THE ENZYMES

Bacterial Carbonic Anhydrases

T0348753

Serial Editors

CLAUDIU T. SUPURAN
Neurofarba Department, Pharmaceutical and Nutraceutical Section, University of Florence, Sesto Fiorentino, Florence; Neurofarba Department, Sezione di Chimica Farmaceutica e Nutraceutica, Universita degli Studi di Firenze, Firenze, Italy

VOLUME FIFTY FIVE

THE ENZYMES
Bacterial Carbonic Anhydrases

Edited by

CLAUDIU T. SUPURAN

Neurofarba Department, Pharmaceutical and Nutraceutical Section, University of Florence, Sesto Fiorentino, Florence; Neurofarba Department, Sezione di Chimica Farmaceutica e Nutraceutica, Universita degli Studi di Firenze, Firenze, Italy

ACADEMIC PRESS

An imprint of Elsevier

Academic Press is an imprint of Elsevier
125 London Wall, London, EC2Y 5AS, United Kingdom
50 Hampshire Street, 5th Floor, Cambridge, MA 02139, United States
525 B Street, Suite 1650, San Diego, CA 92101, United States

First edition 2024

ISBN: 978-0-443-29518-8
ISSN: 1874-6047

For information on all Academic Press publications
visit our website at https://www.elsevier.com/books-and-journals

Publisher: Zoe Kruze
Acquisitions Editor: Leticia Lima
Editorial Project Manager: Palash Sharma
Production Project Manager: Abdulla Sait
Cover Designer: Gopalakrishnan Venkatraman

Typeset by MPS Limited, India

Working together
to grow libraries in
developing countries

www.elsevier.com • www.bookaid.org

Contents

Contributors   xi
Preface   xiii

1. Overview on bacterial carbonic anhydrase genetic families   1

Clemente Capasso and Claudiu T. Supuran

1. Introduction   2
2. Type of BCAs   4
3. Evolutionary divergence of CA classes   15
4. BCAs are key targets in biomedical research and drug development   17
5. BCA and their potential applications in carbon capture technologies   20
6. Conclusions   22
References   23

2. Bacterial α-CAs: a biochemical and structural overview   31

Vincenzo Massimiliano Vivenzio, Davide Esposito,
Simona Maria Monti, and Giuseppina De Simone

1. Introduction   31
2. α-CAs as novel molecular targets to combat antibiotic resistance in   32
   bacteria
3. α-CAs from extremophilic bacteria   43
4. α-CAs from cyanobacteria   51
5. Conclusions and future perspectives   52
References   53

3. Bacterial β-carbonic anhydrases   65

Marta Ferraroni

1. Introduction   66
2. Physiological role of β-carbonic anhydrases in bacteria   67
3. Structural characterization of β-Carbonic anhydrases   73
4. Allostery in β-carbonic anhydrases   78
5. Inhibitor binding to β-carbonic anhydrases   82
6. Catalyic mechanism of β-CAs   84
References   87

4. Bacterial γ-carbonic anhydrases   **93**

Andrea Angeli

1. Introduction   94
2. Catalytic mechanism of bacterial γ-CAs   94
3. The subgroup of γ-CAs: the case of CamH   98
4. γ-Carbonic anhydrases from bacteria   99
5. Crystal structures   102
6. Inhibitors and activators of bacterial γ-CAs   114
7. Conclusion   115
References   116

5. Bacterlal ι-CAs   **121**

Clemente Capasso and Claudiu T. Supuran

1. Introduction   122
2. The journey for identifying the bacterial ι-CA class   123
3. An intriguing aspect of the ι-CAs: they lack metal ion cofactors   132
4. Novel fusion CA gene   135
5. Conclusions   136
References   138

6. Sulfonamide inhibitors of bacterial carbonic anhydrases   **143**

Alessio Nocentini

1. Introduction   143
2. Mechanism of action of sulfonamide inhibitors of bacterial CA   145
    isoforms
3. Inhibition of bacterial CA isoforms with sulfonamide compounds   149
4. Conclusions   184
References   185

7. Non-sulfonamide bacterial CA inhibitors   **193**

Fabrizio Carta

1. Introduction   193
2. Anionic species   194
3. Coumarins   201
4. Phenols   202
5. N-Methyl thiosemicarbazones   203
6. N-Hydroxy ureas   205

7. Miscellaneous   207
8. Conclusions   207
References   208

8. *Helicobacter pylori* CAs inhibition   **213**

Bianca Laura Bernardoni, Concettina La Motta, Simone Carradori,
and Ilaria D'Agostino

1. *Helicobacter pylori*, a bacterium threatening public health: an   213
   overview of the pathogen, its related diseases, and current treatments
2. HpCAs as promising targets for anti-Hp agents: functional role and   215
   kinetics
3. Structural insight into HpCAs   218
4. Modulation of HpCAs for therapeutical purposes: CAI libraries from   219
   the literature
5. Sulfonamido-containing CAIs   221
6. Anion and acid CAIs   228
7. Phenolic CAIs   230
8. Conclusions: how far HpCAs came and future perspectives   232
Acknowledgments   232
References   232

9. *Neisseria gonorrhoeae* carbonic anhydrase inhibition   **243**

Molly S. Youse, Katrina J. Holly, and Daniel P. Flaherty

1. Introduction   243
2. Enzyme genetic families encoded in *Neisseria gonorrhoeae*   244
3. Enzyme inhibition   258
4. Conclusion   272
References   272

10. *Enterococci* carbonic anhydrase inhibition   **283**

Katrina J. Holly, Molly S. Youse, and Daniel P. Flaherty

1. Introduction   283
2. Carbonic anhydrase genetic families encoded in *Enterococci*   284
3. Enzyme inhibition   295
4. Conclusion   307
References   308

11. Carbonic anhydrases in bacterial pathogens   313

Reygan E. Braga, Fares Z. Najar, Chelsea L. Murphy, and
Marianna A. Patrauchan

1. Introduction   313
2. Distribution and sequence conservation in carbonic anhydrases   315
   among bacterial pathogens
3. Importance of carbonic anhydrases in physiology of bacterial   328
   pathogens
4. Role in bacterial virulence and host-pathogen interactions   332
5. Potential in therapeutic implications   334
6. Remaining unknowns and challenges   335
References   336

12. Mycobacterial β-carbonic anhydrases: Molecular biology,   343
role in the pathogenesis of tuberculosis and inhibition
studies

Jenny Parkkinen, Ratul Bhowmik, Martti Tolvanen, Fabrizio Carta,
Claudiu T. Supuran, Seppo Parkkila, and Ashok Aspatwar

1. Introduction   344
2. Computational tools for studying *M. tuberculosis* β-carbonic   350
   anhydrases
3. Molecular biology   355
4. β-CA1 is induced during environmental stress   357
5. β-CA2 is essential for mycobacterial virulence   358
6. β-CA3 serves as a dual-function transmembrane protein   360
7. Inhibition of *M. tuberculosis* β-carbonic anhydrases   361
8. Inhibition with sulfonamides   362
9. Inhibition with dithiocarbamates   368
10. Inhibition with natural products and carboxylic acids   368
11. Inhibition of *M. tuberculosis* in culture   369
12. Summary and future directions   374
Acknowledgments   375
References   375

13. Challenges for developing bacterial CA inhibitors   383
as novel antibiotics

Claudiu T. Supuran

1. Introduction   384
2. CA inhibitors (CAIs) classes   385

3. Drug repurposing of human CAIs as antiinfectives   386
4. Drug design studies of bacterial CAIs   389
5. Vacomycin resistant *Enterococci* CAs   395
6. Conclusions   400
Funding   402
Declaration of interest   402
References   402

Contributors

Andrea Angeli
Neurofarba Department, University of Florence, Sesto Fiorentino, Florence, Italy

Ashok Aspatwar
Faculty of Medicine and Health Technology, Tampere University, Tampere, Finland

Bianca Laura Bernardoni
Department of Pharmacy, University of Pisa, Pisa, Italy

Ratul Bhowmik
Faculty of Medicine and Health Technology, Tampere University, Tampere, Finland

Reygan E. Braga
Department of Microbiology and Molecular Genetics, Oklahoma State University, Stillwater, OK, United States

Clemente Capasso
Department of Biology, Agriculture and Food Sciences, Institute of Biosciences and Bioresources, CNR, Napoli, Italy

Simone Carradori
Department of Pharmacy, G. d'Annunzio University of Chieti-Pescara, Chieti, Italy

Fabrizio Carta
NEUROFARBA Department, Sezione di Scienze Farmaceutiche e Nutraceutiche, University of Florence, Sesto Fiorentino, Florence; Neurofarba Department, Sezione di Chimica Farmaceutica e Nutraceutica, Università degli Studi di Firenze, Firenze, Italy

Giuseppina De Simone
Istituto di Biostrutture e Bioimmagini, CNR, Napoli, Italy

Ilaria D'Agostino
Department of Pharmacy, University of Pisa, Pisa, Italy

Davide Esposito
Istituto di Biostrutture e Bioimmagini, CNR, Napoli, Italy

Marta Ferraroni
Dipartimento di Chimica "Ugo Schiff", Università di Firenze, Sesto Fiorentino, Firenze, Italia

Daniel P. Flaherty
Borch Department of Medicinal Chemistry and Molecular Pharmacology, College of Pharmacy, Purdue University, West Lafayette, IN, United States

Katrina J. Holly
Borch Department of Medicinal Chemistry and Molecular Pharmacology, College of Pharmacy, Purdue University, West Lafayette, IN, United States

Concettina La Motta
Department of Pharmacy, University of Pisa, Pisa, Italy

Simona Maria Monti
Istituto di Biostrutture e Bioimmagini, CNR, Napoli, Italy

Chelsea L. Murphy
Bioinformatics Core, Oklahoma State University, Stillwater, OK, United States

Fares Z. Najar
Bioinformatics Core, Oklahoma State University, Stillwater, OK, United States

Alessio Nocentini
Sezione di Scienze Farmaceutiche, NEUROFARBA Department, University of Florence, Sesto Fiorentino, Florence, Italy

Seppo Parkkila
Faculty of Medicine and Health Technology, Tampere University; Fimlab Ltd. and Tampere University Hospital, Tampere, Finland

Jenny Parkkinen
Faculty of Medicine and Health Technology, Tampere University, Tampere, Finland

Marianna A. Patrauchan
Department of Microbiology and Molecular Genetics, Oklahoma State University, Stillwater, OK, United States

Claudiu T. Supuran
Neurofarba Department, Pharmaceutical and Nutraceutical Section, University of Florence, Sesto Fiorentino, Florence; Neurofarba Department, Sezione di Chimica Farmaceutica e Nutraceutica, Universita degli Studi di Firenze, Firenze, Italy

Martti Tolvanen
Department of Computing, University of Turku, Finland

Vincenzo Massimiliano Vivenzio
Istituto di Biostrutture e Bioimmagini, CNR, Napoli, Italy

Molly S. Youse
Borch Department of Medicinal Chemistry and Molecular Pharmacology, College of Pharmacy, Purdue University, West Lafayette, IN, United States

Preface

The Enzymes book series was initiated by Paul D. Boyer in 1970 and 54 volumes were published since then, with different other scientists acting as series editors over this time span. The present volume is the 55th and I will act as editor of the series, with the commitment of continuing the excellence that characterised the preceding volumes. I take here the opportunity to thank my predecessors editors for their fantastic work which inspired me when I was a student, as a researcher later and even now, as a professor, since many of the volumes in *The Enzymes* constitute reference books in many fields dedicated to our beloved proteins possessing catalytic activity.

The present volume is focused on one of the enzymes which received a lot of attention ever since it has been discovered in 1933, the carbonic anhydrase, and more precisely to the bacterial carbonic anhydrases. These enzymes catalyze the humble but fundamental recation between carbon dioxide and water, generating bicarbonate and protons. This simple reaction may occur without catalysts, but it is so slow at the physiological pH. Thus, there was a stringent biological need that catalysts evolve for facilitating it. And evolution "worked" efficiently, since at least eight times, independently, carbonic anhydrases evolved in organisms all over the phylogenetic tree, in prokaryotes and eukaryotes, with very few known organisms in which they are absent.

Volume 55 contains 13 chapters dedicated to carbonic anhydrases present in bacteria. Of the eight genetically distinct carbonic anhydrase families, the bacteria encode for four of them, the α-, β-, γ- and ɩ-class enzymes. Chapter 1 by Capasso and Supuran presents an overview of these four enzyme classes present in bacteria, their role in the bacterial metabolism, with a general presentation of each class, of their similarities and differences and a thorough phylogenetic analyses of such enzymes from pathogenic as well as non-pathogenic bacteria. A brief discussion of the potential of these enzymes as antibacterial drug targets is also provided. Chapter 2, by Vivenzio et al., presents the bacterial α-carbonic anhydrases, with detailed information on the biochemical and structural characteristics of these enzymes, whereas Chapter 3, by Ferraroni, Chapter 4, by Angeli, and Chapter 5 by Capasso and Supuran do the same for the β-, γ- and ɩ-class bacterial carbonic anhydrases. Chapter 6, by Nocentini, and Chapter 7, by Carta, discuss the sulfonamide and non-sulfonamide bacterial carbonic anhydrase inhibitors, resepectively, with exhaustive presentation of the various classes of inhibitors, which are crucial for their pharmacological applications as anti-infectives. Indeed, the next five chapters present different pathogenic bacteria for which relevant progress has been done ultimately in understanding the role played by their carbonic anhydrases in pathogenicity, invasion, survival in particular niches but also on how to use them as antibacterial drug targets. Chapter 8, by Bernardoni et al., presents *Helicobacter pylori* carbonic anhydrases and their inhibitors, which represents an interesting case study, since a drug used in the past for the management of ulcers, acetazolamide, was subsequently demonstrated to have such beneficial effects due to inhibition of the bacterial enzymes (an α- and a β-carbonic anhydrase are present in *H. pylori*) and not of the human gastric mucosa carbonic anhydrases. Chapters 9 and 10 are both from Flahaerty's group and examine two other pathogenic organisms, *Neisseria gonorrhoeae* and drug resistant *Enterococci*, respectively, for which the carbonic anhydrases were validated in recent years as antiinfective drug targets. Bacteria belonging to species treated in all these chapters developed high degrees of resistance to most clinically used antibiotics in the last decades, and the inhibition of their carbonic anhydrases might thus represent an opportunity to develop new drugs., Chapter 11 by Braga et al. focuses exactly on the different carbonic anhydrases present in pathogenic bacteria and how they can be investigated and validated as drug targets. Chapter 12, by Parkkinen et al., presents in detail the carbonic anhydrases present in *Mycobacterium tuberculosis* and related species (e.g., *M. marinum*), and their inhibition as an approach for obtaining effective antimycobacterial drugs. The last chapter (Chapter 13) by myself, discusses the challenges and opportunities for designing bacterial carbonic anhydrase inhibitors as innovative anti-infective agents. The challenges are many since the α-carbonic anhydrases are present indeed in some bacteria but also in vertebrates, humans included, and as thus the available inhibitors effectively inhibit both of them, leading thus to potential side effects of drugs based on them. However, there are recent examples showing the possibility to obtain compounds which are better inhibitors of the bacterial than of the human enzymes, which are discussed in detail. On the other hand, the β-, γ- and ι-carbonic anhydrases are not present in humans and in these cases, there may be more opportunities to design inhibitors which should not cross-react with the human enzymes, but the field is still in its infancy.

Overall, the book presents updated information on many well-studied bacterial carbonic anhydrases present in relevant human pathogens, a field in great expansion in the last years, which led to the validation of some of these enzymes as drug targets, with interesting drug design studies which are presented for each particular pathogen examined and with the possibility to extend them to other pathogenic bacteria poorly investigated to date.

I am grateful to all authors who contributed to this book, providing insightful, high quality chapters. I also wish to thank Leticia Lima from Elsevier Brasil and the desk editor Palash Sharma for their constant help throughout the project.

CLAUDIU T. SUPURAN

CHAPTER ONE

Overview on bacterial carbonic anhydrase genetic families

Clemente Capasso[a],* and Claudiu T. Supuran[b]

[a]Department of Biology, Agriculture and Food Sciences, Institute of Biosciences and Bioresources, CNR, Napoli, Italy
[b]Neurofarba Department, Pharmaceutical and Nutraceutical Section, University of Florence, Sesto Fiorentino, Firenze, Italy
*Corresponding author. e-mail address: clemente.capasso@ibbr.cnr.it

Contents

1. Introduction   2
2. Type of BCAs   4
   2.1 Bacterial α-CAs   5
   2.2 Bacterial β-CAs   7
   2.3 Bacterial γ-CAs   11
   2.4 Bacterial ι-CAs   13
3. Evolutionary divergence of CA classes   15
4. BCAs are key targets in biomedical research and drug development   17
5. BCA and their potential applications in carbon capture technologies   20
6. Conclusions   22
References   23

Abstract

Bacterial carbonic anhydrases (BCAs, EC 4.2.1.1) are indispensable enzymes in microbial physiology because they facilitate the hydration of carbon dioxide (CO_2) to bicarbonate ions (HCO_3^-) and protons (H^+), which are crucial for various metabolic processes and cellular homeostasis. Their involvement spans from metabolic pathways, such as photosynthesis, respiration, to organic compounds production, which are pivotal for bacterial growth and survival. This chapter elucidates the diversity of BCA genetic families, categorized into four distinct classes (α, β, γ, and ι), which may reflect bacterial adaptation to environmental niches and their metabolic demands. The diversity of BCAs is essential not only for understanding their physiological roles but also for exploring their potential in biotechnology. Knowledge of their diversity enables researchers to develop innovative biocatalysts for industrial applications, including carbon capture technologies to convert CO_2 emissions into valuable products. Additionally, BCAs are relevant to biomedical research and drug development because of their involvement in bacterial pathogenesis and microbial survival within the host. Understanding the diversity and function of BCAs can aid in designing targeted therapeutics that interfere with bacterial metabolism and potentially reduce the risk of infections.

The Enzymes, Volume 55
ISSN 1874-6047, https://doi.org/10.1016/bs.enz.2024.05.004

1. Introduction

Bacterial carbonic anhydrases (BCAs, EC 4.2.1.1) are essential components of microbial physiology, significantly affecting a range of metabolic processes through their critical role in facilitating the hydration of carbon dioxide (CO_2) via the following chemical reaction: $CO_2 + H_2O \rightleftharpoons HCO_3^- + H^+$ [1–4]. This reaction provides bicarbonate ions (HCO_3^-) and protons (H^+), which are indispensable for sustaining numerous biochemical reactions and maintaining cellular homeostasis. For example, in non-photosynthetic bacteria, the release of H^+ helps regulate cytoplasmic pH, thereby ensuring optimal conditions for enzymatic activity and cellular function, whereas HCO_3^- facilitates the synthesis of key molecules such as amino acids, nucleotides, and fatty acids [4–7]. In photosynthetic bacteria, BCAs also play a central role in carbon fixation, where CO_2 is assimilated into organic molecules during the Calvin cycle [8–10]. BCAs provide the necessary substrate for carboxylation reactions catalyzed by enzymes such as ribulose-1,5-bisphosphate carboxylase/oxygenase (RuBisCO) [7,11,12]. Thus, CO_2 hydration represents a crucial step in various metabolic pathways, including photosynthesis, respiration, and production of organic compounds that are vital for cell growth and survival. At the core of the microbial metabolism lies the need for efficient CO_2 management and BCAs play a central role in this process.

The classification system of CAs is based on the use of Greek letters to denote the various distinct families or classes of these enzymes. To date, eight distinct classes of CAs have been identified in all living organisms, each represented by a different Greek letter, including α (alpha), β (beta), γ (gamma), δ (delta), ζ (zeta), η (eta), θ (theta), and ι (iota) [13–15]. CA classes possesses distinct structural features, such as conserved domains, active site configurations, and oligomeric states, which contribute to their classification into separate families or classes. In the realm of BCAs, four distinct classes α, β, γ, and ι have been identified, each with varying distributions across bacterial genomes [1–4,16–19]. Most BCAs contain a metal ion, generally Zn^{2+}, in their active site, which plays a crucial role in catalyzing the reaction. In all CA classes except the ι-CAs, the catalytic reaction typically follows a two-step mechanism. In the first step, the zinc ion coordinates with water molecules to form a hydroxide ion (OH^-), which acts as a nucleophile on CO_2, forming HCO_3^- [12,20–22]. After the formation of HCO_3^-, a water molecule displaces it from the active site. The second step occurs after the displacement of bicarbonate ion. The active site is regenerated through a proton transfer reaction, where a proton from the water molecule is transferred to the zinc-bound hydroxide ion, reforming a water molecule, and completing the catalytic cycle. The rate-determining step in many enzymes is the generation of metal hydroxide species from coordinated water, with some α-CAs achieving high k_{cat}/K_M values ($>10^8$ $M^{-1} \times s^{-1}$), making CAs highly effective catalysts [23–26]. For ι-CAs, the proposed catalytic mechanism (without metal ions) is shown in Scheme 1:

Different CA classes may exhibit variations in enzymatic activity, substrate specificity, and metal cofactor preferences, which further contributes to their classification [13]. Investigating the diversity of BCA classes offers valuable insights into the evolutionary history of bacteria and their capacity to thrive in diverse environmental niches. Bacteria have thrived in a wide range of environments for billions of years, resulting in the evolution of diverse metabolic pathways and enzyme systems that enable them to adapt to their surroundings [27]. The presence of multiple classes of BCAs across bacterial taxa can shed light on the evolutionary history of bacteria and trace the need of these enzymes in various microbial lineages. Bacteria exhibit remarkable adaptability to diverse environmental conditions, including variations in temperature, pH, salinity, and nutrient availability [28]. The presence of different BCA classes in bacteria reflects their adaptation to specific environmental niches and metabolic requirements [28]. Bacteria inhabiting high temperature environments possess

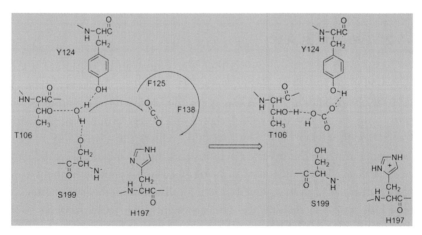

Scheme 1 Catalytic mechanism of CO_2 hydration proposed of ι-CAs: the water molecule acting in the catalytic cycle is activated for the nucleophilic attack on the CO_2 molecule bound in a hydrophobic pocket, by formation of a network of hydrogen bonds with Thr106, Ser199 and Tyr124. The proton transfer is achieved presumably by His197.

BCAs with optimal activity at elevated temperatures, whereas those living in high CO_2 environments may have specialized BCAs for efficient CO_2 hydration [29–33]. Bacteria often exist in complex ecological communities where they interact with other organisms and compete for resources [34]. The diversity of BCA classes may influence bacterial interactions within these communities by affecting carbon metabolism, pH regulation, and nutrient cycling. Understanding how different BCA classes contribute to bacterial ecology provides insights into the dynamics of microbial communities and ecosystem functioning. The diversity and functional implications of BCA classes is crucial not only for elucidating their physiological functions but also for investigating their potential applications in biotechnology. By possessing knowledge of BCA diversity and mechanisms, researchers can contribute to the creation of innovative biocatalysts for industrial purposes [19,35–39]. BCAs have potential applications in carbon capture technologies, where they can be used to capture and convert CO_2 emissions into useful products [19,35–39]. Bacterial CAs are of interest in biomedical research and drug development [4,40]. They play a role in bacterial pathogenesis, where they contribute to virulence and survival within the host environment. Understanding the diversity and functions of bacterial CA classes can aid in the design of targeted therapeutics to inhibit CA activity, thereby disrupting bacterial physiology and potentially mitigating infections.

2. Type of BCAs

In Bacteria, it is common to find a combination of α, β, and γ CA classes, while the ι-class is relatively rare [1–4,16–19]. However, there is considerable variability in the CA classes encoded by bacterial genomes. Some bacteria possess only one or two CA classes, and exceptionally, some lack any CA enzymes altogether [41]. This variability reflects the adaptability of bacterial metabolism and the diverse environmental niches inhabited by different bacterial species. The presence of an N-terminal signal peptide in the α-CAs of Gram-negative bacteria suggests their localization in the periplasmic or extracellular space, facilitating the conversion of CO_2 to bicarbonate [41]. This conversion is vital for metabolic pathways that require CO_2 or bicarbonate as substrates, ensuring microbial survival and its metabolic demands. Conversely, β-, γ- and ι-CAs, with cytoplasmic localization, supply CO_2 for intracellular processes like carboxylase enzymes and pH homeostasis [3,4,41].

Some pathogenic Gram-negative bacteria encode β-, γ- and ι-CAs with signal peptides, indicating potential periplasmic localization akin to α-CAs, suggesting similar roles in supporting microbial physiology. Additionally, the variety of BCAs presents a promising opportunity for the development of selective inhibitors that specifically target BCAs without adversely affecting mammalian homologous enzymes. The genome of pathogenic bacteria frequently encodes BCA classes that are absent in mammals, whose genome encodes only α-CAs [3,4,41]. Moreover, the structural differences between the bacterial and mammalian CA classes provide a basis for designing inhibitors that specifically target bacterial enzymes. This minimizes off-target effects and reduces the risk of adverse reactions in mammalian hosts. The ability to selectively target bacterial CAs opens new avenues for precision medicine approaches for the treatment of bacterial infections. By exploiting the differences in CAs between bacteria and mammals, researchers can develop tailored therapies that specifically target bacterial pathogens while minimizing the risk of collateral damage to host microbiota or mammalian cells. BCAs are characterized by presence of specific conserved domains that define each class. A conserved domain is a recurring unit within a protein's molecular structure or sequence that remains relatively unchanged across different organisms through evolutionary time. These domains often correspond to the specific functions or structural features of proteins. They are identified through sequence and structural analyses, typically using multiple sequences and model alignments.

2.1 Bacterial α-CAs

According to the Conserved Domain Database (CDD) [42], which contains curated and annotated multiple sequence alignment models for protein domains, bacterial α-CAs possess a conserved **COG3338** domain, also known as Cah. This domain is crucial for maintaining the structural integrity and functionality of this class of enzymes. The alignment provided in Fig. 1 provides insights into the conservation and variation of **COG3338** domain across α-CAs from different bacterial species.

Bacterial α-CAs have received relatively less attention compared to their mammalian counterparts. Despite this, their three-dimensional structures closely resemble those of human α-CAs, which have been extensively studied in the presence of acetazolamide (AAZ), a well-known sulfonamide CA inhibitor [2,3]. The structural analysis of the bacterial α-CA revealed a homodimeric arrangement that is stabilized by numerous hydrogen bonds and several hydrophobic interactions [43,44] (Fig. 2).

Halalkalibacterium halodurans 1 MKKYLMGktclVVSLSVMVTACSSAPSTEpvdepsetheetsggaheVHWSYTGDT-GPEHWAELDSEYGACAQGEEQSP 79
Pasteurella multocida 1 MKKSNLF----VLTLGALALTACTSPLKKee---------------thTHWGYTGHE-SPEHWAELSPKFRICGEGKNQTP 61
Sinorhizobium meliloti 1 MERRDFLr--gLALLAACPLCVKTAYAAEg----------------VHWRYEGEE-GPEHWGSLAKENSACSAGSQQSP 60
Vibrio cholerae 1 MKKTTWV----LAMVASMSFGVQAS---------------------EWGYEGEH-APEHWGKVAP---LCAEGKNQSP 49
Nostoc sp. PCC 7120 5 LYRRQLLKilgMSVLGTSFSSCVTSPARAk----------------tVNNGYIGKV-GPEHWGELSPDFALCQIGRKQTP 67
Helicobacter pylori J99 1 MKKYFLI----ALALTASLIGAENT---------------------KWDYKNKEnGPHRWDKLHKDFEVCKSGKSQSP 53
Listeria innocua 1 --------------------MTKEKV---------------------LWGYDEKT-GPEMMGHICSDFEIAHTGKAQSP 37
Mesorhizobium loti 1 MHRRNLIr--aFIALGVCPICAQTARAAS----------------AHWGYGGSV-GPEHWADLDTRNFACSAGRQQSP 59

80 IN--LDKAEAVD---TDTEIQVHYEP-SAFTIKNNGHTIQAETTSDGNTIEIDGKEYTLVQFHFHIPSEHEMEGKNLDMEL 154
62 ID--IKHTIDG---KLAPIKLDYRF-SNVEIVNNGHTIQVDFKEASNRMQLNGKTFTLKQFHFHVPSENLIRGKSFPMEA 135
61 ID--IRGAVKA---DIPELTADWK--SGGTILNNGHTIQVNAAOG-TLRRGD-KSYDLVQYHFHSPSEHFVDGKSFPMEA 131
50 ID--VAQSVEA---DLQPFTLNYQG-QVVGLLNNGHTLQAIVR-GNNPLQIDGKTFQLKQFHFHTPSENLLKGKQFPLEA 122
68 IDiqIADVKDVHssaSQDLLVTNYQP-TALHLINNGKTVQVNYQPG-SYLKYAHQKFELLQFHFHHFSEHRVDGKLYDMEL 145
54 IN--IEHYYHTQ--DKADLQFKYAAsKPKAVFFTHHTLKASFEPT-NHINYRGHDYVLDNVHFHAPMEFLINNKTRPLSA 128
38 VN--IEQADVVK---LKPSTMKFYYKeTDYTIKRIEQSVHVFPHDKEQGLRFNGEYYPLVSFHAHIPAEHLLDGYVYPIEN 113
60 ID--IAGTVKA---DIPRIDISWLK-GGGRMVNNGHTIQINMPEGSTLTRGD-RVYQLAQLHFHAPSEHHVAGMSFPMEA 132

155 HFVHKNEN-DELAVLGVLMKAGEENEELAKLWSKLPAEETEENIsLDESIDLNALLPESKEGFHYNGSLTTPPCSEGVKW 233
136 HFVHADNE-GNLAVLGVLYVLSSENQRLAPIWQNFPQKAGEKYT-LSTAFDPATLIPKKRDYYRFSGSLTTPPCSEGVNW 213
132 HFVHKNAEtGTLGVLGVFLVPGAANSTFASLAEKFPRNPGEESP-LIT-IDPKGLLPSSLSYWTYEGSLTTPPCSEIVDW 209
123 HFVHADEQ-GNLAVVAVMYQVGSENPLLKVLTADMPTK-GNSTQ-LTQGIPLADWIPESKHYYRFNGSLTTPPCSEGVRW 199
146 HLVHRSKS-GDLAVMGIFLQAGAFNPTLQIIWDATPQNQGTDKRiEDINIDASQFLPAQHRFFTYSGSLTTPPCSENVLW 224
129 HFVHKDAK-GRLLVLAIGFEEGKENPNLDPILEGIQKK--QN---LKE-VALDAFLPKSINYYHFNGGLTAPPCTEGVAW 201
114 HFVHEKPDqTTLVMSAWMKIDNTNNVEFKDLPTYFPEVFADFETeREITLDVNEFMPEERVFYTYQGSRTTPPTVEGVTW 193
133 HFVHKDTKsDTLGVLSALLMPGATNNSFAGLAKVFPARPGEEMA-VDE-FDPNGLLPASLGYWTYEGSLTTPPCTENVEW 210

234 TVLSEPITVSQEQIDAFAEIFP--DNHRPVQPWNDRDVYDVI 273
214 LVLKHYDNISQSQVEAFATLMKq-HNNRPVQPINARVIVE-- 252
210 MVAMEAVEVDPGDIKKFTALYs--MNARPALAGNRRYVLSSS 249
200 IVLKEPAHLSNQQEQQLSAVMq--HNNRPVQPHNARLVLQAD 239
225 CVMATPIEASPAQIAKFSQMFP--QNARPVQPLNDRLVIKAT 264
202 FVIREPLEVQAHQLADIHHHHH...FNQRFVQPDYNTVILKSS 243
194 IVLKNAKTLGQEDFTEFEKAIG--NTSRPVQDLNGREITFYN 233
211 MVAMKPVEVDTADIKRFTTLYa--SNARSIRPSNRRFILGLS 250

Fig. 1 Representative sequences of the conserved COG3338 (Cah) domain according to the Conserved Domain Database (CDD). *Red*, same residue; *bold red*, amino acid residues of the catalytic triad; *blue*, conserved residues; *lowercase amino acids*, unaligned residues.

The α-CAs demonstrate activity as both monomers and dimers. The active site is situated within a deep cavity extending from the protein surface to the core of the molecule, exhibiting both hydrophilic and hydrophobic regions [43,44]. The hydrophilic segment facilitates the transfer of protons from zinc-bound water to the solvent, whereas the hydrophobic region is involved in CO_2 binding and ligand recognition. At the base of this cavity lies the catalytic zinc ion, tetrahedrally coordinated by three histidine residues and the nitrogen atom of the sulfonamide moiety of the inhibitor or potentially by a water molecule in the uninhibited enzyme. Notably, bacterial α-CAs display a more compact structure than their mammalian counterparts, primarily because of the absence of certain insertions [3]. This structural difference results in a larger active site in bacterial CAs compared to human enzymes. Additionally, the structures of thermostable CAs, such as SspCA (from *Sulfurihydrogenibium yellowstonense*) and SazCA (from *Sulfurihydrogenibium azorense*), identified in thermophilic bacteria, are characterized by a higher content of secondary structural elements and an increased number of charged residues, all of which contribute to protein thermostability [45,46] (Fig. 3).

It is noteworthy that the crystal structure of TaCA from *Thermovibrio ammonificans* is tetrameric, featuring a central core stabilized by two intersubunit disulfides and a single lysine residue from each monomer, which

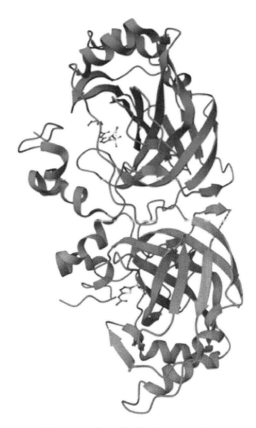

Fig. 2 Crystal structure of the complex of *Helicobacter pylori* α-CA with acetazolamide, a carbonic anhydrase inhibitor. The structure is depicted in two different colors to distinguish between the two monomers that compose the complex.

participates in inter-subunit ionic interactions [47]. It is noteworthy that α-CAs, beyond their primary function of facilitating the reversible hydration of carbon dioxide, have been observed to exhibit additional enzymatic behavior. This includes a tendency for weak and nonspecific esterase activity towards activated esters, as well as cyanamide hydration to urea [48,49]. The discovery of these secondary reactions serves as a notable example of enzyme promiscuity, highlighting the inherent flexibility of enzymes to catalyze reactions beyond their evolutionary design.

2.2 Bacterial β-CAs

β-CAs belong to the **cd00382** domain, which is part of the broader superfamily **cl00391**. Within this superfamily, β-CAs are further divided into four

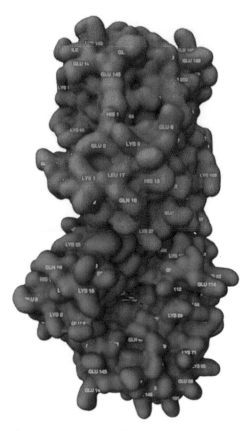

Fig. 3 Gaussian surface representation of the 3D structure of α-CA from the extremophilic bacterium *Sulfurihydrogenibium yellowstonense* YO3AOP1. Ligand (AZA) and surface amino acid labels are included. The two different colors refer to the two monomers.

clades: **cd00883** (beta_CA_cladeA), **cd00884** (beta_CA_cladeB), **cd03378** (beta_CA_cladeC)**, and cd03379** (beta_CA_cladeD). Fig. 4 displays the multiple alignments of bacterial β-CAs belonging to conserved domains **cd00883, cd00884, cd03378**, and **cd03379**, which also include β-CA amino acid sequences from non-bacterial species possessing the same CCD.

β-CAs have evolved unique structural and functional properties that enable them to perform specific roles in bacterial physiology that are distinct from those of α, γ, and ι CAs. X-ray crystallography has been instrumental in elucidating the structural features of various β-CAs derived from bacterial sources including *Escherichia coli*, *Haemophilus influenzae*, *Mycobacterium tuberculosis*, *Salmonella enterica*, and *Vibrio cholerae* [50–53].

cd00883 (beta_CA_cladeA)

Porphyridium purpureum
Escherichia coli
Saccharomyces cerevisiae
Dictyostelium discoideum
Coccomyxa sp. PA
Saccharomyces pombe
Phaeodactylum tricornutum
Caulobacter vibrioides
Ralstonia pseudosolanacearum

cd00884 (beta_CA_cladeB)

Pisum sativum
Synechococcus elongatus
Zea mays
Helicobacter pylori
Drosophila melanogaster
Mesorhizobium loti MAFF303099
Pseudomonas aeruginosa PAO1
Campylobacter jejuni
Arabidopsis thaliana
Caenorhabditis elegans

cd03378 (beta_CA_cladeC)

Mycobacterium tuberculosis
Solibacter usitatus
Salmonella enterica
Nitrobacter hamburgensis
Nostoc sp. PCC 7120
Legionella pneumophila
Gloeobacter violaceus
Leifsonia xyli
Corynebacterium diphtheriae
Nitrosomonas europaea ATCC 19718

cd03379 (beta_CA_cladeD)

Methanothermobacter thermautotrophicus
Mycobacterium tuberculosis H37Rv
Methanothermobacter thermautotrophicus
Methanothermobacter thermautotrophicus DH
Bacillus subtilis
Sulfolobus solfataricus
Neisseria meningitidis MC58
Streptococcus pneumoniae TIGR4
Clostridium perfringens str. 13
Neurospora crassa

Fig. 4 Representative sequences of the conserved domains characteristic of beta_CA_clades according to CDD. *Red*, same residue; *bold red*, amino acid residues of the catalytic triade; *blue*, conserved residues; *lowercase amino acids*, unaligned residues.

These enzymes exhibit a degree of structural conservation in their three-dimensional folds, albeit with variations in their oligomeric states; some exist as dimers while others form tetramers. Notably, all bacterial β-CAs manifest activity as dimers or tetramers, each possessing two or four identical active sites [50] (Fig. 5).

The active sites typically exhibit a characteristic elongated channel morphology, housing a catalytic zinc ion situated at the bottom, tetrahedrally coordinated by two cysteines, one histidine, and one aspartic acid residue, thereby forming what is commonly referred to as the "closed active site." This nomenclature is attributed to the lack of catalytic activity of the enzyme under acidic conditions (pH < 8.3). Interestingly, in this inactive state, bicarbonate ions are bound within a pocket in close proximity to the zinc ion [50]. However, at pH levels exceeding 8.3, a transformation occurs wherein the closed active site transitions to an "open active site," accompanied by the acquisition of catalytic activity. This transition is associated with the displacement of the aspartic acid residue from its position adjacent to the catalytic Zn(II) ion, facilitating the coordination of an incoming water molecule

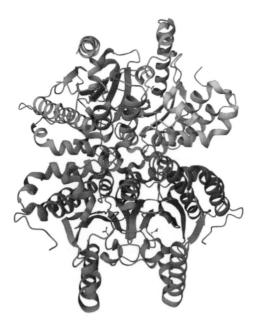

Fig. 5 Crystal structure of β-CAfrom *Vibrio cholerae* in the presence of bicarbonate ions associated with each individual monomer and localized in a noncatalytic binding pocket close to the zinc ion. The structure is depicted in distinct colors, enabling the visualization and analysis of the spatial arrangement of the four monomers.

to the metal ion [50]. This coordinated water molecule, often present as a hydroxide ion, plays a pivotal role in the catalytic activity, as previously elucidated for other CAs. It is intriguing to note that experimental evidence suggests that CA from *Acidithiobacillus thiooxidans* possesses the ability to catalyze the conversion of CS_2 and COS to H_2S and CO_2, albeit with lower efficiency compared to dedicated CS_2 hydrolases [54]. Notably, this enzyme, clustering within the β-CA family, implying that the CA from *A. thiooxidans* and the CS_2 hydrolases evolved from a common ancestral protein and have undergone divergence to acquire specialized functions related to CS_2 and COS conversion. The existence of this secondary catalytic activity underscores the plasticity of enzymatic systems and highlights the potential for functional diversity within the CA families. This phenomenon underscores the dynamic nature of β-CAs and their ability to modulate catalytic activity in response to changes in the local environment. Such structural insights not only deepen our understanding of the molecular mechanisms underlying enzymatic function but also hold potential implications for biotechnological and therapeutic applications targeting CA activity.

2.3 Bacterial γ-CAs

Bacterial γ-CAs are characterized by the presence of either **cd00710** (LbH_gamma_CA domain-containing protein) or **COG0663** (PaaY, carbonic anhydrase or acetyltransferase, isoleucine patch superfamily) domains, which exhibit unique structural features (Fig. 6).

The presence of **cd00710** or **cd00710** domains underscores the evolutionary divergence and functional specialization of γ-CAs within bacterial taxa. The CA Methanosarcina (CAM) from *Methanosarcina thermophila* stands as the archetype of the γ-CA class [55,56]. Structurally, this enzyme adopts a left-handed parallel α-helix fold and crystallizes as a trimer, each containing three zinc-containing active sites situated at the interface between the two monomers. Intriguingly, unlike α-CAs, which feature a histidine residue as a proton shuttle, CAM incorporates a glutamic acid residue for this function. High-resolution crystallography of CAM revealed that Glu89 exhibits two orientations akin to His64 in α-CAs [57]. Subsequent years witnessed the biochemical and structural characterization of other γ-CAs. Remarkably, despite sharing significant structural similarities, these proteins display varying catalytic activities. Specifically, crystallographic analysis of BpscCA from *Burkholderia pseudomallei* revealed a typical trimeric arrangement with active sites at the monomer-monomer interface [58] (Fig. 7).

cd00710 (LbH_gamma_CA domain-containing protein)

```
Methanosarcina thermophila   25 SAPVIDPTAYIDPEASVIGEVTIGANVMVSPMASIRSDEGM------PIFVGDRSNVQDGVVLHALetineegepiedni 98
Xylella fastidiosa 9a5c      34 DLPDVHESAFVDPTAILCGCVIVEAGVFIGPYAVIRADETTwegdikPIRIGIGANIQDGVVIHSKag------------ 101
Nitrosomonas europaea ATCC 19718  9 DQPIVHESAFVDPTAILCGRIVVHENVFIGPYAVIRADEVDatglxmPITVGAHSNIQDGVVIHSKag------------ 76
Desulfitobacterium hafniense Y51  16 KHPKVHHTAYVHPTAVLIGDVRVCENVMICPNVVLRADEGL------TIIIGAGVNIQDGVIHHCLkg------------ 77
Methanosarcina acetivorans C2A    9 QHPRVSKRAWISETAVIIGNISIADYVFVGPNAVLRADEPGs-----SITVQSGCHVQDNVVVHSLsh------------ 71

99 vevdqkeyAVYIGNRVSLAHQSQVHGPAAVGDDTFIGMQAFVFKSKVGHNCVLEPRSAAIGVTIPDGRYIPagMVVTSqaeadkLPEVTDGYaysnTNEAVVYVHVHLAEGYKE 212
102 ------aAVTIGARTSIAHRAIVHGPCTIGERVFIGFNSVLFNCVVGDGCVVRHNAVVDGCDLPPGFYVPstERIGSstdlgrIPRVTVAAs--eFSEDVAQTHNRLVEGYRR 206
77 ------aAVTIGERTSIAHRAIVHGPCTVGPDVVIGFNSVLFNCTIGEGCVVRHNAVVDGCDLPPGFYVPstQRIGPrtdlstIPKVSPKAs--eFSEDVARTNNTLVQGYKH 181
78 ------gSIEIGQDCSIAHCALVHGPAVIGRETFVGFRAIVHNATLGKNSFISHGAMVLNVELPDDSSVPvgKIIQSeaevreLPAIVVQEl--oFKHDVQQIHRELGRGYRE 182
72 ------sDVLVGKNTSLAHSCIVHGPCRIGEGCFIGFGAVVFDCNIGKDTLVLHRSVVRGIDIFSGRIVPdgYVITRqayanaLEPITKEMt--oFKRSVVRANIELVEGYMK 178
```

COG0663 (PaaY, carbonic anhydrase or acetyltransferase, isoleucine patch superfamily)

```
Escherichia coli O157:H7     74 SDVLRPYRDLFPQIGQR--VMIDDSSVVIGDVRLADDVGIWPLVVIRGDVHYVQIGARTNIQDGSMLHVTHKasyn---- 147
Agrobacterium fabrum.        32 MFLYRLA-DRVPQTPAPdrYWIAPDANVIGSVTLGEDVGIWFGATLRGDNEPISVGRGTNIQEGVMVHSDFG-------- 102
Bacillus halodurans          1 --MIFRYSDFSPKVDTT--AFIAESATVIGNVEIGAYTTVLFNAVIRGDEGLIKIGRRCNIQENVMCHLYEQy------- 69
Bacillus halodurans          1 --MLYSYKDKKPKIAEN--VFLADYVTITGDVTIGADSSIWYNTVIRGDVSPTFIGERVNIQDNSVLHQS-P-------- 68
Brucella melitensis          1 MPIYAYN-GHKPQFADReaNWIAPDATLIGKVVVGENAGFWFGAVLAGDNEPITIGADTHVQEQTIMHTDIG-------- 71
Bacillus subtilis            1 --MIYPYKEHKPDIHPT--AFIADNATITGDVVIGEQSSIWFSVVIRGOVAPTRIGNRVNIQDLSCLHQS-Pn------- 68
Clostridium acetobutylicum   1 --MIYSYKDKKPNIHSS--VFIAKSADIIGDVNIDKNSSVWFGAVIRGDSNYIRIGEGTNIQDNSVLHTNYYs------- 69
Caulobacter vibrioides.      1 HTVYSI-G-ATTPTLPAQgeYWIAPSASVMONVILKFNASIWWGAVARGUNDPITIGENSNVQDGSVLHTDLG------- 71
Corynebacterium glutamicum   4 QFLILPFGDKVPRIHES--AWIAPNATIIGDVEIGPDASIFYGVVLRGDVNKITIGARTNVQDNCVLHVDGD------- 73
Campylobacter jejuni         1 -MLIK-FKNHSPKLGQN--VFVAEGAKIIGEIEIGDESSIWFNCVLRADVNFIKIGKRTNIQDLSTVHVHHRefdekgkl 76
```

```
148 -pdqNPLTIGEDVTVGHKVMLHGCTIGNRVLVGMGSILLDGAIVEDDVHIGAGSLVPQNKRLESGYLYLGSFVKQIRPLSDEEKAGLRYSANNYVKWKDEYLDQGNQTQP 256
103 ----FAAVIGDMCTIGHHAIVHGCSIGDNSLIGMGATILNGAKIGNVGANALVTEGKEFPDNSLIVGSPARAIRTLDEDTVAGIRRSAEKYIENWKRFSTDLATIE- 207
70 -----PLVLEDEVSLGHHAIVHGCTLKQGVLIGMGATVLDGAEIGEYSIVGAGSVVPPGKVIPPRSLVLGSPGKVVREVTEQDLAMIEETVETYVRNGQEPKRSOVLERi 174
69 ----tPLVIEDDUVTVGHQVILHSCTIRKEALIGMGSIILDGAEVGEGAFIGAGSLVPQGKKIPANSLAFGRFAKVVHTLTEEDRKGMARIRREYVEKGQYYKQVQQEYRS 174
72 FPLTIGAGCTIGHHAILHGCTIGDHTLIGHGAIVLNGAKVCKGCLICAGTLVKEGMEIPDNELVPGEPARVLRQLDDAAVEYLRASAKVVYEPGHSFFMGRZMEPA-- 175
69 ----xTLLIEDDATIGHQVTLRSAVIRKNALIGMGSIILDGAEIGEGAFIGAGSLVPPGKIIPPGHLAFGRPAKVIRPLTEKIGHKUMQRiKHSNIVEKUQYYKrLQQi---- 17?
70 -----GIGIKNNVTIGHQVILHGCIINSNCIGMGATILDDVKIGEYTIVGANSLITSGKKIPDGVLCMGIPAKVIHELTVDEKLGIDKNAEEHIEMGKKYFK------- 168
72 ----APLTIGANTVIGHMVMLHGCTIGDGSLIGIGSIVLNGAKIGKNCLIGAGALITEGKEIPDNSMVMGAPGKVVREIGEQHAMILQASALHYVENWKRYVRDLKIYE- 176
74 ----APCTLGDDVTVGHHALVHGAVTVGNGTLVGHKSALLSGSHVGAGALIAAGAVVLEGHEIPAKALAAGVPAKVRLLDDAQSQSFIPHAGRYVETSKAQASIAEALGL 179
?? k1aoFPTIIGDDVTIGHNCVIBACVIKNRVLIGNNAVIMDNALIEEDSIVGAGSVVTKGKKFPPRSLILGNPAKFVRELNDEEVSFLKOSALNYVDFKNEFLKDlU---- 182
```

Fig. 6 Representative sequences of the conserved **cd00710** (LbH_gamma_CA domain-containing protein) and **COG0663** (carbonic anhydrase or acetyltransferase, isoleucine patch superfamily) domains, according to the Conserved Domain Database (CDD). *Red*, same residue; *bold red*, amino acid residues of the catalytic triad; *blue*, conserved residues; *lowercase amino acids*, unaligned residues.

Fig. 7 Three-dimensional structure of a γ-CA from the pathogenic bacterium *Burkholderia pseudomallei*. 2-Mercaptoethanol and the zinc ions (gray spheres) are also shown.

Unexpectedly, a molecule of 2-mercaptoethanol was observed in the bicarbonate binding pocket, forming hydrogen bonds with the zinc-bound water molecule. This discovery, coupled with the known inhibition mechanism of several inhibitors of human carbonic anhydrases (hCAs), suggests promising prospects for designing selective inhibitors of BpscCA [58]. Structural alignment of BpscCA with previously characterized γ-CAs highlighted substantial structural conservation. However, several residues crucial for Cam catalytic activity, notably Glu84, were not conserved in BpscCA. Further investigation identified Glu123 as a potential proton shuttle residue in BpscCA, which, similarly to Glu84 in Cam, lacks strict conservation among active γ-CAs [58]. Understanding the molecular features that dictate the catalytic activity of pathogenic CAs is of paramount importance for fine-tuning their enzymatic activity and combating bacterial infections. This knowledge sets the stage for structure-based drug design and facilitattes the development of novel antibacterial agents.

2.4 Bacterial ι-CAs

Bacterial ι-CAs represent a distinct class delineated by **COG4875** and **COG4337** domains, indicating their evolutionary divergence and functional specialization from other CA classes (Fig. 8).

LCIP63 (ι-CA from the diatom *Thalassiosira pseudonana*) and its bacterial homologs (**COG4875** and **COG4337)** exhibit a primary sequence distinct from any previously identified CA class [13,59–61]. Generally, the polypeptide chains of bacterial ι-CAs exhibit a pre-sequence of 19 or more amino acid residues at the N-terminal region and contain one or two repeated domains. The amino acid sequences of these bacterial ι-CAs show homology to proteins annotated as the SgcJ/EcaC oxidoreductase family, whose function remains unknown. Interestingly, proteins within this family share a structural resemblance with the NTF2-like superfamily, which is characterized by a hydrophobic pocket [26,59,61,62]. This pocket is hypothesized to potentially serve as a substrate-binding or catalytically active site, implicating a role in enzymatic function. One notable feature observed across all analyzed and classified bacterial ι-CAs is the presence of a consensus motif within the C-terminal domain. This motif is characterized by the arrangement of specific amino acid residues, namely (H)HHSS, which appears to be a distinctive hallmark of ι-CAs [26,59,62]. This motif's consistent presence across diverse ι-CAs suggests its functional significance and potential role in defining the enzyme's catalytic properties or structural stability. Further investigation into the specific roles and contributions of the

COG4875

```
Xylella fastidiosa 9a5c    2 YRILLSLSLSFFAFPVFAGNVKCYKGSAPLISN--VEEREVAQLFDHWNAALTTGNPNNVAELYAPDAVLLPTMSNQVRS  79
Brucella melitensis        1 -----------------------------mPTKAKVAALFERWNKSLLTGDPAKVVANCAPDAVLLPTLSKKVRY  46
Ralstonia solanacearum     1 MKHAGVVSLMAAAATVLLVSSPASAQTAPAASCtpTTDQQIAALFDRWNDTLRTGDPDKVTANYATDGVLLPTVSNTPRT  80

                          80 SRSEILDYFSHFLALKPQGYINYRKVTLLsKNNALDAGVVTFILTnkDGSkhNVQARYSFTYSKIDGTWLIVNHHSSAMPEGK 162
                          47 TQKEREDYFRGFLKKKPVGHIDSRTIRTG-CNEALDTGTYTFTFG--DGS--KAAARYTFTYGWNGKNWVITSHHSSAMPQ-- 122
                          81 SPGAIRDYFVKFLKGKPQGTIDSRVIKIG-CNIAQDIGTYTFKFA--DGK--AVHARYTYVYEWQNGQWLIAHHHSSAMFERD 158
```

COG4337

```
Caulobacter vibrioides.   34 LYLLsRRsAPRRFHADQEKDMTPRSIIAAGALFAAVATAPAAWASSTYMGALTEAEVLAAQKAWGEALVAISREHEAGGQ 113
Vibrio cholerae            1 -----------------MQRMTRTLSVSALILCGLLPFTTQAADLVVNNNITHEDVLKAQQANGEALIKISDTYQKQGI  62
Nostoc sp. PCC 7120        1 MNLFKPR-ILVLFAATALISGIAIVAQTSVADSGDKITATSSLKTPIVNRAITESEVLAAQKAWGEALVAISTTYDAKGK  79

                         114 AKAKALASTVLDKAYGYNLGPVLFKPTLTMAPQTFRTTKDGALAYFVGGDPNYPMDKGFALQGWRSVAIKNAAIQIHGDVATTTGNVMITDKNGNVTTVDKTWAFKKTDDGAIRIVVHHSSLPYTAK 240
                          63 KAATELANQVLDQAYGYQQGAVLFKPTLASGKQTFRTDTEGALSYFVGNNKSYPQOSGFALKGWKEYRFENAAVYIDGDLALTTGNVFLVNDKGQETVVDKSWAFKKDEDQQQLRIVLHHSSLPYQAK 189
                          80 ASAKALAERVIDDAYGYQFGPVLFKPTLAISPRTFRTTRAGALAYFVGDDKAFFEDKGFALSSWRKVEIKNAAIFITGNTATTMGNVIITDKQGKATTVDKTWQFLKDDRGKLRIITHHSSLPYEQ- 205
```

Fig. 8 Representative sequences of the conserved **COG4875** and **COG4337** domains according to the Conserved Domain Database (CDD). *Red*, same residue; *blue*, conserved residues; *lowercase amino acids*, unaligned residues.

(H)HHSS motif to ι-CA function is warranted to fully comprehend its significance in the context of catalysis, substrate binding, or protein folding [61]. Their association with protein families, such as CaMKII-AD and SgcJ/EcaC oxidoreductase, highlights the complexity of their functional roles and suggests potential novel enzymatic activities beyond traditional CA functions. A recent study by a Japanese research group uncovered novel carbonic anhydrases (CAs) designated as **COG4337**, encoded by the genomes of the eukaryotic microalga *Bigelowiella natans* and cyanobacterium *Anabaena* sp. PCC7120. Interestingly, COG4337 homologs identified in eukaryotic organisms are multidomain proteins comprising up to five domains, whereas homologs in prokaryotic organisms encode a single domain of approximately 160 amino acid residues in length [13]. These newly identified CAs from *B. natans* and *Anabaena* sp. PCC7120, exhibited catalytic activity even in the absence of the metal ion cofactor, a characteristic not observed in previously reported ι-CAs or any other known CAs studied to date [13] (Fig. 9). Their catalytic mechanism has been mentioned in the introduction of the chapter (see Scheme 1).

This revelation challenges the conventional understanding of the catalytic mechanisms of CAs and suggests the presence of unique structural features or alternative catalytic pathways in the newly identified enzymes. The discovery of these novel **COG4337** CAs expands our knowledge of the diversity and function of CAs across different organisms. The crystal structures of the newly identified ι-CAs suggest a unique homodimeric arrangement [13]. Each monomer within the homodimer forms a cone-shaped barrel structure consisting of three α-helices and a four-stranded antiparallel β-sheet. This folding pattern differs significantly from that observed in the other CAs classes. The proposed catalytic mechanism for CO_2 hydration by **COG4337**

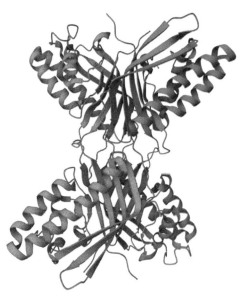

Fig. 9 The structure of ι-CA from the cyanobacterium *Anabaena* sp. PCC7120 complexed with bicarbonate.

proteins involves specific residues in **COG4337** proteins that play essential roles in mediating the hydration of CO_2 despite the absence of metal ions in their active sites [13]. Further experimental validation is required to confirm the proposed mechanism and elucidate the exact details of CO_2 hydration by the **COG4337** proteins.

3. Evolutionary divergence of CA classes

The sequences shown in Figs. 1, 4, 6, and 8 were aligned for phylogenetic analysis using PhyML 3.0 [63]. The resulting dendrogram in Fig. 10 illustrates the significant evolutionary divergence observed among the BCA classes over time. The separation obtained in the dendrogram suggests that each CA family follows a distinct evolutionary path, leading to the formation of well-defined clusters. Each cluster exhibited CAs with potentially divergent features as well as the multifaceted nature of CA evolution. Although all CA classes have a common ancestor, the primary structures of the polypeptide chains of each class differ significantly, leading to low sequence similarity. Additionally, each class possesses characteristic

Fig. 10 Phylogenetic analysis of α-, β-, γ-, and ι-CAs across various species, as obtained from the CDD. The tree was constructed using PhyML 3.0 program. Each name in the dendrogram includes a sequence acronym, conserved domain (CCD), accession number, and CA class. For complete species names, please refer to the alignments provided in Figs. 1, 2, 3, and 4.

folds and structures that are separated from one another. Thus, CA classes have been subjected to a variety of factors that have influenced the evolutionary history of these enzymes, inducing them to evolve independently from one another. Gene duplications, losses, and other evolutionary processes such as genetic drift or horizontal gene transfer may have contributed to the divergence of CAs. This divergence likely led to the development of specialized CA classes adapted to specific biological functions or environmental niches. However, beneath this diversity between the CA classes, there is remarkable consistency in the fundamental mechanism underlying the reversible hydration of CO_2. Convergent evolution refers to the independent evolution of similar traits or features in unrelated species or lineages owing to similar environmental pressures or functional demands [64,65]. Despite their divergent evolutionary trajectories and genetic compositions, CAs collectively exhibit a shared mechanism involving catalysis of the reversible hydration of carbon dioxide to bicarbonate ions and protons. The juxtaposition of sequence and structural diversity with functional conservation in bacterial CA classes exemplifies the complex interplay between evolutionary divergence and functional convergence within the CA superfamily. Each CA class has likely evolved independently to fulfill a specific biochemical function tailored to the physiological needs of the organisms in which they are found, contributing to their survival and adaptation in their respective habitats. The reversible CO_2 hydratase reaction is of paramount importance for various physiological functions across a wide spectrum of organisms. The convergence of this fundamental mechanism across different CA classes underscores the evolutionary plasticity and versatility of this enzyme family for adapting to diverse biological and environmental contexts [65].

4. BCAs are key targets in biomedical research and drug development

To understand the multifaceted nature of pathogenicity, researchers have delved into the intricate role of BCAs as a significant contributor to microbial virulence and antibiotic resistance. The indispensability of CAs for bacterial growth is evident in both non-pathogenic and pathogenic strains. For instance, in *Ralstonia eutropha* [22,66] and *E. coli* [67], CA activity is crucial for growth at ambient CO_2 concentrations to maintain the dissolved inorganic carbon balance [1–3, 68]. In pathogens like

Helicobacter pylori [69–71] and *V. cholerae* [72], CAs facilitate acid acclimatization and induce toxin expression, respectively, highlighting their significance in bacterial survival and virulence. Similarly, *Brucella suis* and *M. tuberculosis* rely on functional CAs for growth, emphasizing the therapeutic potential of targeting bacterial CAs for drug development [17]. Pathogenic microorganisms deploy various strategies to evade host defenses, including pH manipulation facilitated by CA activity, which creates an environment that is conducive to infection [73]. Additionally, CAs play crucial roles in biofilm formation and bacterial proliferation, further complicating the treatment strategies [74]. The role of BCA in microbial life cycles has been extensively investigated using various methodologies. One such approach involves genetic manipulation, in which the deletion of genes encoding bacterial CAs prevents their expression, thus hindering enzyme activity. However, this method has inherent complexities and implementation challenges. Alternatively, employing compounds tailored to inhibit CA enzymes may offer a more accessible avenue. CAIs represent a diverse array of compounds classified into several structural classes, including sulfonamides, sulfamates, coumarins, phenols, and dithiocarbamates [75,76]. Despite their structural diversity, most CAIs share a common binding motif involving the coordination of the zinc-binding group (ZBG). Sulfonamides and their bioisosteres (sulfamates) (Fig. 11), extensively studied as prototypes for understanding inhibitor binding, coordinate with the zinc ion in the active site of the enzyme, thereby interfering with enzyme function through the formation of a stable tetrahedral intermediate [23,77,78].

The sulfonamide binding mode, supported by X-ray crystallographic studies, involves interactions between the sulfonamide group and active site amino acid residues, particularly histidine residues [75]. Hydrophobic and aromatic substituents attached to the sulfonamide nitrogen contribute to inhibitor potency and selectivity by interacting with the hydrophobic pocket in the enzyme active site [1,76,79]. Other classes of CAIs, such as coumarins [80,81], phenols [82], and dithiocarbamates [82,83], exhibit distinct binding patterns and inhibition mechanisms, thereby broadening the chemical variety of CAIs and offering valuable insights into the structural features necessary for effective enzyme inhibition (Fig. 12).

While much research has focused on human CA II, differences in the structure and function between human and bacterial CAs necessitate careful consideration in drug development [84]. Structural disparities, metal cofactor preferences, and substrate specificities between bacterial and human CAs influence inhibitor binding patterns and catalytic mechanisms,

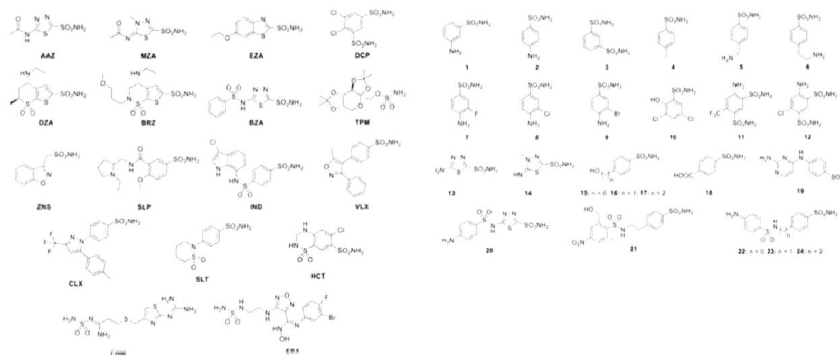

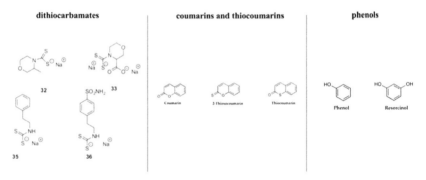

Fig. 11 AAZ-EPA agents employed in clinical practice and commercially accessible sulfonamide derivatives numbered 1–24, utilized as CAIs.

Fig. 12 Dithiocarbamates (compounds with a dithiocarbamate group ($-HNCS_2^-$)), coumarins (compounds incorporating a fused benzene ring and a lactone moiety), phenols (characterized by a hydroxyl group attached to an aromatic ring). Dithiocarbamates coordinate with the zinc ion via one sulfur atom, forming a stable complex. Coumarin derivatives, leveraging hydrophobic interactions and π–π stacking with aromatic residues within the enzyme's active site, these inhibitors exhibit heightened binding affinity and inhibitory potency. Phenols engage in hydrogen bonding and hydrophobic interactions with the active site zinc ion and neighboring amino acid residues.

highlighting the importance of tailored drug design for effective targeting of bacterial infections [85–87]. CAIs disrupt pH-regulating mechanisms, rendering pathogens more susceptible to host defenses and antimicrobial treatments. Furthermore, the diverse mechanisms of CA inhibition offer a promising avenue for drug development with sulfonamide-based CAIs emerging as potent candidates [88,89]. Through drug repurposing and structural insights into CA-inhibitor interactions, researchers aim to develop novel antibacterial agents with improved selectivity and efficacy, offering innovative solutions to combat microbial infections and addressing antibiotic resistance challenges. CAIs have been used to elucidate their effects on the microbial physiology and metabolism [68,75,90]. This process involves selecting a microorganism, conducting genome analysis to identify its CA enzymes, and employing specific inhibitors to disrupt enzyme activity [4,17,74,87,91]. Evaluating CA inhibition in microorganisms involves assessing microbial growth and metabolic processes, shedding light on how these inhibitors influence microbial metabolism [18,92]. Nonetheless, the primary challenge lies in the limited selectivity of CAIs for human enzyme isoforms, necessitating the development of novel inhibitors with enhanced specificity. Recent studies have explored the efficacy of CAIs against bacterial CAs, revealing their potential to inhibit bacterial growth and virulence. Notably, some CAIs, notably sulfonamides, have demonstrated significant inhibitory effects on bacterial CAs and impede bacterial growth in cell culture [93]. However, the translation of in vitro efficacy to in vivo inhibition of microorganisms poses a challenge because of the limited ability of inhibitors to penetrate bacterial cells. To address this issue, researchers have evaluated the capacity of CAIs to interfere with bacterial growth and metabolism using non-pathogenic prokaryotic cells as model microorganisms, as exemplified by studies using *E. coli* [18]. Such endeavors underscore the feasibility of identifying novel chemical alternatives capable of disrupting bacterial metabolism, offering promising avenues for combating drug-resistant pathogens such as vancomycin-resistant enterococci (VRE) and *Neisseria gonorrhoeae* [94–100]. These findings not only provide insights into the intricate relationship between CAs and bacterial metabolism but also offer potential therapeutic strategies against antibiotic-resistant bacterial infections.

5. BCA and their potential applications in carbon capture technologies

CAs offer promising avenues for addressing global warming by facilitating the capture and sequestration of CO_2 from combustion gases. These enzymes, particularly thermophilic variants, remain active under harsh conditions typical of CO_2 capture processes, making them valuable biocatalysts [101–103]. Various techniques, including chemical absorption, have been explored for CO_2 capture. Chemical absorption involves the reaction of CO_2 with a solvent, followed by regeneration to release pure CO_2. This method, which typically employs alkanolamines as absorbents, requires a significant energy input for desorption and solvent pumping. The use of CAs, particularly heat-stable variants from thermophilic bacteria, such as *S. yellowstonense* and *Caminibacter mediatlanticus*, enhances the rate of CO_2 absorption, thereby offering a potential solution to reduce energy costs [104]. Furthermore, the direct conversion of CO_2 to calcium carbonate ($CaCO_3$) catalyzed by CAs is a promising approach for safer CO_2 storage [105]. Recent developments have focused on heterologous expression and immobilization techniques to enhance enzyme stability and efficiency [106–113]. Immobilization of CAs onto solid supports, such as polyurethane foam or hollow fibers, addresses the challenges associated with enzyme stability and repeated usage, enhancing the efficiency of CCS technologies. Recent advancements, including the development of γ-CA nanoassemblies for controlled immobilization, have underscored the potential of CAs to revolutionize CO_2 capture and storage processes [114]. Recombinant thermostable SspCA from *S. yellowstonense* was successfully immobilized onto magnetic Fe_3O_4 nanoparticles (MNP), enabling its recovery and reuse through simple magnet-based separation methods [110]. Moreover, innovative approaches, involving direct anchoring of SspCA onto the outer membrane of *E. coli* cells have shown promising results [101] (Fig. 13). This strategy not only simplifies the enzyme purification and immobilization processes but also enhances enzyme stability

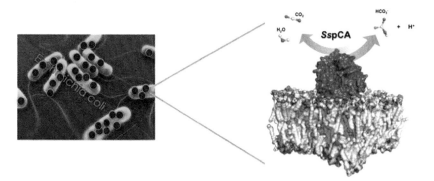

Fig. 13 *Escherichia coli* cells (left side), transformed with a specific construct, expressing SspCA directly immobilized on the outer membrane. This system significantly reduces the cost of producing biocatalyst and enzyme immobilization. The blue dots visible on the left side of the figure represent SspCA on the outer membrane of *E. coli* cells. On the right side of the figure, the description proceeds from bottom to top, detailing the following components: the outer membrane of *E. coli*, the recombinant SspCA (blue color) on the external surface of the bacterial membrane, and the reversible reaction catalyzed by SspCA.

and activity. By harnessing these approaches, the biomimetic capture of CO_2 becomes more cost-effective and environmentally friendly, paving the way for broader biotechnological applications.

6. Conclusions

Bacterial α-, β-, γ-, and ι-CAs exhibit diverse structural features and catalytic activities, essential for various physiological functions and enzymatic behaviors. α-CAs, characterized by the conserved **COG3338** domain, demonstrate structural similarities to mammalian counterparts and possess compact structures with unique active site configurations. β-CAs, divided into distinct clades, exhibit diverse oligomeric states and undergo pH-dependent conformational changes, influencing catalytic activity. γ-CAs, marked by specific conserved domains, display trimeric arrangements and vary in catalytic efficiencies, offering potential as targets for selective inhibitors in pathogenic bacteria. ι-CAs, identified by unique **COG4875** and **COG4337** domains, present novel structural arrangements and catalytic mechanisms, challenging traditional understanding and expanding the diversity of CA functions. The phylogenetic analysis reveals significant evolutionary divergence and specialization among α, β, γ, and ι carbonic anhydrase (CA) classes, indicating the intricate multifaceted nature of CA evolution. Despite their divergent evolutionary trajectories and genetic compositions, CAs collectively exhibit a shared mechanism involving reversible hydration of carbon dioxide. The convergence of this mechanism across different CA classes highlights the interplay between evolutionary divergence and functional convergence, which contributes to the survival and adaptation of organisms in their respective habitats.

BCAs play a significant role in microbial virulence, antibiotic resistance, and overall bacterial physiology. They are essential for the growth, acid acclimatization, toxin expression, and biofilm formation of pathogenic bacteria. Targeting BCAs for drug development offers a promising avenue for combating bacterial infections, with CAIs emerging as potential candidates. The development of selective inhibitors with enhanced efficacy against bacterial CAs remains a challenge. However, recent studies have shown that CAIs have the potential to inhibit bacterial growth and virulence, thereby providing hope for combating antibiotic-resistant pathogens.

BCAs offer promising solutions for mitigating global warming by capturing and sequestering CO_2 from combustion gases. In particular, thermophilic variants exhibit robust activity under the harsh conditions typical of CO_2 capture processes, making them valuable biocatalysts. Recent advancements in heterologous expression and immobilization techniques have enhanced CA stability and efficiency, furthering the potential of CAs for CO_2 capture and storage. These advancements have paved the way for broader biotechnological applications of CAs in combating climate change.

Thus, insights into bacterial CA classes not only deepen our understanding of enzymatic behavior, but also hold promise for biotechnological applications and therapeutic interventions targeting bacterial infections.

References

[1] C. Capasso, C.T. Supuran, An overview of the selectivity and efficiency of the bacterial carbonic anhydrase inhibitors, Curr. Med. Chem. 22 (18) (2015) 2130–2139.

[2] C. Capasso, C.T. Supuran, An overview of the alpha-, beta- and gamma-carbonic anhydrases from Bacteria: can bacterial carbonic anhydrases shed new light on evolution of bacteria? J. Enzym. Inhib. Med. Ch 30 (2) (2015) 325–332.

[3] C.T. Supuran, C. Capasso, An overview of the bacterial carbonic anhydrases, Metabolites 7 (4) (2017) 56–74.

[4] C.T. Supuran, C. Capasso, Biomedical applications of prokaryotic carbonic anhydrases, Expert. Opin. Ther. Pat. 28 (10) (2018) 745–754.

[5] R.S. Kumar, J.G. Ferry, Prokaryotic carbonic anhydrases of Earth's environment, Subcell. Biochem. 75 (2014) 77–87.

[6] K.S. Smith, J.G. Ferry, Prokaryotic carbonic anhydrases, FEMS Microbiol. Rev. 24 (4) (2000) 335–366.

[7] N. Mangan, M. Brenner, Systems analysis of the CO_2 concentrating mechanism in cyanobacteria, Elife 3 (2014) 02043–02059.

[8] R. Occhipinti, W.F. Boron, Role of carbonic anhydrases and inhibitors in acid-base physiology: insights from mathematical modeling, Int. J. Mol. Sci. 20 (15) (2019) 3841–3870.

[9] L.L. Wang, J.J. Liang, Y. Zhou, T. Tian, B.L. Zhang, D.Q. Duanmu, Molecular characterization of carbonic anhydrase genes in Lotus japonicus and their potential roles in symbiotic nitrogen fixation, Int. J. Mol. Sci. 22 (15) (2021) 7766–7780.

[10] L.V.N.S. Khandavalli, T. Lodha, M. Abdullah, L. Guruprasad, S. Chintalapati, V.R. Chintalapati, Insights into the carbonic anhydrases and autotrophic carbon dioxide fixation pathways of high CO_2 tolerant Rhodovulum viride JA756, Microbiol. Res. 215 (2018) 130–140.

[11] F.Y. Gong, G.X. Liu, X.Y. Zhai, J. Zhou, Z. Cai, Y. Li, Quantitative analysis of an engineered CO_2-fixing Escherichia coli reveals great potential of heterotrophic CO_2 fixation, Biotechnol. Biofuels 8 (2015) 86–96.

[12] Y.J. Wang, D.J. Stessman, M.H. Spalding, The CO_2 concentrating mechanism and photosynthetic carbon assimilation in limiting CO_2: how Chlamydomonas works against the gradient, Plant. J. 82 (3) (2015) 429–448.

[13] Y. Hirakawa, M. Senda, K. Fukuda, H.Y. Yu, M. Ishida, M. Taira, et al., Characterization of a novel type of carbonic anhydrase that acts without metal cofactors, BMC Biol. 19 (1) (2021) 105–119.

[14] A. Aspatwar, M.E.E. Tolvanen, H. Barker, L. Syrjänen, S. Valanne, S. Purmonen, et al., Carbonic anhydrases in metazoan model organisms: molecules, mechanisms, and physiology, Physiol. Rev. 102 (3) (2022) 1327–1383.

[15] Y. Hirakawa, Y. Hanawa, K. Yoneda, I. Suzuki, Evolution of a chimeric mitochondrial carbonic anhydrase through gene fusion in a haptophyte alga, FEBS Lett. 596 (23) (2022) 3051–3059.

[16] C. Capasso, Carbonic anhydrases: a superfamily of ubiquitous enzymes, Int. J. Mol. Sci. 24 (8) (2023) 7014–7017.

[17] C. Capasso, C.T. Supuran, Carbonic anhydrase and bacterial metabolism: a chance for antibacterial drug discovery, Expert. Opin. Ther. Pat. (2024) 1–10.

[18] V. De Luca, V. Carginale, C.T. Supuran, C. Capasso, The gram-negative bacterium *Escherichia coli* as a model for testing the effect of carbonic anhydrase inhibition on bacterial growth, J. Enzym. Inhib. Med. Ch 37 (1) (2022) 2092–2098.

[19] S. Del Prete, R. Merlo, A. Valenti, R. Mattossovich, M. Rossi, V. Carginale, et al., Thermostability enhancement of the alpha-carbonic anhydrase from *Sulfurihydrogenibium yellowstonense* by using the anchoring-and-self-labelling-protein-tag system (ASL(tag)), J. Enzyme Inhib. Med. Chem. 34 (1) (2019) 946–954.

[20] E. Kupriyanova, N. Pronina, D. Los, Carbonic anhydrase - a universal enzyme of the carbon-based life, Photosynthetica 55 (1) (2017) 3–19.

[21] K.S. Smith, C. Jakubzick, T.S. Whittam, J.G. Ferry, Carbonic anhydrase is an ancient enzyme widespread in prokaryotes, Proc. Natl Acad. Sci. USA 96 (26) (1999) 15184–15189.

[22] C.S. Gai, J.N. Lu, C.J. Brigham, A.C. Bernardi, A.J. Sinskey, Insights into bacterial CO_2 metabolism revealed by the characterization of four carbonic anhydrases in *Ralstonia eutropha* H16, Amb. Express 4 (1) (2014) 12.

[23] D. Cuffaro, E. Nuti, A. Rossello, An overview of carbohydrate-based carbonic anhydrase inhibitors, J. Enzym. Inhib. Med. Ch 35 (1) (2020).

[24] R. McKenna, S.C. Frost, Overview of the carbonic anhydrase family carbonic anhydrase: mechanism, regulation, links disease, Ind. Appl. 75 (2014) 3–5.

[25] M. Sola, A. Lledos, M. Duran, J. Bertran, Theoretical-study of the catalyzed hydration of CO_2 by carbonic-anhydrase - a brief overview, Nato Adv. Sci. I C-Mat 368 (1992) 263–298.

[26] C.T. Supuran, An overview of novel antimicrobial carbonic anhydrase inhibitors, Expert. Opin. Ther. Tar. 27 (10) (2023) 897–910.

[27] E.T. Granato, T.A. Meiller-Legrand, K.R. Foster, The evolution and ecology of bacterial warfare, Curr. Biol. 29 (11) (2019) R521–R537.

[28] A.K. Wani, N. Akhtar, F. Sher, A.A. Navarrete, J.H.P. Americo-Pinheiro, Microbial adaptation to different environmental conditions: molecular perspective of evolved genetic and cellular systems, Arch. Microbiol. 204 (2) (2022) 144.

[29] A. Akdemir, D. Vullo, V. De Luca, A. Scozzafava, V. Carginale, et al., The extremo-alpha-carbonic anhydrase (CA) from *Sulfurihydrogenibium azorense*, the fastest CA known, is highly activated by amino acids and amines, Bioorg. Med. Chem. Lett. 23 (4) (2013) 1087–1090.

[30] A.M. Alafeefy, H.A. Abdel-Aziz, D. Vullo, A.M. Al-Tamimi, N.A. Al-Jaber, C. Capasso, et al., Inhibition of carbonic anhydrases from the extremophilic bacteria *Sulfurihydrogenibium yellowstonense* (SspCA) and *S. azorense* (SazCA) with a new series of sulfonamides incorporating aroylhydrazone-, [1,2,4]triazolo[3,4-b][1,3,4]thiadiazinyl- or 2-(cyanophenylmethylene)-1,3,4-thiadiazol-3(2H)-yl moieties, Bioorg. Med. Chem. 22 (1) (2014) 141–147.

[31] V. De Luca, S. Del Prete, V. Carginale, D. Vullo, C.T. Supuran, C. Capasso, A failed tentative to design a super carbonic anhydrase having the biochemical properties of the most thermostable CA (SspCA) and the fastest (SazCA) enzymes, J. Enzyme Inhib. Med. Chem. 30 (6) (2015) 989–994.

[32] D. Vullo, V. De Luca, A. Scozzafava, V. Carginale, M. Rossi, C.T. Supuran, et al., Anion inhibition studies of the fastest carbonic anhydrase (CA) known, the extremo-CA from the bacterium *Sulfurihydrogenibium azorense*, Bioorg. Med. Chem. Lett. 22 (23) (2012) 7142–7145.

[33] D. Vullo, V. De Luca, A. Scozzafava, V. Carginale, M. Rossi, C.T. Supuran, et al., The extremo-alpha-carbonic anhydrase from the thermophilic bacterium *Sulfurihydrogenibium azorense* is highly inhibited by sulfonamides, Bioorg. Med. Chem. 21 (15) (2013) 4521–4525.

[34] R.M. Stubbendieck, C. Vargas-Bautista, P.D. Straight, Bacterial communities: interactions to scale, Front. Microbiol. 7 (2016) 1234.

[35] S. Del Prete, V. De Luca, C. Capasso, C.T. Supuran, V. Carginale, Recombinant thermoactive phosphoenolpyruvate carboxylase (PEPC) from *Thermosynechococcus elongatus* and its coupling with mesophilic/thermophilic bacterial carbonic anhydrases (CAs) for the conversion of CO_2 to oxaloacetate, Bioorg. Med. Chem. 24 (2) (2016) 220–225.

[36] F. Migliardini, V. De Luca, V. Carginale, M. Rossi, P. Corbo, C.T. Supuran, et al., Biomimetic CO_2 capture using a highly thermostable bacterial alpha carbonic anhydrase immobilized on a polyurethane foam, J. Enzyme Inhib. Med. Chem. 29 (1) (2014) 146–150.

[37] L. Parri, A. Fort, A. Lo Grasso, M. Mugnaini, V. Vignoli, C. Capasso, et al., Evaluating the efficiency of enzyme accelerated CO(2) capture: chemical kinetics modelling for interpreting measurement results, J. Enzyme Inhib. Med. Chem. 36 (1) (2021) 394–401.

[38] R. Perfetto, S. Del Prete, D. Vullo, G. Sansone, C. Barone, M. Rossi, et al., Biochemical characterization of the native alpha-carbonic anhydrase purified from the mantle of the Mediterranean mussel, *Mytilus galloprovincialis*, J. Enzyme Inhib. Med. Chem. 32 (1) (2017) 632–639.

[39] R. Perfetto, S. Del Prete, D. Vullo, G. Sansone, C.M.A. Barone, M. Rossi, et al., Production and covalent immobilisation of the recombinant bacterial carbonic anhydrase (SspCA) onto magnetic nanoparticles, J. Enzyme Inhib. Med. Chem. 32 (1) (2017) 759–766.

[40] A. Di Fiore, C.T. Supuran, A. Scaloni, G. De Simone, Human carbonic anhydrases and post-translational modifications: a hidden world possibly affecting protein properties and functions, J. Enzym. Inhib. Med. Ch 35 (1) (2020) 1450–1461.

[41] C.T. Supuran, C. Capasso, New light on bacterial carbonic anhydrases phylogeny based on the analysis of signal peptide sequences, J. Enzyme Inhib. Med. Chem. 31 (6) (2016) 1254–1260.

[42] J.Y. Wang, F. Chitsaz, M.K. Derbyshire, N.R. Gonzales, M. Gwadz, S.N. Lu, et al., The conserved domain database in 2023, Nucl. Acids Res. 51 (D1) (2023) D384–D388.

[43] J.K. Modak, Y.C. Liu, M.A. Machuca, C.T. Supuran, A. Roujeinikova, Structural basis for the inhibition of *Helicobacter pylori* alpha-carbonic anhydrase by sulfonamides, PLoS One 10 (5) (2015) e0127149.

[44] P.O. Staff, Correction: structural basis for the inhibition of *Helicobacter pylori* alpha-carbonic anhydrase by sulfonamides, PLoS One 10 (7) (2015) e0132763.

[45] A. Di Fiore, C. Capasso, V. De Luca, S.M. Monti, V. Carginale, C.T. Supuran, et al., X-ray structure of the first;extremo-alpha-carbonic anhydrase', a dimeric enzyme from the thermophilic bacterium *Sulfurihydrogenibium yellowstonense* YO3AOP1, Acta 69 (Pt 6) (2013) 1150–1159.

[46] G. De Simone, S.M. Monti, V. Alterio, M. Buonanno, V. De Luca, M. Rossi, et al., Crystal structure of the most catalytically effective carbonic anhydrase enzyme known SazCA thermophilic bacterium, Bioorg. Med. Chem. Lett. 25 (9) (2015) 2002–2006.

[47] P. James, M.N. Isupov, C. Sayer, V. Saneei, S. Berg, M. Lioliou, et al., The structure of a tetrameric α-carbonic anhydrase from *Thermovibrio ammonificans* reveals a core formed around intermolecular disulfides that contribute to its thermostability, Acta Crystallogr. D. 70 (2014) 2607–2618.

[48] P. Piazzetta, T. Marino, N. Russo, Mechanistic explanation of the weak carbonic anhydrase's esterase activity, Molecules 22 (6) (2017).

[49] F. Briganti, S. Mangani, A. Scozzafava, G. Vernaglione, C.T. Supuran, Carbonic anhydrase catalyzes cyanamide hydration to urea: is it mimicking the physiological reaction? J. Biol. Inorg. Chem. 4 (5) (1999) 528–536.

[50] M. Ferraroni, S. Del Prete, D. Vullo, C. Capasso, C.T. Supuran, Crystal structure and kinetic studies of a tetrameric type II beta-carbonic anhydrase from the pathogenic bacterium *Vibrio cholerae*, Acta 71 (Pt 12) (2015) 2449–2456.

[51] J.D. Cronk, J.A. Endrizzi, M.R. Cronk, J.W. O'Neill, K.Y.J. Zhang, Crystal structure of *E. coli* β-carbonic anhydrase, an enzyme with an unusual pH-dependent activity, Protein Sci. 10 (5) (2001) 911–922.

[52] J.D. Cronk, R.S. Rowlett, K.Y.J. Zhang, C.K. Tu, J.A. Endrizzi, J. Lee, et al., Identification of a novel noncatalytic bicarbonate binding site in eubacterial β-carbonic anhydrase, Biochemistry-Us 45 (14) (2006) 4351–4361.

[53] A.S. Covarrubias, T. Bergfors, T.A. Jones, M. Högbom, Structural mechanics of the pH-dependent activity of β-carbonic anhydrase from, J. Biol. Chem. 281 (8) (2006) 4993–4999.

[54] M.J. Smeulders, A. Pol, H. Venselaar, T.R. Barends, J. Hermans, M.S. Jetten, et al., Bacterial CS2 hydrolases from *Acidithiobacillus thiooxidans* strains are homologous to the archaeal catenane CS2 hydrolase, J. Bacteriol. 195 (18) (2013) 4046–4056.

[55] B.E. Alber, J.G. Ferry, A carbonic-anhydrase from the archaeon *Methanosarcina thermophila*, Proc. Natl Acad. Sci. USA 91 (15) (1994) 6909–6913.

[56] C. Kisker, H. Schindelin, B.E. Alber, J.G. Ferry, D.C. Rees, A left-handed beta-helix revealed by the crystal structure of a carbonic anhydrase from the archaeon *Methanosarcina thermophila*, EMBO J. 15 (10) (1996) 2323–2330.

[57] T.M. Iverson, B.E. Alber, C. Kisker, J.G. Ferry, D.C. Rees, A closer look at the active site of γ-class carbonic anhydrases: high-resolution crystallographic studies of the carbonic anhydrase from, Biochemistry-Us 39 (31) (2000) 9222–9231.

[58] A. Di Fiore, V. De Luca, E. Langella, A. Nocentini, M. Buonanno, S.M. Monti, et al., Biochemical, structural, and computational studies of a y-carbonic anhydrase from the pathogenic bacterium, Comput. Struct. Biotec. 20 (2022) 4185–4194.

[59] A. Nocentini, C.T. Supuran, C. Capasso, An overview on the recently discovered iota-carbonic anhydrases, J. Enzym. Inhib. Med. Ch 36 (1) (2021) 1988–1995.

[60] E.L. Jensen, V. Receveur-Brechot, M. Hachemane, L. Wils, P. Barbier, G. Parsiegla, et al., Structural contour map of the iota carbonic anhydrase from the diatom *Thalassiosira pseudonana* using a multiprong approach, Int. J. Mol. Sci. 22 (16) (2021).

[61] E.L. Jensen, R. Clement, A. Kosta, S.C. Maberly, B. Gontero, A new widespread subclass of carbonic anhydrase in marine phytoplankton, ISME J. 13 (8) (2019) 2094–2106.

[62] S. Del Prete, A. Nocentini, C.T. Supuran, C. Capasso, Bacterial ι-carbonic anhydrase: a new active class of carbonic anhydrase identified in the genome of the Gram-negative bacterium, J. Enzym. Inhib. Med. Ch 35 (1) (2020) 1060–1068.

[63] S. Guindon, J.F. Dufayard, V. Lefort, M. Anisimova, W. Hordijk, O. Gascuel, New algorithms and methods to estimate maximum-likelihood phylogenies: assessing the performance of PhyML 3.0, Syst. Biol. 59 (3) (2010) 307–321.

[64] K. Tomii, Y. Sawada, S. Honda, Convergent evolution in structural elements of proteins investigated using cross profile analysis, BMC Bioinforma 13 (2012).

[65] A.J.M. Ribeiro, I.G. Riziotis, N. Borkakoti, J.M. Thornton, Enzyme function and evolution through the lens of bioinformatics, Biochem. J. 480 (22) (2023) 1845–1863.

[66] E. Schwartz, B. Voigt, D. Zühlke, A. Pohlmann, O. Lenz, D. Albrecht, et al., A proteomic view of the facultatively chemolithoautotrophic lifestyle of *Ralstonia eutropha* H16, Proteomics 9 (22) (2009) 5132–5142.

[67] C. Merlin, M. Masters, S. McAteer, A. Coulson, Why is carbonic anhydrase essential to *Escherichia coli*? J. Bacteriol. 185 (21) (2003) 6415–6424.

[68] C.T. Supuran, Bacterial carbonic anhydrases as drug targets: toward novel antibiotics? Front. Pharmacol. 2 (2011) 1–6.

[69] S. Bury-Mone, G.L. Mendz, G.E. Ball, M. Thibonnier, K. Stingl, Ecobichon, et al., Roles of alpha and beta carbonic anhydrases of *Helicobacter pylori* in the urease-dependent response to acidity and in colonization of the murine gastric mucosa, Infect. Immun. 76 (2) (2008) 497–509.

[70] J.K. Modak, Y.C. Liu, C.T. Supuran, A. Roujeinikova, Structure-activity relationship for sulfonamide inhibition of *Helicobacter pylori* α carbonic anhydrase, J. Med. Chem. 59 (24) (2016) 11098–11109.

[71] D.Scott, E.Marcus, G.Sachs, Inventors carbonic anhydrase inhibitors as drugs to eradicate *Helicobacter pylori* in the mammalian, including human stomach patent US20040204493A1, 2004.

[72] B.H. Abuaita, J.H. Withey, Bicarbonate induces *Vibrio cholerae* virulence gene expression by enhancing ToxT activity, Infect. Immun. 77 (9) (2009) 4111–4120.

[73] J. Pieters, Evasion of host cell defense mechanisms by pathogenic bacteria, Curr. Opin. Immunol. 13 (1) (2001) 37–44.

[74] C. Campestre, V. De Luca, S. Carradori, R. Grande, V. Carginale, A. Scaloni, et al., Carbonic anhydrases: new perspectives on protein functional role and inhibition in, Front. Microbiol. 12 (2021).

[75] C.T. Supuran, How many carbonic anhydrase inhibition mechanisms exist? J. Enzym. Inhib. Med. Chem. 31 (3) (2016) 345–360.

[76] D. Cuffaro, E. Nuti, A. Rossello, An overview of carbohydrate-based carbonic anhydrase inhibitors, J. Enzyme Inhib. Med. Chem. 35 (1) (2020) 1906–1922.

[77] S.P. Gupta, Quantitative structure-activity relationships of carbonic anhydrase inhibitors, Prog. Drug. Res. 60 (2003) 171–204.

[78] C.S. Hewitt, N.S. Abutaleb, A.E.M. Elhassanny, A. Nocentini, X.F. Cao, D.P. Amos, et al., Structure-activity relationship studies of acetazolamide-based carbonic anhydrase inhibitors with activity against, ACS Infect. Dis. 7 (7) (2021) 1969–1984.

[79] S. Galati, D. Yonchev, R. Rodríguez-Pérez, M. Vogt, T. Tuccinardi, J. Bajorath, Predicting isoform-selective carbonic anhydrase inhibitors via machine learning and rationalizing structural features important for selectivity, ACS Omega 6 (5) (2021) 4080–4089.

[80] M. Kaur, S. Kohli, S. Sandhu, Y. Bansal, G. Bansal, Coumarin: a promising scaffold for anticancer agents, Anti-Cancer Agent. Me 15 (8) (2015) 1032–1048.

[81] F. Annunziata, C. Pinna, S. Dallavalle, L. Tamborini, A. Pinto, An overview of coumarin as a versatile and readily accessible scaffold with broad-ranging biological activities, Int. J. Mol. Sci. 21 (13) (2020) 4618–4698.

[82] J.Y. Winum, C.T. Supuran, Recent advances in the discovery of zinc-binding motifs for the development of carbonic anhydrase inhibitors, J. Enzym. Inhib. Med. Ch 30 (2) (2015) 321–324.

[83] T.O. Ajiboye, T.T. Ajiboye, R. Marzouki, D.C. Onwudiwe, The versatility in the applications of dithiocarbamates, Int. J. Mol. Sci. 23 (3) (2022) 1317–1352.

[84] D.P. Flaherty, M.N. Seleem, C.T. Supuran, Bacterial carbonic anhydrases: under-exploited antibacterial therapeutic targets, Fut. Med. Chem. 13 (19) (2021) 1619–1622.

[85] R. McKenna, C.T. Supuran, Carbonic anhydrase inhibitors drug design carbonic anhydrase: mechanism, regulation, links disease, Ind. Appl. 75 (2014) 291–323.

[86] M. Mustafa, J.Y. Winum, The importance of sulfur-containing motifs in drug design and discovery, Expert. Opin. Drug. Dis. 17 (5) (2022) 501–512.

[87] A. Nocentini, A. Angeli, F. Carta, J.Y. Winum, R. Zalubovskis, S. Carradori, et al., Reconsidering anion inhibitors in the general context of drug design studies of modulators of activity of the classical enzyme carbonic anhydrase, J. Enzym. Inhib. Med. Ch 36 (1) (2021) 561–580.

[88] J. Kaur, X. Cao, N.S. Abutaleb, A. Elkashif, A.L. Graboski, A.D. Krabill, et al., Optimization of acetazolamide-based scaffold as potent inhibitors of vancomycin-resistant enterococcus, J. Med. Chem. 63 (17) (2020) 9540–9562.

[89] N.S. Abutaleb, A. Elkashif, D.P. Flaherty, M.N. Seleem, In vivo antibacterial activity of Acetazolamide, Antimicrob. Agents Ch 65 (4) (2021).

[90] C.T. Supuran, Carbonic anhydrase versatility: from pH regulation to CO(2) sensing and metabolism, Front. Mol. Biosci. 10 (2023) 1326633.

[91] A. Nocentini, C. Capasso, C.T. Supuran, Carbonic anhydrase inhibitors as novel antibacterials in the era of antibiotic resistance: where are we now? Antibiotics-Basel 12 (1) (2023) 12010142–12010154.

[92] A. Shah, B.J. Eikmanns, Transcriptional regulation of the β-type carbonic anhydrase gene bca by RamA in Corynebacterium glutamicum, PLoS One 11 (4) (2016) 0154382–0154399.

[93] L. Kergoat, P. Besse-Hoggan, M. Leremboure, J. Beguet, M. Devers, F. Martin-Laurent, et al., Environmental concentrations of sulfonamides can alter bacterial structure and induce diatom deformities in freshwater biofilm communities, Front. Microbiol. 12 (2021).

[94] N.S. Abutaleb, A.E.M. Elhassanny, A. Nocentini, C.S. Hewitt, A. Elkashif, B.R. Cooper, et al., Repurposing FDA-approved sulphonamide carbonic anhydrase inhibitors for treatment of Neisseria gonorrhoeae, J. Enzym. Inhib. Med. Ch 37 (1) (2022) 51–61.

[95] S. Giovannuzzi, N.S. Abutaleb, C.S. Hewitt, F. Carta, A. Nocentini, M.N. Seleem, et al., Dithiocarbamates effectively inhibit the alpha-carbonic anhydrase from Neisseria gonorrhoeae, J. Enzym. Inhib. Med. Ch 37 (1) (2022) 1–8.

[96] C.S. Hewitt, N.S. Abutaleb, A.E.M. Elhassanny, A. Nocentini, X. Cao, D.P. Amos, et al., Structure-activity relationship studies of acetazolamide-based carbonic anhydrase inhibitors with activity against Neisseria gonorrhoeae, ACS Infect. Dis. 7 (7) (2021) 1969–1984.

[97] A. Nocentini, C.S. Hewitt, M.D. Mastrolorenzo, D.P. Flaherty, C.T. Supuran, Anion inhibition studies of the alpha-carbonic anhydrases from Neisseria gonorrhoeae, J. Enzym. Inhib. Med. Ch 36 (1) (2021) 1061–1066.

[98] G.S. Chilambi, Y.H. Wang, N.R. Wallace, C. Obiwuma, K.M. Evans, Y. Li, et al., Carbonic anhydrase inhibition as a target for antibiotic synergy in enterococci, Microbiol. Spectr. 11 (4) (2023) e0396322.

[99] A. Palmieri, M. Martinelli, A. Pellati, F. Carinci, D. Lauritano, C. Arcuri, et al., Prevalence of enterococci and vancomycin resistance in the throat of non-hospitalized individuals randomly selected in Central Italy, Antibiotics (Basel) 12 (7) (2023).

[100] M.E. de Kraker, V. Jarlier, J.C. Monen, O.E. Heuer, N. van de Sande, H. Grundmann, The changing epidemiology of bacteraemias in Europe: trends from the European Antimicrobial Resistance Surveillance System, Clin. Microbiol. Infect. 19 (9) (2013) 860–868.

[101] S. Del Prete, R. Perfetto, M. Rossi, F.A.S. Alasmary, S.M. Osman, Z. AlOthman, et al., A one-step procedure for immobilising the thermostable carbonic anhydrase (SspCA) on the surface membrane of, J. Enzym. Inhib. Med. Ch 32 (1) (2017) 1120–1128.

[102] F. Migliardini, V. De Luca, V. Carginale, M. Rossi, P. Corbo, C.T. Supuran, et al., Biomimetic CO_2 capture using a highly thermostable bacterial α-carbonic anhydrase immobilized on a polyurethane foam, J. Enzym. Inhib. Med. Ch 29 (1) (2014) 146–150.

[103] D. Vullo, V. De Luca, A. Scozzafava, V. Carginale, M. Rossi, C.T. Supuran, et al., The first activation study of a bacterial carbonic anhydrase (CA). The thermostable α-CA from YO3AOP1 is highly activated by amino acids and amines, Bioorg. Med. Chem. Lett. 22 (20) (2012) 6324–6327.

[104] A. Di Fiore, V. Alterio, S.M. Monti, G. De Simone, K. D'Ambrosio, Thermostable carbonic anhydrases in biotechnological applications, Int. J. Mol. Sci. 16 (7) (2015) 15456–15480.

[105] T. Sharma, A. Sharma, C.L. Xia, S.S. Lam, A.A. Khan, S. Tripathi, et al., Enzyme mediated transformation of CO_2 into calcium carbonate using purified microbial carbonic anhydrase, Env. Res. 212 (2022).

[106] H. Rasouli, K. Nguyen, M.C. Iliuta, Recent advancements in carbonic anhydrase immobilization and its implementation in CO_2 capture technologies: a review, Sep. Purif. Technol. 296 (2022).

[107] R. Merlo, S. Del Prete, A. Valenti, R. Mattossovich, V. Carginale, C.T. Supuran, et al., An AGT-based protein-tag system for the labelling and surface immobilization of enzymes on E. coli outer membrane, J. Enzym. Inhib. Med. Ch 34 (1) (2019) 490–499.

[108] S. Peirce, M.E. Russo, V. De Luca, C. Capasso, M. Rossi, G. Olivieri, et al., Immobilization of carbonic anhydrase for biomimetic CO_2 capture, N. Biotechnol. 31 (2014) S20–S21.

[109] S. Peirce, M.E. Russo, V. De Luca, C. Capasso, M. Rossi, G. Olivieri, et al., Immobilization of carbonic anhydrase for biomimetic CO_2 capture in a slurry absorber as cross-linked enzyme aggregates (CLEA), Chem. Eng. Trans. 43 (2015) 259–264.

[110] R. Perfetto, S. Del Prete, D. Vullo, G. Sansone, C.M.A. Barone, M. Rossi, et al., Production and covalent immobilisation of the recombinant bacterial carbonic anhydrase (SspCA) onto magnetic nanoparticles, J. Enzym. Inhib. Med. Ch 32 (1) (2017) 759–766.

[111] M.E. Russo, C. Capasso, A. Marzocchella, P.I. Salatino, Immobilization of carbonic anhydrase for CO_2 capture and utilization, Appl. Microbiol. Biot. 106 (9-10) (2022) 3419–3430.

[112] M.E. Russo, S. Peirce, R. Perfetto, C. Capasso, M. Rossi, A. Marzocchella, et al., Immobilization of carbonic anhydrase for enhancement of CO_2 reactive absorption, N. Biotechnol. 44 (2018) S44 -S44.

[113] M.E. Russo, S. Scialla, V. De Luca, C. Capasso, G. Olivieri, A. Marzocchella, Immobilization carbonic anhydrase biomim. CO_2 capture, Chem. Eng. Trans. 32 (2013) 1867–1872.

[114] H. Bose, T. Satyanarayana, Microbial carbonic anhydrases in biomimetic carbon sequestration for mitigating global warming: prospects and perspectives, Front. Microbiol. 8 (2017).

CHAPTER TWO

Bacterial α-CAs: a biochemical and structural overview

Vincenzo Massimiliano Vivenzio, Davide Esposito, Simona Maria Monti*, **and Giuseppina De Simone**

Istituto di Biostrutture e Bioimmagini, CNR, Napoli, Italy
*Corresponding author. e-mail address: simonamaria.monti@cnr.it

Contents

1.	Introduction	31
2.	α-CAs as novel molecular targets to combat antibiotic resistance in bacteria	32
3.	α-CAs from extremophilic bacteria	43
	3.1 α-CAs from extremophiles: biotechnological applications	49
4.	α-CAs from cyanobacteria	51
5.	Conclusions and future perspectives	52
	References	53

Abstract

Carbonic anhydrases belonging to the α-class are widely distributed in bacterial species. These enzymes have been isolated from bacteria with completely different characteristics including both Gram-negative and Gram-positive strains. α-CAs show a considerable similarity when comparing the biochemical, kinetic and structural features, with only small differences which reflect the diverse role these enzymes play in Nature. In this chapter, we provide a comprehensive overview on bacterial α-CA data, with a highlight to their potential biomedical and biotechnological applications.

1. Introduction

Among the eight different carbonic anhydrase (CA) families, the α-one is the most populated with representative members in many living organisms such as animals and higher plants, algae, protozoa, fungi and bacteria [1]. Members of this family differ in catalytic activity, biological function and oligomeric arrangement. However, despite these differences, they all share a common structural fold characterized by a central ten-stranded β-sheet, surrounded by α-helices, 3_{10}-helices and additional β-strands. The active site is placed in a wide cavity at the base of which a catalytic zinc ion is positioned, being coordinated by three conserved histidine residues and a solvent molecule [2].

The Enzymes, Volume 55
ISSN 1874-6047, https://doi.org/10.1016/bs.enz.2024.07.001

Bacterial species contain α-, β-, γ- and ι-CAs; however, the different distribution of these enzymes is a characteristic unique to each bacterium [3,4]. In general, β- and γ-CAs are the most commonly present, while ι- and α-CAs are less common. Until recently, α-CAs were primarily found in Gram-negative bacteria, where they have a periplasmic or extracellular localization thanks to the presence of an N-terminal signal peptide [3,5]. However, more recent reports have shown that the genomes of the probiotic Gram positive bacteria *Lactobacillus delbrueckii* and *L. rhamnosus* GG [6,7] as well as the two vancomycin-resistant *enterococci* (VRE), *F. faecium* and *E. faecalis*, also encode active α-CAs [8–10].

Although in the past most studies on α-CAs focused on human iso-forms, due to their involvement in a range of physiological and patholo-gical processes, there has recently been renewed interest in bacterial α-CAs. Indeed, important applications can arise from the functional, biochemical, and structural study of these enzymes, both in the biomedical field, through the design of new antibacterial drugs targeting α-CAs from pathogenic bacteria [11,12], and in the biotechnological field through the use of these enzymes in carbon capture, utilization, and storage processes [13]. On this basis, a comprehensive overview of the literature data available on bacterial α-CAs is here provided, highlighting the physiological role these enzymes have in the organisms where they are, along with their biochemical and structural features, and eventually their biomedical and/or biotechnological application.

2. α-CAs as novel molecular targets to combat antibiotic resistance in bacteria

Resistance to antibiotics is a worldwide problem that is responsible for more deaths per year than HIV/AIDS or malaria. According to the latest study by the European Centre for Disease Prevention and Control, only in Europe, more than 35,000 people die each year for this emerging phenomenon [14], clearly indicating the necessity of drugs with novel mechanism of action.

In recent decades, important advancements in genome sequencing and high-throughput screening technologies have made it possible to identify new virulence factors or proteins essential for bacterial life; among these, CAs are emerging as new possible targets for the development of new antimicrobial drugs [12,15]. In this paragraph, biochemical and structural data on α-CAs from pathogenic bacteria including *Vibrio cholerae, Helicobacter pylori,* VRE *E. faecium* and *E. faecalis,* and *Neisseria gonorrhoeae* will be summarized along with experimental evidence, when available, of their potential utility as antibacterial drug targets.

V. cholerae is a Gram–negative bacterium that lives in aquatic environments and is the causative agent of cholera [16,17]. It colonizes the upper small intestine where sodium bicarbonate is present at a high concentration. According to World Health Organization (WHO) every year there are between 1.3 to 4.0 million infections of cholera, resulting in 21,000–143,000 deaths worldwide [18]. Over 200 serogroups have been identified, but only those belonging to the O1 and O139 classes are causative of epidemics and pandemics [16,17]. The infection of a host initiates a complex regulatory cascade that results in production of ToxT, a regulatory protein that directly activates transcription of the genes encoding cholera toxin (CT), toxin-coregulated pilus (TCP), and other virulence genes [17,19,20].

Current research suggests that CAs may play an important role in the induction of *V. cholerae* virulence by modulating intracellular bicarbonate levels since (i) bicarbonate stimulates virulence gene expression by enhancing ToxT activity; (ii) ethoxyzolamide, a potent CA inhibitor, inhibits bicarbonate-mediated virulence induction; (iii) no bicarbonate transporters have been reported in *V. cholerae* [16]. Interestingly, genome inspection of *V. cholerae* resulted in the identification of three different CAs belonging to the α-, β-, and γ-classes, named VchαCA, VchβCA and VchγCA. These enzymes could be considered suitable targets for the development of new therapeutics against bacterial colonization. In the following, the main biochemical and structural features of VchαCA will be summarized.

VchαCA was cloned, expressed in *E. coli* and characterized the first time in 2012 [21]. As evident from Table 1, this enzyme has a significant catalytic activity, being more active than the human isoform CA I or the *H. pylori* α-class CA (HpαCA). VchαCA is about half as active as hCA II, one of the best catalysts known in nature [21]. This protein appeared to be stable at thermal treatment, showing activity even at 70 °C [22]. Extensive investigations testing wide panels of inhibitors have been performed by Supuran's group which identified strong nanomolar inhibitors [21,23–29]. Promising inhibition studies have been conducted with amides and sulfonamides incorporating imidazole moieties which not only exerted inhibitory capacity in the nM range between 5.4 and 39.2 nM, but also showed a pronounced selectivity towards VchαCA compared to hCA I and

Table 1 Kinetic and structural features of α-CAs from pathogenic bacteria compared to human CA I and CA II.

Enzyme name	Bacterium	Kinetic constants	PDB code	Quaternary structure	References
VchαCA	V. cholerae	k_{cat}: 8.23×10^5 s^{-1} k_{cat}/K_m: 7.0×10^7 M^{-1} s^{-1}	N.A.*	No experimental data available	[21]
HpαCA	H. pylori	k_{cat}: 2.4×10^5 s^{-1} k_{cat}/K_m: 1.4×10^7 M^{-1} s^{-1} measured at pH 8.9 and 25 °C	4XFW 4YGF 5TV3 5TT8 5TT3 5TUO	Monomer ↔ Dimer	[35,43–45]
NgαCA	N. gonorrhoeae	k_{cat}: 1.1×10^6 s^{-1} k_{cat}/K_m: 5.5×10^7 M^{-1} s^{-1} measured at pH 9.0 and 25 °C	1KOP 1KOQ 8DPC 8DYQ 8DQF 8DR2 8DRB	Dimer	[61,69,70]
VREαCAs	E. faecium E. faecalis	N.A.	N.A.	No experimental data available	—
hCA I	Homo sapiens	k_{cat}: 2.00×10^5 s^{-1} k_{cat}/K_m: 5.0×10^7 M^{-1} s^{-1}	3LXE	Monomer	[165,166]
hCA II	Homo sapiens	k_{cat}: 1.40×10^6 s^{-1} k_{cat}/K_m: 1.5×108 M^{-1} s^{-1}	1CA2	Monomer	[47,167]

*N.A., Not available.

Kinetic parameters refer to the CO_2 hydration reaction and are measured at pH 7.5 and 20 °C unless otherwise stated.

hCA II [30]. Additional compounds tested include N'-aryl-N-hydroxy-ureas (Kis ranging from 97.5 nM to 7.26 μM) [31], simple phenols and phenolic acids (Ki between 0.6 and 76 μM) [32], and coumarins (Kis ranging from 39.8 to 438.7 μM) [33]. So far, investigation carried out using this isoform did not lead to the resolution of 3D structure.

$H.\ pylori$ is a gram-negative bacterium responsible for different gastric diseases, including gastritis and peptic ulcer disease [34]. Even though the bacterium needs elevated levels of carbon dioxide for its growth, genomic data revealed the absence of anaplerotic carbon dioxide fixation mechanisms suggesting the presence of players for the reversible conversion between CO_2 and HCO_3^-. Indeed, $H.\ pylori$ has two genes encoding different forms of CA enzymes, namely an α- and a β-CA (HpαCA and HpβCA), which have a different subcellular localization, being periplasmic the first and cytoplasmic the second [34]. The joint action of these two enzymes along with urease are essential for generating the NH_3/NH_4^+ and CO_2/HCO_3^- pairs. These pairs help to regulate the periplasmic and cytoplasmic pH of $H.\ pylori$, keeping it near neutral in the stomach's highly acidic environment, which is crucial for the survival and growth of the bacterium gastric niche [35–37]. It has also been shown that HpαCA is overexpressed at low pH in the presence of urea [38,39], and that its gene is regulated when the periplasmic pH changes but the cytosolic one remains constant [40]. All these data strongly suggest that HpαCA and HpβCA are potential targets for the development of novel anti-$H.\ pylori$ drugs.

HpαCA has been also experimented in biotechnological applications involving its expression on the surface of $E.\ coli$ for CO_2 capture and mineralization [41,42].

HpαCA was first cloned and expressed in 2001 in a truncated form lacking the N-terminal signal peptide [43]. The enzyme showed a moderate CA activity (Table 1). Its crystallographic structure was determined both in the unbound form and in complex with several sulfonamide inhibitors [35,44,45], revealing the classical fold of α-CAs, characterized by a central ten-stranded twisted β-sheet surrounded by several helices and additional β-strands (Fig. 1). The active site is positioned within a conical-shaped cavity hosting the catalytic Zn^{2+} ion at the bottom. An intramolecular disulfide bond (Cys45–Cys195) that is conserved in nearly all bacterial α-CA structures as well as in some human isoforms stabilizes the structure [46] (Figs. 2 and 3). The HpαCA overall fold is very similar to that of human CAs [2]. Fig. 4 reports its structural superimposition with hCA II (PDB code 1CA2) [47] showing that most of the secondary

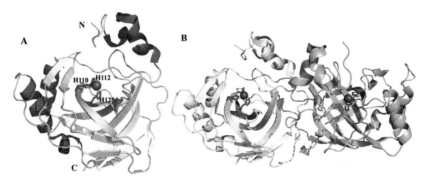

Fig. 1 (A) Cartoon representation of HpαCA monomer (PDB: 4YGF) [35]. Helices are colored in red and β-strands in yellow. The zinc ion (sphere) and the three coordinating histidines are also shown in red. (B) Cartoon representation of HpαCA dimer. The two monomers are coloured in yellow and light blue, respectively; the zinc ion (sphere) and the three coordinating histidines are represented in red.

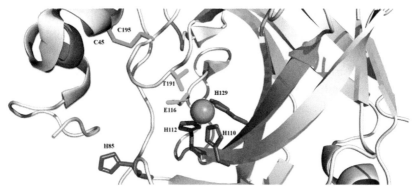

Fig. 2 Zoom into the active site of HpαCA (PDB: 4YGF). The zinc ion is represented as an orange sphere, the three zinc-coordinating histidines as purple-red sticks, the proton shuttle histidine in blue, the gatekeeper residues in green, and the intramolecular disulfide bridge between Cys45 and Cys195 in cyan.

structure elements present in the human enzyme are retained also in HpαCA. The structural differences between the two enzymes are mainly due to the length of the loop regions, which in HpαCA are generally shorter (Figs. 3 and 4). As evident from Fig. 3, all the amino acid residues important for catalytic activity are conserved and have similar structural positions in the human and bacterial enzyme.

Although in all cases HpαCA crystallizes as a dimer [35,44,45] (Fig. 1B), conflicting data are available on the quaternary structure of the enzyme in

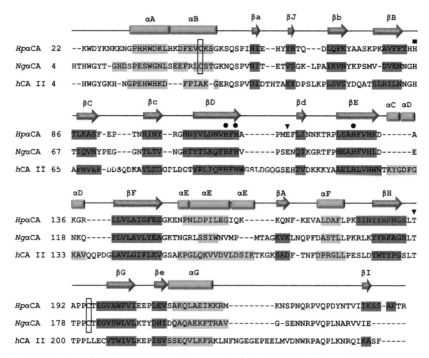

HpαCA   22   --KWDYKNKENGPHRWDKLHKDFEVCKSGKSQSPINIE--HYYHTQ---DLQFKYAASKPKAVFFTHH

NgαCA   4   HTHWGYT-GHDSPESWGNLSEEFRICSTGKNQSPVNIT--ETVSGK-LPAIKVNYKPSMV-DVENNGH

hCA II   4   --HWGYGKH-NGPEHWHKD---FPIAK-GERQSPVDIDTHTARYDPSLKPLSVSYDQATSLRILNNGH

βC   βc   βD   βd   βE   αC αD

HpαCA   86   TLKASF-EP---TNHINY--RGHDYVLDNVHFHA------PMEFLINNKTRPLSAHFVHKD------A

NgαCA   67   TIQVNYPEG---GNTLTV--NGRTYTLKQFHFHV------PSENQIKGRTFPMEAHFVHLD------E

hCA II   65   AFNVLE-DDSQDKAVLKGGPLDGTYPLIQFHFHWGSLDGQGSEHTVDKKKYAAELHLVHWNTKYGDFG

αD   βF   αE   αE   αE   βA   αF   βH

HpαCA   136   KGR-----LLVLAIGFEEGKENPNLDPILEGIQK------KQNF-KEVALDAFLPKSINYYHPNGSLT

NgαCA   118   NKQ-----PLVLAVLYEAGKTNGRLSSIWNVMP---MTAGKVKLNQPFDASTLLPKRLKYYRFAGSLT

hCA II 133   KAVQQPDGLAVLGIFLKVGSAKPGLQKVVDVLDSIKTKGKSADF-TNFDPRGLLPESLDYWTYPGSLT

βG   βe   αG   βI

HpαCA   192   APPCTEGVAWFVIEEPLEVSAKQLAEIKKRM---------KNSPNQRPVQPDYNTVIIKSS-AFTR

NgαCA   178   TPPCTEGVSWLVLKTYDHIDQAQAEKFTRAV---------G-SENNRPVQPLNARVVIE-------

hCA II 200   TPPLLECVTWIVLKEPISVSSEQVLKFRKLNFNGEGEPEELMVDNWRPAQPLKNRQIKASF-----

Fig. 3 Structure-based sequence alignment of HpαCA, NgαCA and hCA II. hCA II secondary structure elements [47] are shown schematically: helices are represented as green cylinders and β-strands as red arrows. Helices and β-strand regions for all enzymes are highlighted in green and red, respectively. Sequence numbering refers to the PDB structures: HpαCA (4YGF) [35], NgαCA (1KOP) [70], and hCA II (1CA2) [47]. The zinc coordinating histidines (His94, His96, His119, hCA II numbering) are indicated with the symbol •, gatekeeper residues (Glu106, Thr199) with the symbol ▼, the proton shuttle residue (His64) with the symbol ■ and the cysteines involved into the intra-molecular disulfide bridge are boxed (Cys45, Cys195, HpαCA numbering).

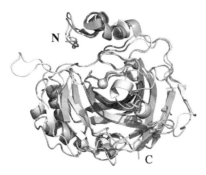

Fig. 4 Structural superimposition between HpαCA in light blue (4YGF) [35] and hCA II in yellow (1CA2) [47].

solution [44,48], suggesting the presence of a monomer/dimer equilibrium dependent on physiological conditions (ionic strength, pH) (Table 1) [44]. Interestingly, a careful structural comparison of the HpαCA/inhibitor complexes [35,45] with the corresponding complexes with hCA II revealed that the bacterial enzyme presents an additional small pocket in the active site (absent in hCA II) that functions as an alternative binding site for sulfonamides [45]. These differences could be utilized for the design of inhibitors selective for HpαCA, which could lead to novel anti-*H. pylori* drugs [45].

The first inhibition studies on HpαCA were carried out by Supuran's group using a panel of sulfonamides, which showed nM inhibition properties, buy were completely devoid of selectivity for the bacterial enzyme with respect to the human counterparts [49]. Subsequently several classes of inhibitors were investigated [50–54]. Among these, it is worth of note famotidine (Fig. 5), an antiulcer drug incorporating a sulfamide motif. This molecule showed a nM inhibition activity against HpαCA, suggesting that its effective anti-ulcer effects may not only stem from its role as an H2-receptor antagonist but also from its inhibition of the bacterial CAs [51]. It was hypothesized that by targeting CAs of *H. pilory*, famotidine could potentially disrupt the survival of the pathogen in the gastric environment [51]. Another interesting study, carried out by Benito and coworkers [55], reported the design and synthesis of hybrid molecules containing the EGFR inhibitor erlotinib and sulfonamide CA inhibitors (CAIs) (Fig. 6), with the aim of inhibiting simultaneously EGFR and HpαCA. Some of these compounds showed strong and selective HpαCA inhibitory activity as well antibacterial activity against the *H. pylori* strain ATCC with moderate minimum inhibitory

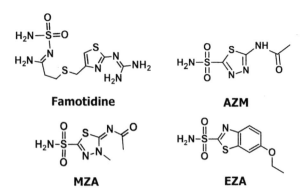

Fig. 5 Chemical structures of famotidine, acetazolamide (AZM), methazolamide (MZA), and ethoxzolamide (EZA).

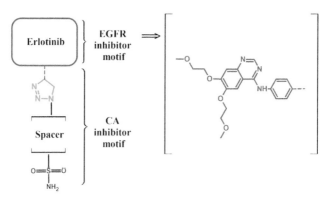

Fig. 6 Schematic scaffold of hybrid molecules containing the EGFR inhibitor erlotinib and the sulfonamide CA inhibitor motif.

concentrations [55]. Finally, also the sulfonamide drugs acetazolamide (AZM), methazolamide (MZA), and ethoxzolamide (EZA) (Fig. 5) were shown to possess antimicrobial activity against several *H. pylori* laboratory and clinical strains [56].

Neisseria gonorrhoeae, also known as the gonococcus, is a Gram-negative bacterium causative of gonorrhea, a sexually transmitted disease [57]. It is capable to colonize the vaginal, oral, and anal mucosae [58–60] and can survive in these environments by expressing several factors able to evade the immune system. According to the World Health Organization (WHO), the emergence of antibiotic-resistance in *N. gonorrhoeae* is making the treatment of gonorrhea increasingly challenging, necessitating the rational use of antibiotics and the search for new molecular targets. Recent advancements have demonstrated promising results targeting *N. gonorrhoeae* α-CA (NgαCA) using CAIs; however, to date the role of NgαCA in pathogenesis has not yet been elucidated [61–66].

Beyond its potential utilization as antibacterial drug target, NgαCA has also been studied for CO_2 sequestration processes. In a first study the enzyme was cloned in fusion with a cellulose binding domain (CBD), resulting in a chimeric enzyme of particular interest due to its binding affinity for cellulose and retained CA activity. This could serve as a starting point for improved technology to capture CO_2 from flue gases [67]. Subsequently, NgαCA's ability to mineralize CO_2 into $CaCO_3$, both in free form and when expressed in the periplasm of *E. coli*, was investigated [68], highlighting the recombinant NgαCA protein as a potentially effective and economical biocatalyst for practical CO_2 conversion system.

The α-CA from *N. gonorrhoeae* (NgαCA) is the first enzyme cloned and characterized from pathogenic bacteria [11] showing at pH 9.0 and 25 °C a high CO_2 hydrase activity [69] (Table 1). Subsequently, the crystallographic structure of the protein was determined by Lindskog's group [70], both in the native form and in complex with acetazolamide, showing again an homodimeric structure, well superimposable with the dimeric structure of HpαCA [35,44,45] (Fig. 7), and the typical fold of α-CAs, stabilized by the intramolecular disulfide bond between the two conserved cysteines C28 and C181 (Fig. 3) [70–72]. Like above described for HpαCA, despite the long evolutionary distance between hCA II and NgαCA, most of the secondary structure is conserved, the biggest differences being in the length of the loops which are considerably longer in the human form compared to the bacterial one. The observed similarity with hCA II includes also the amino acid residues in the active site including the zinc ligands, the "gate keeper" residues and the proton shuttle [70]. More recently, the structures of NgαCA adducts with several thiadiazolesulfonamide inhibitors have also been determined, highlighting the binding mechanisms of these compounds [61].

In vitro inhibition studies on the recombinant protein were performed using a panel of anions [73], phenols [32], coumarins [33], dithiocarbamates and monotiocarbamates [64] identifying some effective non-classical inhibitors.

The first *in vitro* studies on the bacterial growth inhibition using CAIs date back to the late 1960s [74], but only in the 2020 s there were observed important advances of these promising results, thanks to investigations

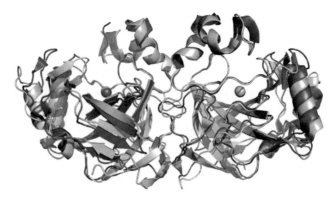

Fig. 7 Structure superposition of the dimers of HpαCA in light blue (4YGF) [35] and NgαCA in purple-red (1KOP) [70].

carried out by Flaherty and co-workers [61–64]. In particular, the synthesis and testing of a panel of AZM-based CAIs against several isolates of *N. gonorrhoeae*, allowed the selection of two efficient and non-toxic compounds, namely compounds **1** and **2** (Fig. 8), which displayed minimum inhibitory concentration (MIC) values as low as $0.25\,\mu g/mL$ equating to an 8- to 16-fold improvement in antigonococcal activity compared to AZM. Interestingly, the FDA approved drug EZA (Fig. 5) also showed to be efficient in inhibiting the growth of *N. gonorrhoeae* displaying MIC against a panel of *N. gonorrhoeae* isolates of $0.125\,\mu g/mL$. These evidences strongly encourage further studies to investigate EZA for its potential translation into an effective anti *N. gonorrhoeae* agent [63]. Finally, the *in vivo* validation of AZM usage was recently given in a mouse model of gonococcal infection, significantly reducing *N. gonorrhoeae* levels in infected vaginal tissues [65]. In agreement with these studies a patent has been filed and published by Purdue Pharma describing CAIs for the treatment of *N. gonorrhea* infection [66].

Enterococci are Gram-positive, facultatively anaerobic oval-shaped cocci that form chains of varying lengths. They are sturdy and versatile, capable of surviving harsh conditions such as high salt concentrations and a wide temperature range (from $10\,°C$ to over $45\,°C$). VRE are one of the leading causes of nosocomial infections in the United States causing diseases in hospitalized and immunosuppressed patients [75,76]. Among the *Enterococcus* genus, the species *E. faecium* and *E. faecalis* were the first VRE isolated [77], associated to infections like infective endocarditis, urinary tract infections, and intra-abdominal infections [78–80]. The genomes of both species encode for α-CAs, but only that of *E. faecium* (EfαCA) has been recombinantly expressed and purified [9].

Flaherty's group carried out very interesting investigations in the field, with the aim to explore drug repurposing as a strategy to discover new anti-VRE agents [81–84]. In fact, repurposing FDA-approved drugs, which have well-documented toxicology and pharmacology profiles, is an appealing approach to reduce time, cost, and risk associated with antimicrobial drug development.

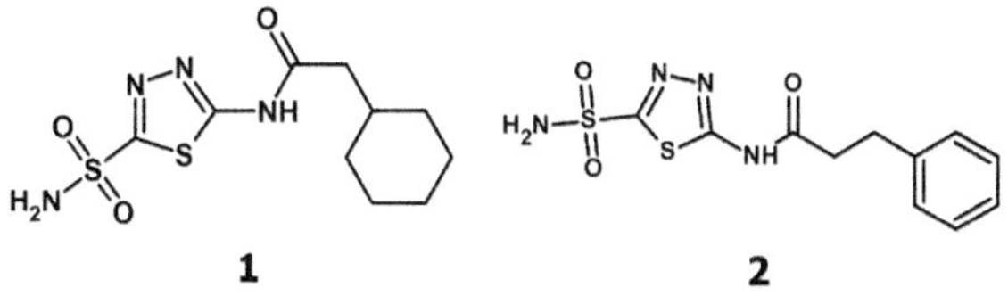

Fig. 8 Chemical structure of AZM derivatives **1** and **2** with anti-gonococcal properties.

FDA-approved CAIs were then assayed as a novel class of potent anti-enterococcal agents. AZM showed greater reductions in VRE CFU counts compared to linezolid (LZD), the current drug of choice for VRE infections, in mouse models of VRE colonization and systemic infection. Additionally, AZM has a well-documented safety profile and can be administered to humans in doses up to 4 g/day, with highly favorable pharmacokinetic properties [85]. Therefore, AZM represents a promising new treatment option for VRE infections. Starting from AZM, through structure—activity relationship optimization several compounds, among which **3**, **4** and **5** (Fig. 9), were developed with improved activity with respect to the lead AZM against a panel of *E. faecium* and *E. faecalis* isolates [8]. The mechanisms of action of these molecules was not clear, but it was speculated that they could target the α- and γ-CAs of *E. faecium*, namely EfαCA and EfγCA [8,9]. To test this hypothesis, the inhibitory properties of the entire series of AZM derivatives against these two CAs were evaluated, showing effective inhibition of the bacterial enzymes, even though a correlative trends between EfCAs inhibition and improved antibacterial potency towards VRE was not observed [9]. Continuing with these studies, the same group of Authors evaluated the *in vivo* efficacy of compounds **3**, **4** and **5** in the gastrointestinal tract decolonization of VRE in an animal model of the disease [86]. Overall, investigated molecules showed a varied antimicrobial potency and intestinal permeability, which influenced their efficacy in VRE gut decolonization. Notably, the exanimated CAIs demonstrated superior VRE decolonization efficacy compared to LZD [86].

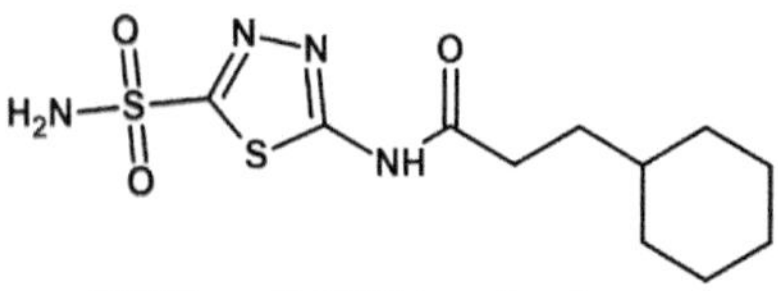

MIC = 0.007µg/mL

3

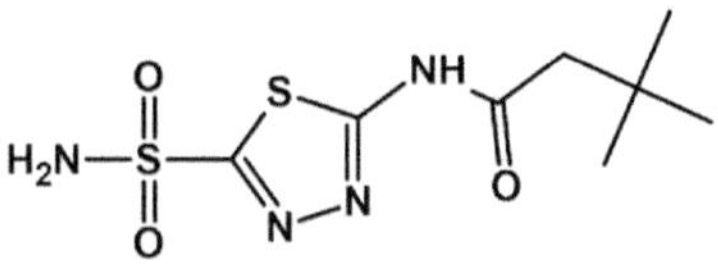

MIC = 0.03µg/mL

5

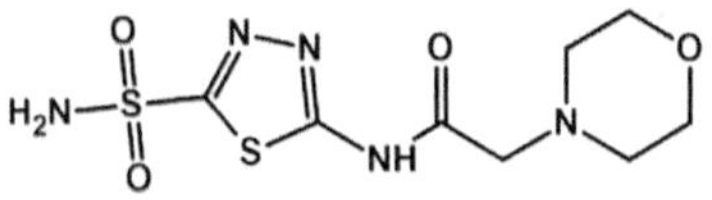

MIC = 1µg/mL

4

Fig. 9 Chemical structures of acetazolamide derivatives **3**, **4** and **5** with anti-VRE properties. MIC values are also indicated.

A very interesting investigation, carried out recently, revealed that an *E. faecalis* clinical strain from an infected patient with a mutation disrupting the α-CA enzyme made the bacteria hypersensitive to aminoglycoside antibiotics. Further investigation revealed that this α-CA disruption led to increased gentamicin accumulation in *E. faecalis* through two mechanisms encompassing enhanced proton motive force-powered uptake and increased membrane permeability. Additionally, the FDA-approved AZM could sensitize various *E. faecalis* strains to gentamicin [10]. These findings suggest that the readily available FDA approved CAIs could serve as effective partner antibiotics for treating *E. faecalis* infections and should be considered as a supplement to existing therapies.

3. α-CAs from extremophilic bacteria

Extremophilic bacteria survive in conditions that would be lethal for most other organisms, such as extreme temperatures, high acidity, salinity, or pressure [87]. The proteins they produce are extremo-proteins, stable and active under extreme conditions, and if produced recombinantly, they can be used for a wide range of biotechnological purposes [88]. In the last decade, the genomes of many microorganisms living in extreme environments have been sequenced and deposited in Genome Data Banks, and thanks to bioinformatic tools based on sequence homology, the identification of genes encoding α-CAs has been possible. *Thermovibrio ammonificans, Hydrogenovibrio crunogenus, Sulfurihydrogenibium yellowstonense, Sulfurihydrogenibium azorense, Persephonella marina, Photobacterium profundum* and *Hydrogenvibrio marinus* are examples of gram-negative bacteria living in extreme environments for which α-CAs have been identified, characterized, and proposed for biotechnological applications. Table 2 summarizes the main biochemical and structural features of these enzymes, whereas Fig. 10 reports their multiple sequence alignment, which highlights the conservation of the zinc binding histidines, the proton shuttle, and the gatekeeping residues. All the sequences contain a signal peptide at the N-terminus, which suggests their periplasmic or extracellular localization in the host. The α-CAs from *H. crunogenus* (TcrαCA) [89] and *H. marinus* (HmCA) [90–92], contain an additional sequence at the N-terminus which will be discussed below in the case of HmCA [90]. As shown in Table 2 all the enzymes are active with k_{cat}/K_m values for the CO_2 hydration reaction ranging from $1.1 * 10^7$ M^{-1} s^{-1} in the case of the α-CA from *H. crunogenus* [93], hereafter indicated as TcrαCA, to $3.5 * 10^8$ $M^{-1}s^{-1}$ in the case of the α-CA

Table 2 Kinetic and structural features of extremophilic bacteria α-CAs.

Enzyme name	Bacterium	Kinetic constants	PDB code	Quaternary structure	References
TaαCA	T. ammonificans	k_{cat}: 9.6×10^5 s^{-1} k_{cat}/K_m: 2.5×10^7 M^{-1} s^{-1} measured at pH 8.5 and 25 °C	4C3T 4COQ 4UOV	Tetramer	[98,104]
TcrαCA	H. crunogenus	k_{cat}/K_m: $1.1 * 10^7$ M^{-1} s^{-1} measured at pH 7.5 and 25 °C	4XZ5	dimer	[89,108]
SspαCA	S. yellowstonense	k_{cat}: 9.35×10^5 s^{-1} k_{cat}/K_m: 1.1×10^8 M^{-1} s^{-1} measured at pH 7.5 and 20 °C	4G7A	dimer	[46,113]
SazαCA	S. azorense	k_{cat}: 4.4×10^6 s^{-1} k_{cat}/K_m: 3.5×10^8 M^{-1} s^{-1} measured at pH 7.5 and 20 °C	4X5S	dimer	[94,99]
PmαCA	P. marina	k_{cat}: 4.0×105 s^{-1} k_{cat}/K_m: 3.9×107 M^{-1} s^{-1} measured at pH 8.5 and 25 °C	6IM0 6IM3	dimer	[100,104]
PprαCA	P. profundum	N.A.	5HPJ	Monomer ↔ Dimer	[101]
HmαCA	H. marinus	k_{cat}/K_m: 2.0×10^7 M^{-1} s^{-1} measured at pH 8.5 and 25 °C	N.A.*	No experimental data available	[90]

*N.A., Not available
Kinetic parameters refer to the CO$_2$ hydration reaction.

```
hCA II      ------------------------------------------------------------   0
TcraCA      MK--KRFSFIFIF----LVALPLYSA--NNVA APLIDLGAEAKKQAQKSAATQSAVPEKESA   54
HmaCA       MKALKRLLIAFISFGVCLVAPQGWSAAVQ HSNAPLIDLGAEMKKQ-HKEAAPEGAAPAQGKA   61
SspaCA      MR--KILISAVLV----LSSISIS------------------------------------   18
SazaCA      MK--KFILS-ILS----LSIVSIAGEHAILQKN---------------------------   26
PpraCA      MK--TRTLH-VLA----FAMVTTW------ATT---------------------------   20
TaaCA       MK--RVLVT-LGA----VAALATG------AVA---------------------------   20
PmaCA       MK--KFVVG-LSS----LVLATSS------------------------------------   17

hCA II      --------------------MSHHWGYGKHNGPEHWHK---DFPIAK-GERQSPVDIDTHTA   38
TcraCA       TKVAEKQKEPEEKAKPEP-K KPPHWGYFGEEGPQYWGELAPEFST CKTGKNQSPINLKPQTA   115
HmaCA        PAAEAKK---EEAPKPKPVV HNPHWSYSGEEGPDHWGDLSPDYAT CKTGKNQSPINLMADDA   120
SspaCA      -------------------FAEHEWSYEGEKGPEHWAQLKPEFFW CKL-KNQSPINIDKKYK   60
SazaCA      -------------------AEVHHWSYEGENGPENWAKLNPEYFW CNL-KNQSPVDISDNYK   68
PpraCA       -------------------SYAAEWSYTGEHGTEHWGD---SFAT CAEGVNQTPIDINQTT-   59
TaaCA       -------------------GGGAHWSYSGSIGPEHWGDLSPEYLM CKIGKNQSPIDINSADA   63
PmaCA       -------------------FAGGGWSYHGEHGPEHWGDLKDEYIM CKIGKNQSPVDINRIV-   59
                                                         *

hCA II      KYDFSLKPLQVQVDQATSLRILNNGHAFNVEFDDSQDKAVLKGGPLDGTYRLIQFHFHWGSL   100
TcraCA      VGTTSLPGFDVYYRETA-LKLINNGHTLQVNIPL-GSYIKING----HRYELLQYHFH----   167
HmaCA       VGTTSLPGFDVHYRDTV-LKVINNGHTLQANVPL-GSYIKIKN----QRYELLQYHFH----   172
SspaCA      V-KANLPKLNLYYKTAKESEVVNNGHTIQINIKE-DNTLNYLG----EKYQLKQFHFH----   112
SazaCA      V-HAKLEKLHINYNKAVNPEIVNNGHTIQVNVLE-DFKLNIKG----KEYHLKQFHFH----   120
PpraCA      --QAELAPLHLDYEGQV-TELVMNGHTIQANLTG-KNTLTVDG----KTFELHLQFHFH----   109
TaaCA       V-KACLAPVSVYYVSDA-KYVVNNGHTIKVVMGG-RGYVVVDG----KRFYLKQFHFH----   114
PmaCA       --DAKLKPIKIEYRAGA-TKVLNNGHTIKVSYEP-GSYIVVDG----IKFELKQFHFH----   109
                                 ***                                  **

hCA II      DGQGSEHTVDKKKYAAELHLVHWNTKYGDFGKAVQQPDGLAVLGIFLKVG-SAKPGLQKVVD   161
TcraCA      --TPSEHQRDGFNYPMEMHLVH-KDGDGN----------LAVIAILFQEG-EENETLAKLMS   215
HmaCA       --TPSEHQLNGFNYPMELHLVH-RDGRGH----------YLVIGILFREG-KENDALQTILN   220
SspaCA      --TPSEHTIEKKSYPLEIHFVH-KTEDGK----------ILVVGVMAKLG-KTNKELDKILN   160
SazaCA      --APSEHTVNGKYYPLEMHLVH-KDKDGN----------IAVIGVFFKEG-KANPELDKVFK   168
PpraCA      --TPSENYLKGKQYPLEAHFVH-ATDKGE----------LAVVAVMFDFGPRSNNELTTLLA   158
TaaCA       --APSEHTVNGKHYPFEAHFVH-LDKNGN----------ITVLGVFVKG-KENPELEKVWR   162
PmaCA       --APSEHKLKGQHYPFEAHFVH-ADKHGN----------LAVIGVFFKEG-RENPILEKIWK   157
                                   ▲

hCA II      VLDSIKTKGKSADFTNFDPRGLLPESLDYWTYPGSLTTPPLLECVTWIVLKEPISVSSEQVL   223
TcraCA      FLPQTLKKQEIHESVKIHPAKFFPADKKFYKYSGSLTTPP CSEGVYWMVFKQPIQASVTQLE   277
HmaCA       HLPKKVGKQEIFNGIEFNPNVFFPESKKFFKYSGSLTTPP CTEGVYWMVFKQPIEASAEQLE   282
SspaCA      VAPAEEGEK-ILDK-NLNLNNLIPKDKRYMTYSGSLTTPP CTEGVRWIVLKKPISISKQQLE   220
SazaCA      NALKEEGSK-VFDG-SININALLPPVKNYYTYSGSLTTPP CSEGVRWFVMQEPQTSSKAQTE   228
PpraCA      SIPSK-GQTVELKE-ALNPADLLPRDREYYRFNGSLTTPP CSEGVRWFVMQEPQTSSKAQTE   218
TaaCA       VMPEEPGQKRHLTA-RIDPEKLLPENRDYYRYSGSLTTPP CSEGVRWIVFKEPVEMSREQLE   223
PmaCA       VMPENAGEEVKLAH-KINAEDLLPKDRDYYRYSGSLTTPP CSEGVRWIVMEEEMEMSKEQIE   218
                                                     ▲

hCA II      KFRKLNFNGEGEPEELMVDNWRPAQPLKNRQIKASFK---------------   260
TcraCA      KMHEY---------LG-SNARPVQRQNARTLLKSWPDR---NRANTVYEFY   315
HmaCA       KMNEL---------MG-ANARPVQDLEARSLLKSWSNPKNDSQDHRYYQYY   323
SspaCA      KLKSV---------MVNPNNRPVQEINSRWIIEGF--------------   246
SazaCA      LFKSI---------MKHNNNRPTQPINSRYILESN---------------   254
PpraCA      KLQAV---------MG-NNARPLQPLNARLILE-----------------   241
TaaCA       KFRKV---------MGFDNNRPVQPLNARKVMK-----------------   247
PmaCA       KFRKI---------MGGDTNRPVQPLNARMIMEK----------------   243
```

Fig. 10 Amino acid sequence alignment of hCA II, TcraCA, HmaCA, PpraCA, TaaCA, PmaCA, SspaCA, and SazaCA. The N-terminal signal peptide of bacterial α-CAs is highlighted in yellow, the disordered regions of TcraCA and HmaCA are boxed in blue, and the strictly conserved amino acids are in bold. The proton shuttle residue His64 (hCA II numbering) is shown in bold blue (indicated by *), the gatekeeper residues (Glu106, Thr199) are in bold magenta (indicated by ▲), the zinc-coordinating residues (His94, His96, His119) are in bold red (indicated by •) and Cys residues involved into the intra-disulfide bond in bacterial α-CAs are boxed.

from *S. azorense,* named SazαCA, which is among the most active α-CAs characterized to date, even more active than hCA II [94–97]. Except for HmαCA, the crystallographic structure of all the enzymes reported in Table 2 has been determined revealing the typical features of bacterial α-CAs above described [46,89,98–101] and a dimeric arrangement with the exception of TaαCA which presents a peculiar tetrameric quaternary structure [98].

In the following, a more detailed description of the enzymes listed in Table 2 will be provided.

T. ammonificans is a thermophilic, anaerobic, and chemolithoautotrophic bacterium, isolated from the walls of a deep-sea hydrothermal vent chimney. It has an optimal growth condition at 75 °C, 2% w/v NaCl, and pH 5.5 [102,103]. Accordingly, TaαCA, an α-CA identified in this bacterium, exhibits an outstanding thermostability, being stable up to 80 °C, and high thermoactivity, showing a continuous increase in esterase activity as the temperature increases up to 95 °C [104]. The crystallographic structure of the enzyme was determined in the native form and in complex with two bound inhibitors [98], showing a tetrameric arrangement stabilized by inter-subunit disulfide bonds and ionic interactions (Fig. 11), which were suggested to be the main determinants of the TaαCA thermostability. Subsequent molecular dynamics studies allowed to design variants of the enzyme with increased stability [105]. In particular, the variants N140G and T175P tested *in vitro*

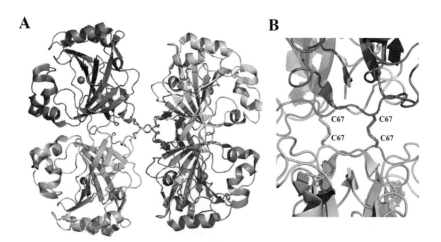

Fig. 11 Tetrameric structure of TaαCA (PDB: 4UOV) [98]. (A) Ribbon representation of TaαCA, the four monomers are differently coloured (blue, green, red, and pink). (B) Central core of the tetramer, the monomers are linked by two inter-molecular disulfide bonds involving Cys67 residues (PDB 4UOV numbering).

demonstrated a 3-fold higher half-life then wild type at 60 °C. Surprisingly these studies, in contrast with the previously reported hypothesis [98], demonstrated that the inter-molecular disulfide bonds were not essential for the enzyme's thermostability [106].

H. crunogenus, previously known as *Thiomicrospira crunogena*, is a chemolithoautotrophic sulfur-oxidizing bacterium, isolated from deep-sea hydrothermal vents [107]. Its genome encodes the periplasmic TcrαCA (Table 2), which was inhibited by anions (Kis in a range from 90 μM to 92 mM) [108,109] and sulfonamides (Kis in a range from 2.7 to 981 nM) [93]. The enzyme presented a dimeric quaternary structure and was demonstrated to be active and thermostable over a wide range of pH values [89]. TcrαCA functional role is still to be determined; indeed, the transcription levels of its gene were not affected by pH and dissolved inorganic carbon (DIC) and the enzyme did not appear to play a role in DIC uptake and fixation [93]. McKenna's group determined the crystallographic structure of the enzyme, which like other bacterial α-CAs, was rather compact with deletions in surface loops and helical structures compared to the human analogues. The compact structure, the presence of an intramolecular disulfide bond and the dimeric arrangement were identified as the molecular determinants responsible of the enzyme thermostability [89].

A very intriguing α-CA, named SspαCA, was identified in *S. yellowstonense*. *S. yellowstonense* is a chemolithoautotrophic bacterium isolated from the calcite hot springs in Yellowstone National Park, which can live at temperatures exceeding 110 °C and at hydrogen sulfide (H_2S) concentrations ranging from 1 to 100 μM [110,111]. SspαCA is a highly active enzyme (Table 2), characterized by an exceptional thermal stability, since it was still active after incubation at 100 °C for 3 h [46,112]. It was inhibited by anions (Kis in a range from 4 μM to 84.6 mM) [113] and sulfonamides (Kis in a range from 4.5 to 876 nM) [114,115]. Activation constants against a panel of amino acids and amines were also assessed (Kas in a range from 1 nM to 5.13 μM) [116]. SspαCA crystallographic structure was determined by our group in 2013 [46], suggesting that an increased structural compactness, together with an increased number of charged residues on the protein surface and a greater number of ionic networks, were the key factors involved in the higher thermostability of this enzyme compared to its mesophilic homologues.

S. azorense is a bacterial species related to *S. yellowstonense,* isolated from terrestrial hot springs at Furnas, São Miguel Island, Azores [117]. Its genome encodes for an α-CA named SazαCA, an interesting enzyme which was shown to possess complementary properties with respect to SspαCA. Indeed, although thermostability studies demonstrated that SazαCA was less stable with respect to SspαCA, its kinetic characterization revealed that it was one of the most active enzymes of the α-CA family (Table 2) [94]. SazαCA maintains its catalytic activity at 90 °C [95] and is strongly inhibited by a wide range of anions (Kis ranging from 3 μM to 32 mM) [118] and a large panel of sulfonamides (Kis ranging from 0.77 to 639 nM) [95,115]. Moreover, SazαCA is particularly activated by a panel of amino acids and amines (Kas ranging from 7 nM to 1.15 μM)[119]. The crystal structure determination allowed the molecular understanding of the marked activity of this enzyme, which was ascribed to the presence of two His residues on the rim of the active site cavity, which were hypothesized to influence the pKa of the His residue which works as proton shuttle [99].

PmαCA is an α-CA identified in *P. marina,* a microaerophilic, and chemolithoautotrophic bacterium, isolated from deep-sea hydrothermal vents in the Pacific Ocean, which has an optimum growth temperature of 73 °C and can use elemental sulfur, thiosulfate, or hydrogen as electron donors and oxygen or nitrate as electron acceptors [120]. Initially, PmαCA was cloned and expressed in *Bacillus subtilis* by Novozyme, which patented the technology [121]. Subsequently, it was expressed in *E. coli* and kinetically characterized (Table 2), showing an optimal catalytic activity at 95 °C [104]. The thermal stability of PmαCA, measured by differential scan fluorimetry, exceeds 85 °C. Similarly to the other bacterial α-CAs, it is a homodimer [100,104,122] with a more compact structure compared to hCA II. Also in this case, the presence of an intramolecular disulfide bond between Cys44 and Cys197, gives rigidity and thermostability [100].

Continuing this brief excursus on the α-CAs identified in extremophilic bacteria, PprαCA [123,124] from the halophilic bacterium *P. profundum* SS9 definitely needs to be mentioned. Differently from the bacteria described above, *P. profundum* SS9 lives in deep-sea environments at low temperatures, between 0 and 25 °C, showing an optimal growth at 15 °C and 28 MPa pressure. Accordingly, the PprαCA exhibited maximal catalytic activity at psychrophilic temperatures with substantial decrease in activity in the mesophilic and thermophilic range. The enzyme was also characterized by salt-dependent thermotolerance and catalytic activity under extreme halophilic conditions and was able to catalyze the CO_2 hydration reaction in both acidic and alkaline pH values [125]. The structural characterization of the protein highlighted the presence of a chloride ion at the dimeric interface that is unique to PprαCA compared to the other extremophilic α-CAs above described [101]. In addition, SEC analysis of the enzyme oligomeric status indicate a monomer/dimer equilibrium. Further investigation is necessary to clarify the exact reason why the enzyme exists in solution as a mixture of monomers and dimers [101].

Finally, *H. marinus* is a halophilic marine bacterium, which has an optimal growth at 37 °C and in presence of 0.5 M NaCl [126]. Sequencing of its genome revealed the presence of a gene encoding for a peculiar α-CA named HmαCA. Indeed, this enzyme, beyond the classical signal peptide for periplasmic localization, possesses at the N-terminus an unusual intrinsically disordered sequence which strongly influences its solubility [90]. As evident in Fig. 10, this sequence is present only in HmαCA and TcrαCA. Given its features, a solubility tag for the expression of recombinant proteins has been developed from it (NEXT-tag) [91,92]. Recombinant HmαCA showed catalytic efficiency comparable to other bacterial α-CAs (Table 2) and was only poorly inhibited by anionic inhibitors because of its halotolerance. Moreover, it was stabilized by alkali metal salts in a concentration-dependent manner. An uneven distribution of electrostatic potential and the localized negative charge on the protein surface were identified as the main players of its remarkable halotolerance [90].

3.1 α-CAs from extremophiles: biotechnological applications

In the last century, due to anthropogenic activities, CO_2 levels in the atmosphere have increased enormously, rising from 280 ppm before the Industrial Revolution to 400 ppm in 2015, and finally to 425 ppm in February 2024, having important consequences on global warming [127,128]. In this context, the development of effective methods for the reduction of atmospheric CO_2 has become of enormous importance [129,130].

Carbon capture technologies refer to a series of strategies aimed at reducing atmospheric CO_2 emissions. In this context, α-CAs from extremophilic organisms have been extensively studied, and in this paragraph, the most interesting results will be summarized.

Among the above analyzed extremo-α-CAs, TaαCA was proposed as a promising biological catalyst for CO_2 sequestration in terms of both activity and stability. Indeed, the enzyme was able to efficiently accelerate $CaCO_3$ formation from gaseous CO_2 at both mesophilic (30 °C) and thermophilic (60 °C) temperatures [104].

Extensive investigations for utilization in carbon capture processes and other biotechnological applications were carried out also on SspαCA. In particular, a preliminary study highlighted the long-term resistance to high temperatures of the free SspαCA, as in all capture processes based on absorption/vacuum–desorption cycles [131]. Subsequent studies, focused on the covalently immobilized enzyme on a polyurethane (PU) foam surface, showed that SspαCA retained its catalytic activity and long-term stability. A bioreactor containing the "PU-immobilized enzyme" as shredded foam was then successfully tested to verify its CO_2 capturing capability under conditions close to those of a power plant application [132], opening an interesting scenario for biotechnological applications of this enzyme. In following studies, SspαCA was expressed on the *E. coli* outer membrane (OM) fused with an OM anchoring domain [133] and characterized as a potential promoter of the CO_2 absorption rate in carbonate/bicarbonate alkaline buffer at operating conditions relevant for industrial CO_2 capture processes. Results showed a substantial similarity between the performances of free SspαCA and of OM-SspαCA anchored on membrane debris [134]. The OM–SspαCA complex fragments were also studied to evaluate the effect on growth and photo-synthetic efficiency of the freshwater microalga *Chlorella sorokiniana* [135]. Other types of SspαCA immobilization were also investigated, such as on magnetic nanoparticles [136], on supported ionic liquid membranes (SILMs) [137], and on polyacrylonitrile (PAN) and polyethylene terephthalate (PET) nanofibers [138]. Interestingly, thermostability and CO_2 mineralization capacity were markedly improved when SspαCA was co-expressed in *E. coli* with TrxA and the molecular chaperone GroELS [139].

The enzyme strictly related to SspαCA, namely SazαCA, was also proposed for several biotechnological applications [42,140–142] and its immobilization on several supports largely investigated [96,143].

HmαCA is another enzyme that has been widely investigated to evaluate its potential utilization in carbon capture biotechnological applications [144–148]. In 2020, Jo and coworkers reported the use of this enzyme for the construction of a highly active and thermostable whole-cell biocatalyst through its periplasm expression in *E. coli* [144]; subsequently, the obtained biocatalyst was immo-bilized on a polyurethane foam matrix and applied to a designed reactor, demonstrating its feasibility for practical applications [145]. Nano-biocatalysts systems for CO_2 capture were also generated through the bacterial production and *in vitro* self-assembly of HmαCA fused to the coat protein of potato virus Y (CP_{PVY}) [147]. Finally, HmαCA was immobilized also on diatom biosilica, showing activity comparable to the free enzyme [146].

4. α-CAs from cyanobacteria

Cyanobacteria are unicellular prokaryotic and photoautotrophic Gram-negative organisms [149] that have been subjected to selective pressures due to variations in atmospheric CO_2 concentrations during Earth's atmosphere evolution [150]. The transition from a CO_2-rich to a CO_2-poor atmosphere led cyanobacteria to develop alternative carbon capture systems where CAs play a key role [151–153]. In cyanobacteria genomes, β- and γ-CAs are predominant, while α-CAs are poorly represented. In this paragraph, we will focus on the most well-characterized cyanobacterial α-CAs.

The first characterized α-CAs from cyanobacteria belong to *Anabaena cylindrica* PCC 7120 and *Synechococcus elongatus* PCC 7942 [154]. The expression of both the enzymes, hereafter indicated as AcαCA and ScαCA, is regulated by CO_2 concentration, being highest in cells grown at elevated CO_2 levels. They possess an N-terminal signal peptide for extracellular localization and show significant sequence similarity to hCAs, including conservation of residues important for catalytic activity (Fig. 12) [154].

```
hCA II    --------------------------------MSHHWGYGKHNGPEHWHKDFP----IAK    24
ScαCA     ----MRRRSLLAALGGSCIGWLGS-----SQPVWASADWDYSRRRGPRQWAKLDPAYAICQQ    53
AcαCA     MSSTLYRRQLLKLLGMSVLGTSFS-SCVTSPARAKTVNWGYIGKVGPEHWGELSPDFALCQI    61
CyαCA     MKSTK----VIHWILVCLL-FSFSTISVVLGAEKTTHSWGYEGDNNPSQWGRLSAEYETCAM    57

hCA II    GERQSPVDIDTHTAKYD-PSLKPLSVSYDQATSLRILNNGHAFNVEFDDSQDKAVLKGGPLD    85
ScαCA     GRQQSPINLTGQPDRTP-------LDYRDRPFKGILQQAPHSLRIDCPAGNGFW-----EAG   103
AcαCA     GRKQTPIDLQIADVKDVHSSSQDLLVTNYQPTALHLINNGKTVQVNYQPGSYLK-----YAH   118
CyαCA     GLFQSPIHLTPSNLIR----QDGKIETHYQPTPLVINNNGHTIQINYEPGSYAM-----ING   110

hCA II    GTYRLIQFHFHWGSLDGQGSEHTVDKKKYAAELHLVHWNTKYGDFGKAVQQPDGLAVLGIFL   147
ScαCA     TFYELLQFHFHT------PSEHQHQGQRFPAEIHFVHRSD----------RDQLAVVGVFL   148
AcαCA     QKFELLQFHFHH------FSEHRVDGKLYDMELHLVHRSK----------SGDLAVMGIFL   163
CyαCA     KKYELKQFHFHT------PSEHSLNGKKSDMELHLVHQNE----------KGNIAIVGVFI   155
                    • •            ▲                                •
hCA II    KVGSAKPGLQKVVDVLDSIKTKGKSADFTNFDPRGLLPESLDYWTYPGSLTTPPLLECVTWI   209
ScαCA     AAGDRPLPILDTL--LAVPPSTDNQLLSTAIQPTDLMPRDRTVWRYSGSLTTPPCSEPVLWR   208
AcαCA     QAGAFNPTLQIIWDATPQNQGTDKRIEDINIDASQFLPAQHRFFTYSGSLTTPPCSENVLWC   225
CyαCA     EAGKANSTVTTIWNHLSETEGIQEV-KEQIINASDFIANNASYFHYQGSLTTPPCSENVIWN   216
                                                        ▲
hCA II    VLKEPISVSSEQVLKFRKLNFNGEGEPEELMVDNWRPAQPLKNRQIKASFK---          260
ScαCA     VCDRPLFVARQQLRQLRQR-----------LGMNARPLQA-------------          237
AcαCA     VMATPIEASPAQIAKFSQM-----------FPQNARPVQPLNDRLVIEAI----          264
CyαCA     VMREPLEASRQQIDQFMKL-----------YPMNARPIQEHNNRLVELIINQSS          259
```

Fig. 12 Amino acid sequence alignment of hCA II, ScαCA, AcαCA and CyαCA. The N-terminal signal peptide of bacterial α-CAs is highlighted in yellow, the conserved amino acids are in bold, the gatekeeper residues (Glu106, Thr199, hCA II numbering) are in bold magenta (indicated by ▲), the zinc-coordinating residues (His94, His96, His119) are in bold red (indicated by •) and the proton shuttle residue His64 is blue boxed (not conserved in AcαCA).

Despite its identification in 1997, the catalytic activity of the recombinant AcαCA was only evaluated in 2009 by overexpressing it in *E. coli* BL21(DE3) strain. The measured catalytic activity was 21.8 U mg^{-1} protein compared to a non-detectable CA activity observed in the control strain [155]. On the other hand, the catalytic activity of ScαCA was measured for the first time only in 2018 [156]. Interestingly, the enzyme showed high catalytic activity only when expressed in a bacterial system that supports optimal disulfide bond formation, indicating that, in this case, the conserved bacterial α-CA intramolecular disulfide bond is fundamental for the catalytic activity [156]. Additionally, since its expression levels are relatively low, it is supposed that it is not involved in the CO_2 concentration mechanism (CCM), but it acts as a sensor of environmental CO_2 levels [153,156–158].

In 2019 another cyanobacterial α-CA was identified and characterized, namely α-CA form *Cyanothece* sp. ATCC 51142 (CyαCA) [159]. After observation of CA activity in intact cells of the cyanobacterium, the gene encoding for CyαCA was cloned and heterologously expressed in *E. coli*. The recombinant protein exhibited high specific activity (~1.1 * 10^4 WAU/mg protein) and this activity was not redox regulated unlike ScαCA [159]. Immunological analysis using specific antibodies against the recombinant protein demonstrated the presence of CyαCA in the periplasmic space of *Cyanothece*. Authors hypothesized that CyαCA is most likely involved in the regulation of inorganic carbon homeostasis in *Cyanothece* sp. ATCC 51142 under inorganic carbon limiting conditions. However further investigation into the functional role of this enzyme is still necessary.

Finally, two further cyanobacterial α-CAs have been identified in alkalophilic cyanobacteria as *Microcoleus chthonoplates* (homotypic synonym of *Coleofasciculus chthonoplastes*) and *Rhabdoderma lineare* [160]. These two α-CAs are localized in the glycocalyx membrane and are likely involved into photosynthetic assimilation of inorganic carbon, $CaCO_3$ deposition during mineralization of cyanobacterial cells [161,162] and into prevention of CO_2 leakage from cells by converting it back to HCO_3^-. As for α-CA from *S. elongatus* PCC 7942, also *R. lineare* α-CA may play a role as a sensor for the CO_2 level in the external medium [160,163,164].

5. Conclusions and future perspectives

In this chapter we have summarized the key features of α-CAs identified in bacteria with a particular focus on the functional, biochemical and structural data obtained so far. Reviewed studies highlighted some common features among the analyzed enzymes: (i) they present the typical α-CA fold, but with a more compact structure compared to their human counterparts; (ii) they show by crystallographic studies the common dimeric arrangement defined as "bacterial dimer" [44–46,70,89,99–101], with the exception of TaαCA which has a tetrameric structure constituted by two "bacterial dimers" [98]. Interestingly, a different situation is observed in solution, where some of these enzymes are present as an equilibrium of the monomeric and dimeric forms. Further investigations are necessary to understand a putative functional role of the equilibrium.

Most studies on bacterial α-CAs have been carried out on enzymes identified in extremophiles, with the main aim to use them in CO_2 capture and storage processes [46]. However, the recent validation of the α-CAs from pathogenic bacteria as drug targets has greatly reignited interest in this class of enzymes [11,12]. In this context, of particular interest is the repurposing of FDA-approved drugs as treatments for interfering with the life cycle of pathogenic species [81–84]. However, many studies still need to be carried out to realize the full potential of CA inhibition as novel strategy to develop new antibacterial drugs. Considering this, we believe that there will be a substantial increase in studies on this topic in the coming years.

References

[1] E. Langella, A. Di Fiore, V. Alterio, S.M. Monti, G. De Simone, K. D'Ambrosio, α-CAs from photosynthetic organisms, Int. J. Mol. Sci. 23 (2022).

[2] V. Alterio, A. Di Fiore, K. D'Ambrosio, C.T. Supuran, G. De Simone, Multiple binding modes of inhibitors to carbonic anhydrases: how to design specific drugs targeting 15 different isoforms? Chem. Rev. 112 (2012) 4421–4468.

[3] C. Capasso, C.T. Supuran, An overview of the alpha-, beta- and gamma-carbonic anhydrases from Bacteria: can bacterial carbonic anhydrases shed new light on evolution of bacteria? J. Enzyme Inhib. Med. Chem. 30 (2015) 325–332.

[4] A. Nocentini, C. Capasso, C.T. Supuran, Carbonic anhydrase inhibitors as novel antibacterials in the era of antibiotic resistance: where are we now? Antibiotics 12 (2023).

[5] C.T. Supuran, C. Capasso, An overview of the bacterial carbonic anhydrases, Metabolites 7 (2017).

[6] C.X. Li, X.C. Jiang, Y.J. Qiu, J.H. Xu, Identification of a new thermostable and alkali-tolerant α-carbonic anhydrase from Lactobacillus delbrueckii as a biocatalyst for CO_2 biomineralization, Bioresour. Bioprocess. 2 (2015) 1–9.

[7] L.J. Urbański, S. Bua, A. Angeli, R.Z. Emameh, H.R. Barker, M. Kuuslahti, et al., The production and biochemical characterization of α-carbonic anhydrase from Lactobacillus rhamnosus GG, Appl. Microbiol. Biotechnol. 106 (2022) 4065–4074.

[8] J. Kaur, X. Cao, N.S. Abutaleb, A. Elkashif, A.L. Graboski, A.D. Krabill, et al., Optimization of acetazolamide-based scaffold as potent inhibitors of vancomycin-resistant *Enterococcus*, J. Med. Chem. 63 (2020) 9540–9562.

[9] W. An, K.J. Holly, A. Nocentini, R.D. Imhoff, C.S. Hewitt, N.S. Abutaleb, et al., Structure-activity relationship studies for inhibitors for vancomycin-resistant *Enterococcus* and human carbonic anhydrases, J. Enzyme Inhib. Med. Chem. 37 (2022) 1838–1844.

[10] G.S. Chilambi, Y.-H. Wang, N.R. Wallace, C. Obiwuma, K.M. Evans, Y. Li, et al., Carbonic anhydrase inhibition as a target for antibiotic synergy in Enterococci, Microbiol. Spectr. (2023) 11.

[11] C.T. Supuran, Bacterial carbonic anhydrases as drug targets: toward novel antibiotics? Front. Pharmacol. 2 (2011).

[12] C.T. Supuran, An overview of novel antimicrobial carbonic anhydrase inhibitors, Expert. Opin. Ther. Targets 27 (2023) 897–910.

[13] C. Bhagat, P. Dudhagara, S. Tank, Trends, application and future prospectives of microbial carbonic anhydrase mediated carbonation process for CCUS, J. Appl. Microbiol. 124 (2018) 316–335.

[14] European Centre for Disease Prevention and Control. 35000 annual deaths from antimicrobial resistance in the EU/EEA, 2022. https://www.ecdc.europa.eu/en/news-events/eaad-2022-launch (accessed June 25, 2024).

[15] D.P. Flaherty, M.N. Seleem, C.T. Supuran, Bacterial carbonic anhydrases: under-exploited antibacterial therapeutic targets, Future Med. Chem. 13 (2021) 1619–1622.

[16] B.H. Abuaita, J.H. Withey, Bicarbonate induces Vibrio cholerae virulence gene expression by enhancing ToxT activity, Infect. Immun. 77 (2009) 4111–4120.

[17] D.A. Montero, R.M. Vidal, J. Velasco, S. George, Y. Lucero, L.A. Gómez, et al., Vibrio cholerae, classification, pathogenesis, immune response, and trends in vaccine development, Front. Med. 10 (2023).

[18] World Health Organization. Cholera. https://www.who.int/health-topics/cholera#tab=tab_1 (accessed June 25, 2024).

[19] S. Almagro-Moreno, K. Pruss, R.K. Taylor, Intestinal colonization dynamics of Vibrio cholerae, PLoS Pathog. 11 (2015) e1004787.

[20] T. Ramamurthy, R.K. Nandy, A.K. Mukhopadhyay, S. Dutta, A. Mutreja, K. Okamoto, et al., Virulence regulation and innate host response in the pathogenicity of Vibrio cholerae, Front. Cell Infect. Microbiol. 10 (2020) 572096.

[21] S. Del Prete, S. Isik, D. Vullo, V. De Luca, V. Carginale, A. Scozzafava, et al., DNA cloning, characterization, and inhibition studies of an α-carbonic anhydrase from the pathogenic bacterium vibrio cholerae, J. Med. Chem. 55 (2012) 10742–10748.

[22] S. Del Prete, V. De Luca, A. Scozzafava, V. Carginale, C.T. Supuran, C. Capasso, Biochemical properties of a new α-carbonic anhydrase from the human pathogenic bacterium, Vibrio cholerae, J. Enzyme Inhib. Med. Chem. 29 (2014) 23–27.

[23] S. Del Prete, D. Vullo, V. De Luca, V. Carginale, S.M. Osman, Z. Alothman, et al., Comparison of the sulfonamide inhibition profiles of the α-, β- and γ-carbonic anhydrases from the pathogenic bacterium Vibrio cholerae, Bioorg Med. Chem. Lett. 26 (2016) 1941–1946, https://doi.org/10.1016/J.BMCL.2016.03.014

[24] F. Carta, S.M. Osman, D. Vullo, Z. AlOthman, S. Del Prete, C. Capasso, et al., Poly (amidoamine) dendrimers show carbonic anhydrase inhibitory activity against α-, β-, γ- and η-class enzymes, Bioorg. Med. Chem. 23 (2015) 6794–6798.

[25] A. Angeli, G. Abbas, S. Del Prete, F. Carta, C. Capasso, C.T. Supuran, Acyl selenoureido benzensulfonamides show potent inhibitory activity against carbonic anhydrases from the pathogenic bacterium Vibrio cholerae, Bioorg. Chem. 75 (2017) 170–172.

[26] A. Angeli, M. Pinteala, S.S. Maier, B.C. Simionescu, A. Milaneschi, G. Abbas, et al., Evaluation of thio- and seleno-acetamides bearing benzenesulfonamide as inhibitor of carbonic anhydrases from different pathogenic bacteria, Int. J. Mol. Sci. 21 (2020).

[27] A. Bonardi, A. Nocentini, R. Cadoni, S. Del Prete, P. Dumy, C. Capasso, et al., Benzoxaboroles: new potent inhibitors of the carbonic anhydrases of the pathogenic bacterium Vibrio cholerae, ACS Med. Chem. Lett. 11 (2020) 2277–2284.

[28] F. Mancuso, A. Angeli, V. De Luca, F. Bucolo, L. De Luca, C. Capasso, et al., Synthesis and biological evaluation of sulfonamide-based compounds as inhibitors of carbonic anhydrase from Vibrio cholerae, Arch. Pharm. Weinh. 355 (2022) 2200070.

[29] K. Demir-Yazıcı, Ö. Güzel-Akdemir, A. Angeli, C.T. Supuran, A. Akdemir, Novel indole-based hydrazones as potent inhibitors of the α-class carbonic anhydrase from pathogenic bacterium Vibrio cholerae, Int. J. Mol. Sci. 21 (2020).

[30] D. De Vita, A. Angeli, F. Pandolfi, M. Bortolami, R. Costi, R. Di Santo, et al., Inhibition of the α-carbonic anhydrase from Vibrio cholerae with amides and sulfonamides incorporating imidazole moieties, J. Enzyme Inhib. Med. Chem. 32 (2017) 798–804.

[31] E. Berrino, M. Bozdag, S. Del Prete, F.A.S. Alasmary, L.S. Alqahtani, Z. AlOthman, et al., Inhibition of α-, β-, γ-, and δ-carbonic anhydrases from bacteria and diatoms with N'-aryl-N-hydroxy-ureas, J. Enzyme Inhib. Med. Chem. 33 (2018) 1194–1198.

[32] S. Giovannuzzi, C.S. Hewitt, A. Nocentini, C. Capasso, G. Costantino, D.P. Flaherty, et al., Inhibition studies of bacterial α-carbonic anhydrases with phenols, J. Enzyme Inhib. Med. Chem. 37 (2022) 666–671.

[33] S. Giovannuzzi, C.S. Hewitt, A. Nocentini, C. Capasso, D.P. Flaherty, C.T. Supuran, Coumarins effectively inhibit bacterial α-carbonic anhydrases, J. Enzyme Inhib. Med. Chem. 37 (2022) 333–338.

[34] L.C. Chirica, C. Petersson, M. Hurtig, B.H. Jonsson, T. Borén, S. Lindskog, Expression and localization of α- and β-carbonic anhydrase in Helicobacter pylori, Biochim. Biophys. Acta - Proteins Proteom. 1601 (2002) 192–199.

[35] J.K. Modakh, Y.C. Liu, M.A. Machuca, C.T. Supuran, A. Roujeinikova, Structural basis for the inhibition of Helicobacter pylori α-carbonic anhydrase by sulfonamides, PLoS One 10 (2015).

[36] E.A. Marcus, A.P. Moshfegh, G. Sachs, D.R. Scott, The periplasmic alpha-carbonic anhydrase activity of Helicobacter pylori is essential for acid acclimation, J. Bacteriol. 187 (2005) 729–738.

[37] D.R. Scott, E.A. Marcus, Y. Wen, S. Singh, J. Feng, G. Sachs, Cytoplasmic histidine kinase (HP0244)-regulated assembly of urease with UreI, a channel for urea and its metabolites, CO₂, NH3, and NH4+, is necessary for acid survival of Helicobacter pylori, J. Bacteriol. 192 (2010) 94–103.

[38] Y. Wen, E.A. Marcus, U. Matrubutham, M.A. Gleeson, D.R. Scott, G. Sachs, Acid-adaptive genes of Helicobacter pylori, Infect. Immun. 71 (2003) 5921–5939.

[39] E.A. Marcus, G. Sachs, Y.I. Wen, J. Feng, D.R. Scott, Role of the Helicobacter pylori sensor kinase ArsS in protein trafficking and acid acclimation, J. Bacteriol. 194 (2012) 5545–5551.

[40] Y. Wen, J. Feng, D.R. Scott, E.A. Marcus, G. Sachs, Involvement of the HP0165-HP0166 two-component system in expression of some acidic-pH-upregulated genes of Helicobacter pylori, J. Bacteriol. 188 (2006) 1750–1761.

[41] L.H. Fan, N. Liu, M.R. Yu, S.T. Yang, H.L. Chen, Cell surface display of carbonic anhydrase on Escherichia coli using ice nucleation protein for CO₂ sequestration, Biotechnol. Bioeng. 108 (2011) 2853–2864.

[42] Y. Zhu, Y. Liu, M. Ai, X. Jia, Surface display of carbonic anhydrase on Escherichia coli for CO₂ capture and mineralization, Synth. Syst. Biotechnol. 7 (2021) 460–473.

[43] L.C. Chirica, B. Elleby, S. Lindskog, Cloning, expression and some properties of α-carbonic anhydrase from Helicobacter pylori, Biochim. Biophys. Acta—Protein Struct. Mol. Enzymol. 1544 (2001) 55–63.

[44] M.E. Compostella, P. Berto, F. Vallese, G. Zanotti, Structure of α-carbonic anhy-drase from the human pathogen Helicobacter pylori, Acta Crystallogr. Sect. F, Struct. Biol. Commun. 71 (2015) 1005–1011.

[45] J.K. Modak, Y.C. Liu, C.T. Supuran, A. Roujeinikova, Structure-activity relation-ship for sulfonamide inhibition of Helicobacter pylori α-carbonic anhydrase, J. Med. Chem. 59 (2016) 11098–11109.

[46] A. Di Fiore, C. Capasso, V. De Luca, S.M. Monti, V. Carginale, C.T. Supuran, et al., X-ray structure of the first 'extremo-α-carbonic anhydrase', a dimeric enzyme from the thermophilic bacterium Sulfurihydrogenibium yellowstonense YO3AOP1, Acta Crystallogr. Sect. D. 69 (2013) 1150–1159.

[47] A.E. Eriksson, T.A. Jones, A. Liljas, Refined structure of human carbonic anhydrase II at 2.0 Å resolution, Proteins Struct. Funct. Bioinforma. 4 (1988) 274–282.

[48] J.K. Modak, S.A. Revitt-Mills, A. Roujeinikova, Cloning, purification and preliminary crystallographic analysis of the complex of Helicobacter pylori α-carbonic anhydrase with acetazolamide, Acta Crystallogr. Sect. F. Struct. Biol. Cryst. Commun. 69 (2013) 1252–1255.

[49] I. Nishimori, D. Vullo, T. Minakuchi, K. Morimoto, S. Onishi, A. Scozzafava, et al., Carbonic anhydrase inhibitors: cloning and sulfonamide inhibition studies of a car-boxyterminal truncated alpha-carbonic anhydrase from Helicobacter pylori, Bioorg. Med. Chem. Lett. 16 (2006) 2182–8.

[50] A. Maresca, D. Vullo, A. Scozzafava, C.T. Supuran, Inhibition of the alpha- and beta-carbonic anhydrases from the gastric pathogen Helycobacter pylori with anions, J. Enzyme Inhib. Med. Chem. 28 (2013) 388–391.

[51] A. Angeli, M. Ferraroni, C.T. Supuran, Famotidine, an antiulcer agent, strongly inhibits Helicobacter pylori and human carbonic anhydrases, ACS Med. Chem. Lett. 9 (2018) 1035–1038.

[52] A. Angeli, M. Pinteala, S.S. Maier, S. Del Prete, C. Capasso, B.C. Simionescu, et al., Inhibition of bacterial α-, β- and γ-class carbonic anhydrases with selenazoles incor-porating benzenesulfonamide moieties, J. Enzyme Inhib. Med. Chem. 34 (2019) 244.

[53] R. Grande, S. Carradori, V. Puca, I. Vitale, A. Angeli, A. Nocentini, et al., Selective inhibition of Helicobacter pylori carbonic anhydrases by carvacrol and thymol could impair biofilm production and the release of outer membrane vesicles, Int. J. Mol. Sci. 22 (2021).

[54] M.M. Rahman, A. Tikhomirova, J.K. Modak, M.L. Hutton, C.T. Supuran, A. Roujeinikova, Antibacterial activity of ethoxzolamide against Helicobacter pylori strains SS1 and 26695, Gut Pathog. 12 (2020).

[55] G. Benito, I. D'agostino, S. Carradori, M. Fantacuzzi, M. Agamennone, V. Puca, et al., Erlotinib-containing benzenesulfonamides as anti- Helicobacter pylori agents through carbonic anhydrase inhibition, Future Med. Chem. 15 (2023) 1865–1883.

[56] J.K. Modak, A. Tikhomirova, R.J. Gorrell, M.M. Rahman, D. Kotsanas, T.M. Korman, et al., Anti-Helicobacter pylori activity of ethoxzolamide, J. Enzyme Inhib. Med. Chem. 34 (2019) 1660–1667.

[57] S.J. Quillin, H.S. Seifert, Neisseria gonorrhoeae host adaptation and pathogenesis, 2018 164, Nat. Rev. Microbiol. 16 (2018) 226–240.

[58] J.S. Lee, H.Y. Choi, J.E. Lee, S.H. Lee, B.S. Oum, Gonococcal keratoconjunctivitis in adults, Eye 16 (2002) 646–649 165 2002.

[59] R.C. Noble, R.M. Cooper, B.R. Miller, Pharyngeal colonisation by Neisseria gonorrhoeae and Neisseria meningitidis in black and white patients attending a venereal disease clinic, Br. J. Vener. Dis. 55 (1979) 14–19.

[60] C.S. Danby, L.A. Cosentino, L.K. Rabe, C.L. Priest, K.C. Damare, I.S. Macio, et al., Patterns of extragenital chlamydia and gonorrhea in women and men who have sex with men reporting a history of receptive anal intercourse, Sex. Transm. Dis. 43 (2016) 105.

[61] A.K. Marapaka, A. Nocentini, M.S. Youse, W. An, K.J. Holly, C. Das, et al., Structural characterization of thiadiazolesulfonamide inhibitors bound to Neisseria gonorrhoeae α-carbonic anhydrase, ACS Med. Chem. Lett. 14 (2022) 103–109.

[62] C.S. Hewitt, N.S. Abutaleb, A.E.M. Elhassanny, A. Nocentini, X. Cao, D.P. Amos, et al., Structure-activity relationship studies of acetazolamide-based carbonic anhydrase inhibitors with activity against Neisseria gonorrhoeae, ACS Infect. Dis. 7 (2021) 1969–1984.

[63] N.S. Abutaleb, A.E.M. Elhassanny, A. Nocentini, C.S. Hewitt, A. Elkashif, B.R. Cooper, et al., Repurposing FDA-approved sulphonamide carbonic anhydrase inhibitors for treatment of Neisseria gonorrhoeae, J. Enzyme Inhib. Med. Chem. 37 (2022) 51–61.

[64] S. Giovannuzzi, N.S. Abutaleb, C.S. Hewitt, F. Carta, A. Nocentini, M.N. Seleem, et al., Dithiocarbamates effectively inhibit the α-carbonic anhydrase from Neisseria gonorrhoeae, J. Enzyme Inhib. Med. Chem. 37 (2022) 1–8.

[65] N.S. Abutaleb, A.E.M. Elhassanny, M.N. Seleem, In vivo efficacy of acetazolamide in a mouse model of Neisseria gonorrhoeae infection, Microb. Pathog. 164 (2022).

[66] D. Flaherty, M. Seleem, Inventors. Carbonic Anhydrase Inhibitors for Treatment of Neisseria Gonorrhoeae Infection. United States Patent US 2022213047A1, July 7, 2022.

[67] Z. Liu, P. Bartlow, R.M. Dilmore, Y. Soong, Z. Pan, R. Koepsel, et al., Production, purification, and characterization of a fusion protein of carbonic anhydrase from Neisseria gonorrhoeae and cellulose binding domain from Clostridium thermocellum, Biotechnol. Prog. 25 (2009) 68–74.

[68] I.G. Kim, B.H. Jo, D.G. Kang, C.S. Kim, Y.S. Choi, H.J. Cha, Biomineralization-based conversion of carbon dioxide to calcium carbonate using recombinant carbonic anhydrase, Chemosphere 87 (2012) 1091–1096.

[69] L.C. Chiricǎ, B. Elleby, B.H. Jonsson, S. Lindskog, The complete sequence, expression in Escherichia coli, purification and some properties of carbonic anhydrase from Neisseria gonorrhoeae, Eur. J. Biochem. 244 (1997) 755–760.

[70] S. Huang, Y. Xue, E. Sauer-Eriksson, L. Chirica, S. Lindskog, B.H. Jonsson, Crystal structure of carbonic anhydrase from Neisseria gonorrhoeae and its complex with the inhibitor acetazolamide, J. Mol. Biol. 283 (1998) 301–310.

[71] L.-G. Mårtensson, M. Karlsson, U. Carlsson, Dramatic stabilization of the native state of human carbonic anhydrase II by an engineered disulfide bond, Biochemistry 41 (2002) 15867–15875.

[72] M. Karlsson, L.G. Mårtensson, C. Karlsson, U. Carlsson, Denaturant-assisted formation of a stabilizing disulfide bridge from engineered cysteines in nonideal conformations, Biochemistry 44 (2005) 3487–3493.

[73] A. Nocentini, C.S. Hewitt, M.D. Mastrolorenzo, D.P. Flaherty, C.T. Supuran, Anion inhibition studies of the α-carbonic anhydrases from Neisseria gonorrhoeae, J. Enzyme Inhib. Med. Chem. 36 (2021) 1061–1066.

[74] A. Forkman, A.B. Laurell, The effect of carbonic anhydrase inhibitor on the growth of Neisseriae, Acta Pathol. Microbiol. Scand. 65 (1965) 450–456.

[75] C.A. Arias, B.E. Murray, The rise of the Enterococcus: beyond vancomycin resistance, Nat. Rev. Microbiol. 10 (2012) 266–278.

[76] D.M. Sievert, P. Ricks, J.R. Edwards, A. Schneider, J. Patel, A. Srinivasan, et al., Antimicrobial-resistant pathogens associated with healthcare-associated infections: summary of data reported to the National Healthcare Safety Network at the Centers for Disease Control and Prevention, 2009-2010, Infect. Control. Hosp. Epidemiol. 34 (2013) 1–14.

[77] D.A. Robinson, D. Falush, E.J. Feil (Eds.), Bacterial Population Genetics in Infectious Disease, Wiley-Blackwell, 2010.

[78] J.M. Miro, J.M. Pericas, A. Del Rio, A new era for treating enterococcus faecalis endocarditis ampicillin plus short-course gentamicin or ampicillin plus ceftriaxone: That is the question!, Circulation 127 (2013) 1763–1766.

[79] E. Fiore, D. Van Tyne, M.S. Gilmore, Pathogenicity of Enterococci, Microbiol. Spectr. 7 (2019).

[80] M. Golob, M. Pate, D. Kušar, U. Dermota, J. Avberšek, B. Papić, et al., Antimicrobial resistance and virulence genes in Enterococcus faecium and Enterococcus faecalis from humans and retail red meat, Biomed. Res. Int. 2019 (2019).

[81] N.S. Abutaleb, A.E.M. Elhassanny, D.P. Flaherty, M.N. Seleem, In vitro and in vivo activities of the carbonic anhydrase inhibitor, dorzolamide, against vancomycin-resistant enterococci, PeerJ 9 (2021) e11059.

[82] A. AbdelKhalek, N.S. Abutaleb, H. Mohammad, M.N. Seleem, Repurposing ebselen for decolonization of vancomycin-resistant enterococci (VRE), PLoS One 13 (2018).

[83] A. AbdelKhalek, N.S. Abutaleb, K.A. Elmagarmid, M.N. Seleem, Repurposing auranofin as an intestinal decolonizing agent for vancomycin-resistant Enterococci, Sci. Rep. 8 (2018) 1–9 81 2018.

[84] H. Mohammad, A. AbdelKhalek, N.S. Abutaleb, M.N. Seleem, Repurposing niclosamide for intestinal decolonization of vancomycin-resistant Enterococci, Int. J. Antimicrob. Agents 51 (2018) 897–904.

[85] N.S. Abutaleb, A. Elkashif, D.P. Flaherty, M.N. Seleem, In vivo antibacterial activity of acetazolamide, Antimicrob. Agents Chemother. 65 (2021), https://doi.org/10.1128/AAC.01715-20

[86] N.S. Abutaleb, A. Shrinidhi, A.B. Bandara, M.N. Seleem, D.P. Flaherty, Evaluation of 1,3,4-thiadiazole carbonic anhydrase inhibitors for gut decolonization of vancomycin-resistant Enterococci, ACS Med. Chem. Lett. 14 (2023) 487–492.

[87] N. Merino, H.S. Aronson, D.P. Bojanova, J. Feyhl-Buska, M.L. Wong, S. Zhang, et al., Living at the extremes: extremophiles and the limits of life in a planetary context, Front. Microbiol. 10 (2019) 447668.

[88] N. Raddadi, A. Cherif, D. Daffonchio, M. Neifar, F. Fava, Biotechnological applications of extremophiles, extremozymes and extremolytes, Appl. Microbiol. Biotechnol. 99 (2015) 7907–7913.

[89] N.A. Díaz-Torres, B.P. Mahon, C.D. Boone, M.A. Pinard, C. Tu, R. Ng, et al., Structural and biophysical characterization of the α-carbonic anhydrase from the gammaproteobacterium Thiomicrospira crunogena XCL-2: insights into engineering thermostable enzymes for CO_2 sequestration, Acta Crystallogr. D. Biol. Crystallogr 71 (2015) 1745–1756.

[90] B.H. Jo, S.K. Im, H.J. Cha, Halotolerant carbonic anhydrase with unusual N-terminal extension from marine Hydrogenovibrio marinus as novel biocatalyst for carbon sequestration under high-salt environments, J. CO_2 Util. 26 (2018) 415–424.

[91] B.H. Jo, An intrinsically disordered peptide tag that confers an unusual solubility to aggregation-prone proteins, Appl. Env. Microbiol. 88 (2022).

[92] I.S. Hwang, J.H. Kim, B.H. Jo, Enhanced production of a thermostable carbonic anhydrase in Escherichia coli by using a modified NEXT tag, Molecules 26 (2021) 5830 2021;26:5830.

[93] D. Vullo, A. Bhatt, B.P. Mahon, R. McKenna, C.T. Supuran, Sulfonamide inhibition studies of the α-carbonic anhydrase from the gammaproteobacterium Thiomicrospira crunogena XCL-2, TcruCA, Bioorg. Med. Chem. Lett. 26 (2016) 401–405.

[94] V. De Luca, D. Vullo, A. Scozzafava, V. Carginale, M. Rossi, et al., An α-carbonic anhydrase from the thermophilic bacterium Sulphurihydrogenibium azorense is the fastest enzyme known for the CO_2 hydration reaction, Bioorg. Med. Chem. 21 (2013) 1465–1469.

[95] D. Vullo, V. De Luca, A. Scozzafava, V. Carginale, M. Rossi, C.T. Supuran, et al., The extremo-α-carbonic anhydrase from the thermophilic bacterium Sulfurihydrogenibium azorense is highly inhibited by sulfonamides, Bioorg. Med. Chem. 21 (2013) 4521–4525.

[96] W.D. Fu, C.J. Hsieh, C.J. Hu, C.Y. Yu, Entrapment of carbonic anhydrase from Sulfurihydrogenibium azorense with polyallylamine-mediated biomimetic silica, Bioresour. Technol. Rep. 20 (2022) 101217.

[97] V. De Luca, S. Del Prete, V. Carginale, D. Vullo, C.T. Supuran, C. Capasso, A failed tentative to design a super carbonic anhydrase having the biochemical properties of the most thermostable CA (SspCA) and the fastest (SazCA) enzymes, J. Enzyme Inhib. Med. Chem. 30 (2015) 989–994.

[98] P. James, M.N. Isupov, C. Sayer, V. Saneei, S. Berg, M. Lioliou, et al., The structure of a tetrameric α-carbonic anhydrase from Thermovibrio ammonificans reveals a core formed around intermolecular disulfides that contribute to its thermostability, Acta Crystallogr. D. Biol. Crystallogr. 70 (2014) 2607–2618.

[99] G. De Simone, S.M. Monti, V. Alterio, M. Buonanno, V. De Luca, M. Rossi, et al., Crystal structure of the most catalytically effective carbonic anhydrase enzyme known, SazCA from the thermophilic bacterium Sulfurihydrogenibium azorense, Bioorg. Med. Chem. Lett. 25 (2015) 2002–2006.

[100] S. Kim, J. Sung, J. Yeon, S.H. Choi, M.S. Jin, Crystal structure of a highly thermostable α-carbonic anhydrase from Persephonella marina EX-H1, Mol. Cell 42 (2019) 460–469.

[101] V. Somalinga, G. Buhrman, A. Arun, R.B. Rose, A.M. Grunden, A high-resolution crystal structure of a psychrohalophilic α–carbonic anhydrase from Photobacterium profundum reveals a unique dimer interface, PLoS One 11 (2016) e0168022.

[102] C. Vetriani, M.D. Speck, S.V. Ellor, R.A. Lutz, V. Starovoytor, Thermovibrio ammonificans sp. nov., a thermophilic, chemolithotrophic, nitrate-ammonifying bacterium from deep-sea hydrothermal vents, Int. J. Syst. Evol. Microbiol. 54 (2004) 175–181.

[103] D. Giovannelli, S.M. Sievert, M. Hügler, S. Markert, D. Becher, T. Schweder, et al., Insight into the evolution of microbial metabolism from the deep-branching bacterium, thermovibrio ammonificans, Elife 6 (2017).

[104] B.H. Jo, J.H. Seo, H.J. Cha, Bacterial extremo-α-carbonic anhydrases from deep-sea hydrothermal vents as potential biocatalysts for CO_2 sequestration, J. Mol. Catal. B Enzym. 109 (2014) 31–39.

[105] R. Parra-Cruz, C.M. Jäger, P.L. Lau, R.L. Gomes, A. Pordea, Rational design of thermostable carbonic anhydrase mutants using molecular dynamics simulations, J. Phys. Chem. B 122 (36) (2018) 8526.

[106] R. Parra-Cruz, P.L. Lau, H.S. Loh, A. Pordea, Engineering of Thermovibrio ammonificans carbonic anhydrase mutants with increased thermostability, J. CO_2 Util. 37 (2020) 1–8.

[107] K.M. Scott, S.M. Sievert, F.N. Abril, L.A. Ball, C.J. Barrett, R.A. Blake, et al., The genome of deep-sea vent chemolithoautotroph Thiomicrospira crunogena XCL-2, PLoS Biol. 4 (2006) e383.

[108] B.P. Mahon, N.A. Díaz-Torres, M.A. Pinard, C. Tu, D.N. Silverman, K.M. Scott, et al., Activity and anion inhibition studies of the α-carbonic anhydrase from Thiomicrospira crunogena XCL-2 Gammaproteobacterium, Bioorg. Med. Chem. Lett. 25 (2015) 4937–4940.

[109] B.P. Mahon, A. Bhatt, D. Vullo, C.T. Supuran, R. McKenna, Exploration of anionic inhibition of the α-carbonic anhydrase from Thiomicrospira crunogena XCL-2 gammaproteobacterium: a potential bio-catalytic agent for industrial CO_2 removal, Chem. Eng. Sci. 138 (2015) 575–580.

[110] S. Nakagawa, Z. Shataih, A. Banta, T.J. Beveridge, Y. Sako, A.L. Reysenbach, Sulfurihydrogenibium yellowstonense sp. nov., an extremely thermophilic, facultatively heterotrophic, sulfur-oxidizing bacterium from Yellowstone National Park, and emended descriptions of the genus Sulfurihydrogenibium, Sulfurihydrogenibium subterraneum and Sulfurihydrogenibium azorense, Int. J. Syst. Evol. Microbiol. 55 (2005) 2263–2268.

[111] K. Kubo, K. Knittel, R. Amann, M. Fukui, K. Matsuura, Sulfur-metabolizing bacterial populations in microbial mats of the Nakabusa hot spring, Japan, Syst. Appl. Microbiol. 34 (2011) 293–302.

[112] C. Capasso, V. De Luca, V. Carginale, R. Cannio, M. Rossi, Biochemical properties of a novel and highly thermostable bacterial α-carbonic anhydrase from Sulfurihydrogenibium yellowstonense YO3AOP1, J. Enzyme Inhib. Med. Chem. 27 (2012) 892–897.

[113] V. De Luca, D. Vullo, A. Scozzafava, V. Carginale, M. Rossi, C.T. Supuran, et al., Anion inhibition studies of an α-carbonic anhydrase from the thermophilic bacterium Sulfurihydrogenibium yellowstonense YO3AOP1, Bioorg. Med. Chem. Lett. 22 (2012) 5630–5634.

[114] D. Vullo, V. De Luca, A. Scozzafava, V. Carginale, M. Rossi, C.T. Supuran, et al., The alpha-carbonic anhydrase from the thermophilic bacterium Sulfurihydrogenibium yellowstonense YO3AOP1 is highly susceptible to inhibition by sulfonamides, Bioorg. Med. Chem. 21 (2013) 1534–1538.

[115] A.M. Alafeefy, H.A. Abdel-Aziz, D. Vullo, A.M.S. Al-Tamimi, N.A. Al-Jaber, C. Capasso, et al., Inhibition of carbonic anhydrases from the extremophilic bacteria Sulfurihydrogenibium yellostonense (SspCA) and S. azorense (SazCA) with a new series of sulfonamides incorporating aroylhydrazone-, [1,2,4]triazolo[3,4-b][1,3,4] thiadiazinyl- or 2-(cyanophenylmethylene)-1,3,4-thiadiazol-3(2H)-yl moieties, Bioorg. Med. Chem. 22 (2014) 141–147.

[116] D. Vullo, V. De Luca, A. Scozzafava, V. Carginale, M. Rossi, C.T. Supuran, et al., Anion inhibition studies of the fastest carbonic anhydrase (CA) known, the extremo-CA from the bacterium Sulfurihydrogenibium azorense, Bioorg. Med. Chem. Lett. 22 (2012) 7142–7145.

[117] P. Aguiar, T.J. Beveridge, A.L. Reysenbach, Sulfurihydrogenibium azorense, sp. nov., a thermophilic hydrogen-oxidizing microaerophile from terrestrial hot springs in the Azores, Int. J. Syst. Evol. Microbiol. 54 (2004) 33–39.

[118] D. Vullo, V. De Luca, A. Scozzafava, V. Carginale, M. Rossi, C.T. Supuran, et al., The first activation study of a bacterial carbonic anhydrase (CA). The thermostable α-CA from Sulfurihydrogenibium yellowstonense YO3AOP1 is highly activated by amino acids and amines, Bioorg. Med. Chem. Lett. 22 (2012) 6324–6327.

[119] A. Akdemir, D. Vullo, V.De Luca, A. Scozzafava, V. Carginale, M. Rossi, et al., The extremo-α-carbonic anhydrase (CA) from Sulfurihydrogenibium azorense, the fastest CA known, is highly activated by amino acids and amines, Bioorg. Med. Chem. Lett. 23 (2013) 1087–1090.

[120] D. Götz, A. Banta, T.J. Beveridge, A.I. Rushdi, B.R.T. Simoneit, A.L. Reysenbach, Persephonella marina gen. nov., sp. nov. and Persephonella guaymasensis sp. nov., two novel, thermophilic, hydrogen-oxidizing microaerophiles from deep-sea hydrothermal vents, Int. J. Syst. Evol. Microbiol. 52 (2002) 1349–1359.

[121] M.S. Borchert, Inventor, Novozymes As, Owners. Heat-Stable Persephonella Carbonic Anhydrases and Their Use. United States Patent US 2013/0149771 A1. June 13, 2013.

[122] B.K. Kanth, S.Y. Jun, S. Kumari, S.P. Pack, Highly thermostable carbonic anhydrase from Persephonella marina EX-H1: its expression and characterization for CO_2-sequestration applications, Process. Biochem. 49 (2014) 2114–2121.

[123] S. Campanaro, A. Vezzi, N. Vitulo, F.M. Lauro, M. D'Angelo, F. Simonato, et al., Laterally transferred elements and high pressure adaptation in Photobacterium profundum strains, BMC Genomics 6 (2005) 1–15.

[124] F.M. Lauro, E.A. Eloe-Fadrosh, T.K.S. Richter, N. Vitulo, S. Ferriera, J.H. Johnson, et al., Ecotype diversity and conversion in Photobacterium profundum strains, PLoS One 9 (2014) e96953.

[125] V. Somalinga, E. Foss, A.M. Grunden, V. Somalinga, E. Foss, A.M. Grunden, Biochemical characterization of a psychrophilic and halotolerant α–carbonic anhydrase from a deep-sea bacterium, *Photobacterium profundum*, AIMS Microbiol. 9 (2023) 540–553 3540 2023.

[126] H. Nishihara, Y. Igarashi, T. Kodama, Hydrogenovibrio marinus gen. nov., sp. nov., a marine obligately chemolithoautotrophic hydrogen-oxidizing bacterium, Int. J. Syst. Bacteriol. 41 (1991) 130–133.

[127] Carbon Dioxide. https://climate.nasa.gov/vital-signs/carbon-dioxide/?intent=121 (accessed June 24, 2024).

[128] Y. Yang, T. Zhou, M. Chong, M. Xie, N. Shi, T. Liu, et al., Recent advances in organic waste pyrolysis and gasification in a CO_2 environment to value-added products, J. Environ. Manage. 356 (2024).

[129] Y. Mao, H. Wang, F. Jiang, F. Justino, D. Bromwich, et al., A comparison of low carbon investment needs between China and Europe in stringent climate policy scenarios, Environ. Res. Lett. 14 (2019) 054017.

[130] K.R. Smith, A. Woodward, D. Campbell-Lendrum, D.D. Chadee, Y. Honda, Q. Liu, et al., Human health: impacts, adaptation, and co-benefits, Clim. Chang. 2014 Impacts, Adapt. Vulnerability Part. A Glob. Sect. Asp. (2014) 709–754.

[131] M.E. Russo, G. Olivieri, C. Capasso, V. De Luca, A. Marzocchella, P. Salatino, et al., Kinetic study of a novel thermo-stable α-carbonic anhydrase for biomimetic CO_2 capture, Enzyme Microb. Technol. 53 (2013) 271–277.

[132] F. Migliardini, V. De Luca, V. Carginale, M. Rossi, P. Corbo, C.T. Supuran, et al., Biomimetic CO_2 capture using a highly thermostable bacterial α-carbonic anhydrase immobilized on a polyurethane foam, J. Enzyme Inhib. Med. Chem. 29 (2014) 146–150.

[133] S. Del Prete, R. Perfetto, M. Rossi, F.A.S. Alasmary, S.M. Osman, Z. AlOthman, et al., A one-step procedure for immobilising the thermostable carbonic anhydrase (SspCA) on the surface membrane of Escherichia coli, J. Enzyme Inhib. Med. Chem. 32 (2017) 1120–1128.

[134] S. Fabbricino, S. Del Prete, M.E. Russo, C. Capasso, A. Marzocchella, P. Salatino, In vivo immobilized carbonic anhydrase and its effect on the enhancement of CO_2 absorption rate, J. Biotechnol. 336 (2021) 41–49.

[135] G. Salbitani, S. Del Prete, F. Bolinesi, O. Mangoni, V. De Luca, V. Carginale, et al., Use of an immobilised thermostable α-CA (SspCA) for enhancing the metabolic efficiency of the freshwater green microalga Chlorella sorokiniana, J. Enzyme Inhib. Med. Chem. 35 (2020) 913–920.

[136] R. Perfetto, S. Del Prete, D. Vullo, G. Sansone, C.M.A. Barone, M. Rossi, et al., Production and covalent immobilisation of the recombinant bacterial carbonic anhydrase (SspCA) onto magnetic nanoparticles, J. Enzyme Inhib. Med. Chem. 32 (2017) 759–766.

[137] M.Y. Abdelrahim, C.F. Martins, L.A. Neves, C. Capasso, C.T. Supuran, I.M. Coelhoso, et al., Supported ionic liquid membranes immobilized with carbonic anhydrases for CO_2 transport at high temperatures, J. Memb. Sci. 528 (2017) 225–230.

[138] S.S.W. Effendi, C.Y. Chiu, Y.K. Chang, I.S. Ng, Crosslinked on novel nanofibers with thermophilic carbonic anhydrase for carbon dioxide sequestration, Int. J. Biol. Macromol. 152 (2020) 930–938.

[139] W.S.S. Effendi, S.I. Tan, W.W. Ting, I.S. Ng, Enhanced recombinant Sulfurihydrogenibium yellowstonense carbonic anhydrase activity and thermostability by chaperone GroELS for carbon dioxide biomineralization, Chemosphere 271 (2021).

[140] S. Del Prete, V. De Luca, C. Capasso, C.T. Supuran, V. Carginale, Recombinant thermoactive phosphoenolpyruvate carboxylase (PEPC) from ThermoSynechococcus elongatus and its coupling with mesophilic/thermophilic bacterial carbonic anhydrases (CAs) for the conversion of CO_2 to oxaloacetate, Bioorg. Med. Chem. 24 (2016) 220–225.

[141] J. Hou, X. Li, M.B. Kaczmarek, P. Chen, K. Li, P. Jin, et al., Accelerated CO_2 hydration with thermostable Sulfurihydrogenibium azorense carbonic anhydrase-chitin binding domain fusion protein immobilised on chitin support, Int. J. Mol. Sci. 20 (2019) 1494 2019;20:1494.

[142] M. Kumari, J. Lee, D.W. Lee, I. Hwang, High-level production in a plant system of a thermostable carbonic anhydrase and its immobilization on microcrystalline cellulose beads for CO_2 capture, Plant. Cell Rep. 39 (2020) 1317–1329.

[143] C.J. Hsieh, J.C. Cheng, C.J. Hu, C.Y. Yu, Entrapment of the fastest known carbonic anhydrase with biomimetic silica and its application for co2 sequestration, Polymers (Basel) 13 (2021) 2452.

[144] B.H. Jo, H. Moon, H.J. Cha, Engineering the genetic components of a whole-cell catalyst for improved enzymatic CO_2 capture and utilization, Biotechnol. Bioeng. 117 (2020) 39–48.

[145] H. Moon, S. Kim, B.H. Jo, H.J. Cha, Immobilization of genetically engineered whole-cell biocatalysts with periplasmic carbonic anhydrase in polyurethane foam for enzymatic CO_2 capture and utilization, J. CO_2 Util. 39 (2020) 101172.

[146] S. Kim, K.Il Joo, B.H. Jo, H.J. Cha, Stability-controllable self-immobilization of carbonic anhydrase fused with a silica-binding tag onto diatom biosilica for enzymatic CO_2 capture and utilization, ACS Appl. Mater. Interfaces 12 (2020) 27055–27063.

[147] S. Wi, I.S. Hwang, B.H. Jo, Engineering a plant viral coat protein for in vitro hybrid self-assembly of CO_2-capturing catalytic nanofilaments, Biomacromolecules 21 (2020) 3847–3856.

[148] H. Sun, J.H. Han, Y.W. Jo, S.O. Han, J.E. Hyeon, Increased thermal stability of the carbonic anhydrase enzyme complex for the efficient reduction of CO_2 through cyclization and polymerization by peptide bonding, Process. Biochem. 120 (2022) 195–201.

[149] R.P. Sinha, D.P. Häder, UV-protectants in cyanobacteria, Plant. Sci. 174 (2008) 278–289.

[150] P.W. Crockford, Y.M. Bar On, L.M. Ward, R. Milo, I. Halevy, The geologic history of primary productivity, Curr. Biol. 33 (2023) 4741–4750.e5.

[151] D.E. Canfield, M.T. Rosing, C. Bjerrum, Early anaerobic metabolisms, Philos. Trans. R. Soc. B Biol. Sci. 361 (2006) 1819–1836.

[152] C.F. Demoulin, Y.J. Lara, L. Cornet, C. François, D. Baurain, A. Wilmotte, et al., Cyanobacteria evolution: insight from the fossil record, Free. Radic. Biol. Med. 140 (2019) 206–223.

[153] K.S. Smith, J.G. Ferry, Prokaryotic carbonic anhydrases, FEMS Microbiol. Rev. 24 (2000) 335–366.

[154] E. Soltes-Rak, M.E. Mulligan, J.R. Coleman, Identification and characterization of a gene encoding a vertebrate-type carbonic anhydrase in cyanobacteria, J. Bacteriol. 179 (1997) 769–774.

[155] D. Wang, Q. Li, W. Li, J. Xing, Z. Su, Improvement of succinate production by overexpression of a cyanobacterial carbonic anhydrase in Escherichia coli, Enzyme Microb. Technol. 45 (2009) 491–497.

[156] E.V. Kupriyanova, M.A. Sinetova, V.S. Bedbenov, N.A. Pronina, D.A. Los, Putative extracellular α-class carbonic anhydrase, EcaA, of Synechococcus elongatus PCC 7942 is an active enzyme: a sequel to an old story, Microbiology 164 (2018) 576–586.

[157] A.K.C. So, H.G.C. Van Spall, J.R. Coleman, G.S. Espie, Catalytic exchange of 18O from 13C18O-labelled CO_2 by wild-type cells and ecaA, ecaB, and ccaA mutants of the cyanobacteria Synechococcus PCC7942 and Synechocystis PCC6803, Can. J. Bot. 76 (1998) 1153–1160.

[158] A.K.C. So, G.S. Espie, Cyanobacterial carbonic anhydrases, Can. J. Bot. 83 (2011) 721–734.

[159] E.V. Kupriyanova, M.A. Sinetova, K.S. Mironov, G.V. Novikova, L.A. Dykman, M.V. Rodionova, et al., Highly active extracellular α-class carbonic anhydrase of Cyanothece sp. ATCC 51142, Biochimie 160 (2019) 200–209.

[160] E.V. Kupriyanova, N.V. Lebedeva, M.V. Dudoladova, L.M. Gerasimenko, S.G. Alekseeva, N.A. Pronina, et al., Carbonic anhydrase activity of alkalophilic cyanobacteria from Soda Lakes, Russ. J. Plant. Physiol. 50 (2003) 532–539.

[161] E.V. Kupriyanova, A.G. Markelova, N.V. Lebedeva, L.M. Gerasimenko, G.A. Zavarzin, N.A. Pronina, Carbonic anhydrase of the alkaliphilic cyanobacterium Microcoleus chthonoplastes, Microbiology 73 (2004) 255–259.

[162] E. Kupriyanova, A. Villarejo, A. Markelova, L. Gerasimenko, G. Zavarzin, G. Samuelsson, et al., Extracellular carbonic anhydrases of the stromatolite-forming cyanobacterium Microcoleus chthonoplastes, Microbiology 153 (2007) 1149–1156.

[163] M.V. Dudoladova, A.G. Markelova, N.V. Lebedeva, N.A. Pronina, Compartmentation of α- and β-carbonic anhydrases in cells of halo- and alkalophilic cyanobacteria Rhabdoderma lineare, Russ. J. Plant. Physiol. 51 (2004) 806–814.

[164] M.V. Dudoladova, E.V. Kupriyanova, A.G. Markelova, M.P. Sinetova, S.I. Allakhverdiev, N.A. Pronina, The thylakoid carbonic anhydrase associated with photosystem II is the component of inorganic carbon accumulating system in cells of halo- and alkaliphilic cyanobacterium Rhabdoderma lineare, Biochim. Biophys. Acta 1767 (2007) 616–623.

[165] S.M. Monti, G. De Simone, N.A. Dathan, M. Ludwig, D. Vullo, A. Scozzafava, et al., Kinetic and anion inhibition studies of a β-carbonic anhydrase (FbiCA 1) from the C4 plant Flaveria bidentis, Bioorg. Med. Chem. Lett. 23 (2013) 1626–1630.

[166] V. Alterio, S.M. Monti, E. Truppo, C. Pedone, C.T. Supuran, G. De Simone, The first example of a significant active site conformational rearrangement in a carbonic anhydrase-inhibitor adduct: the carbonic anhydrase I–topiramate complex, Org. Biomol. Chem. 8 (2010) 3528–3533.

[167] G. De Simone, A. Di Fiore, E. Truppo, E. Langella, D. Vullo, C.T. Supuran, et al., Exploration of the residues modulating the catalytic features of human carbonic anhydrase XIII by a site-specific mutagenesis approach, J. Enzyme Inhib. Med. Chem. 34 (2019) 1506–1510.

CHAPTER THREE

Bacterial β-carbonic anhydrases

Marta Ferraroni*

Dipartimento di Chimica "Ugo Schiff", Università di Firenze, Sesto Fiorentino, Firenze, Italia
*Corresponding author. e-mail address: marta.ferraroni@unifi.it

Contents

1. Introduction   66
2. Physiological role of β-carbonic anhydrases in bacteria   67
3. Structural characterization of β-Carbonic anhydrases   73
4. Allostery in β-carbonic anhydrases   78
5. Inhibitor binding to β-carbonic anhydrases   82
6. Catalyic mechanism of β-CAs   84
References   87

Abstract

β-Carbonic anhydrases (β-CA; EC 4.2.1.1) are widespread zinc metalloenzymes which catalyze the interconversion of carbon dioxide and bicarbonate. They have been isolated in many pathogenic and non-pathogenic bacteria where they are involved in multiple roles, often related to their growth and survival.

β-CAs are structurally distant from the CAs of other classes. In the active site, located at the interface of a fundamental dimer, the zinc ion is coordinated to two cysteines and one histidine.

β-CAs have been divided in two subgroups depending on the nature of the fourth ligand on the zinc ion: class I have a zinc open configuration with a hydroxide ion completing the metal coordination, which is the catalytically active species in the mechanism proposed for the β-CAs similar to the well-known of α-CAs, while in class II an Asp residue substitute the hydroxide. This latter active site configuration has been showed to be typical of an inactive form at pH below 8. An Asp-Arg dyad is thought to play a key role in the pH-induced catalytic switch regulating the opening and closing of the active site in class II β-CAs, by displacing the zinc-bound solvent molecule. An allosteric site well-suited for bicarbonate stabilizes the inactive form. This bicarbonate binding site is composed by a triad of well conserved residues, strictly connected to the coordination state of the zinc ion. Moreover, the escort site is a promiscuous site for a variety of ligands, including bicarbonate, at the dimer interface, which may be the route for bicarbonate to the allosteric site.

The Enzymes, Volume 55
ISSN 1874-6047, https://doi.org/10.1016/bs.enz.2024.05.009

1. Introduction

β–CA is a metalloenzyme that catalyzes the reversible hydration of carbon dioxide to bicarbonate releasing a proton in the process. The β–CAs are broadly distributed in nature and they have been found in species from all three domains of life, with representatives in several bacteria, in the thermophilic archaeon *Methanobacterium thermoautotrophicum*, in yeast (*Saccharomyces cerevisiae*), in green (*Chlamydomonas reinhardtii*) and red (*Porphyridium purpureum*) algae, in a variety of higher plants and in invertebrates [1–3].

CAs have been cloned and characterized in many pathogenic and non-pathogenic bacteria. The first bacterial CA to be recognized as a β–CA was the product of the CynT gene of *Escherichia coli* in 1992 [4]. Many bacterial β–CAs are now known, including enzymes from common pathogens such as *Helicobacter pylori, Mycobacterium tuberculosis, Vibrio cholerae* and *Salmonella typhimurium*. β–CAs are also known in cyanobacteria (*Synechocystis* PCC6803) and in the carboxysomes of chemoautotrophic bacteria (*Halothiobacillus neapolitanus*) [3].

Among the eight genetic families of CAs which have been discovered to date (α, β, γ, δ, ζ, η, θ, ι) only the α–CAs, β–CAs, γ–CAs and ι–CAs are present in bacteria with a predominance of the β–CAs and γ–CAs (Table 1). α–CAs have been detected only in a few of the bacteria examined, only Gram–negative bacteria, and none in the Archaea domain [5].

Kinetics studies of plant and bacterial β–CAs show that both subclasses of enzyme are highly efficient catalysts for the CO_2 hydration reaction, with kcat values on the order of 10^5 s^{-1} and kcat/Km values on the order of 10^8 M^{-1} s^{-1} [6,7]. These values approach that of the human Carbonic anhydrase isoform II (hCA II), among the fastest of known enzymes [8].

In plants, algae and cyanobacteria, β–CAs are required for transport and maintenance of CO_2 and HCO_3^- concentrations for photosynthesis, and similarly, in prokaryotes they are involved in the maintenance of internal pH and CO_2 and HCO_3^- concentrations required for biosynthetic reactions and also in degradation of xenobiotics (e.g. cyanate in *Escherichia coli*), as well as in the survival of intracellular pathogens within their host.

To date, unlike γ– and δ–CAs that utilize iron [9,10] and cadmium [11] respectively, all other CAs included β–CAs have been shown to be zinc metalloenzymes.

Here, we review the bacterial β–CAs with a particular focus on the aspects related to the physiological roles, structure and catalytic mechanism.

2. Physiological role of β-carbonic anhydrases in bacteria

Interestingly, many prokaryotic genomes contain multiple genes encoding CAs and some even contain genes from more than one family (Table 1) [2]. The wide distribution and multiple occurrences of CAs in bacteria suggest a fundamental physiological significance of these enzymes.

Often their activity is connected with the survival of the microbes since the reaction catalyzed by CAs is essential for supporting numerous physiological functions involving dissolved inorganic carbon. In particular, the β-CAs play an essential role in facilitating aerobic growth of microbes at low partial pressures of CO_2 by providing endogenous $HCO3^-$.

For example, in non-pathogenic bacteria such as Escherichia coli and Ralstonia eutropha, it has been demonstrated in vivo that the bacterial growth at an ambient CO_2 concentration is dependent on CA activity. Two different β-CAs have been identified in E. coli. The first to be characterized was CynT [4]. The gene encoding CynT is transcribed as part of an operon, the cyn operon, which is induced in the presence of cyanate, along with the gene encoding cyanase, an enzyme that catalyzes the conversion of cyanate into ammonia and carbon dioxide. The role of CynT is to catalyze the hydration of the CO_2 generated by cyanase into $HCO3^-$, thus preventing CO_2 loss by rapid diffusion from the cell and depletion of the $HCO3^-$ required for further degradation of cyanate or for other metabolic processes. Hence, CynT is essential in providing sufficient levels of cellular $HCO3^-$ for growth with cyanate as the sole nitrogen source [12].

Besides, during normal growth E. coli requires a supply of bicarbonate/ CO_2 as a metabolic substrate for the biosynthesis of fatty acids and other various small molecules and in central metabolism. The low concentration of CO_2 in air and its rapid diffusion from the cell mean that insufficient bicarbonate is spontaneously made in vivo to meet metabolic and biosynthetic needs. Merlin et al. calculated that demand for bicarbonate is 10^3- to 10^4-fold greater than that provided by the uncatalyzed intracellular reaction and that enzymatic conversion of CO_2 to bicarbonate is therefore necessary for the growth of E. coli at the CO_2 concentration characteristic of air. This role is accomplished by a second β-CA (YadF) discovered in E. coli, which was demonstrated being essential for the aerobic microorganism growth [13].

The facultative chemoautotroph Ralstonia eutropha produces a homolog of YadF, Can, which has been characterized [14] and found to be essential for growth of the wild type in the presence of low CO_2 concentrations. In

Table 1 Distribution of CA families in Gram-negative and positive bacteria (for which the genome was cloned).

Gram negative	CA family		
Escherichia coli	–	β	γ
Haemophilus influenzae	–	β	–
Brucella suis	–	β	γ
Salmonella enterica	–	β	–
Vibrio cholerae	α	β	γ
Sulfurihydrogenibium yellowstonense	α	–	γ
Porphyromonas gingivalis	–	β	γ
Sulfurihydrogenibium azorense	α	–	γ
Neisseria gonorrhoeae	α	–	–
Helicobacter pylori	α	β	γ
Ralstonia eutropha	α	β	γ
Burkholderia pseudomallei	–	β	γ
Gram positive	**CA family**		
Mycobacterium tuberculosis	–	β	γ
Clostridium perfringens	–	β	γ
Streptococcus pneumoniae	–	β	γ
Bacillus subtilis	–	β	γ
Leifsonia xyli	–	β	γ
Staphylococcus aureus	–	–	γ
Enterococcus faecalis	–	–	γ

fact, it was verified that a *Ralstonia eutropha* strain with a mutation in Can and another mutant generated by deletion of the gene encoding for Can are associated with a high CO_2 requiring phenotype and growth of the bacterial mutants was observed only in condition of elevated CO_2 concentrations [14].

The genome sequence in *Corynebacterium glutamicum* revealed two genes encoding a putative β-type and γ-type CA. Internal deletion of the chromosomal bca gene encoding for the β-CA resulted in a phenotype showing severely reduced growth under atmospheric conditions that was restored under conditions of elevated CO_2 or heterologous expression of *P. purpureum* CA. Instead, the γ-CA was found dispensable for normal growth under ordinary atmospheric conditions [15].

More interesting is the in vivo evidence concerning the involvement of CAs for the growth of pathogenic bacteria. For example, the genome of *Helicobacter pylori*, a Gram-negative bacterium colonizing the human gastric mucosa, encodes a pleriplasmic α-CA and a cytoplasmic β-CA [16] which seem to be fundamental in the resistance of the bacterium to the acidic environment.

H. pylori plays a major role in the pathogenesis of chronic gastritis, which, depending on environmental factors, can evolve into peptic ulcer disease and gastric cancer. Among bacteria, *H. pylori* has the unique ability to grow in the stomach, where conditions are highly acidic. Therefore, the pathogen has evolved specialized processes for survival that maintain the cytoplasmic pH close to neutrality. Urease and CAs are the two enzymatic systems used by the microbe for growing in this extreme conditions. Bicarbonate is produced in large amounts by the urease reaction of urea degradation and *H. pylori* is additionally exposed to bicarbonate that is continuously secreted by gastric epithelial cells. Thus, the interconversion of bicarbonate by CA enzymes could contribute significantly to urease-mediated acid resistance and adaptation to the gastric environment. The two CA of *H. pylori* have attracted considerable interest since these enzymes were inhibited in vitro by several sulfonamide and sulfamate inhibitors and therefore have been proposed to be alternative therapeutic targets for the management of patients infected by drug-resistant *H. pylori* [17].

Three putative carbonic anhydrases are encoded by the pathogenic bacterium *V. cholerae*, the causative agent of cholera. They belong to three distinct classes: an α-CA, a β-CA and a γ-CA [18] which play a role in virulence gene induction. In fact, bicarbonate induces the expression of the major *V. cholerae* virulence factors by enhancing the activity of ToxT, a regulatory protein that directly activates the transcription of the genes encoding the cholera toxin [19,20]. Thus, *V. cholerae* utilizes the CA system to accumulate bicarbonate in its cells, suggesting a pivotal role of these metalloenzymes in microbial virulence.

Similarly, two β-CA from *B. suis*, a Gram-negative coccobacillus responsible for brucellosis, and three β-CAs (Rv1284, Rv3588c, and Rv3273) from *M. tuberculosis*, the causative agent of tuberculosis, were demonstrated being essential for the growth of the corresponding microbes [21–23]. Analogously, the genome of *S. enterica serovar Typhimurium*, a Gram-negative bacterium causing gastroenteritis, also encodes for a β-CA [24] which is highly expressed during the bacterial infection, as demonstrated by in vivo gene expression studies [25].

In the genome of *Pseudomonas aeruginosa* PAO1, an opportunistic human pathogen causing life threatening infections, three genes, PAO102, PA2053 and PA4676, encoding β-CAs were identified. The three CAs are differentially expressed in PAO1 cells, supporting the hypothesis that they may play different physiological roles in *P. aeruginosa* [26]. PsCA1, the most active and abundant, plays a role in PAO1 survival at low partial pressure of CO_2 in ambient air, similarly to the β-CA YadF in *E. coli* and Can in *R. eutropha*. Furthermore, a more recent study of the same group suggested that PsCAs are involved in the formation of extracellular calcium deposits by $CaCO_3$ precipitation. Analyses of the deletion mutants lacking one, two or all three PsCA genes showed a reduction of calcium salt deposition [27]. This ability of the pathogen may increase its virulence by contributing to soft tissue calcification or stone formation during bacterial infections which was observed, for example, in late stages of cystic fibrosis or primary ciliary dyskinesia.

β-CAs have a cytoplasmic localization, while all bacterial α-CAs known to date are characterized by the presence of an N-terminal signal peptide, essential for the translocation across the cytoplasmic membrane, suggesting a periplasmic or extracellular location. A role in the acclimatization of the microorganisms to a hostile environment has been proposed for the procaryotic α-CAs (as it is the case for *Helicobacter pylori*) [5]. Recently, it has been noted that the primary structure of β-CAs identified in the genome of some pathogenic Gram-negative bacteria, such as Helicobacter pylori, *Vibrio cholerae, Neisseria gonorrhoeae* and *Streptococcus salivarius*, presents a secretory signal peptide at the N-terminal part. Hence, it has been suggested that the β-CAs of Gram-negative bacteria characterized by the presence of a signal peptide might exhibit a periplasmic localization and a role similar to that described for the α-CAs [28].

Cyanobacteria are the most abundant photosynthetic organisms on Earth that, together with plants, are responsible of most of the global CO_2 fixation into biomass. In cyanobacteria and many chemoautotrophic

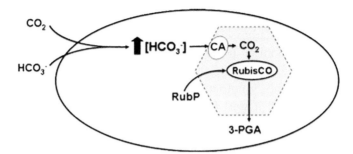

Fig. 1 The CO_2 concentrating mechanism (CCM) of cyanobacteria The model shows the two stages of the CCM: 1. Accumulation of an intracellular bicarbonate pool via ATP- and NADH-dependent pumps; 2. dehydration of bicarbonate to the RubisCO substrate, CO_2, by the carboxysomal CA. Subsequent fixation of CO_2 onto ribulose-1,5-bispho- sphate (RubP) by RubisCO yields two molecules of 3-phosphoglycerate (3-PGA). *Reproduced with permission from reference G.C. Cannon, S. Heinhorst, C.A. Kerfeld, Carboxysomal carbonic anhydrases: structure and role in microbial CO_2 fixation, Biochim. Biophys. Acta. 1804 (2) (2010) 382–392.*

bacteria carboxysomes are polyhedral specialized microcompartments, enveloped by a proteinaceous shell, in which carbon fixation is catalyzed by ribulose 1,5-bisphosphate carboxylase/oxygenase (RuBisCO). RuBisCO is a rather inefficient enzyme with very low affinity for its second substrate CO_2 and low CO_2 versus O_2 (the competing substrate) specificity. In order to overcome these catalytic limitations, cyanobacteria possess a complex mechanism known as CO_2 concentrating mechanism, which is crucial for their growth and survival (Fig. 1). Bicarbonate is transported across the membrane cell through a variety of ATP- and NADH-dependent CO_2 and HCO3⁻ pumps to create a cytosolic pool of inorganic carbon that is estimated to be three orders of magnitude higher than that outside the cell [29]. The cyanobacterial genome sequences do not appear to code for a CA that is expressed in the cytosol, as to prevent the dissipation of this pool the CA activity should be absent from the region where this pool is accu- mulated [30]. Then, bicarbonate diffuses inside the carboxysome where it is converted to CO_2 by a specific CA prior to fixation. It is also possible that the carboxysomal shell provides a barrier to the escape of CO_2 from RuBisCO. This mechanism results in the accumulation of CO_2 in the vicinity of RuBisCO at concentration high enough to promote efficient rates of fixation [31]. The operation of the CO_2 concentrating mechanisms requires the intervention of multiple CAs as there are several steps where CO_2 and HCO_3^- require interconversion.

Carboxysomes are found in two varieties based on the RuBisCO type encapsulated [32]. α-Carboxysomes contain a β-CA, CsoSCA, which is so divergent from canonical β-carbonic anhydrases that it was originally thought to be the founding member of a new (ε) class of Carbonic anhydrases. CsoSCA of the cyanobacterium *H. neapolitanus* has been structurally characterized [33] and it was shown to be an integral component of the microcompartment shell. The protein is composed of three domains: the N-terminal domain is entirely α-helical, forming an up–down four-helix bundle. The N-terminal domain is located at an isolated extremity of the protein, suggesting the possibility that it might function to anchor CsoSCA to the carboxysomal shell or to molecules of RuBisCO. The second domain, the catalytically active domain, has an α/β fold common to all β-CAs and also coordinate a catalytic zinc ion similarly to classical CAs. The C-terminal domain is structurally similar to the catalytic domain and is spatially related to it by a two-fold rotation axis. The similarity likely reflects an ancient gene duplication and fusion followed by divergence (see also below).

Cyanobacteria like *Synechocystis sp.* PCC6803 and *Synechococcus elongatus* PCC7942 contain β-carboxysomes which have CcaA as a component, a CA with high homology with the β-class and most closely related to the plant type [31]. In *Synechococcus* PCC7942 and *Synechocystis* PCC6803, CcaA has been shown to be essential for efficient CO_2 fixation, and mutants defective in the ccaA gene encoding CcaA require high CO_2 levels for growth since are unable to generate enough CO_2 in the carboxysomes to sustain normal rates of photosynthesis [34,35]. This protein is recruited to the carboxysome via interactions with CcmM, which is a γ-CA homolog with enzymatic activity in many, but not all cyanobacteria. Catalytic characterization showed CcaA to be among the least active β-CAs characterized to date, with activity comparable with CcmM, which it either complements or replaces in β-carboxysomes.

From the structural characterization of CcaA from *Synechocystis sp.* PCC 6803 [36] it was found that CcaA has a novel trimer of dimers organization, in contrast to the dimers, tetramers, and octamers previously seen with β-CAs. It is hypothesized that the trimeric N-terminus of CcmM could interact by aligning the threefold symmetry axes and recruiting CcaA to the carboxysome.

CcaA has a characteristic C-terminal extension of approximately 80 amino acids of unknown function. This region is not found in other β-CAs and appears to be important for catalytic activity and intermolecular interactions and it is also thought to interact with the carboxysomal shell proteins.

3. Structural characterization of β-Carbonic anhydrases

The first reported X-ray structure of a β-Carbonic anhydrase was that of the red marine alga *Porphyridium purpureum* in 2000 [37]. The crystal structures of the β-CA from the dicotyledonous plant *Pisum sativum* [38] and that of the bacterial YadF from *Escherichia coli* [39] shortly followed. At the time of writing eighteen crystal structures of β-CAs have been already deposited in the PDB of which nine are of bacterial enzymes (Table 2).

The β-CAs are structurally distant from the other CAs of different classes. The general architecture of β-CA monomer is an α/β fold, being composed of three sequential components: an N-terminal arm, a conserved zinc-binding core, and a C-terminal subdomain that is dominated by α-helices. The zinc binding-core resembles the classical nucleotide binding Rossmann fold motif, composed of a central parallel β-sheet composed of four β-strands and the connecting segments between them, including two helices that pack against either side of the sheets (Fig. 2).

Comparison of the YadF monomer with those of the previously determined of the plant and red alga β-CA structures showed significant differences in the C-terminal and N-terminal subdomains. Instead, the bacterial β-CAs share a very similar fold (Fig. 3).

The oligomeric state is variable inside the β-class with tetramers (or pseudotetramer as in *P. purpureum* and *H. neapolitanus* β-CAs) and octamers being the most usual arrangement, whereas the fundamental catalytic unit is a dimer. The N-terminal extension and zinc-binding core mediate formation of the dimer. The orientation and spacing of the two N-terminal arms in the dimer create the sides of a broad and deep groove in the structure and, although zinc binding is wholly mediated by residues within a given monomer, interactions at the dimer interface contribute to the local environment at the metal-binding site. This dimeric association of monomers appears to be common to all β-CA structures determined so far, whereas variation in the C-terminal subdomain contributes to the diverse higher-order oligomerization.

A particular case are the β-CAs from *Porphyridium purpureum* and *Halothiobacillus neapolitanus* [33,37]. In fact the gene encoding for the two CAs has been formed by duplication and fusion of a primordial β-CA gene and the monomer is composed of two internally repeating domains, each similar to that of other β-CAs, forming a pseudodimer. However, there is 70% sequence identity between the N- and C-terminal domains of *P. purpureum* CA but only 11% identity between the two domains of

Table 2 Crystallographic structures of bacterial β-CAs (in bold) and of other β-CA from plants and algae cited in the text.

Source, protein name	Class	Zinc coordination	Oligomeric state	Resolution (Å)	PDB code	References
Porphyridium purpureum	Type II	Closed	Pseudotetramer	2.2	1DDZ	[37]
Pisum sativum	Type I	Open	Octamer	1.9	1EKJ	[38]
Escherichia coli, YadF	Type II	Closed	Tetramer	2.0	1I6P	[39]
Methanobacterium thermoautotrophicum	Type I	Open	Dimer	2.1	1G5C	[44]
Mycobacterium tuberculosis, Rv3588c	Type II	Closed	Dimer	1.7	1YM3	[42]
Mycobacterium tuberculosis, Rv1284	Type I	Open	Tetramer	2.0	1YLK	[42]
Haemophilus influenzae	Type II	Closed	Tetramer	2.3	2A8C	[41]
Halothiobacillus neapolitanus, CsoSCA	Type I	Open	Pseudotetramer	2.2	2FGY	[33]
Vibrio cholerae	Type II	Closed	Tetramer	1.9	5CXK	[40]
Pseudomonas aeruginosa, psCA3	Type II	Open	Dimer	1.9	4RXY	[51]
Synechocystis sp. PCC 6803	Type I	Open	Hexamer	1.4	5SWC	[36]
Burkholderia pseudomallei	Type II	Closed	Tetramer	2.7	6YJN	[43]

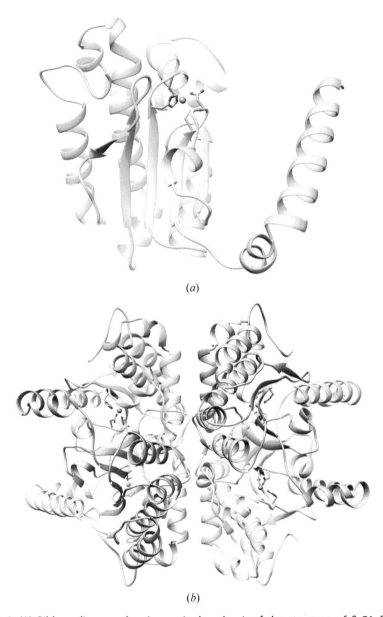

(a)

(b)

Fig. 2 (A) Ribbon diagram showing a single subunit of the structure of β-CA from *Vibrio cholerae*. The active-site zinc is also represented with its ligands. (B) Ribbon diagram showing the tetrameric arrangement of β-CA from *Vibrio cholerae*. *Reproduced with permission of the International Union of Crystallography from reference M. Ferraroni, S. Del Prete, D. Vullo, C. Capasso, C.T. Supuran, Crystal structure and kinetic studies of a tetrameric type II β-carbonic anhydrase from the pathogenic bacterium Vibrio cholerae, Acta Crystallogr. D. Biol. Crystallogr. 71 (Pt 12) (2015) 2449–2456.*

Fig. 3 Superposition of a monomer of β-CA from *Vibrio cholerae* (orange) with that of β-CA from *E. coli* (gold) and *H. influenzae* (light blue). The catalytic zinc ion is represented as a magenta sphere. *Reproduced with permission of the International Union of Crystallography from reference M. Ferraroni, S. Del Prete, D. Vullo, C. Capasso, C.T. Supuran, Crystal structure and kinetic studies of a tetrameric type II β-carbonic anhydrase from the pathogenic bacterium Vibrio cholerae, Acta Crystallogr. D. Biol. Crystallogr. 71 (Pt 12) (2015) 2449–2456.*

Halothiobacillus neapolitanus CA and the C-terminal domain of this enzyme has diverged so greatly that it has lost the ligands for the catalytic zinc ion.

The active site of β-CAs is located at one end of the central β-sheet at the interface of one dimer. Interestingly, the active sites of α- and β-CAs exhibit certain architectural similarities. In both classes, the active sites are divided into two regions, one hydrophobic and the other hydrophilic, while the β-CAs have a much shallower active site, constructed by dimer formation.

The zinc coordination of β-CAs is unique within the CA superfamily. The zinc ion is coordinated by two cysteines and one histidine differently for example from the 3His coordination of α-CAs and γ-CAs. In the X-ray structures of β-CAs the fourth zinc coordination site was found occupied by a water molecule/hydroxide ion or a substrate analogue/inhibitor but unexpectedly some of them showed also a different arrangement with an Asp as fourth ligand.

After that, the β-CAs have been classified in two distinct subgroups based on the presence or not of the Aspartate residue inside the zinc coordination sphere. Within this classification class I β-CAs are characterized by an "open" active site, whereas that of class II have a unique zinc coordination geometry, where the water molecule has been replaced by the Asp residue, resulting in a "closed" Cys2HisAsp coordination sphere. Examples of this second class of β-CAs are the enzymes from *Porphyridium purpureum* [37], *Escherichia coli* [39], *Vibrio cholerae* [40], *Haemophiles influenzae* [41], *Mycobacterium tuberculosis* Rv3588c [42], and *Burkholderia pseudomallei* [43] (Table 2).

In class I β-CAs, exemplified by the enzymes from *Pisum sativum* [38], *Methanobacterium thermoautotrophicum* [44] *Halothiobacillus neapolitanus* [33] and *Mycobacterium tuberculosis* Rv1284 [42], the zinc bound Aspartate forms strong hydrogen bond interactions with a nearby Arginine in all structures solved to date, which liberate the fourth coordination position at the Zn^{2+} ion, which coordinates an incoming water molecule (Fig. 4). Together with the three zinc ligands, these two residues Asp44 and Arg46 (from here on the numbering of YadF and of most of the bacterial β-CAs is used unless otherwise specified) are the only strictly conserved in the β-CA active sites, suggesting a role for them inside the catalytic mechanism [33]. Site-directed mutagenesis of the equivalent residues in the *M. thermoautotrophicum* enzyme confirmed that the two residues play an essential role in the reaction [45].

Only the active site setup with three amino acid coordination to the zinc (open active site) was consistent with the catalytic mechanism proposed for the β-CAs, based on the kinetic studies, like the ping-pong mechanism used by the α-CAs [46]. This mechanism includes the attack of a catalytically

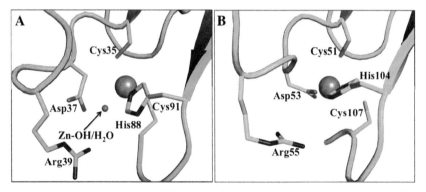

Fig. 4 (A) Active site geometry of the type I β-CA, Rv1284 (PDB ID: 1YLK) isolated from *Mycobacterium tuberculosis* exhibiting the 'open' configuration. (B) Active site geometry of the type II β-CA, Rv3588c (PDB ID: 1YM3) isolated from *Mycobacterium tuberculosis* exhibiting the 'closed' configuration. *Reproduced with permission from reference M.A. Pinard, S.R. Lotlikar, C.D. Boone, D. Vullo, C.T. Supuran, M.A. Patrauchan, et al., Structure and inhibition studies of a type II beta-carbonic anhydrase psCA3 from Pseudomonas aeruginosa, Bioorg Med. Chem. 23 (15) (2015) 4831–4838.*

active zinc-bound hydroxide on the carbon dioxide substrate and subsequent proton transfer from a new water ligand [38]. Moreover, the lack of a zinc coordinated hydroxide ion in class II β-CAs had to be reconciled with the finding that members of both classes are kinetically active. Class I β-CAs show catalytic activity at pH from as low as 6.5 to greater than 9.0 [47], while class II β-CAs have a pH dependent activity. In fact, members of this class are active at pH above 8.0 but their activity drops abruptly at lower pH, with the enzymes becoming inactive at neutral pH [39,48].

Ref. [37] described an alternative mechanism in which the Asp44 (Asp405 in the *Porphyridium purpureum* β-CA) functioned as a base that activates a water molecule, with the resulting hydroxide which displaces the aspartyl group as the fourth zinc ligand in the next steps of the reaction.

Later, this inconsistency led to the suggestion that the Aspartate may be involved in activity regulation of the enzyme by somehow changing between the accessible and blocked states [39].

Covarrubias *et al.* firstly proved that the β-CA from *Mycobacterium tuberculosis Rv3588c* is able to adopt both an accessible and a blocked form of the active site [48]. Interestingly, the structure of the protein complex with thiocyanate showed the small inorganic anion displacing Asp44 (Asp53 in β-CA from *Mycobacterium tuberculosis* Rv3588c) from its zinc coordinating position to become involved in hydrogen bonds with Arg46 (Arg55 in β-CA from *Mycobacterium tuberculosis* Rv3588c). More recently the active site open configuration has also been observed in the D44N mutant of *H. influenzae* β-CA (HICA) [49]. The only substantial structural or conformational difference between the active site of the *H. influenzae* β-CA mutant D44N and that of a class I β-CA like *Pisum sativum* CA is the position of the side chain of Arg46. In PSCA, Arg46 interacts with Asp44 to orient it as a hydrogen bond acceptor for the water/hydroxide ion bound to the zinc ion, whereas in HICA-D44N they do not interact [49] (Fig. 5). These structural data strongly suggested that the two variants of β-CA represent an alternate conformation of the protein and led to propose a regulatory role for Asp44 by functioning as an on/off switch.

4. Allostery in β-carbonic anhydrases

A bicarbonate binding site, distinct from the zinc coordination site at which the dehydration of bicarbonate is expected to occur, was discovered by Cronk et al. in *E. coli* and *H. influenzae* β-CAs [41] (Fig. 6). The

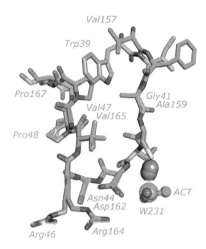

Fig. 5 Alignment of allosteric and active-site residues of PSCA (PDB 1EKJ) and HICA–D44N (PDB 3E1V). Carbon atoms of PSCA are colored green and HICA–D44N carbon atoms are colored orange. Zinc ions are depicted as gray spheres. An acetate ion (ACT) bound to PSCA is depicted as a ball and stick model. Water-231, bound near the zinc ion in HICA–D44N, is depicted as a transparent sphere. Some key residues are labeled: green labels are for PSCA, and orange labels are for HICA–D44N. *Reproduced with permission from reference R.S. Rowlett, C. Tu, J. Lee, A.G. Herman, D.A. Chapnick, S.H. Shah, et al., Allosteric site variants of Haemophilus influenzae beta-carbonic anhydrase, Biochemistry, 48 (26) (2009) 6146–6156.*

bicarbonate site is in the vicinity of both the active site and the dimerization interface and, relative to the zinc ion, is on the opposite side of the Asp44 ligand [37,38].

The non catalytic site for bicarbonate binding seems to be perfectly tailored for accommodating this ion, as other isostructural ions such as acetate or nitrate would not be able to achieve all the hydrogen bonding interactions in which bicarbonate takes part simultaneously. In fact, the side chains of Trp39, Arg64, and Tyr181, as well as two water molecules, donate hydrogen bonds to the bicarbonate ion. Moreover, a critical hydrogen bond interaction with the carbonyl of Val47 assists in discrimination between $HCO3^-$ and other anions that cannot act as a hydrogen bond donor. The triad of residues Trp-Arg-Tyr is fully conserved in the class II β-CAs, while β-CAs of the other class lack key elements of the non-catalytic bicarbonate binding site described above. The binding of bicarbonate in this site is strictly connected to the coordination state of the zinc ion. In fact, bicarbonate expels the side chain of Val47, which causes a reorganization of the 44–48 loop and results in the

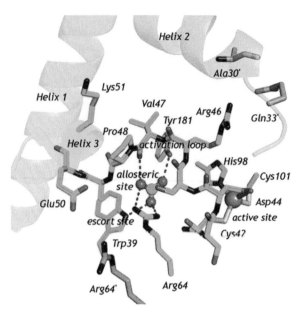

Fig. 6 The non-catalytic bicarbonate-binding site, which includes the coordinating residues Trp39, Arg64, and Tyr 181. The active site contains the metal ligands Cys42, Cys101, and His98 as well as (in the T-state) Asp44. Hydrogen bonds formed by the bicarbonate ions to enzyme residues are also depicted as dashed lines. *Reproduced with permission from reference K.M. Hoffmann, H.R. Million-Perez, R. Merkhofer, H. Nicholson, R.S. Rowlett, Allosteric reversion of Haemophilus influenzae β-carbonic anhydrase via a proline shift, Biochemistry 54 (2) (2015) 598–611.*

disruption of the Asp44–Arg46 dyad, allowing Asp44 to displace the zinc-bound water and to bind directly to the zinc ion.

These findings and kinetic evidences that bicarbonate is an inhibitor as well as substrate of HICA [41] opened the possibility that bicarbonate can function as a regulator, activating an allosteric switch that regulates the equilibrium between the inactive form (also called T-state) and active form of the enzyme (R-state) (Scheme 1).

A bicarbonate was also found serendipitously bound to the non-catalytic pocket in the structure of *Vibrio cholerae* β-CA [40], reinforcing the hypothesis that the bicarbonate binding to this allosteric site near the zinc ion is a distinctive feature of type II β-CAs and plays a significant role in the functionality of this enzymes.

Comparing the structure of *Pisum sativum* β-CA and that of the mutant HICA-D44N, both endowed with an open active site, Rowlett et al. identified three key structural elements that may distinguish

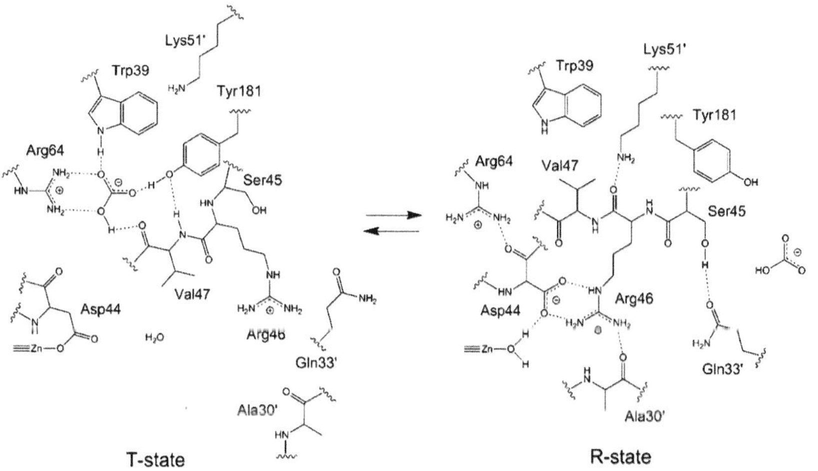

Scheme 1 Schematic representation of active site and non-catalytic bicarbonate binding site interactions in the hypothesized active (R-State) and inactive (T-State) conformations of β-CA from *H. influenzae. Reproduced with permission from reference K.M. Hoffmann, H.R. Million-Perez, R. Merkhofer, H. Nicholson, R.S. Rowlett, Allosteric reversion of Haemophilus influenzae β-carbonic anhydrase via a proline shift, Biochemistry 54 (2) (2015) 598–611.*

between allosteric and non-allosteric β-CAs. These structural differences were recognized as the substitution of Ala for Gly41, the substitution of Val for Trp39, and the repositioning of Pro48 to a position one residue later in the amino acid sequence in *Pisum sativum* CA that resulted in the reorientation of the proline ring, so that it protruded into the allosteric site rather than pointing away from it, as in the wild-type HICA [49] (Fig. 5).

Later, HICA has been engineered in the allosteric site region to resemble the non-allosteric *Pisum sativum* enzyme and to find critical features of allostery. However, the variants G41A, V47A were found to be similar to wild-type HICA in their overall structure and catalytic ability toward CO_2 hydration, showing kcat/Km values similar to that of the wild-type enzyme and exhibiting a similar dramatic decrease in catalytic activity at pH < 8.0 [47]. Since these single variants were insufficient to revert HICA to a non-allosteric enzyme the mutant P48S/A49P HICA was constructed to mimic the proline shift [50]. This double variant adopted an active site conformation nearly identical to that of non-allosteric β-carbonic anhydrase (R-state) for one chain, whereas the other in the asymmetric unit associated in a biologically relevant oligomer, conserved the T-state of the wild-type enzyme, showing for the first time the two distinct active site conformations simultaneously present in an allosteric β-CA.

Another bicarbonate binding site has been found in the structures of HICA-V47A, HICA-Y181F and HICA-G41A variants near the dimer-ization interface and the allosteric site (Fig. 7) [47]. It has been suggested that this site, defined by Arg64 and Glu50 from each of the two protein chains at the dimer interface, represents the route of bicarbonate to the allosteric binding site and it has been designated as the escort site. In the absence of bicarbonate in the structures of the same variants a sulfate ion binds in the escort site. The competition of the two anions, sulfate and bicarbonate, for this site has been proposed as a possible explanation of the observed activation effect of HICA by sulfate ions in vitro. Thus, it seems likely that sulfate ion acts as a competitive inhibitor for bicarbonate ion at the escort site, hindering access of bicarbonate to produce the inactive T state of the enzyme. It has also argued that the binding of sulfate to HICA cannot be physiologically relevant, as the sulfate concentrations needed for the activation effect are higher than intracellular concentration of sulfate ions. However, also other anions such as phosphates have been found bound in the escort site and they may function as an allosteric effector of HICA as well.

Interestingly, the crystal structure of PsCA3, a β-CA from *Pseudomonas aeruginosa*, solved at pH 8.3 showed an open conformation of the active site with a water molecule coordinated to the zinc ion and the Asp44 inter-acting with Arg46 [51]. PsCA3 is a type II β-CA as it shows the highest sequence identity with type II β-CAs and the Trp-Arg-Tyr triad, char-acteristic of the allosteric bicarbonate binding site, is conserved in the enzyme. However, bicarbonate was not found in the bicarbonate binding site as the enzyme is in the 'open' configuration, but the presence of this conserved site suggests bicarbonate may also play a similar role in the allosteric regulation of this β-CA. Nevertheless, the same group reported that soaking crystals of PsCA3 in solutions at pH < 6 has never produced the expected pH–induced structural switch [52].

5. Inhibitor binding to β-carbonic anhydrases

Since bacteria gain constant resistance toward antibiotics, it is imperative that novel targets be evaluated for the development of more sensitive and specific drugs. In many pathogenic bacteria involved in

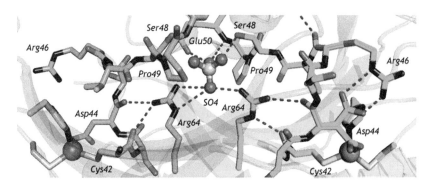

Fig. 7 Asymmetric hydrogen-bonding network around the anion binding (escort) site in the HICA-P48S/A49P variant. Key residues of chain A (green) and chain B (orange) are depicted as sticks. Sulfate is depicted as ball and stick. Zinc ions are gray spheres. Selected hydrogen-bonding interactions are depicted by magenta dashes. *Reproduced with permission from reference K.M. Hoffmann, H.R. Million-Perez, R. Merkhofer, H. Nicholson, R.S. Rowlett, Allosteric reversion of Haemophilus influenzae β-carbonic anhydrase via a proline shift, Biochemistry 54 (2) (2015) 598–611.*

human infections such as *Vibrio cholerae, Brucella suis, Salmonella enterica, Pseudomonas aeruginosa*, and *Helicobacter pylori*, it has been demonstrated that β-CAs are essential for the life cycle of the microorganism and that the conversion of CO_2 to bicarbonate plays a role in virulence-gene induction [53]. Hence, it cannot be excluded that interference with their activity may be exploited therapeutically to obtain antibiotics with a different mechanism of action. Moreover, β-CAs represent a promising target for developing specific inhibitors due to the lack of β-CAs in humans and no similarity between bacterial β-CAs and human α-CAs. Nevertheless, structural studies aimed at obtaining detailed information on the binding mode of inhibitors to the active site of bacterial β-CAs are scarce.

The inhibition by anions and sulfonamides, well-known inhibitors of α-CA, has been investigated in plant β-CAs [7] and more recently also in bacterial β-CAs such as *H. pylori* [54], *Francisella tularensis* [55], *Clostridium perfringes* [56], *E. coli* [57] and *Staphylococcus aureus* [58].

At the time of writing, the work of Murray et al. is the unique report of a structural study investigating the binding mode of inhibitors to the β-CAs [59]. The authors studied the binding of anions, such as sulfamide, tiocyanate, imidazole and 4-methylimidazole, to the β-CA from *P. aeruginosa* PsCA3 and compared the binding of these small molecules to those in the complexes with hCA II. Each of the four anions inhibits PsCA3 with Kis in the µM range. Conversely, the imidazoles are activators of hCA II [60,61]. All the inhibitors bind to PsCA3 coordinating the zinc ion and displacing the zinc-bound water molecule and other one or two of the four ordered waters present in the active site. Each inhibitor binds the zinc with a nitrogen atom except for 4-methylimidazole that coordinates the metal with the methyl group. They also form hydrogen bonds with residues Asp44 and Gln33' (from an adjacent monomer). The 4-methylimidazole also forms a π-stacking interaction with the phenyl ring of Phe83 and a π–CH2 interaction with Phe61 (Fig. 8).

Comparison of the PsCA3 complexes with the analogous complexes of hCA II showed marked differences in the binding of thiocyanate and 4-methylimidazole. In the hCA II complex with thiocyanate the anion binds to the zinc as in the complex with PsCA3, but the zinc maintains the coordination with the zinc-bound water, adopting an overall trigonal bipyramidal geometry. In the hCAII complex the 4-methylimidazole binds the active site zinc through one aromatic nitrogen atoms, like imidazole.

A structural comparison was also done between the crystal structure of psCA3 and representative of β-CA crystal structures from various organisms to identify conserved and variable regions near the active site [59]. From this characterization, it was found that the region where the small anions are bound is highly conserved across the different β-CA forms which indicate nonspecific β-CA inhibition. However, this mapping also revealed a non-conserved region that could be targeted for the development of β-CA selective inhibitors.

6. Catalyic mechanism of β-CAs

Most of the kinetic properties observed for the β-CAs [62–64] and the structural organization of the active site are consistent with a catalytic mechanism similar to the well-established ping-pong mechanism utilized by the α-CAs [46]. In this mechanism the catalytically active entity is a zinc-bound water molecule, the fourth ligand of the tetrahedrally coordinated zinc, which ionizes to a hydroxide ion and attacks CO_2. The resulting zinc-coordinated $HCO3^-$ ion is displaced from the metal ion by a solvent molecule. The rate determining step of the reaction mechanism is the proton transfer from the zinc-bound water to the buffer molecules in the reaction medium [7].

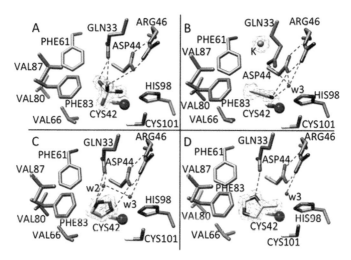

Fig. 8 Stick representation of the active site of psCA3 in complex with (A) sulfamide, (B) thiocyanate, (C) imidazole, and (D) 4-methylimidazole. Hydrogen bonds are represented as red dashes. Catalytic zinc is shown as a magenta sphere. Shown as a blue mesh is the 2 F_o–F_c density contoured at 1.4 σ. *Reproduced with permission from reference A.B. Murray, M. Aggarwal, M. Pinard, D. Vullo, M. Patrauchan, C.T. Supuran, et al., Structural mapping of anion inhibitors to β-carbonic anhydrase psCA3 from Pseudomonas aeruginosa, ChemMedChem 13 (19) (2018) 2024–2029.*

The mechanism of proton transfer is thought to be different between the two CA classes [63]. In α-CAs, this function is conducted by a well-ordered network of water molecules and a histidine proton shuttle residue [65,66]. In fact, most of the fifteen human α-CA isoforms known to date have a His residue at position 64 which is actively involved in the proton shuttle and is responsible for converting the zinc-bound water molecule to hydroxide ion [67]. Initially, it was proposed that in the β-CAs a buffer molecule could act in a direct proton transfer from the zinc-coordinated water [62,63]. Later, site-directed mutagenesis of β-CAs has led to the widely accepted view that a combination of residues is responsible for proton shuttling, and the identity of the residues differs among β-CAs. For example, it has been proposed in β-CA from *Arabidopsis thaliana* that residues His216 and Tyr212, conserved in the plant β-CAs, are both important for efficient proton transfer in analogy with His64 in α-CAs [68]. Tyr212 (Tyr83 in β-CA from *V. cholerae* and *E. coli*) is conserved also in the β-CAs from the Vibrio genus and in other bacterial enzymes. Also, the role of His208, was investigated by site-directed mutagenesis of *P. sativum* β-CA (His216 in the *A. thaliana* β-CA, not conserved in the bacterial β-CAs) as possible internal intramolecular proton acceptor. The results of these studies indicated that this residue is not crucial for the catalytic mechanism [69]. Subsequently, Kimber et al. showed that His208 is not in the vicinity of the zinc-bound water position [38], whereas identified Tyr205 (the counterpart of Tyr212 in *A. thaliana* and Tyr83 in the bacterial β-CAs) as a possible route for the proton passage to bulk solvent.

Of the residues in the active site, only the three zinc ligands and the Asp/Arg pair are conserved in the diverse β-CA sequences [2]. The role of Asp44 has been hypothesized as that of a 'gatekeeper' residue, in analogy with Thr199 in α-CAs, which prevents non-protonated atoms from binding effectively to the zinc ion. It also should ensure that the hydroxide ion is correctly oriented on the zinc for nucleophilic attack and, through the hydrogen bond it provides, help to lower the hydroxide pKa [38,49]. Also, by site-directed mutagenesis a crucial role has been suggested for Gln151 and Gly224 (Gln 33 and Gly103 in the bacterial β-CAs), in stabilization of the transition state during the catalysis [70]. The two residues interact with the acetate ion, an analogue of bicarbonate, bound to the zinc ion in the active site of the *P. sativum* β-CA structure [38].

Aggarwal et al. have determined the structure of the complex of PsCA3 with the substrate CO_2 by pressurized cryocooled crystallography. They also compared the PsCA3 complex with that of hCA II, the only other CA to have CO_2 captured in its active site [46,52]. As in the case of hCA II, the CO_2 displaces a water, that in the unbound state lies in the hydrophobic pocket occupied by CO_2 and is stabilized by a short strong hydrogen bond with the zinc-bound water. Despite the lack of structural similarity between psCA3 and hCA II, the CO_2 binding orientation relative to the zinc-bound solvent is identical, with the CO_2 central carbon aligned and distanced 2.8 Å from the zinc-bound water, which is stabilized by a H-bond with Thr199 in HCA II and Asp44 in psCA3 (Fig. 9).

The CO_2 resides within the hydrophobic cleft formed by Val66, Gly102, Gly103, Phe61', Phe83', Val87', and Leu88' (from an adjacent monomer). In addition, the CO_2 is further stabilized by H-bonds with the side chain oxygen of Gln33', and the backbone amide nitrogen of Gly103. A H-bond between CO_2 and Gln33 is particularly interesting as it supports the previously hypothesized importance of this residue.

Another CO_2 molecule was found at the dimer interface in the so-called escort site, where it is completely surface "inaccessible", sandwiched between symmetry-paired residues Ala49, Val62, and Arg64. Moreover,

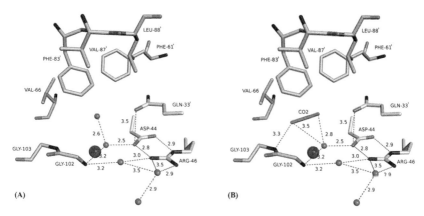

Fig. 9 Hydrogen bond network in the psCA3 active site: (A) open without CO_2 and (B) with CO_2. *Reproduced with permission from reference M. Aggarwal, T.K. Chua, M.A. Pinard, D.M. Szebenyi, R. McKenna, Carbon dioxide "trapped" in a β-carbonic anhydrase, Biochemistry 54 (43) (2015) 6631–6638.*

CO_2 is stabilized by two paired H-bonds with the side chains of Arg64 (from each monomer). As already mentioned, this pocket has been shown to be a promiscuous site for a variety of ligands, including bicarbonate, sulfate, and phosphate ions and probably involved in the allosteric switching between active and inactive form of the type II β-CAs.

References

[1] K.S. Smith, C. Jakubzick, T.S. Whittam, J.G. Ferry, Carbonic anhydrase is an ancient enzyme widespread in prokaryotes, Proc. Natl. Acad. Sci. U. S. A. 96 (1999) 15184–15189.

[2] K.S. Smith, J.G. Ferry, Prokaryotic carbonic anhydrases, FEMS Microbiol. Rev. 24 (2000) 335–366.

[3] R.S. Rowlett, Structure and catalytic mechanism of the beta-carbonic anhydrases, Biochim. Biophys. Acta, Proteins Proteomics 1804 (2010) 362–373.

[4] M.B. Guilloton, J.J. Korte, A.F. Lamblin, J.A. Fuchs, P.M. Anderson, Carbonic anhydrase in *Escherichia coli* a product of the cyn operon, J. Biol. Chem. 267 (1992) 3731–3734.

[5] C. Capasso, C.T. Supuran, An overview of the alpha-, beta- and gamma-carbonic anhydrases from Bacteria: can bacterial carbonic anhydrases shed new light on evolution of bacteria? J. Enzyme Inhib. Med. Chem. 30 (2) (2015) 325–332.

[6] R.S. Rowlett, C. Tu, M.M. McKay, J.R. Preiss, R.J. Loomis, K.A. Hicks, et al., Kinetic characterization of wild-type and proton transfer-impaired variants of beta-carbonic anhydrase from Arabidopsis thaliana, Arch. Biochem. Biophys. 404 (2002) 197–209.

[7] R.S. Rowlett, M.R. Chance, M.D. Wirt, D.E. Sidelinger, J.R. Royal, M. Woodroffe, et al., Kinetic and structural characterization of spinach carbonic anhydrase, Biochemistry 33 (1994) 13967–13976.

[8] D.N. Silverman, S. Lindskog, The catalytic mechanism of carbonic anhydrase: implications of a rate-limiting protolysis of water, Acc. Chem. Res. 21 (1) (1988) 30–36.

[9] B.C. Tripp, C.B. Bell 3rd, F. Cruz, C. Krebs, J.G. Ferry, A role for iron in an ancient carbonic anhydrase, J. Biol. Chem. 279 (8) (2004) 6683–6687.

[10] S.R. MacAuley, S.A. Zimmerman, E.E. Apolinario, C. Evilia, Y.M. Hou, J.G. Ferry, et al., The archetype γ-class carbonic anhydrase (Cam) contains iron when synthesized in vivo, Biochemistry 48 (5) (2009) 817–819.

[11] Y. Xu, L. Feng, P.D. Jeffrey, Y. Shi, F.M. Morel, Structure and metal exchange in the cadmium carbonic anhydrase of marine diatoms, Nature 452 (7183) (2008) 56–61.

[12] M.B. Guilloton, A.F. Lamblin, E.I. Kozliak, M. Gerami-Nejad, C. Tu, D. Silverman, et al., A physiological role for cyanate-induced carbonic anhydrase in Escherichia coli, J. Bacteriol. 175 (5) (1993) 1443–1451.

[13] C. Merlin, M. Masters, S. McAteer, A. Coulson, Why is carbonic anhydrase essential to Escherichia coli? J. Bacteriol. 185 (21) (2003) 6415–6424.

[14] B. Kusian, D. Sültemeyer, B. Bowien, Carbonic anhydrase is essential for growth of Ralstonia eutropha at ambient CO_2 concentrations, J. Bacteriol. 184 (18) (2002) 5018–5026.

[15] S. Mitsuhashi, J. Ohnishi, M. Hayashi, M. Ikeda, A gene homologous to beta-type carbonic anhydrase is essential for the growth of Corynebacterium glutamicum under atmospheric conditions, Appl. Microbiol. Biotechnol. 63 (5) (2004) 592–601.

[16] L.C. Chirica, C. Petersson, M. Hurtig, B.H. Jonsson, T. Boren, S. Lindskog, Expression and localization of α- and β-carbonic anhydrase in Helicobacter pylori, Biochim. Biophys. Acta 1601 (2) (2002) 192–199.

[17] I. Nishimori, T. Minakuchi, K. Morimoto, S. Sano, S. Onishi, H. Takeuchi, et al., Carbonic anhydrase inhibitors: DNA cloning and inhibition studies of the alpha-carbonic anhydrase from Helicobacter pylori, a new target for developing sulfonamide and sulfamate gastric drugs, J. Med. Chem. 49 (2006) 2117–2126.

[18] J.F. Heidelberg, J.A. Eisen, W.C. Nelson, R.A. Clayton, M.L. Gwinn, R.J. Dodson, et al., DNA sequence of both chromosomes of the cholera pathogen Vibrio cholerae, Nature 406 (6795) (2000) 477–483.

[19] B.H. Abuaita, J.H. Withey, Bicarbonate induces Vibrio cholerae virulence gene expression by enhancing ToxT activity, Infect. Immun. 77 (9) (2009) 4111–4120.

[20] M. Cobaxin, H. Martínez, G. Ayala, J. Holmgren, A. Sjöling, J. Sánchez, Cholera toxin expression by El Tor Vibrio cholerae in shallow culture growth conditions, Microb. Pathog. 66 (2014) 5–13.

[21] I. Nishimori, T. Minakuchi, A. Maresca, F. Carta, A. Scozzafava, C.T. Supuran, The β-carbonic anhydrases from Mycobacterium tuberculosis as drug targets, Curr. Pharm. Des. 16 (29) (2010) 3300–3309.

[22] S. Köhler, S. Ouahrani-Bettache, J.-Y. Winum, Brucella suis carbonic anhydrases and their inhibitors: towards alternative antibiotics? J. Enzyme Inhib. Med. Chem. 32 (1) (2017) 683–687.

[23] A. Aspatwar, J.Y. Winum, F. Carta, C.T. Supuran, M. Hammaren, M. Parikka, et al., Carbonic anhydrase inhibitors as novel drugs against mycobacterial β-carbonic anhydrases: an update on in vitro and in vivo studies, Molecules 23 (11) (2018) 2911.

[24] I. Nishimori, T. Minakuchi, D. Vullo, A. Scozzafava, C.T. Supuran, Inhibition studies of the β-carbonic anhydrases from the bacterial pathogen Salmonella enterica serovar Typhimurium with sulfonamides and sulfamates, Bioorg. Med. Chem. 19 (16) (2011) 5023–5030.

[25] C. Rollenhagen, D. Bumann, Salmonella enterica highly expressed genes are disease specific, Infect. Immun. 74 (3) (2006) 1649–1660.

[26] S.R. Lotlikar, S. Hnatusko, N.E. Dickenson, S.P. Choudhari, W.L. Picking, M.A. Patrauchan, Three functional β-carbonic anhydrases in *Pseudomonas aeruginosa* PAO1: role in survival in ambient air, Microbiology (Read.) 159 (8) (2013) 1748–1759.

[27] S.R. Lotlikar, B.B. Kayastha, D. Vullo, S.S. Khanam, R.E. Braga, A.B. Murray, et al., *Pseudomonas aeruginosa* β-carbonic anhydrase, psCA1, is required for calcium deposition and contributes to virulence, Cell Calcium 84 (2019) 102080.

[28] C.T. Supuran, C. Capasso, New light on bacterial carbonic anhydrases phylogeny based on the analysis of signal peptide sequences, J. Enzyme Inhib. Med. Chem. 31 (6) (2016) 1254–1260.

[29] G.D. Price, Inorganic carbon transporters of the cyanobacterial CO_2 concentrating mechanism, Photosynth. Res. 109 (2011) 47–57.

[30] M.R. Badger, D.T. Hanson, G.D. Price, Evolution and diversity of CO_2 concentrating mechanisms in cyanobacteria, Funct. Plant. Biol. 29 (2002) 161–173.

[31] G.C. Cannon, S. Heinhorst, C.A. Kerfeld, Carboxysomal carbonic anhydrases: structure and role in microbial CO_2 fixation, Biochim. Biophys. Acta 1804 (2) (2010) 382–392.

[32] M.R. Badger, The roles of carbonic anhydrases in photosynthetic CO_2 concentrating mechanisms, Photosynth. Res. 77 (2–3) (2003) 83–94.

[33] M.R. Sawaya, G.C. Cannon, S. Heinhorst, S. Tanaka, E.B. Williams, T.O. Yeates, et al., The structure of beta-carbonic anhydrase from the carboxysomal shell reveals a distinct subclass with one active site for the price of two, J. Biol. Chem. 281 (11) (2006) 7546–7555.

[34] A.K.-C.M. So, J. McKay, G.S. Espie, Characterization of a mutant lacking carboxysomal carbonic anhydrase from the cyanobacterium *Synechocystis* PCC6803, Planta 214 (2002) 456–467.

[35] H. Fukuzawa, E. Suzuki, Y. Komukai, S. Miyachi, A gene homologous to chloroplast carbonic anhydrase (icfA) is essential to photosynthetic carbon dioxide fixation by *Synechococcus* PCC7942, Proc. Natl. Acad. Sci. U. S. A. 89 (1992) 4437–4441.

[36] L.D. McGurn, M. Moazami-Goudarzi, S.A. White, T. Suwal, B. Brar, J.Q. Tang, et al., The structure, kinetics and interactions of the β-carboxysomal β-carbonic anhydrase, CcaA, Biochem. J. 473 (24) (2016) 4559–4572.

[37] S. Mitsuhashi, T. Mizushima, E. Yamashita, M. Yamamoto, T. Kumasaka, H. Moriyama, et al., X-ray structure of beta-carbonic anhydrase from the red alga, Porphyridium purpureum, reveals a novel catalytic site for CO_2 hydration, J. Biol. Chem. 275 (8) (2000 Feb 25) 5521–5526, https://doi.org/10.1074/jbc.275.8.5521

[38] M.S. Kimber, E.F. Pai, The active site architecture of *Pisum sativum* beta-carbonic anhydrase is a mirror image of that of alpha-carbonic anhydrases, EMBO J. 19 (7) (2000) 1407–1418.

[39] J.D. Cronk, J.A. Endrizzi, M.R. Cronk, J.W. O'neill, K.Y. Zhang, Crystal structure of E. coli beta-carbonic anhydrase, an enzyme with an unusual pH-dependent activity, Protein Sci. 10 (5) (2001) 911–922.

[40] M. Ferraroni, S. Del Prete, D. Vullo, C. Capasso, C.T. Supuran, Crystal structure and kinetic studies of a tetrameric type II β-carbonic anhydrase from the pathogenic bacterium *Vibrio cholerae*, Acta 71 (Pt 12) (2015) 2449–2456.

[41] J.D. Cronk, R.S. Rowlett, K.Y. Zhang, C. Tu, J.A. Endrizzi, J. Lee, et al., Identification of a novel noncatalytic bicarbonate binding site in eubacterial beta-carbonic anhydrase, Biochemistry 45 (14) (2006) 4351–4361.

[42] A.S. Covarrubias, A.M. Larsson, M. Högbom, J. Lindberg, T. Bergfors, C. Björkelid, et al., Structure and function of carbonic anhydrases from *Mycobacterium tuberculosis*, J. Biol. Chem. 280 (19) (2005) 18782–18789.

[43] A. Angeli, M. Ferraroni, M. Pinteala, S.S. Maier, B.C. Simionescu, F. Carta, et al., Crystal structure of a tetrameric type II β-carbonic anhydrase from the pathogenic bacterium *Burkholderia pseudomallei*, Molecules 25 (10) (2020) 2269.

[44] P. Strop, K.S. Smith, T.M. Iverson, J.G. Ferry, D.C. Rees, Crystal structure of the "cab"-type beta class carbonic anhydrase from the archaeon Methanobacterium thermoautotrophicum, J. Biol. Chem. 276 (13) (2001) 10299–10305.

[45] K.S. Smith, C. Ingram-Smith, J.G. Ferry, Roles of the conserved aspartate and arginine in the catalytic mechanism of an archaeal β-class carbonic anhydrase, J. Bacteriol. 184 (2002) 4240–4245.

[46] J.F. Domsic, B.S. Avvaru, C.U. Kim, S.M. Gruner, M. Agbandje-McKenna, D.N. Silverman, et al., Entrapment of carbon dioxide in the active site of carbonic anhydrase II, J. Biol. Chem. 283 (45) (2008) 30766–30771.

[47] R.S. Rowlett, K.M. Hoffmann, H. Failing, M.M. Mysliwiec, D. Samardzic, Evidence for a bicarbonate "escort" site in Haemophilus influenzae beta-carbonic anhydrase, Biochemistry 49 (2010) 3640–3647.

[48] A.S. Covarrubias, T. Bergfors, T.A. Jones, M. Högbom, Structural mechanics of the pH-dependent activity of beta-carbonic anhydrase from Mycobacterium tuberculosis, J. Biol. Chem. 281 (8) (2006) 4993–4999.

[49] R.S. Rowlett, C. Tu, J. Lee, A.G. Herman, D.A. Chapnick, S.H. Shah, et al., Allosteric site variants of Haemophilus influenzae beta-carbonic anhydrase, Biochemistry 48 (26) (2009) 6146–6156.

[50] K.M. Hoffmann, H.R. Million-Perez, R. Merkhofer, H. Nicholson, R.S. Rowlett, Allosteric reversion of Haemophilus influenzae β-carbonic anhydrase via a proline shift, Biochemistry 54 (2) (2015) 598–611.

[51] M.A. Pinard, S.R. Lotlikar, C.D. Boone, D. Vullo, C.T. Supuran, M.A. Patrauchan, et al., Structure and inhibition studies of a type II beta-carbonic anhydrase psCA3 from Pseudomonas aeruginosa, Bioorg. Med. Chem. 23 (15) (2015) 4831–4838.

[52] M. Aggarwal, T.K. Chua, M.A. Pinard, D.M. Szebenyi, R. McKenna, Carbon dioxide "trapped" in a β-carbonic anhydrase, Biochemistry 54 (43) (2015) 6631–6638.

[53] C. Capasso, C.T. Supuran, Anti-infective carbonic anhydrase inhibitors: a patent and literature review, Expert. Opin. Ther. Pat. 23 (2013) 693–704.

[54] S. Morishita, I. Nishimori, T. Minakuchi, S. Onishi, H. Takeuchi, Y. Sugiura, et al., Cloning, polymorphism, and inhibition of beta-carbonic anhydrase of Helicobacter pylori, J. Gastroenterol. 43 (2008) 849–857.

[55] S. Del Prete, D. Vullo, S.M. Osman, Z. AlOthman, W.A. Donald, J.Y. Winum, et al., Anion inhibitors of the β-carbonic anhydrase from the pathogenic bacterium responsible of tularemia, Francisella tularensis, Bioorg Med. Chem. 25 (17) (2017) 4800–4804.

[56] D. Vullo, R.S.S. Kumar, A. Scozzafava, J.G. Ferry, C.T. Supuran, Sulphonamide inhibition studies of the β-carbonic anhydrase from the bacterial pathogen Clostridium perfringens, J. Enzyme Inhib. Med. Chem. 33 (1) (2018) 31–36.

[57] S. Del Prete, V. De Luca, A. Nocentini, A. Scaloni, M.D. Mastrolorenzo, C.T. Supuran, et al., Anion inhibition studies of the beta-carbonic anhydrase from Escherichia coli, Molecules 25 (11) (2020) 2564.

[58] L.J. Urbanski, D. Vullo, S. Parkkila, C.T. Supuran, An anion and small molecule inhibition study of the β-carbonic anhydrase from Staphylococcus aureus, J. Enzyme Inhib. Med. Chem. 36 (1) (2021) 1088–1092.

[59] A.B. Murray, M. Aggarwal, M. Pinard, D. Vullo, M. Patrauchan, C.T. Supuran, et al., Structural mapping of anion inhibitors to β-carbonic anhydrase psCA3 from Pseudomonas aeruginosa, ChemMedChem 13 (19) (2018) 2024–2029.

[60] F. Briganti, S. Mangani, P. Orioli, A. Scozzafava, G. Vernaglione, C.T. Supuran, Carbonic anhydrase activators: X-ray crystallographic and spectroscopic investigations for the interaction of isozymes I and II with histamine, Biochemistry 36 (34) (1997) 10384–10392.

[61] M. Ilies, M.D. Banciu, M.A. Ilies, A. Scozzafava, M.T. Caproiu, C.T. Supuran, Carbonic anhydrase activators: design of high affinity isozymes I, II, and IV activators, incorporating tri-/tetrasubstituted-pyridinium-azole moieties, J. Med. Chem. 45 (2) (2002) 504–510.

[62] I.M. Johansson, C. Forsman, Kinetic studies of pea carbonic anhydrase, Eur. J. Biochem. 218 (1993) 439–446.

[63] I.M. Johansson, C. Forsman, Solvent hydrogen isotope effects and anion inhibition of CO_2 hydration catalysed by carbonic anhydrase from *Pisum sativum*, Eur. J. Biochem. 224 (1994) 901–907.

[64] K.S. Smith, N.J. Cosper, C. Stalhandske, R.A. Scott, J.G. Ferry, Structural and kinetic characterization of an archaeal beta-class carbonic anhydrase, J. Bacteriol. 182 (2000) 6605–6613.

[65] C.K. Tu, D.N. Silverman, C. Forsman, B.H. Jonsson, S. Lindskog, Role of histidine-64 in the catalytic mechanism of human carbonic anhydrase II studied with a site-specific mutant, Biochemistry 28 (19) (1989) 7913–7918.

[66] D.N. Silverman, R. Mckenna, Solvent-mediated proton transfer in catalysis by carbonic anhydrase, Acc. Chem. Res. 40 (8) (2007) 669–675.

[67] M. Aggarwal, C.D. Boone, B. Kondeti, R. McKenna, Structural annotation of human carbonic anhydrases, J. Enzyme Inhib. Med. Chem. 28 (2) (2013) 267–277.

[68] R.S. Rowlett, C. Tu, M.M. McKay, J.R. Preiss, R.J. Loomis, K. Hicks, et al., Kinetic characterization of wild-type and proton transfer-impaired variants of beta-carbonic anhydrase from *Arabidopsis thaliana*, Arch. Biochem. Biophys. 404 (2002) 197–209.

[69] H. Björkbacka, I.M. Johansson, C. Forsman, Possible roles for His 208 in the active-site region of chloroplast carbonic anhydrase from *Pisum sativum*, Arch. Biochem. Biophys. 361 (1) (1999) 17–24.

[70] R.S. Rowlett, C. Tu, P.S. Murray, J.E. Chamberlin, Examination of the role of Gln-158 in the mechanism of CO_2 hydration catalyzed by beta-carbonic anhydrase from *Arabidopsis thaliana*, Arch. Biochem. Biophys. 425 (1) (2004) 25–32.

CHAPTER FOUR

Bacterial γ-carbonic anhydrases

Andrea Angeli*

Neurofarba Department, University of Florence, Sesto Fiorentino, Florence, Italy
*Corresponding author. e-mail address: andrea.angeli@unifi.it

Contents

1. Introduction   94
2. Catalytic mechanism of bacterial γ-CAs   94
3. The subgroup of γ-CAs: the case of CamH   98
4. γ-Carbonic anhydrases from bacteria   99
5. Crystal structures   102
   5.1 *Methanosarcina thermophila*   102
   5.2 *Pyrococcus horikoshii*   105
   5.3 *Escherichia coli*   106
   5.4 *Brucella abortus*   108
   5.5 *ThermoSynechococcus elongatus*   110
   5.6 *Thermus thermophilus*   111
   5.7 *Burkholderia pseudomallei*   112
6. Inhibitors and activators of bacterial γ-CAs   114
   6.1 Inhibition mechanism   114
   6.2 Activation studies   114
7. Conclusion   115
References   116

Abstract

Carbonic anhydrases (CAs) are a ubiquitous family of zinc metalloenzymes that catalyze the reversible hydration of carbon dioxide to bicarbonate and protons, playing pivotal roles in a variety of biological processes including respiration, calcification, acid-base balance, and CO_2 fixation. Recent studies have expanded the understanding of CAs, particularly the γ-class from diverse biological sources such as pathogenic bacteria, extremophiles, and halophiles, revealing their unique structural adaptations and functional mechanisms that enable operation under extreme environmental conditions. This chapter discusses the comprehensive catalytic mechanism and structural insights from X-ray crystallography studies, highlighting the molecular adaptations that confer stability and activity to these enzymes in harsh environments. It also explores the modulation mechanism of these enzymes, detailing how different modulators interact with the active site of γ-CAs. Comparative analyzes with other CA classes elucidate the evolutionary trajectories and functional diversifications of these enzymes. The synthesis of this knowledge not only sheds light on the fundamental aspects of CA biology but also opens new avenues for therapeutic and

The Enzymes, Volume 55
ISSN 1874-6047, https://doi.org/10.1016/bs.enz.2024.05.002

industrial applications, particularly in designing targeted inhibitors for pathogenic bacteria and developing biocatalysts for industrial processes under extreme conditions. The continuous advancement in structural biology promises further insights into this enzyme family, potentially leading to novel applications in medical and environmental biotechnology.

1. Introduction

Carbonic anhydrases (CAs) are fundamental metalloenzymes catalyzing the reversible hydration of carbon dioxide to bicarbonate and proton across all life forms, reflecting their essential role in biological processes like pH regulation and CO_2 transport [1]. Spanning eight classes (α, β, γ, δ, ζ, η, ι and θ), CAs exhibit remarkable examples of convergent evolution, with no significant sequence or structural similarities, suggesting they do not share a common ancestor [2–6]. γ-CAs, found extensively in Bacteria and Archaea and uniquely in photosynthetic eukaryotes, represent the most ancient class, highlighting their foundational role in early life forms [7]. These enzymes share a unique structural motif: the left-handed parallel β-helix, marked by a triple helical face pattern. This motif is essential for their function, involving hexapeptide repeats crucial for enzyme stability and activity [8,9]. The functional γ-CAs were found as homotrimers, with active sites situated between monomer interfaces, reflecting a case of convergent evolution due to similar catalytic needs across different enzyme classes, particularly in their metal coordination strategies. In addition, their ubiquitous nature and diverse functional roles, from maintaining internal pH to facilitating CO_2 transport for photosynthesis, underscore the enzyme universal importance. Despite the majority of research focusing on mammalian α-class isozymes, the γ-CAs, with their unique metal ion requirements and evolutionary significance, offer intriguing insights into ancient life physiology and enzyme functionality.

2. Catalytic mechanism of bacterial γ-CAs

The first bacterial γ-CA to be identified was sourced from the archaeal domain, specifically within the anaerobic microorganism *Methanosarcina thermophila* marking it as the prototype of the γ-class, with the acronym Cam [10]. To date, it remains the most well-characterized and thoroughly understood member within this class of CAs and, like other CA classes, γ-CA catalyzes the hydration of carbon dioxide. The catalytic mechanism of this class exhibits similarities to the extensively studied α-class [11], following in a common two-step ping-pong reaction, outlined by the equations provided below, where E represents enzyme residues, M denotes the metal, and B the buffer:

$$E - M^{2+} - OH^- \underset{}{\overset{CO_2}{\rightleftharpoons}} E - M^{2+} - HCO_3^- \underset{}{\overset{H_2O}{\rightleftharpoons}} E - M^{2+} - OH_2 + HCO_3^-$$

(1)

$$E - M^{2+} - OH_2 \underset{BH^+}{\overset{B}{\rightleftharpoons}} E - M^{2+} - OH^-$$

(2)

The initial step of the reaction mechanism involves a zinc-bound hydroxide ion with a nucleophilic attack on CO_2 and, subsequently, the replacement of the metal-associated bicarbonate with a water molecule (Eq. 1). Following this, the mechanism undergoes a step where the zinc-bound hydroxide is regenerated. This involves removing a proton from the metal-associated water and transferring it to a buffer, as illustrated in the Eq. 2, where B serves as an external proton acceptor or an enzyme residue [12,13]. During this process, the zinc ion functions as a Lewis acid reducing the pK_a of water.

Additionally, studies indicate that the efficiency of CO_2 hydration by Cam, as measured by k_{cat}/K_m and k_{cat} values, increases with pH, hinting at a hydroxyl group nucleophilic attack on CO_2. Two specific pK_a values were observed, with the first pK_a ranging between 6.7 and 6.9 and the second pK_a between 8.2 and 8.4, suggesting that the first pK_a correlates with the ionization of zinc-bound water [11]. However, this is where the parallels with the other classes cease, as the active site residues surrounding Cam metal ion significantly differ. Various studies have explored the roles of these distinct residues in Cam, drawing comparisons to key residues in human CA II for functional insights [14–17].

In terms of the metal ion, recombinant Cam typically undergoes production in a zinc-supplemented medium and is aerobically purified [10]. Nevertheless, anaerobic purification from *Escherichia coli* reveals the presence of iron in the active site, indicating its potential physiological relevance *in vivo* [18]. This iron-bound form of Cam displays a threefold increase in activity compared to its zinc counterpart, although it exhibits sensitivity to oxygen and hydrogen peroxide, leading to the oxidation of Fe^{2+} to Fe^{3+}, which is then lost from the active site [19]. These observations led to the hypothesis that iron might serve as the physiologically significant metal in Cam. This raises the possibility that iron could act as a metal cofactor across all γ-CAs and potentially in other classes as well. Furthermore, subsequent experiments reconstituting apo-Cam with various metals revealed that while ions like Cu^{2+}, Mn^{2+}, Ni^{2+}, and Cd^{2+} can substitute zinc in the active site, they result in significantly lower kinetic parameters compared to Zn-Cam [18]. Interestingly, cobalt-reconstituted Cam demonstrated even greater activity than Zn-Cam [11].

This γ-CA was characterized by its moderate enzymatic velocity with a k_{cat} near 10^5 s^{-1}, but with notable CO_2 hydrase activity.

This activity suggests the involvement of an ionizable, dynamic residue acting as a proton shuttle. Investigations into Cam catalytic mechanism have utilized site-directed mutagenesis of active site residues and detailed analysis of the wild-type Cam crystal structure via X-ray diffraction, leading to the proposition of a specific catalytic mechanism depicted in Fig. 1.

From X-ray Observations, was identified Glu84 in dual conformations and in one of them, it seemingly shares a proton with Glu62. This observation prompted the suggestion that Glu84 and Glu62 collectively perform a role similar to that of His64 in human CA II, transferring protons from the zinc-associated water to the solvent [20]. Furthermore, mutagenesis of Glu84 confirmed its critical role as a proton shuttle, mirroring His64 function in hCA II [14]. Replacing Glu84 with Asp and His yielded functional enzymes, while substitutions with Ala and other non-ionizable residues significantly reduced activity. This reduction could be compensated by imidazole, which acts as an external proton donor and acceptor, facilitating proton transfer [21]. The replacement of Glu62 led to reduced enzymatic efficiency, albeit less so with Glu62Asp variants, indicating Glu62 involvement in CO_2 hydration and a preference for a carboxylate at position 62 for catalysis. While Thr199 and Glu106 in hCA II serve to have a "door keeper" function, ensure only protonated ligands bind to the Zn (II) site, a similar role for Glu62 in Cam has been proposed [22]. Despite lower activity in Glu62 variants compared to wild-type, the orientation of Glu62 towards Glu84 in the enzyme structure suggests it might contribute to proton transport [14]. Other significant residues, Asn73, Gln75, and Asn202, have been implicated in the enzyme interaction with HCO_3^-, positioned within hydrogen-bonding distance to bicarbonate oxygen [16]. These residues are considered potential contributors to positioning the metal-bound hydroxide and CO_2 for nucleophilic attack or stabilizing the transition state. Variants replacing these residues showed decreased efficiency compared to the wild type, particularly those replacing Gln75, which also demonstrated an increase in pK_a related to the ionization of the

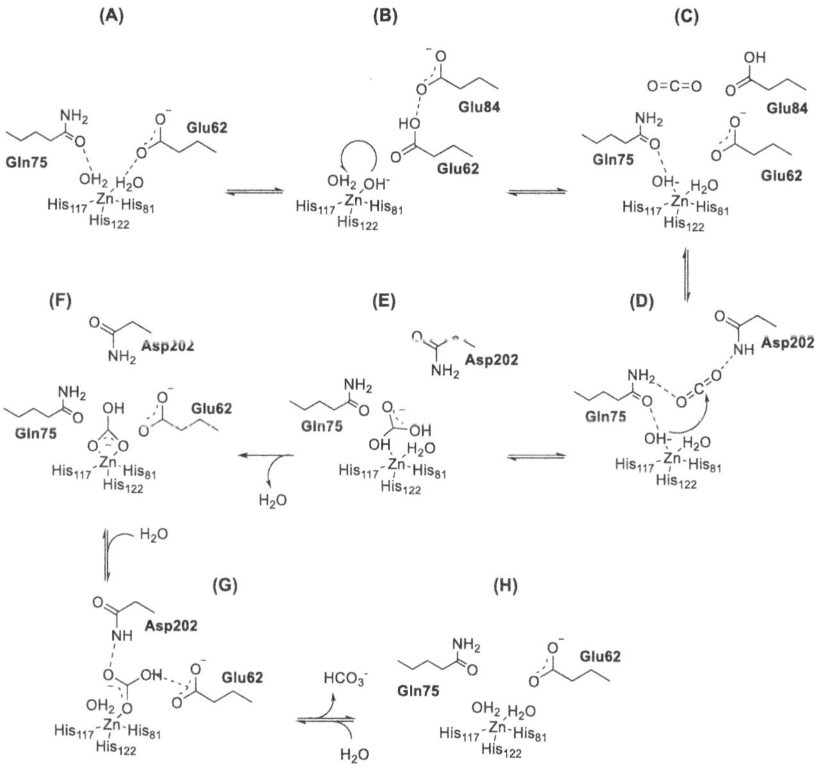

Fig. 1 The reaction mechanism proposed for Zn-Cam. (A) Zn^{2+} is coordinated by two water molecules. (B) Glu62 extracts a proton from a water molecule, which then extracts a proton from the adjacent water. (C–F) $Zn-OH^-$ attacks the CO_2 resulting in a bound HCO_3^- that displaces a water molecule. The HCO_3^- undergoes a bidentate transition state where the proton either rotates or transfers to the nonmetal-bound oxygen of the HCO_3^- (G) Glu62 hydrogen bonds with hydroxyl HCO_3^-, destabilizing it. Incoming water further destabilizes the HCO_3^- by replacing one of the bound oxygens. (H) The second incoming water completely displaces the HCO_3^- resulting in product removal and regeneration of the active site.

metal–bound hydroxide. This supports Gln75 essential role in orienting the metal–bound hydroxide for CO_2 attack and aiding the metal in lowering its pK_a. Asn73 variants exhibited slight decreases in efficiency, hinting at a role in arranging Gln75 for optimal catalytic interaction, akin to the function of Thr199 and Glu106 in hCA II. Analysis of Asn202 variants revealed significant decreases in activity, affirming its proposed function in CO_2 polarization and enhancing attack by the metal–bound hydroxide.

Arg59, positioned at the juncture of monomers, plays a critical role in both the formation and stabilization of the trimeric structure. This residue is not only vital for the enzyme catalytic function, affecting the CO_2 hydration process (with observed reductions in both k_{cat} and k_{cat}/K_m upon mutation), but its functionality can also be partially restored with the addition of positively charged guanidinium compounds [15]. Among the mutations, Arg59Lys is notable for retaining some level of activity, underscoring the necessity of a positive charge at this specific site. As the sole positively charged residue in the active site of Cam, located merely 6Å away from the metal ion, Arg59 is speculated to contribute to CO_2 conversion and/or the release of reaction products. Additionally, two other residues, Trp19 and Tyr200, located further from the catalytic metal at distances of 10Å and 12Å respectively, have been subject to structural and kinetic studies. These residues are unique in that they challenge the notion that the active sites of γ-class CAs significantly diverge from those in the α-class, as they occupy positions comparable to Trp5 and Tyr7 in human CA II [17]. Analysis of Trp19 variants reveals that while Trp19 is not indispensable, it plays a significant role in the CO_2 to bicarbonate conversion and the proton transfer steps. The mutagenesis of Tyr200 has been observed to expedite intramolecular proton transfer. This observation draws parallels to the enhanced proton transfer seen with the Tyr7Phe mutation in human CA II, suggesting that such acceleration could be linked to the configuration of structured water molecules within the enzyme's active site [23].

3. The subgroup of γ-CAs: the case of CamH

From the sequence analysis all γ-CAs share conservation in metal ligand residues (His81, His117, His122), residues crucial for catalysis (Gln75 and Gln73), and those vital for maintaining the active site structure (Arg59, Asp76, and Asp61). However, a significant portion of γ-CAs distinguishes itself by not including Asn202, Glu62, and an acidic loop with Glu84, which is a critical proton shuttle in Cam. The CamH variant from *M. thermophila* falls into this specific subgroup of γ-CAs. Produced recombinantly, CamH forms a homotrimer and exhibits carbonic anhydrase activity [24]. Kinetic studies reveal that CamH follows a conventional two-step metal hydroxide mechanism, although with k_{cat} and k_{cat}/K_m values considerably lower than Cam when aerobically purified from *E. coli* supplemented with zinc (Zn-CamH), showing a zinc/iron ratio of 0.71 to 0.15 per monomer. Conversely, anaerobically purified CamH from *E. coli*

(Fe-CamH), with iron supplementation, showed equivalent iron levels with minimal zinc presence. Notably, Fe-CamH effective k_{cat} was significantly reduced compared to Fe-Cam but saw an increase with added imidazole, while its k_{cat}/K_m was similar to Fe-Cam. These findings suggest CamH lacks a direct proton shuttle equivalent to Cam Glu84. Additionally, Fe-CamH activity decreases upon exposure to air or hydrogen peroxide, highlighting iron potential role in the active site and suggesting iron as the physiological metal cofactor *in vivo*, as seen in Cam.

Cam and CamH are theorized to play roles in the metabolic pathway converting acetate to methane in *M. thermophila*. Interestingly, Cam possesses 34 additional *N*-terminal residues indicative of signal peptides found in secretory proteins, implying an extracellular or periplasmic location [24]. In contrast, CamH genetic sequence lacks these additional residues, suggesting a cytosolic presence. Given their proposed cellular localizations, Cam and CamH are believed to facilitate bicarbonate exchange with acetate during growth on acetate, underscoring their distinct yet complementary functions in the organism metabolism [25].

4. γ-Carbonic anhydrases from bacteria

To date, four classes of CAs (α, β, γ and ι) encoded by bacterial genomes have been identified and, an examination of both Gram-positive and Gram-negative bacterial genomes, revealed a diverse pattern of CAs distribution [26]. Specifically, certain bacteria may possess CAs from only a single class, while others encode for two or even three different CA classes. Notably, specific bacteria, including *Lactobacillus reuteri*, *Akkermansia muciniphila*, and *Prevotella intermedia*, are characterized by the presence of solely γ-CA class enzymes [27].

Moreover, a number of bacterial species, particularly some Gram-negative bacteria such as those from the *Buchnera* and *Rickettsia* genera, appear to lack genes encoding for CAs. These bacteria are typically found in environments with high CO_2 concentrations, a condition that often correlates with a genomic reduction including the loss of CA-encoding genes [28].

The study of CA enzymes in pathogenic bacteria has emerged as a promising field in the development of antibiotics that operate through novel mechanisms. Indeed, given the essential role of CAs in the life cycle of many pathogenic organisms, targeting these enzymes can significantly disrupt the growth and viability of the pathogens, presenting a strategic approach to combat bacterial infections.

Analyzing the evolutionary lineage of β- and γ-CAs through phylogenetic trees has revealed that these enzymes form two distinct clusters, underscoring their significant phylogenetic divergence. It is suggested that γ-CA amino acid sequences act as evolutionary intermediates, potentially giving rise to the β-CAs [27]. Regarding the enzymatic activity of CAs, the presence of a Zn^{2+} ion cofactor, coordinated by three protein-derived amino acid residues, is generally essential [29]. The enzyme catalytic cycle is facilitated by a fourth ligand, typically a water molecule or hydroxide ion, serving as the nucleophile. Notably, γ-CAs show a dependence on Fe^{2+} for their activity but can also function with Zn^{2+} or Co^{2+} ions bound.

For γ-CAs to fulfill their catalytic roles effectively, trimer formation is essential. Each monomer within the trimer is marked by a series of tandemly-repeated hexapeptides that are fundamental for establishing the left-handed β-helix structure characteristic of these enzymes [30]. γ-CAs predominantly reside within the cytoplasm, playing vital roles in internal cellular processes like CO_2/HCO_3^- transport and pH regulation [31]. Recent findings have highlighted the presence of a short, putative signal peptide at the N-terminus of some γ-CAs from Gram-negative bacteria, suggesting additional functionalities and localization patterns.

γ-CAs have been successfully cloned and purified from various bacteria, including pathogenic strains such as *Porphyromonas gingivalis*, *Vibrio cholerae*, *E. coli* and *Burkholderia pseudomallei*, as well as from Antarctic bacteria like *Pseudoalteromonas haloplanktis*, *Colwellia psychrerythraea*, and *Nostoc* commune. These advancements underscore the diverse roles and evolutionary significance of γ-CAs across different bacterial species and the kinetic parameters of the CO_2 hydration reaction catalyzed by these CAs are reported in Table 1.

All these enzymes demonstrate notable CO_2 hydration activity, although at a moderate level. In contrast, Antarctic bacteria, known for their extremophilic nature, prosper in environments with temperatures below 15°C and produce enzymes that exhibit enhanced catalytic efficiency at colder temperatures. This attribute has recently attracted attention for its potential in industrial processes that require catalysts functioning optimally at low temperatures.

Comparative analyzes reveal that all the enzymes reported in Table 1 exhibit catalytic activities ranging from moderate to high, consistently surpassing the archetypal enzyme Cam. Sequence comparisons underscore

Table 1 Kinetic parameters for the CO_2 hydration reaction catalyzed by the γ-CAs. Inhibition constant (K_I) for the sulfonamide inhibitor acetazolamide (AAZ) are also shown.

Organism	Acronym	k_{cat} (s^{-1})	k_{cat}/K_m (M^{-1} s^{-1})	K_I AAZ (nM)	References
Methanosarcina thermophila	Zn–Cam	$6.1\ 10^4$	$8.7\ 10^5$	63	[10,11,14]
M. thermophila	Co–Cam	$1.0\ 10^5$	$1.76\ 10^6$	143	[10,11,14]
M. thermophila	Fe–Cam	$2.4\ 10^{5a}$	$5.4\ 10^{6a}$	–	[10,11,14]
M. thermophila	Zn–CamH	$4.9\ 10^2$	$1.6\ 10^5$	–	[15,17]
M. thermophila	Fe–CamH	$4.0\ 10^{4a}$	$1.3\ 10^{7a}$	–	[15,17]
Vibrio cholerae	VhCA	$7.4\ 10^5$	$6.4\ 10^7$	473	[32,33]
Porphyromonas gingivalis	PgiCA	$4.1\ 10^5$	$5.4\ 10^7$	324	[34,35]
Burkholderia pseudomallei	BpsCA	$5.3\ 10^5$	$2.5\ 10^7$	149	[36,37,38]
Colwellia psychrerythraea	CpsCA	$6\ 10^5$	$4.7\ 10^7$	502	[39,40]
Pseudoalteromonas haloplanktis	PhaCA	$1.4\ 10^5$	$1.9\ 10^7$	403	[41,42]
Nostoc commune	NcoCA	$9.5\ 10^5$	$8.3\ 10^7$	75.8	[43,44]
Nostoc sp. PCC 7120	NstCcmM209a	$2.2\ 10^4$	$4.1\ 10^6$	–	[45]
Escherichia coli	EcoCAc	$5.7\ 10^5$	$6.9\ 10^6$	248	[46]
Mammaliicoccus sciuri	MscCAγ	$6.2\ 10^5$	$9.5\ 10^6$	245	[47]

[a]Effective k_{cat} (k_{cat}^{eff}) and k_{cat}/K_m^{eff} values, obtained by dividing apparent k_{cat} and k_{cat}/K_m values by the molar ratio of metal to monomer.

the conservation of metal ion ligands (His81, His117, and His122, according to the Cam nomenclature) across these enzymes. Furthermore, the critical residues for catalysis, namely Asn73, Gln75, and Arg59, maintain strict conservation. However, Asn202, Glu62, and Glu84, the latter identified as a proton shuttle residue in Cam, are either absent or show variable conservation among these enzymes.

5. Crystal structures

5.1 *Methanosarcina thermophila*

As mentioned above, Cam represents the prototype enzyme of γ-class CAs and it was discovered by Ferry approximately two decades ago from *M. thermophila*, an anaerobic methane-producing microorganism [48]. The elucidation of Cam crystal structure in 1996 revealed that a single monomer is composed of seven complete rotations of a left-handed parallel β-helix, capped by a brief α-helix, and succeeded by another C-terminal α-helix aligned antiparallel to the β-helix axis (Fig. 2A) [48]. The β-helix faces are made up of parallel β-strands, each harboring two or three residues, and are linked to the next face via a 120-degree turn, giving the β-helix an equilateral triangular cross-section. Within each β-helix rotation, two type II β-turns are located between the first and second strands and between the second and third strands, finalized by a loop that connects the third strand to the first strand of the ensuing rotation. The helix core is

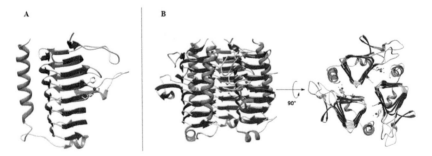

Fig. 2 (A) Ribbon representation of the Cam monomer. β-Strands are shown as *curved arrows* in violet and α-helices as ribbon in red. The overall fold is a left-handed β-helix (PDB: 1QRG). The active site Zn(II) ion (colored in magenta), His81 and His122, and the two metal-bound waters are also shown. (B) View of the Cam trimer along the molecular threefold axis.

characterized by hydrophobic interactions among aliphatic side chains from corresponding positions in adjacent helix rotations (Fig. 2A).

The enzyme active form is a homotrimer (Fig. 2B), formed by aligning three left-handed β-helices parallel to each other. Three amino acids, Arg59, Asp61, and Asp76, play crucial roles in the trimer formation. Arg59 engages in salt bridge interactions with Asp61 from the same monomer and Asp76 from a different monomer. These residues are highly conserved across all proteins with homology to Cam. The active site is situated at the juncture between two subunits (Fig. 2B). Within this site, a Zn(II) ion is coordinated by His81 and His122 from adjacent turns of one monomer and by His117 from a neighboring monomer, forming a distorted trigonal bipyramidal geometry with the inclusion of two water molecules. These water molecules are positioned to form hydrogen bonds with the side chains of Gln75 and Glu62 [49]. Although there is no structural homology between the γ- and α-classes of CAs, Cam and human CA II, the most thoroughly studied human variant, both exhibit a 3-His metal coordination motif.

The coordination of water molecules to the metal ion in Cam contrasts with that observed in hCA II, where a single water molecule fulfills the tetrahedral coordination of either a zinc or cobalt ion. Specifically, the structure of cobalt-substituted Cam (Co-Cam) reveals an octahedral coordination geometry that includes three water molecules (Fig. 3B). Among these, only one water molecule parallels the water coordinating zinc in the Zn-Cam variant (Fig. 3A) [49–51]. Interestingly, substituting

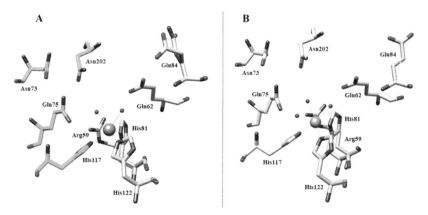

Fig. 3 Metal coordination and active site residues in (A) Zn-Cam (PDB: 1QRG) and (B) Co-Cam (PDB: 1QQ0). Residues belonging to different subunits are depicted using different color.

the metal with cobalt enhances this isoform activity, despite Co-Cam exhibiting a higher coordination number than Zn-Cam. This increase in activity due to a higher coordination number differs from the pattern seen in hCA II, where an increased coordination number of the metal ion typically leads to a reduced turnover rate [11,49,52].

In Zn-Cam, HCO_3^- adopts a bidentate binding approach to the Zn(II) ion, displacing the two metal-associated water molecules. This ion is additionally secured through a hydrogen bond formed between its unco-ordinated oxygen and both oxygens of the Glu62 side chain (Fig. 4A). In contrast, in Co-Cam, HCO_3^- binds in a monodentate manner to cobalt ion, taking the place of one water molecule and another from the active site network of water molecules. It also establishes a hydrogen bond with Glu62 side chain (Fig. 4B). The manner in which bicarbonate interacts with Zn-Cam and Co-Cam diverges, particularly because the water molecule replaced by HCO_3^- in the complex Co-Cam-bicarbonate is distinct from those involved in Zn-Cam hydration process.

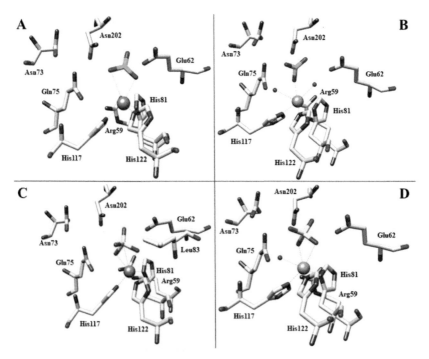

Fig. 4 Metal coordination in (A) Zn-Cam bound to HCO_3^-; (B) Co-Cam bound to HCO_3^-; (C) Zn-Cam bound to SO_4^-; (D) Co-Cam bound to SO_4^-.

Exploring anion inhibition reveals that, contrary to initial findings indicating sulfate does not inhibit Cam activity [53], sulfate direct interaction with the Zn^{2+} ion in the Cam enzyme suggests otherwise. This interaction, which leads to the displacement of water molecules coordinating with zinc, maintains the enzyme distorted trigonal bipyramidal structure, mimicking inhibitor-like behavior. Notably, a hydrogen bond formed between a sulfate oxygen and Glu62 implies a proton exchange with this residue, possibly raising its pK_a value (Fig. 4C).

Shifting focus to the Co-Cam complex with sulfate unveils additional layers of complexity. Here, sulfate adopts a bidentate mode of binding to cobalt, substituting for specific water molecules while leaving one water molecule intact. This pattern of interaction is similar to that seen in the Zn-Cam complex, involving contact with Glu62 side chain. The Co-Cam-sulfate complex, however, distinguishes itself by forming two hydrogen bonds with Glu62, as opposed to the singular bond in the Zn-Cam interaction. This arrangement suggests an elaborate proton-sharing dynamic, possibly supported by Glu62 in a protonated state, and highlights nuanced differences in sulfate interaction with these metal-specific variants of the Cam enzyme (Fig. 4D) [49].

5.2 *Pyrococcus horikoshii*

The determination of a second γ-CA crystal structure in 2008 by Jeyakanthan et al. revealed insights from the OT3 strain of the hyperthermophilic archaeon *Pyrococcus horikoshii* (Cap). This marked a significant advancement following the initial structural insights into γ-CAs. A noticeable distinction between Cap and the previously characterized Cam is the presence of shorter loop regions in Cap [54]. Differences extend to the structural organization between the C-terminal α-helix and the β-helix. Unlike Cam, which features a flexible loop followed by a small α-helix, Cap displays a more structured β-strand, distinguished by a classical β-bulge. Similar to Cam, Cap also forms a trimer, with the active site located at the interface between monomers, where it coordinates a zinc ion with three histidine residues (His 65 and His 87 from one monomer and His 82 from another) along with two water molecules (Fig. 5A).

An intriguing aspect of Cap structure is the identification of a secondary metal-binding site, characterized by the presence of four calcium ions at the monomer interfaces. Notably, one calcium ion is centrally located within the trimer, hexacoordinated by Ser40 and Asn61 from each monomer, suggesting a role in facilitating substrate entry and product exit, akin to the

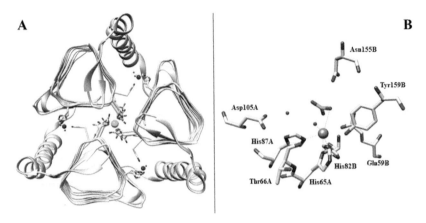

Fig. 5 (A) Ribbon representation of the Zn-Cap trimer; metal ions are shown as spheres. The central Ca^{2+} ion is in green, the Zn^{2+} ions in magenta. The zinc-coordinating histidines are also shown. The central Ca^{2+} ion is coordinated to Ser40 and Asn61 from all three chains (PDB: 1V3W); (B) Metal ion coordination in the structure of Cap-HCO_3^- complex (PDB: 1V67).

zinc ion function (Fig. 5A). The catalytic site of Cap contrasts markedly with that of Cam in terms of residue conservation. The crucial residues for catalysis in Cam, such as Glu62, Glu84, and Asn202, are substituted in Cap with Ile47, His68, and Tyr159, respectively. His68 is hypothesized to serve as a proton shuttle, mirroring the role of Glu84 in Cam and His64 in hCA II, although Cap CA activity remains unreported. Furthermore, Tyr159, substituting Asn202 in Cap, is positioned near the zinc ion, illustrating a notable deviation from Cam. When bicarbonate is bound within Cap structure (Fig. 5B), it attaches in a monodentate manner to the zinc ion, replacing a single water molecule. This differs from the bidentate binding observed with bicarbonate in Cam. A bicarbonate oxygen atom engages in a dual hydrogen-bond with Gln59 (equivalent to Gln75 in Cam), showcasing the structural and functional diversity within the γ-CA family.

5.3 *Escherichia coli*

E. coli expresses two classes of CA, specifically β-CAs with YadF and CynT, but it also encodes several gene products such as CaiE, PaaY, and YrdA, which are closely related to the γ-group CAs. Notably, the YrdA gene is conserved across multiple bacteria and encodes a protein that typically comprises at least 180 amino acids, featuring hexapeptide-repeat motifs. This protein sequence may also extend by up to 70 amino acids at either the N- or C-terminus and has been crystallized in two different forms [55].

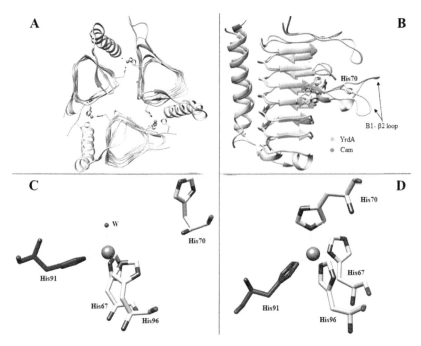

Fig. 6 (A) Ribbon representation of the YrdA trimer; metal ions are shown as spheres. The central Zn^{2+} ions are in gray. The zinc-coordinating histidines are also shown (PDB: 3TIS). (B) Differences in the β1-β2 loops between superposed YrdA and Cam. The main-chain atoms of the two superposed structures are displayed as differently colored ribbon diagrams (blue for Cam and light gray for YrdA). The residues involved in coordinating a metal ion (*gray ball*), and some residues involved in forming the pocket of the active site are represented by stick models for YrdA. (C) Open conformation. The zinc ion is coordinated by three His residues originating from two subunits that are differentiated by light gray and magenta carbon colors. (D) Closed conformation. The zinc ion is coordinated by four His residues. The subunits are differentiated using light gray and magenta carbon colors.

Structurally, YrdA shares similarities with other γ-CAs, though it notably lacks significant CA activity (Fig. 6A). The protein variability mainly lies in the loops between the β1 and β2 strands, and between the β10 and β11 strands (regions where Cam includes the acidic residue Glu84). The β1-β2 loop in YrdA is shorter than in Cam, containing only three amino acids that form a brief turn. This loop contributes to the formation of YrdA active site through interactions between the side chain of the bulky Tyr8 and Trp171 located in the C-terminal α-helix of an adjacent monomer (Fig. 6B).

In the proposed active site, a Zn(II) ion is coordinated in the typical 3-His arrangement found in γ-CAs. However, in six of the nine independent molecules from the two YrdA crystal structures, His70 in the β10-β11 loop coordinates the zinc, forming nearly tetrahedral coordination described as the closed conformation (Fig. 6D). In the remaining three subunits, the zinc coordination sphere is completed by a water molecule. Each of the nine molecules in the asymmetric units of the two structures displays varied conformations for the β10-β11 loop, aligned with the positional shifts of His70 and the rotational conformation of Trp171 seems partly dependent on the positioning of His70. When His70 is distant from the active site, the β10-β11 loop is also remote, leaving YrdA active site in an open state (Fig. 6C). Conversely, when Trp171 is nearer, the active site assumes a closed state (Fig. 6D). The noncanonical zinc coordination in the closed state of YrdA closely mirrors the unique coordination seen in type II β-CAs from organisms like *Haemophilus influenzae*, *E. coli*, and *V. cholerae*, where a water molecule is replaced by an Asp side chain [56–58].

Despite these structural insights, YrdA specific function remains unclear, although the flexibility of the β10-β11 loop suggests it may process a larger substrate than the typical CO_2 processed by other CAs. Notably, significant CA activity has not been detected for YrdA, nor can it compensate for the loss of CA activity in *E. coli* when β-CA (CynT) is silenced [59]. Additionally, YrdA lacks residues equivalent to Glu62, Glu84, and Asn202 found in Cam, further differentiating its potential functional role.

5.4 *Brucella abortus*

The crystal structure of RicA (Rab2 interacting conserved protein A), is related to γ-CA-like proteins from *Brucella abortus* presents a fascinating example of structural conservation and functional adaptation among γ-CAs. RicA crystallizes as a homotrimer, each monomer adopting a left-handed β-helix capped by a C-terminal α-helix, forming a classic γ-CA fold (Fig. 7A) [60]. Notably, the structural analysis revealed three zinc-binding sites per trimer each one coordinated by three histidines (His67 and His89 from one monomer and His84 from and adjacent monomer) and located at the interfaces of the monomers.

This configuration is consistent with the structural organization observed in other γ-CAs, where the active sites are typically formed at similar monomer-monomer interfaces. Despite the structural similarities to known γ-CAs, RicA does not exhibit typical carbonic anhydrase activity, suggesting a divergence in function that underscores the protein role in

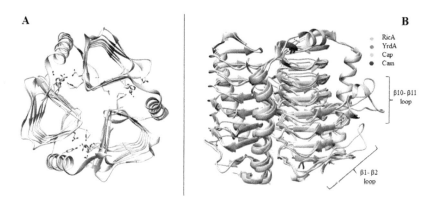

Fig. 7 (A) Ribbon representation of the RicA trimer; metal ions are shown as spheres. The central Zn^{2+} ions are in gray. The zinc-coordinating histidines are also shown (PDB: 4N27). (B) Differences in the β1-β2 and β10-β11 loops between superposed RicA, YrdA, Cap and Cam. The main-chain atoms of the two superposed structures are displayed as differently colored ribbon diagrams (light gray for RicA blue for YrdA, green for Cap and magenta for Cam).

Brucella pathogenesis rather than CO_2 hydration. The absence of carbonic anhydrase activity in RicA might be attributed to its unique active site environment, potentially lacking crucial residues that are necessary for catalytic activity found in other γ-CAs. When compared to other structurally characterized γ-CAs such as those from *M. thermophila* (Cam and CamH), *P. horikoshii* (Cap), and even non-active homologs like YrdA from *E. coli*, RicA shows high structural homology with some distinct differences that may inform its functional divergence (Fig. 7B). For instance, the zinc coordination in RicA, while similar, does not support catalytic activity, even if the residues important for the activity are conserved in RicA with the exception of Trp19 (present in the β1-β2 loop), Glu62, Asp76, Glu84, Tyr200 and Asn202. Residue Asp76 is conserved in all known γ-CA and γ-CA-related structures; the general consequence of the conservative substitution to glutamic acid at this position in RicA is unknown. Residues including Glu62, Glu84 and Tyr200 are also variable at equivalent positions of other active γ-carbonic anhydrases, suggesting that they are not generally required for γ-carbonic anhydrase activity. An asparagine residue at position 202 (Asn202) is found exclusively in the active γ-CA enzymes, and may thus play an important role in CA activity [60]. The active sites of Cam and CamH are known for their robust enzymatic activity, which contrasts starkly with RicA lack of activity. This could be indicative of evolutionary adaptation where RicA has repurposed the ancestral γ-CA fold for a specific role in the pathogen-host interaction, particularly in its binding to the host Rab2 protein, influencing intracellular trafficking processes. The structure-function relationship in RicA provides an excellent case study of how proteins within the γ-CA family can evolve new functions while retaining structural elements characteristic of the family. This phenomenon is seen in RicA ability to bind zinc at traditional γ-CA active sites yet diverge functionally to interact with host cellular machinery (a crucial factor in the pathogenicity of *B. abortus*).

5.5 *ThermoSynechococcus elongatus*

Cyanobacteria are prolific photosynthetic organisms on Earth, playing a crucial role alongside plants in global CO_2 fixation into biomass. Within these organisms, specialized microcompartments called carboxysomes enhance carbon fixation, primarily catalyzed by ribulose 1,5-bisphosphate carboxylase/oxygenase (RuBisCO). Despite RuBisCO critical role, it is notoriously inefficient with a low affinity for CO_2, its secondary substrate. To address these catalytic challenges, cyanobacteria employ a CO_2 concentrating mechanism (CCM), which is vital for their survival and growth. This system involves the active transport of bicarbonate into the cell, which then diffuses into carboxysomes and is converted back to CO_2 by a specific carbonic anhydrase (CA) before being fixed by RuBisCO. This process effectively elevates CO_2 concentrations around RuBisCO, facilitating more efficient fixation rates. Many β-cyanobacteria, such as *Synechocystis* sp. PCC 6803 and *Synechococcus elongatus* PCC 7942 [61,62], encode a carboxysomal CA known as CcaA, which is closely related to the β-class of CAs. However, some, like the thermophilic β-cyanobacterium *Thermo-Synechococcus elongatus* BP-1, do not possess a CcaA homolog. Instead, the role of carboxysomal CA in these bacteria is suggested to be performed by the N-terminal domain of another carboxysomal component, CcmM, which shows sequence homology to the γ-CAs. The structure of two variants of the N-terminal domain of the recombinantly expressed CcmM from *T. elongatus* BP-1, comprising residues 1–193 (TeCcmM193) and 1–209 (TeCcmM209), has been elucidated [63]. The monomer of CcmM209 structurally resembles the left-handed β-helix typical of Cam, capped at the C-terminus by a short α-helix. Additional α-helices follow, contributing to the complex architecture that facilitates enzyme functionality (Fig. 8A).

A key feature in these structures is the formation of a trimeric channel, obstructed centrally by closely packed Arg121 residues. A chloride ion

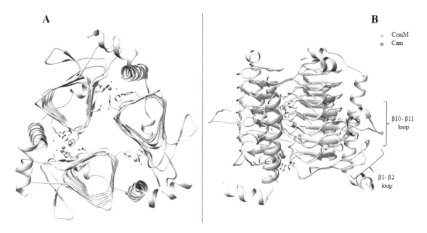

Fig. 8 (A) Ribbon representation of the CcmM trimer; metal ions are shown as spheres. The central Zn^{2+} ions are in gray. The zinc-coordinating histidines are also shown (PDB: 3KWC). (B) Differences in the β1-β2 and β10-β11 loops between superposed CcmM and Cam. The main-chain atoms of the two superposed structures are displayed as differently colored ribbon diagrams (light gray for CcmM and blue for Cam).

found in this structure is coordinated by the guanidinium groups from each monomer. Significantly, the zinc coordination and all catalytic site residues are conserved with Cam structure, with the exception of the Glu84 residue, which is replaced by a Gln in CcmM209. Differences between the two structures are primarily localized to the length of the β1- β2 loop, the helical turns of the α-helices, and the residues forming the subsequent loops and helices which influence the overall structural dynamics and enzyme activity (Fig. 8B). TeCcmM209 exhibits slightly enhanced activity compared to its wild-type counterpart, with activity notably dependent on the oxidation state of a disulfide bond. Conversely, the TeCcmM193 variant shows a lack of ordered structure in certain regions, correlating with a decrease in CA activity. This structural instability, particularly in the absence of the αC helix, leads to a disordered and reorganized trimer, impacting the active site structure and enzymatic function.

5.6 *Thermus thermophilus*

The γ-class CA from *Thermus thermophilus* HB8 (γ-TtCA) exhibits a thermostable molecular structure, crucial for understanding its adaptation to high-temperature environments. The crystal structure reveals that γ-TtCA adopts the typical γ-CA fold consisting of a left-handed β-helix coupled with a C-terminal α-helix [64]. This configuration aligns well with

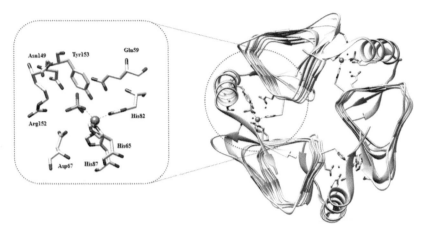

Fig. 9 Ribbon representation of the γ-TtCA trimer; metal ions are shown as spheres. The central Zn^{2+} ions are in gray and phosphate radical in orange. The zinc-coordinating histidines and key amino acid residues are also shown (PDB: 6IVE).

other known c-CAs, albeit with some notable adaptations which might be an evolutionary response to its thermophilic nature. In γ-TtCA, zinc is tetrahedrally coordinated by three histidines and a phosphate radical, a slight deviation from the typical water molecule found in other γ-CAs (Fig. 9). This modification could be indicative of structural adaptations enhancing enzyme stability and function at high temperatures.

When compared with other γ-CAs, γ-TtCA shares the common structural theme of the β-helix and α-helix configuration but differs significantly in the loop regions and terminal helices. These differences are particularly notable in the half-replacement of the C-terminal α-helix with a long loop, which is not observed in the other structures. This unique feature might influence the stability and possibly the substrate specificity or enzymatic activity of γ-TtCA under extreme conditions. Moreover, the γ-TtCA structure is unique not just for its thermostability but also for its lack of significant carbonic anhydrase activity, which is unusual for γ-CAs. The replacement of the C-terminal α-helix with a long loop might contribute to this lack of activity, suggesting a possible evolutionary trade-off between stability and catalytic function.

5.7 Burkholderia pseudomallei

The crystal structure of γ-BpsCA, a c-carbonic anhydrase from the pathogenic bacterium *B. pseudomallei*, offers critical insights into its potential as a novel antibacterial drug target. The structure reveals a seven-

A   B

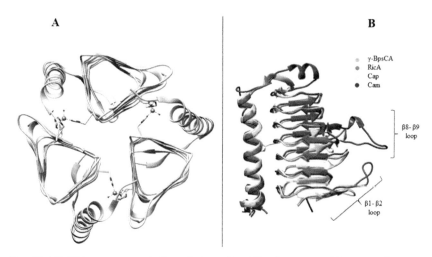

Fig. 10 (A) Ribbon representation of the γ-BpscCA trimer; metal ions are shown as spheres. The central Zn^{2+} ions are in gray. The zinc-coordinating histidines are also shown (PDB: 7ZW9); (B) Structural superposition of γ-BpscCA to related γ-CA family proteins showing loop region variability. γ-BpscCA is colored in light gray, Cam in magenta, RicA in blue, Cap in yellow.

turn left-handed parallel β-helix typical of γ-CAs, followed by an anti-parallel β-strand and a long α-helix, indicative of the enzyme functional adaptation. γ-BpsCA crystallizes in a trimeric form, with each monomer contributing to the overall stability and catalytic potential of the enzyme [65]. Central to the γ-BpsCA function is the zinc ion in the active site, coordinated by three histidine residues and a water molecule or hydroxide ion, which is vital for the enzyme catalytic action of converting CO_2 to bicarbonate. This configuration is highly conserved among γ-CAs and plays a crucial role in their enzymatic activity. The positioning of the zinc ion and its coordinating residues aligns well with what is observed in other γ-CAs, emphasizing the evolutionary conservation of these critical elements (Fig. 10A). Comparative analysis with other γ-CAs such as RicA from *B. abortus*, Cap from *P. horikoshii*, and Cam from *M. thermophila* highlights both similarities and unique adaptations. The overall fold comprising a β-helix flanked by α-helices is conserved, but variations exist particularly in loop regions which can influence substrate access and binding (Fig. 10B).

Indeed, γ-BpsCA shares a high degree of structural homology with Zn-Cap and RicA, particularly in the core elements of the catalytic machinery. Unique features to γ-BpsCA are specific loop conformations and the absence of an "acidic loop" which is present in Cam and some other γ-CAs [63]. These differences may affect the enzyme substrate specificity and resistance to inhibitors, which is critical for designing targeted antibacterial agents. Finally, the structure of γ-BpsCA reveals a potential for high thermal stability and robust enzymatic activity, which is characteristic of γ-CAs from thermophilic organisms. This feature is crucial for its function in the harsh environments that *B. pseudomallei* might encounter.

6. Inhibitors and activators of bacterial γ-CAs

6.1 Inhibition mechanism

The inhibition of γ-CAs, especially the Cam isoform, has been thoroughly explored, utilizing a diverse classes of inhibitors including inorganic anions, sulfonamides, and sulfamates. These compounds highlight the varied inhibitory landscape across different γ-CAs. For instance, the γ-CA from *P. gingivalis* is particularly sensitive to thiocyanate, cyanide, azide, and hydrogen sulfide [66]. Extensive studies with a broad spectrum of sulfonamides and sulfamates reveal varied inhibition profiles when compared to well-characterized human isoforms such as hCA I and II, although all enzymes displayed sensitivity to acetazolamide. However, γ-CAs generally show less susceptibility to sulfonamide inhibition than α-CAs, like hCA II, though some compounds demonstrated low nanomolar affinities [34–44]. Anion inhibition studies on zinc- and cobalt-substituted Cam (Zn-Cam and Co-Cam) indicated distinct inhibitory effects [53]. Zn-Cam was notably inhibited by hydrogen sulfite and cyanate with inhibition constants ranging from 50 to 90 mM, whereas fluoride, chloride, and sulfate had minimal effects. In contrast, Co-Cam was more susceptible to inhibition by carbonate, nitrate, and bicarbonate, with inhibition constants of 9, 90, and 100 mM, respectively. These differences can be attributed to the unique coordination geometries preferred by zinc and cobalt; zinc typically favors tetrahedral or trigonal bipyramidal configurations, while cobalt generally coordinates with five or six ligands. This explains why monodentate anions more effectively inhibit zinc enzymes, while bidentate anions have a stronger impact on cobalt enzymes.

6.2 Activation studies

Activation studies using typical CA activators such as amino acids and amines have been shown to increase the turnover rates of γ-CAs without altering K_m [67]. Although there are currently no crystallographic structures to detail their activation mechanism, it has been proposed that these activators function similarly to mammalian isoforms. For Cam, it is suggested that the activator replaces the proton shuttle residue in facilitating the transfer of protons from the metal-bound water molecule to the external buffer. Both Zn-Cam and Co-Cam exhibit moderate activation by several compounds, though Co-Cam generally shows better activation compared to Zn-Cam. Recent studies have explored the activation of γ-CAs from a range of bacteria, including pathogens like *V. cholerae* and *B. pseudomallei*, as well as extremophiles such as *P. haloplanktis* and *C. psychrerythraea*. These investigations utilized both natural and synthetic amino acids and amines, uncovering unique activation profiles for each species [68–70]. These enzymes were robustly activated by the compounds tested, with *L*-tryptophan and 4-amino-*L*-phenylalanine being the most effective activators for *B. pseudomallei* and *V. cholerae* respectively, showing activation constants in the low nanomolar range.

7. Conclusion

The investigation of γ-CAs from a variety of sources, including pathogens such as *B. pseudomallei*, *P. gingivalis*, and *V. cholerae*, alongside extremophiles, has significantly enhanced our comprehension of the structural and functional diversity within this enzyme family. Structural studies highlight that these enzymes typically consist of three identical subunits, each adopting a left-handed β-helix structure with core elements characteristic of CAs, including zinc-binding histidine residues. However, notable variations are primarily found in the loops connecting the strands and in the C-terminal regions. These structural differences not only allow the enzymes to operate effectively in varied environmental conditions but also influence their catalytic efficiency and sensitivity to inhibitors.

Phylogenetic studies indicate that most γ-CAs do not possess the acidic loop and the proton shuttle residue (Glu84), which are crucial for the carbonic anhydrase activity observed in the prototype isoform, Cam. Within the γ-class, residues such as Arg59, Asp61, and Asp76 are highly conserved, playing essential roles in trimer formation and maintaining the integrity of the active site, while residues like Gln75 and Asn73 show less conservation.

Interestingly, certain γ-CAs, like those from *P. horikoshii* and YrdA from *E. coli*, exhibit no CO_2 hydrase activity. For these proteins, functions other than traditional CO_2 hydration or activation through interactions with other molecules have been suggested.

The structural variances, particularly in loop regions and active sites, present opportunities to develop more selective CA inhibitors. These inhibitors could serve as novel anti-infective agents with potentially fewer side effects in therapeutic applications.

The continuous exploration of γ-CA structural and functional diversity not only deepens our fundamental scientific knowledge but also drives the creation of innovative solutions addressing key environmental and health challenges.

References

[1] C.T. Supuran, Carbonic anhydrases: novel therapeutic applications for inhibitors and activators, Nat. Rev. Drug. Discov. 7 (2008) 168–181.

[2] V. De Luca, A. Angeli, V. Mazzone, C. Adelfio, V. Carginale, A. Scaloni, et al., Heterologous expression and biochemical characterisation of the recombinant β-carbonic anhydrase (MpaCA) from the warm-blooded vertebrate pathogen malassezia pachydermatis, J. Enzyme Inhib. Med. Chem. 37 (2022) 62–68.

[3] A. Angeli, L.J. Urbański, V.P. Hytönen, S. Parkkila, C.T. Supuran, Activation of the β-carbonic anhydrase from the protozoan pathogen *Trichomonas vaginalis* with amines and amino acids, J. Enzyme Inhib. Med. Chem. 36 (2021) 758–763.

[4] A. Angeli, M. Pinteala, S.S. Maier, S. Del Prete, C. Capasso, B.C. Simionescu, et al., Inhibition of bacterial α-, β- and γ-class carbonic anhydrases with selenazoles incorporating benzenesulfonamide moieties, J. Enzyme Inhib. Med. Chem. 34 (2019) 244–249.

[5] A. Angeli, M. Pinteala, S.S. Maier, S. Del Prete, C. Capasso, B.C. Simionescu, et al., Inhibition of α-, β-, γ-, δ-, ζ- and η-class carbonic anhydrases from bacteria, fungi, algae, diatoms and protozoans with famotidine, J. Enzyme Inhib. Med. Chem. 34 (2019) 644–650.

[6] C.T. Supuran, Structure and function of carbonic anhydrases, Biochem. J. 473 (2016) 2023–2032.

[7] C.T. Supuran, C. Capasso, An overview of the bacterial carbonic anhydrases, Metabolites 7 (2017) 56.

[8] C.T. Supuran, C. Capasso, Biomedical applications of prokaryotic carbonic anhydrases, Expert. Opin. Ther. Pat. 28 (2018) 745–754.

[9] C. Capasso, C.T. Supuran, Bacterial, fungal and protozoan carbonic anhydrases as drug targets, Expert. Opin. Ther. Targets 19 (2015) 1689–1704.

[10] B.E. Alber, J.G. Ferry, Characterization of heterologously produced carbonic anhydrase from *Methanosarcina thermophila*, J. Bacteriol. 178 (1996) 3270–3274.

[11] B.E. Alber, C.M. Colangelo, J. Dong, C.M. Sta°lhandske, T.T. Baird, C. Tu, et al., Kinetic and spectroscopic characterization of the gamma-carbonic anhydrase from the methanoarchaeon *Methanosarcina thermophila*, Biochemistry 38 (1999) 13119–13128.

[12] D.N. Silverman, S. Lindskog, The catalytic mechanism of carbonic anhydrase: implications of a ratelimiting protolysis of water, Acc. Chem. Res. 21 (1988) 30–36.

[13] S. Lindskog, Structure and mechanism of carbonic anhydrases, Pharmacol. Ther. 74 (1997) 1–20.

[14] B.C. Tripp, J.G. Ferry, A structure-function study of a proton transport pathway in the gamma-class carbonic anhydrase from *Methanosarcina thermophila*, Biochemistry 39 (2000) 9232–9240.

[15] S. Zimmerman, J.F. Domsic, C. Tu, A.H. Robbins, R. McKenna, D.N. Silverman, et al., Role of Trp19 and Tyr200 in catalysis by the γ-class carbonic anhydrase from *Methanosarcina thermophila*, Arch. Biochem. Biophys. 529 (2013) 11–17.

[16] B.C. Tripp, C. Tu, J.G. Ferry, Role of arginine 59 in the γ-class carbonic anhydrases, Biochemistry 41 (2002) 669–678.

[17] S.A. Zimmerman, J.G. Ferry, Proposal for a hydrogen bond network in the active site of the prototypic γ-class carbonic anhydrase, Biochemistry 45 (2006) 5149–5157.

[18] B.C. Tripp, C.B. Bell 3rd, F. Cruz, C. Krebs, J.G. Ferry, A role for iron in an ancient carbonic anhydrase, J. Biol. Chem. 279 (2004) 6683–6687.

[19] S.R. MacAuley, S.A. Zimmerman, E.E. Apolinario, C. Evilia, Y.M. Hou, J.G. Ferry, et al., The archetype γ-class carbonic anhydrase (Cam) contains iron when synthesized in vivo, Biochemistry 48 (2009) 817–819.

[20] C.K. Tu, D.N. Silverman, C. Forsman, B.H. Jonsson, S. Lindskog, Role of histidine 64 in the catalytic mechanism of human carbonic anhydrase II studied with a site-specific mutant, Biochemistry 28 (1989) 7913–7918.

[21] D. Duda, C. Tu, M. Qian, P. Laipis, M. Agbandje-McKenna, D.N. Silverman, et al., Structural and kinetic analysis of the chemical rescue of the proton transfer function of carbonic anhydrase II, Biochemistry 40 (2001) 1741–1748.

[22] A. Liljas, K Ha kansson, BH Jonsson, Y Xue, Inhibition and catalysis of carbonic anhydrase. Recent crystallographic analyses, Eur. J. Biochem. 219 (1994) 1–10.

[23] C.M. Maupin, M.G. Saunders, I.F. Thorpe, R. McKenna, D.N. Silverman, G.A. Voth, Origins of enhanced proton transport in the Y7F mutant of human carbonic anhydrase II, J. Am. Chem. Soc. 130 (2008) 11399–11408.

[24] S.A. Zimmerman, J.-F. Tomb, J.G. Ferry, Characterization of CamH from *Methanosarcina thermophila*, founding member of a subclass of the g class of carbonic anhydrases, J. Bacteriol. 192 (2010) 1353–1360.

[25] J.G. Ferry, The gamma class of carbonic anhydrases, Biochim. Biophys. Acta 1804 (2010) 374–381.

[26] C. Capasso, C.T. Supuran, An overview of the alpha-, beta- and gamma-carbonic anhydrases from Bacteria: can bacterial carbonic anhydrases shed new light on evolution of bacteria? J. Enzyme Inhib. Med. Chem. 30 (2015) 325–332.

[27] A. Amedei, C. Capasso, G. Nannini, C.T. Supuran, Microbiota, bacterial carbonic anhydrases, and modulators of their activity: links to human diseases? Mediators Inflamm. 2021 (2021) 6926082.

[28] K. Ueda, H. Nishida, T. Beppu, Dispensabilities of carbonic anhydrase in proteobacteria, Int. J. Evol. Biol. 2012 (2012) 324549.

[29] G.M. Buzás, C.T. Supuran, The history and rationale of using carbonic anhydrase inhibitors in the treatment of peptic ulcers. In memoriam Ioan Puşcaş (1932-2015), J. Enzyme Inhib. Med. Chem. 31 (2016) 527–533.

[30] X. Fu, L.J. Yu, L. Mao-Teng, L. Wei, C. Wu, M. Yun-Feng, Evolution of structure in gamma-class carbonic anhydrase and structurally related proteins, Mol. Phylogenet Evol. 47 (2008) 211–220.

[31] C.T. Supuran, C. Capasso, Antibacterial carbonic anhydrase inhibitors: an update on the recent literature, Expert. Opin. Ther. Pat. 30 (2020) 963–982.

[32] S. Del Prete, D. Vullo, V. De Luca, V. Carginale, S.M. Osman, Z. AlOthman, et al., Comparison of the sulfonamide inhibition profiles of the α-, β- and γ-carbonic anhydrases from the pathogenic bacterium *Vibrio cholerae*, Bioorg. Med. Chem. Lett. 26 (2016) 1941–1946.

[33] A. Angeli, S. Del Prete, W.A. Donald, C. Capasso, C.T. Supuran, The γ-carbonic anhydrase from the pathogenic bacterium *Vibrio cholerae* is potently activated by amines and amino acids, Bioorg. Chem. 77 (2018) 1–5.

[34] S. Del Prete, D. Vullo, V. De Luca, V. Carginale, A. Scozzafava, C.T. Supuran, et al., A highly catalytically active γ-carbonic anhydrase from the pathogenic anaerobe *Porphyromonas gingivalis* and its inhibition profile with anions and small molecules, Bioorg. Med. Chem. Lett. 23 (2013) 4067–4071.

[35] C.T. Supuran, C. Capasso, Carbonic anhydrase from *Porphyromonas gingivalis* as a drug target, Pathogens 6 (2017) 30.

[36] S. Del Prete, D. Vullo, P. Di Fonzo, S.M. Osman, Z. AlOthman, W.A. Donald, et al., Sulfonamide inhibition profile of the γ-carbonic anhydrase identified in the genome of the pathogenic bacterium *Burkholderia pseudomallei* the etiological agent responsible of melioidosis, Bioorg. Med. Chem. Lett. 27 (3) (2017) 490–495.

[37] D. Vullo, S. Del Prete, S.M. Osman, Z. AlOthman, C. Capasso, W.A. Donald, et al., *Burkholderia pseudomallei* γ-carbonic anhydrase is strongly activated by amino acids and amines, Bioorg. Med. Chem. Lett. 27 (2017) 77–80.

[38] D. Vullo, S. Del Prete, S.M. Osman, F.A.S. Alasmary, Z. AlOthman, W.A. Donald, et al., Comparison of the amine/amino acid activation profiles of the β- and γ-carbonic anhydrases from the pathogenic bacterium *Burkholderia pseudomallei*, J. Enzyme Inhib. Med. Chem. 33 (2018) 25–30.

[39] D. Vullo, V. De Luca, S. Del Prete, V. Carginale, A. Scozzafava, S.M. Osman, et al., Sulfonamide inhibition studies of the γ-carbonic anhydrase from the Antarctic bacterium *Colwellia psychrerythraea*, Bioorg. Med. Chem. Lett. 26 (2016) 1253–1259.

[40] V. De Luca, D. Vullo, S. Del Prete, V. Carginale, S.M. Osman, Z. AlOthman, et al., Cloning, characterization and anion inhibition studies of a γ-carbonic anhydrase from the Antarctic bacterium *Colwellia psychrerythraea*, Bioorg. Med. Chem. 24 (2016) 835–840.

[41] D. Vullo, V. De Luca, S. Del Prete, V. Carginale, A. Scozzafava, C. Capasso, et al., Sulfonamide inhibition studies of the γ-carbonic anhydrase from the Antarctic bacterium *Pseudoalteromonas haloplanktis*, Bioorg. Med. Chem. Lett. 25 (2015) 3550–3555.

[42] A. Angeli, S. Del Prete, S.M. Osman, Z. AlOthman, W.A. Donald, C. Capasso, et al., Activation studies of the γ-carbonic anhydrases from the Antarctic Marine bacteria *Pseudoalteromonas haloplanktis* and *Colwellia psychrerythraea* with amino acids and amines, Mar. Drugs 17 (2019) 238.

[43] V. De Luca, S. Del Prete, V. Carginale, D. Vullo, C.T. Supuran, C. Capasso, Cloning, characterization and anion inhibition studies of a γ-carbonic anhydrase from the *Antarctic cyanobacterium* Nostoc commune, Bioorg. Med. Chem. Lett. 25 (2015) 4970–4975.

[44] V. De Luca, S. Del Prete, D. Vullo, V. Carginale, P. Di Fonzo, S.M. Osman, et al., Expression and characterization of a recombinant psychrophilic γ-carbonic anhydrase (NcoCA) identified in the genome of the Antarctic cyanobacteria belonging to the genus Nostoc, J. Enzyme Inhib. Med. Chem. 31 (2016) 810–817.

[45] C. de Araujo, D. Arefeen, Y. Tadesse, B.M. Long, G.D. Price, R.S. Rowlett, et al., Identification and characterization of a carboxysomal γ-carbonic anhydrase from the cyanobacterium Nostoc sp. PCC 7120, Photosynth. Res. 121 (2014) 135–150.

[46] S. Del Prete, S. Bua, C.T. Supuran, C. Capasso, *Escherichia coli* γ-carbonic anhydrase: characterisation and effects of simple aromatic/heterocyclic sulphonamide inhibitors, J. Enzyme Inhib. Med. Chem. 35 (2020) 1545–1554.

[47] V. De Luca, S. Giovannuzzi, C.T. Supuran, C. Capasso, May sulfonamide inhibitors of carbonic anhydrases from *Mammaliicoccus sciuri* prevent antimicrobial resistance due to gene transfer to other harmful staphylococci? Int. J. Mol. Sci. 23 (2022) 13827.

[48] C. Kisker, H. Schindelin, B.E. Alber, J.G. Ferry, D.C. Rees, A left-handed β-helix revealed by the crystal structure of a carbonic anhydrase from the archaeon *Methanosarcina thermophila*, EMBO J. 15 (1996) 2323–2330.

[49] T.M. Iverson, B.E. Alber, C. Kisker, J.G. Ferry, D.C. Rees, A closer look at the active site of g-class carbonic anhydrases: high-resolution crystallographic studies of the carbonic anhydrase from *Methanosarcina thermophila*, Biochemistry 39 (2000) 9222–9231.

[50] A.E. Eriksson, T.A. Jones, A. Liljas, Refined structure of human carbonic anhydrase II at 2.0 Å resolution, Proteins 4 (1988) 274–282.

[51] K. Hakansson, A. Wehnert, A. Liljas, X-ray analysis of metal-substituted human carbonic anhydrase II derivatives, Acta Crystallogr. D50 (1994) 93–100.

[52] K.A. Kogut, R.S. Rowlett, A Comparison of the mechanisms of CO_2 hydration by native and Co^{2+}- substituted carbonic anhydrase, J. Biol. Chem. 262 (1987) 16417–16424.

[53] A. Innocenti, S. Zimmerman, J.G. Ferry, A. Scozzafava, C.T. Supuran, Carbonic anhydrase inhibitors. Inhibition of the zinc and cobalt gamma-class enzyme from the archaeon *Methanosarcina thermophila* with anions, Bioorg. Med. Chem. Lett. 14 (2004) 3327–3331.

[54] J. Jeyakanthan, S. Rangarajan, P. Mridula, S.P. Kanaujia, Y. Shiro, S. Kuramitsu, et al., Observation of a calcium-binding site in the gamma-class carbonic anhydrase from *Pyrococcus horikoshii*, Acta 64 (2008) 1012–1019.

[55] H.M. Park, J.H. Park, J.W. Choi, J. Lee, B.Y. Kim, C.H. Jung, et al., Structures of the γ-class carbonic anhydrase homologue YrdA suggest a possible allosteric switch, Acta 68 (2012) 920–926.

[56] J.D. Cronk, R.S. Rowlett, K.Y.J. Zhang, C. Tu, J.A. Endrizzi, J. Lee, et al., Identification of a novel noncatalytic bicarbonate binding site in eubacterial beta-carbonic anhydrase, Biochemistry 45 (2006) 4351–4361.

[57] J.D. Cronk, J.A. Endrizzi, M.R. Cronk, J.W. O'neill, K.Y. Zhang, Crystal structure of *E. coli* beta-carbonic anhydrase, an enzyme with an unusual pH-dependent activity, Protein Sci. 10 (2001) 911–922.

[58] M. Ferraroni, S. Del Prete, D. Vullo, C. Capasso, C.T. Supuran, Crystal structure and kinetic studies of a tetrameric type II β-carbonic anhydrase from the pathogenic bacterium *Vibrio cholerae*, Acta Crystallogr. D71 (2015) 2449–2456.

[59] C. Merlin, M. Masters, S. McAteer, A. Coulson, Why is carbonic anhydrase essential to *Escherichia coli*? J. Bacteriol. 185 (2003) 6415–6424.

[60] J. Herrou, S. Crosson, Molecular structure of the *Brucella abortus* metalloprotein RicA, a Rab2-binding virulence effector, Biochemistry 52 (2013) 9020–9028.

[61] A.K. So, M. John-McKay, G.S. Espie, Characterization of a mutant lacking carboxysomal carbonic anhydrase from the cyanobacterium Synechocystis PCC6803, Planta 214 (2002) 456–467.

[62] H. Fukuzawa, E. Suzuki, Y. Komukai, S. Miyachi, A gene homologous to chloroplast carbonic anhydrase (icfA) is essential to photosynthetic carbon dioxide fixation by Synechococcus PCC7942, Proc. Natl Acad. Sci. U S A 89 (1992) 4437–4441.

[63] K.L. Peña, S.E. Castel, C. de Araujo, G.S. Espie, M.S. Kimber, Structural basis of the oxidative activation of the carboxysomal gamma-carbonic anhydrase, CcmM, Proc. Natl Acad. Sci. USA 107 (2010) 2155–2160.

[64] W. Wang, Y. Zhang, L. Wang, Q. Jing, X. Wang, X. Xi, et al., Molecular structure of thermostable and zinc-ion-binding γ-class carbonic anhydrases, Biometals 32 (2019) 317–328.

[65] A. Di Fiore, V. De Luca, E. Langella, A. Nocentini, M. Buonanno, S.M. Monti, et al., Biochemical, structural, and computational studies of a γ-carbonic anhydrase from the pathogenic bacterium *Burkholderia pseudomallei*, Comput. Struct. Biotechnol. J. 20 (2022) 4185–4194.

[66] S. Del Prete, D. Vullo, V. De Luca, V. Carginale, A. Scozzafava, C.T. Supuran, et al., A highly catalytically active γ-carbonic anhydrase from the pathogenic anaerobe *Porphyromonas gingivalis* and its inhibition profile with anions and small molecules, Bioorg. Med. Chem. Lett. 23 (2013) 4067–4071.

[67] A. Innocenti, S.A. Zimmerman, A. Scozzafava, J.G. Ferry, C.T. Supuran, Carbonic anhydrase activators: activation of the archaeal beta-class (Cab) and gamma-class (Cam) carbonic anhydrases with amino acids and amines, Bioorg. Med. Chem. Lett. 18 (2008) 6194–6198.

[68] A. Angeli, S. Del Prete, W.A. Donald, C. Capasso, C.T. Supuran, The γ-carbonic anhydrase from the pathogenic bacterium *Vibrio cholerae* is potently activated by amines and amino acids, Bioorg. Chem. 77 (2018) 1–5.

[69] A. Stefanucci, A. Angeli, M.P. Dimmito, G. Luisi, S. Del Prete, C. Capasso, et al., Activation of β- and γ-carbonic anhydrases from pathogenic bacteria with tripeptides, J. Enzyme Inhib. Med. Chem. 33 (2018) 945–950.

[70] A. Angeli, S. Del Prete, S.M. Osman, Z. AlOthman, W.A. Donald, C. Capasso, et al., Activation studies of the γ-carbonic anhydrases from the Antarctic Marine bacteria *Pseudoalteromonas haloplanktis* and *Colwellia psychrerythraea* with amino acids and amines, Mar. Drugs 17 (2019) 238.

CHAPTER FIVE

Bacterial ι-CAs

Clemente Capasso[a,*] and Claudiu T. Supuran[b]

[a]Department of Biology, Agriculture and Food Sciences, Institute of Biosciences and Bioresources, CNR, Napoli, Italy
[b]Neurofarba Department, Pharmaceutical and Nutraceutical Section, University of Florence, Sesto Fiorentino, Firenze, italy
*Corresponding author. e-mail address: clemente.capasso@ibbr.cnr.it

Contents

1. Introduction   122
2. The journey for identifying the bacterial ι-CA class   123
   2.1 Unraveling the discovery of iota-CAs   124
   2.2 CA activity and metal ion cofactor of the new class   128
   2.3 Biochemical properties of the ι-CAs from *Burkholderia territorii*   129
3. An intriguing aspect of the ι-CAs: they lack metal ion cofactors   132
4. Novel fusion CA gene   135
5. Conclusions   136
References   138

Abstract

Recent research has identified a novel class of carbonic anhydrases (CAs), designated ι-CA, predominantly found in marine diatoms, eukaryotic algae, cyanobacteria, bacteria, and archaea genomes. This class has garnered attention owing to its unique biochemical properties and evolutionary significance. Through bioinformatic analyses, LCIP63, a protein initially annotated with an unknown function, was identified as a potential ι-CA in the marine diatom *Thalassiosira pseudonana*. Subsequent biochemical characterization revealed that LCIP63 has CA activity and its preference for manganese ions over zinc, indicative of evolutionary adaptation to marine environments. Further exploration of bacterial ι-CAs, exemplified by *Burkholderia territorii* ι-CA (BteCAι), demonstrated catalytic efficiency and sensitivity to sulfonamide and inorganic anion inhibitors, the classical CA inhibitors (CAIs). The classification of ι-CAs into two variant types based on their sequences, distinguished by the COG4875 and COG4337 domains, marks a significant advancement in our understanding of these enzymes. Structural analyses of COG4337 ι-CAs from eukaryotic microalgae and cyanobacteria thereafter revealed a distinctive structural arrangement and a novel catalytic mechanism involving specific residues facilitating CO_2 hydration in the absence of metal ion cofactors, deviating from canonical CA behavior. These findings underscore the biochemical diversity within the ι-CA class and highlight its potential as a target for novel antimicrobial agents. Overall, the elucidation of ι-CA properties and mechanisms advances our knowledge of carbon metabolism in diverse organisms and underscores the complexity of CA evolution and function.

The Enzymes, Volume 55
ISSN 1874-6047, https://doi.org/10.1016/bs.enz.2024.05.003

1. Introduction

Carbonic anhydrases (CAs, EC 4.2.1.1) constitute a superfamily of enzymes that are renowned for their exceptional versatility and ubiquity across all domains of life. Their primary function is to expedite the reversible hydration of carbon dioxide (CO_2) into bicarbonate (HCO_3^-) and a proton (H^+), which is a crucial reaction fundamental to numerous metabolic pathways essential for growth, reproduction, and environmental adaptation [1–3]. The intrinsic rate of spontaneous CO_2 hydration, standing at 0.15 s^{-1}, falls short of meeting the rapid demand for CO_2 and HCO_3^-, which are vital for sustaining complex metabolic processes across organisms [4–7]. However, CA activity dramatically elevates this reaction rate, accelerating it from 100,000 to one million times per second (k_{cat} ranging between 10^4 and 10^6 s^{-1}), thus positioning CAs as one of nature's fastest biocatalysts [7–10]. Across various organisms, CAs play pivotal roles in a myriad of physiological processes, including respiration, photosynthesis, pH and CO_2 homeostasis, electrolyte secretion, bone resorption, and calcification. For instance, in mammals, they aid in the conversion of CO_2 to bicarbonate, facilitating its removal from the cells and subsequent exhalation [7]. In plants, CAs participate in the conversion of bicarbonate ions to CO_2, thereby optimizing the efficiency of enzymes such as RuBisCO, which is involved in photosynthesis [10–12]. Additionally, CAs are integral to processes such as coral calcification, where they facilitate the formation of calcium carbonate and contribute to the structural integrity of corals [13–17]. In bacteria, CAs serve as the primary pathway for the rapid acquisition and regulation of CO_2, carbonic acid (H_2CO_3), HCO_3^-, and carbonate (CO_3^{2-}). Gram-negative bacteria use periplasmic CAs to prevent CO_2 loss, highlighting the significance of these enzymes in bacterial metabolism [4–7,18–21]. The CA superfamily comprises several structurally distinct enzymes categorized into eight different classes or families (α, β, γ, δ, ζ, η, θ, and ι), indicating the diversity of enzymes involved in the reversible CO_2 hydratase reaction (see Chapter 1 of this book for more details) [1–3]. Despite their phylogenetic diversity, these classes exhibit convergent evolution, sharing similar catalytic functions and displaying distinct structural characteristics [12,22–24]. Notably, some CA classes demonstrate esterase activity, broadening their functional repertoire beyond that of CO_2 hydration [25,26]. Structurally, the CA classes exhibit diverse folding patterns and oligomeric states, with each class displaying unique features that are crucial for catalytic activity. Additionally, certain classes exhibit affinity for specific (and sometimes diverse) metal-ion cofactors, further diversifying their biochemical properties [12,22–24]. In fact, till the crystal structure of ι-CAs has been reported (see discussion later in the text) [1], it has been generally considered that all CAs need metal ions for their catalytic activity on CO_2/bicarbonate as substrates [6,7].

2. The journey for identifying the bacterial ι-CA class

Recently, considerable emphasis has been placed on the enzymes produced by bacteria that facilitate the reversible hydration of CO_2. This increased focus indicates a growing appreciation for the significance of bacterial CAs (BCAs) in a variety of biological processes [27–29]. With growing concerns over climate change and carbon emissions, there is increased interest in understanding the biological processes involved in CO_2 metabolism [30–32]. Bacteria play a crucial role in carbon cycling in various environments, including soils, oceans, and freshwater ecosystems [33,34]. BCAs play a significant role in regulating carbon fluxes within microbial communities and have a profound impact on ecosystem dynamics. They are vital for carbon fixation and recycling and are essential for controlling atmospheric CO_2 levels and mitigating the effects of climate change [33,34]. BCAs can be utilized as biocatalysts for CO_2 capture, conversion, and bioremediation of carbon-rich waste streams [35–38]. Engineering BCAs with optimized properties may enhance the efficiency and applicability of these processes [39]. CAs are implicated in bacterial virulence. Some pathogenic bacteria, including *Mycobacterium tuberculosis* [19] and *Helicobacter pylori* [40–42], encode CAs that are essential for their survival and adaptation to the host environment. Understanding the role of these enzymes in bacterial physiology and pathogenesis can inform the development of antimicrobial strategies and provide insights into the fundamental biological processes essential for the survival of organisms across different domains of life [7,19,43–47]. Finally, studying BCAs provides insight into the evolutionary history and diversity of these enzymes across different taxa. Comparative genomics and phylogenetic analyses have elucidated the evolutionary relationships among BCAs and their adaptations to diverse ecological niches [7,19,43–47]. Thus, BCAs reflect their importance in microbial physiology, environmental processes, pathogenesis, biotechnology, and evolutionary biology. Understanding the properties and functions of these enzymes will expand our knowledge of carbon metabolism in bacterial systems and their broader implications in various research fields and applications.

Up to date, only four CA classes have been identified in the bacterial genome (α, β, γ, δ and ι) [4–7,18–21]. Each family of BCAs is described in detail in Chapter 1, which offers a comprehensive understanding of the structural and functional diversity within the CA enzyme family. Here, we embark on a journey delving into the revelation of the ι-CA class, which stands as the latest identified CA family encoded within the bacterial genome. This opens a window for a fascinating world of microbial CAs. The unveiling of the ι-CA class signifies a pivotal juncture in the ongoing saga of CA discovery, representing the latest addition to the pantheon of bacterial CA families. Central to the success of unearthing the ι-CA class is the indispensable role played by bioinformatic methods [48–51]. These tools serve as proverbial lighthouses that guide researchers through the vast expanse of genetic data, enabling the discernment of patterns, motifs, and evolutionary relationships that elude traditional experimental approaches. The identification of this novel CA family not only expands our understanding of microbial enzymology, but also catalyzes further inquiry into the functional diversity and evolutionary dynamics of CAs across the bacterial kingdom.

2.1 Unraveling the discovery of iota-CAs

The journey into the ι-CA class began in 2017 [52], when Clement et al. using bioinformatic tools identified in the genome of marine diatom *Thalassiosira pseudonana* an unknown protein classified as Low-CO_2-inducible protein of 63 kDa, indicated with the acronym LCIP63 and belonging to the **COG4875** family protein (Fig. 1).

As shown in Figs. 1 and 2, LCIP63 as well as its homologs in other diatom species demonstrated substantial similarity to the conserved calcium-calmodulin protein kinase II association domains (CaMKII-AD) that have been identified across various species [53]. For instance, LCIP63 resulted in a multidomain protein, characterized by four repeated domains, each resembling CaMKII-AD, showing an endoplasmic reticulum signal peptide consisting of 22 amino acid residues and a chloroplast signal peptide consisting of 34 amino acid residues (Fig. 1). CaMKII-AD belongs to the NTF2-like protein superfamily, which shares a common fold initially identified in rat Nuclear Transport Factor 2 (NTF2). These proteins, which exhibit the common structural motif NTF2, are characterized by a hydrophobic pocket that may serve as a putative substrate-binding or catalytically active site. Interestingly, a conserved (H)HHSS motif at the C-terminus of each domain underscores the evolutionary conservation of LCIP63 across diverse taxa [52,53]. Moreover,

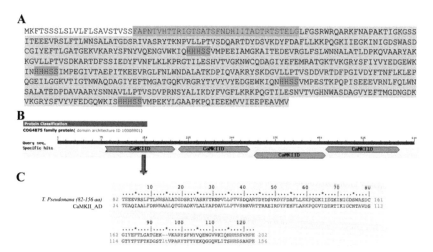

A

MKFTSSSLSLVLFLSAVSTVSSFAPNTVHTTRIGTSATSFNDHIITADTRTSTELGLFGSRWRQARKFNAPAKTIGKGSS
ITEEEVRSLFTLWNSALATGDSRIVASRYTKNPVLLPTVSDQARTDYDSVKDYFDAFLLKKPQGKIIEGKINIGDSWASD
CGIYEFTLGATGEKVKARYSFNYVQENGVWKIQHHHSSVMPEEIAMGKAITEDEVRGLFSLWNNALATLDPKQVAARYAK
KGVLLPTVSDKARTDFSSIEDYFVNFLKLKPRGTILESHVTVGKNWCQDAGIYEFEMRATGKTVKGRYSFIYVYEDGEWK
INNHHHSSIMPEGIVTAEPITKEEVRGLFNLWNDALATKDPIQVAKRYSKDGVLLPTVSDDVRTDFPGIVDYFTNFLKLEP
QGEILGGKVTIGTNWAQDAGIYEFTMGATGQKVRGRYTYVYVYEDGEWKIQNHHHSSVMPESTKPQPISEEEVRNLFQLWN
SALATEDPDAVAARYSNNAVLLPTVSDVPRNSYALIKDYFVGFLKRKPQGTILESNVTVGHNWASDAGVYEFTMGDNGDK
VKGRYSFVYVFEDGQWKISHHHSSVMPEKYLGAAPKPQIEEEMVVIEEPEAVMV

B

Protein Classification
COG4875 family protein(domain architecture ID 10008901)

Query seq.
Specific hits
CaMKIID   CaMKIID   CaMKIID   CaMKIID

C

```
                    10        20        30        40        50        60        70        80
            ....*....|....*....|....*....|....*....|....*....|....*....|....*....|....*....|
T. Pseudonana (82-156 aa)  82 TEEEVRSLFTLWNSALATGDSRIVASRYTKNPVLLPTVSDQARTDYDSVKDYFDAFLLKKPQGKIIEGKINIGDSWASDC 161
          CaMKII_AD        34 TEAQIAALFDRNNAALQTGDADKVLALYAPDAVLLPTVSNKVRTTRAEIRDYFEHFLAKKPQGVIDERTIKIGCNTAVDS 113

                    90       100       110       120
            ....*....|....*....|....*....|....*....|...
           162 GIYEFTLGATGEK---VKARYSFNYVQENGVWKIQHHHSSVMPE 202
           114 GTYTFTFTKDGSTltVPARYTFTYEKQGGQWLITSHHSSAMPE 156
```

Fig. 1 This figure provides a comprehensive visual representation delineating the key components pertinent to the LCIP63 amino acid sequence. It comprises three distinct parts: (A) depicting the amino acid sequence of LCIP63 with distinct colors for the signal peptide, chloroplast signal, and conserved motif HHHSS or HHS, highlighted in yellow, green, and light blue, respectively; (B) presenting a graphical synopsis elucidating conserved domains discerned within the queried sequence (LCIP63); and (C) delineating the alignment between the initial LCIP63 domain (residues 82–202) and CaMKII_AD, with notable emphasis on amino acid identity highlighted in red. Part A provides a detailed description of the amino acid sequence of LCIP63. Part B facilitates the identification of conserved motifs critical for the functional elucidation of LCIP63. Part C accentuates regions of amino acid identity between the initial LCIP63 domain and the CaMKII_AD (with a red arrow indicating this).

sequence comparisons between LCIP63 and known CA subclasses revealed a very low degree of identity, suggesting that LCIP63 represents a distinct CA subclass, denoted by the Greek letter iota (ι) [53].

Clement et al. demonstrated that LCIP63 is rapidly and substantially upregulated in cells exposed to low CO_2 concentrations [52–54]. Conversely, its expression is downregulated under conditions of phosphorus, nitrogen, silicon, or iron limitation, aligning with a diminished demand for carbon within the cellular milieu. This nuanced regulatory profile not only underscores LCIP63's responsiveness to fluctuating environmental conditions, but also hints at its potential significance in mediating carbon acquisition and utilization pathways. Furthermore, the primary sequence structure of LCIP63, characterized by the presence of an endoplasmic reticulum signal peptide and a chloroplast transit peptide, suggests its localization within the chloroplast compartment [52–54]. This spatial association with the chloroplast, coupled with its responsiveness to

Fig. 2 Multiple alignments of LCIP63 with the amino acid sequences of homologous enzymes from different diatom species. *LCIP63*, protein from *Thalassiosira pseudonana*; *Srobusta* , unknown protein from *Seminavis robusta*, GenBank: CAB9511325.1; *Bsp*, hypothetical protein BSKO_11674 from *Bryopsis* sp. KO-2023, GenBank: GMH43740.1; *Ctenuissimus*, hypothetical protein CTEN210_10685 from *Chaetoceros tenuissimus*, GenBank: GFH54209.1; *Toceanica*, hypothetical protein THAOC_16311 from *Thalassiosira oceanica*, GenBank: EJK63051.1; *Fsolaris*, hypothetical protein FisN_1Lh071 from *Fistulifera solaris*, GenBank: GAX21243.1; *Ninconspicua*, calcium/calmodulin dependent protein kinase II Association-domain containing protein from *Nitzschia inconspicua*, GenBank: KAG7370291.1.

CO_2 concentration fluctuations, posits LCIP63 as a potential participant in the Carbon Concentrating Mechanism (CCM) of *T. pseudonana*. Based on these observations, it has been hypothesized that LCIP63 may represent a hitherto undescribed CA entity, substantiated by its upregulation under low CO_2 concentrations, a hallmark feature shared with other canonical CAs. By modulating its expression in response to changes in CO_2 availability and nutrient limitations, LCIP63 likely contributes to the fine-tuning of carbon assimilation pathways, thereby optimizing cellular carbon acquisition efficiency under varying environmental conditions [52–54]. Furthermore, its subcellular localization within the chloroplast suggests direct involvement in the enzymatic processes governing CO_2 hydration and bicarbonate transport, thereby bolstering the resilience of *T. pseudonana* in nutrient-deficient and carbon-limiting environments [52–57].

Remarkably, LCIP63 was conspicuously absent in animals and land plants [52–57], and it was predominantly observed in several bacterial classes and cyanobacteria (Fig. 3). Prokaryotic homolog genes encode a single domain consisting of approximately 170 amino acids (see Chapter 1 of this book). However, despite its prevalence, the sequence of LCIP63 has not been traditionally recognized as a CA, resulting in its annotation in the

A

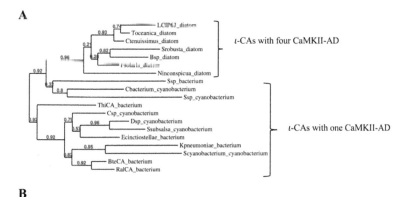

B

BteCA_bacterium
```
mkyvaagsvv iviactlfvp lpasaqnqsa aacvpttepq iaalfdrwnn slrtgdpdkv
vanyavdgvl lptvsnkprt npaairdyfv kflkgkpqgt vnkriikigc niaqdigiyt
fkfgdgktvq arytfvyewq ngkwliahhh ssampekeec de
```

Fig. 3 (Panel A) Phylogenetic analysis of ι-CAs across various species, including diatoms, cyanobacteria, and bacteria. The tree was constructed using the PhyML 3.0 program. Each branch point represents a common ancestor shared by organisms diverged from that point. Variations in branch length indicate the degree of genetic difference or evolutionary distance between organisms. The numbers associated with the branches typically represent bootstrap support values, providing insights into the robustness of the inferred relationships. (Panel B) ι-CA from *Burkholderia territorii* as an example of the amino acid sequence of a bacterial enzyme belonging to this new class, with the signal peptide and the HHHSS motif highlighted in yellow and light blue, respectively. The accession numbers of the i-CAs from the diatoms under consideration are listed in the legend of Fig. 2. Here are the identifiers for the other species used in the phylogenetic analysis: *Scyanobacterium, Synechococcaceae cyanobacterium*, GenBank: NJM10519.1; *Ssubsalsa, Spirulina subsalsa*, WP_017303260.1; *Dsp, Desertifilum* sp., GenBank: MCD8487683.1; *Ssp, Synechococcus* sp. HK05, GenBank: MBV2351208.1; *Csp, Cronbergia* sp. UHCC 0137, WP_323286419.1; *Cbacterium, Cyanobacteria bacterium*, GenBank: MCX5952458.1; *Csp, Cyanobium* sp. PLM2.Bin73, GenBank: TVS06472.1; *Kpneumoniae, Klebsiella pneumoniae*, WP_064115041.1; *Ecinctiostellae, Ephemeroptericola cinctiostellae*, WP_114562659.1; *Ssp, Saccharothrix* sp., WP_053720260.1; *ThiCA, Thiotrichales bacterium*, WP_ OYX05505.1; *BteCA, B. territorii*, WP_063553346.1; *RalCA, Ralstonia solanacearum*, WP_089190700.1.

NCBI databases either as having an unknown function or, in the case of prokaryote homologs, as members of the "SgcJ/EcaC oxidoreductase family."

The results depicted in Fig. 3 (Panel A) present a dendrogram illustrating the evolutionary relationships among ι-CAs from various organisms, including diatoms, cyanobacteria, and bacteria. Distinct clusters and branching patterns emerged, signifying genetic divergence among species. Diatoms form a discrete and cohesive cluster, indicating a shared evolutionary history characterized by conserved genetic traits and close phylogenetic relationships. All diatom ι-CAs considered were characterized by four CaMKII-AD motifs. In contrast, ι-CAs from cyanobacteria and bacteria exhibit a more intertwined arrangement, lacking clear delineations between their respective lineages. It is noteworthy that the cyanobacteria and bacteria considered here contained one CaMKII-AD motif (Fig. 3, Panel B). This mingling suggests a complex evolutionary landscape potentially influenced by frequent genetic exchange, horizontal gene transfer, or convergent evolutionary processes. Additionally, bootstrap support values associated with the branches provide insights into the robustness of inferred relationships, enhancing our understanding of microbial evolution and the intricate interplay between divergent and convergent evolutionary forces.

2.2 CA activity and metal ion cofactor of the new class

In 2019, Jensen et al. cloned, overexpressed, and purified LCIP63 [53]. Experimental exploration of LCIP63's biochemical properties sheds further light on its identity and unveils intriguing facets of its enzymatic functionality. Thus, despite having a primary sequence distinct from known CAs, LCIP63 exhibits CA activity, and the peculiar feature of LCIP63 lies in its preference for Mn^{2+} ions over Zn^{2+}, suggesting a remarkable evolutionary adaptation tailored to marine environments [53]. However, the presence of the metal ion in the protein has not been determined by atomic absorption (AA) or other techniques: simply the CO_2 hydrase activity of the protein was measured in the presence of various metal salts by Clement et al., who observed the highest activity in the presence of Mn(II), thus concluding that these enzymes contain manganese and not zinc in their active site [53]. This prompted the authors to designate a new CA subclass, ι-CA. This divergence in primary sequence but retention of functional similarity is not unprecedented in protein biology, as exemplified by proteins from diverse families sharing overlapping 3D structures, despite low sequence identity [58]. Similarly, within the realm of CAs, a growing number of subclasses with disparate sequences yet shared catalytic and esterase activities further underscore the intricate relationship between sequences and functions [59]. The metal dependency of LCIP63 CA activity was revealed through treatment with the metal ion-chelating agent EDTA, which led to a significant reduction in CA activity to approximately 64% [53]. The subsequent addition of zinc ions failed to restore the activity, contrary to expectations considering the role of zinc as a cofactor in many CA enzymes. Similarly, cadmium and cobalt, alternative cofactors for certain CA enzymes, failed to rescue LCIP63 activity, suggesting unique metal ion requirements for LCIP63 catalytic function. Intriguingly, CA activity was fully restored by the addition of Mn^{2+} ions, underscoring the pivotal role of Mn^{2+} as a cofactor for LCIP63. However, LCIP63's preference for manganese as a cofactor presents a notable deviation from the norm. This preference for manganese is not unprecedented in biological systems, as evidenced by the prototype γ-CA from *Methanosarcina thermophila*, which can substitute Fe^{2+} for Zn^{2+}, and the cambialistic ζ-CA (CDCA1) from *T. weissflogii*, which utilizes both Zn^{2+} and Cd^{2+} [53,60–62]. LCIP63's reliance on manganese for catalysis is further supported by the identification of potential manganese binding sites in its sequence, suggesting an evolutionary adaptation tailored to the marine environment's trace metal availability. In addition to its metal dependency, LCIP63 was sensitivity to sulfonamide CA inhibitors, with a substantial reduction in CA activity upon treatment with acetazolamide [59]. The concurrence of esterase activity with CA activity further bolsters LCIP63's identity as a functional CA, as esterase activity is considered a reliable indicator of CA functionality, at least for α-class enzymes [59].

2.3 Biochemical properties of the ι-CAs from *Burkholderia territorii*

Building upon the research conducted mainly by Clement and Jensen [53] on the functional properties, evolutionary relationships, and biochemical pathways associated with LCIP63 in *T. pseudonana*, Capasso and Supuran decided to investigate the biochemical properties of a bacterial ι-CAs to comprehend its enzymatic activity, inhibition, and structure [63,64]. Thus, in 2020, it was decided to heterologously express the bacterial ι-CA encoded by the genome of *Burkholderia territorii* that is a species of bacteria within the genus Burkholderia, belonging to the phylum Proteobacteria [65]. It has been found in various environmental samples, particularly in the soil and rhizosphere environments. The species name "territorii" refers to its initial isolation from soil samples in the Northern Territory of Australia [65]. Similar to other members of the Burkholderia genus, *B. territorii* has versatile metabolism and can degrade a wide range of organic compounds. It possesses metabolic pathways that enable it to utilize various carbon sources, contributing to its adaptability to different environments [65]. While the specific relationship between CAs and *B. territorii* may not have been extensively studied, understanding the presence and function of ι-CA in this bacterium could provide insights into its metabolic capabilities, environmental adaptation, and potential applications in agriculture or bioremediation. Capasso and Supuran demonstrated the catalytic efficiency of the *B. territorii* ι-CA (BteCAι acronym) using the stopped flow technique. Here, we want stress that Jensen et al. measured the enzymatic activity of the *T. pseudonana* ι-CA measuring the Wilbur-Anderson Units (WAU). Capasso and Supuran reported a k_{cat} of 3.0×10^5 s^{-1} and k_{cat}/K_M of 3.9×10^7 M^{-1} s^{-1} for the hydration of CO_2 to bicarbonate and protons [63,64]. Interestingly, the addition of Zn^{2+} or Ca^{2+} to the culture media during enzyme expression in *E. coli* yielded a catalytically active enzyme, whereas the addition of Mn^{2+} resulted in either diminished enzyme activity or the enzyme incorporating zinc, possibly because of traces of this ion present as impurities in the reagents used. The determination of zinc ions by atomic absorption per polypeptide chain revealed a 1:1 ratio [63,64]. These findings led to the proposition that the bacterial ι-CA is likely a Zn^{2+}-containing enzyme rather than Mn^{2+}-containing, as reported for diatom ι-CA from *T. pseudonana*.

Furthermore, Capasso and Supuran found that BteCAι was sensitive to inhibition by substituted benzene-sulfonamides and various clinically licensed sulfonamide-, sulfamate-, and sulfamide-type drugs, which are well-studied CA Inhibitors (CAIs) [66]. The inhibition profile of BteCAι reveals several key observations: firstly, several benzene-sulfonamides exhibited inhibition constants lower than 100 nM; secondly, its behavior differed from that of the other α, β, and γ-CAs; thirdly, the clinically utilized CAIs showed a micromolar affinity for inhibition [66]. This comprehensive characterization lays the groundwork for identifying compounds that can effectively and selectively inhibit bacterial ι-CA. Such selective inhibition, distinct from that of other protein isoforms present in the host, holds considerable promise for the development of novel antimicrobial agents. Additionally, the investigation of other classes of CAIs, such as inorganic anions and small molecules revealed that sulfamic acid, stannate, phenylarsonic acid, phenylboronic acid, and sulfamide were the most effective inhibitors, with inhibition constant values ranging from 6.2 to 94 mM [67]. Additionally, diethyldithiocarbamate, tellurate, selenate, bicarbonate, and cyanate exhibited sub-millimolar inhibitory activities, with K_I values ranging from 0.71 to 0.94 mM. Halides (excluding iodide), thiocyanate, nitrite, nitrate, carbonate, bisulphite, sulfate, hydrogensulfide, peroxydisulfate, selenocyanate, fluorosulfonate, and trithiocarbonate displayed K_I values in the range of 3.1–9.3 mM [67]. The differential inhibitory potencies observed highlight the importance of understanding the structural and chemical determinants underlying enzyme-inhibitor interactions. Such insights not only deepen our understanding of the catalytic mechanism of BteCAι but also lay the groundwork for the rational design of selective inhibitors with therapeutic potential against bacterial infections.

The CA superfamily exhibits an intriguing characteristic: the ability to bind molecules termed "activators" within the middle-exit region of the active site [68–71]. These activators, known as CAA (CA activators), encompass a range of compounds, including biogenic amines such as histamine, serotonin, and catecholamines, as well as amino acids, oligopeptides, and small proteins [69–71]. Through various analytical techniques including electronic spectroscopy, X-ray crystallography, and kinetic measurements, it has been shown that CAAs operate distinctively from traditional substrate binding [72]. Specifically, the CAAs did not directly influence the binding of CO_2 to the CA active site. Instead, they facilitate the rate-determining step of the catalytic process, expediting the proton transfer from the active site to the surrounding environment. This acceleration of proton transfer ultimately leads to an overall enhancement of the catalytic turnover of the enzyme [72,73]. Importantly, this augmentation was achieved without altering the substrate affinity of the enzyme, as reflected by consistent K_M values. Capasso and Supuran also reported the activation of bacterial BteCAι [68]. Their findings revealed that certain amino acids and amines, including proteinogenic derivatives, histamine, dopamine, and serotonin, exhibit efficient activating properties towards BteCAι, with activation constants in the range of 3.9–13.3 mM [68]. Additionally, L-Phe, L-Asn, L-Glu, and some pyridyl-alkylamines demonstrate a weaker activating effect on BteCAι, with K_A values ranging from 18.4 to 45.6 mM [68]. Comparison of these results with those obtained for other bacterial CA classes underscores a diverse regulatory mechanism within the CA superfamily.

3. An intriguing aspect of the ι-CAs: they lack metal ion cofactors

Recently, Hirakawa et al. identified novel CAs, designated as COG4337, within the genomes of the eukaryotic microalga *Bigelowiella natans* and cyanobacterium *Anabaena* sp. PCC7120 [74]. Classification of ι-CAs into two variant types based on their sequences represents a significant advancement in our understanding of these enzymes. One variant type was characterized by COG4875 domains, whereases the other variant comprised COG4337 domains (Fig. 4).

This classification highlights the diversity within the ι-CA family and suggests potential functional differences between the two variant types. In eukaryotic organisms, COG4337 homologs encode multidomain proteins comprising up to five domains, whereas prokaryotic homologs encode a single domain consisting of approximately 170 amino acid residues. The CAs from *B. natans* and *Anabaena* sp. PCC7120, denoted here as BnaCA and AspCA respectively, exhibited the HHSS consensus motif typical of ι-CAs (Fig. 4, COG4337 protein family) [74]. BnaCA and AspCA CO_2 hydration activities were assessed by determining the Wilburn–Anderson Units (WAU). Although the enzyme activity of BnaCA and AspCA was

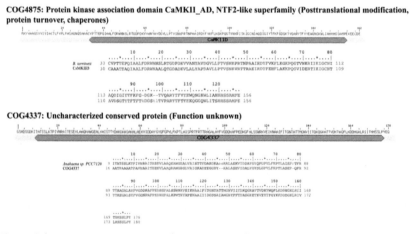

Fig. 4 Schematic representation of COG4875 and COG4337 ι-CAs, which provide a visual comparison of their primary structural features. The ι-CA from *B. territorii* represents COG4875, while the amino acid sequence of *Anabaena* sp. PCC7120 was used for COG4337. Amino acid identities are highlighted in red, aiding the identification of regions of similarity and divergence. Notably, both protein families exhibited a consensus sequence (H)HHSS.

found to be 7 and 37 times lower, respectively, than that of mammalian CAs, it fell within the same range as θ-CA from *Phaeodactylum tricornutum* and ι-CA from *T. pseudonana* [74]. Notably, a surprising discovery was made regarding the catalytic activity of both BnaCA and AspCA. These enzymes exhibited enzymatic activity even in the absence of a metal ion cofactor, which is unlike other reported ι-CAs and known CAs investigated thus far [74]. X-ray crystallographic structures were obtained for the COG4337 proteins in the presence of bicarbonate and the anion inhibitor iodide [74]. Here, as example we report the crystal structure of the ι-CA from cyanobacterium complexed with HCO_3 (Fig. 5).

The crystal structures of the newly identified ι-CAs reveal a distinctive homodimeric arrangement, characterized by each monomer forming a cone-shaped barrel structure comprised of three α-helices and a four-stranded antiparallel β-sheet. This folding pattern sets ι-CAs apart from other CA classes and underscores their unique structural organization. Proposed catalytic mechanisms for CO_2 hydration in ι-CAs involve specific residues within COG4337 proteins, facilitating hydration despite the absence of metal ions in their active sites. In contrast to typical CAs, where the initial step involves the deprotonation of active site water to form an OH^- ion, serving as a nucleophile for CO_2 hydration, ι-CAs exhibit a divergent mechanism (Fig. 6).

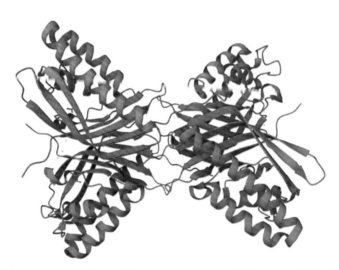

Fig. 5 Ribbon representation of the crystal structure (PDB entry, 7C5V) of the ι-CA from the cyanobacterium *Anabaena* sp. PCC7120 complexed with HCO_3^-.

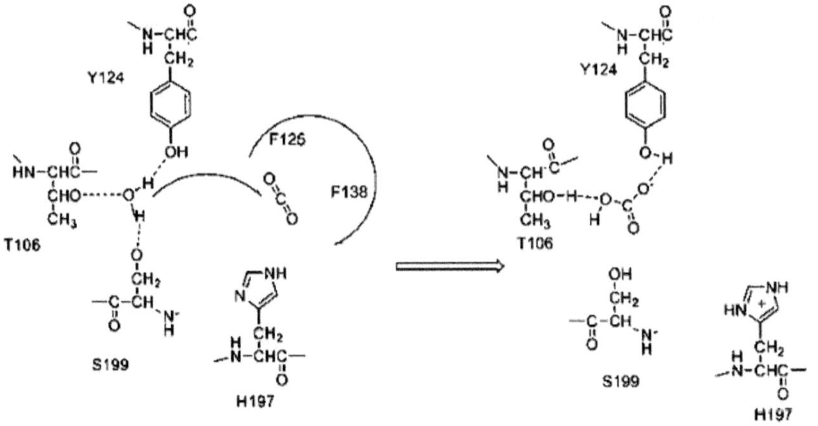

Fig. 6 Proposed catalytic mechanism for CO_2 hydration in ι-CAs. The catalytic cycle involves the activation of a water molecule for nucleophilic attack of the CO_2 molecule, which is bound in a hydrophobic pocket. This activation is facilitated by the formation of a hydrogen-bond network involving Thr106, Ser199, and Tyr124. Proton transfer is presumed to be achieved by a Tyr or His197.

Within the ι-CA structure, a cavity nestled within the cone-shaped barrel accommodates both the substrate, such as bicarbonate, and the inhibitor, for example, the iodide ion, which the authors used to determine the structure complexed with it. This cavity contains a hydrophilic domain, characterized by residues such as Thr, Ser, and His, juxtaposed with a hydrophobic domain featuring residues like Trp and Phe. The residues particularly around the bicarbonate/iodide region, exhibit a high degree of conservation within the COG4337 domains [74]. The deprotonation process within ι-CAs appears to be mediated by specific hydroxyl groups of Thr106 and Tyr124 within the cavity. These hydroxyl groups, positioned in proximity to both bicarbonate and iodide ions, likely facilitated deprotonation of active site water. Additionally, the involvement of the main chain nitrogen of Thr106 and the hydroxyl group of Ser199 in hydrogen-bonding interactions further supports this deprotonation mechanism. Interestingly, whereas histidine traditionally functions as a proton–shuttle residue in other carbonic anhydrases (CAs), in ι-CAs, it is likely that Tyr residues (124 or 176) assume this role instead of His197. This alteration in the proton-shuttle residue is probably due to their orientation towards the cavity and their close proximity to the protein surface, mirroring observations in other CA variants. Furthermore, the positioning of CO_2 within a hydrophobic pocket near Phe125/138 residues suggests a plausible CO_2 hydration site within the ι-CA structure. This arrangement aligns with previous observations of canonical CAs, emphasizing a conserved aspect in the catalytic process despite nuanced differences in the reaction mechanism. In summary, while retaining some similarities to canonical CAs, ι-CAs exhibit distinctive attributes that underscore their biochemical properties. It was also hypothesized that metal-free ι-CAs possibly lack the reversible dehydration reaction of bicarbonate to CO_2, a unique characteristic distinct from other metallo-CAs catalyzing both CO_2 hydration and bicarbonate dehydration reactions [64,74]. However, further experimental validation is necessary to confirm this mechanism and elucidate the exact details of CO_2 hydration by the COG4337 proteins.

4. Novel fusion CA gene

Recent studies have shown that both bacterial and eukaryotic COG4337-type ι-CAs exhibit remarkable characteristics, they are metal-independent CAs. This intriguing property was confirmed using X-ray crystallography and biochemical assays. Unlike many other known CAs that typically require metal ions as cofactors for their catalytic activity, COG4337-type ι-CAs can function effectively in the absence of metal ions. This finding challenges the traditional understanding of CA enzymology and opens new avenues for research on the catalytic mechanisms and evolutionary origins of these enzymes. Furthermore, COG4337-type ι-CA genes have been identified in a wide range of eukaryotic algae, including chlorarachniophytes, dinoflagellates, and haptophytes. This broad distribution across diverse algal groups suggests that COG4337-type ι-CAs play important roles in various biological processes in different algal species. Understanding the functional significance of these enzymes in algae could provide insights into their ecological and physiological adaptations, as well as their potential applications in biotechnology and environmental science. Overall, the identification and characterization of COG4337-type ι-CAs represent a significant advancement in the field of CA research. The recent discovery of a novel gene encoding both β-CA and ι-CA in the transcriptome data of the haptophyte *Isochrysis galbana* (strain CCMP1323) represents a significant finding in CA research [75]. This fusion gene, unique to *I. galbana* among the haptophytes, suggests a fascinating evolutionary event specific to this lineage. The presence of both the β-CA and ι-CA domains within the same gene implies a gene fusion event that likely occurred within the restricted haptophyte lineage. Recombinant protein assays have revealed intriguing functional characteristics of this fusion enzyme. While the C-terminal ι-CA region retains its enzymatic activity by catalyzing the hydration of CO_2, the N-terminal β-CA region no longer exhibits enzymatic activity. This difference in activity between the two regions suggests functional specialization following a gene fusion event. Notably, haptophytes typically possess mitochondrion-localized β-CAs and plastid-localized ι-CAs, reflecting the compartmentalization of CA activity within different cellular organelles. The fusion of β-CA and ι-CA domains in *I. galbana* represents an intermediate stage in the evolutionary transition from mitochondrial β-CA to plastid-localized ι-CA within this haptophyte species [75]. This discovery sheds light on the dynamic nature of the evolution and adaptation of CAs in haptophytes. This highlights the plasticity of genetic mechanisms underlying enzyme diversification and functional specialization in response to environmental and physiological factors [75]. Further investigation into the regulation, expression patterns, and biochemical properties of this fused CA in *I. galbana* will deepen our understanding of its role in carbon acquisition, photosynthesis, and overall cellular physiology. Additionally, comparative genomic and functional studies across haptophyte species may elucidate the evolutionary forces driving the emergence and retention of this fusion gene, and its implications for haptophyte biology.

5. Conclusions

The discovery of LCIP63 in 2019 represents a significant departure from the conventional understanding concerning the functionality of CA enzymes. This protein, characterized by an unconventional primary sequence, challenged established paradigms regarding CA structure and function. Despite its atypical composition, LCIP63 exhibits CA activity, a function historically associated with proteins possessing specific structural motifs and metal ions within their active sites. The peculiar feature of LCIP63 lies in its preference for manganese ions over zinc, suggesting a remarkable evolutionary adaptation tailored to marine environments, although the definitive presence of this metal ion in the protein has not been determined by using a strong technique, such as X-ray crystallography or atomic absorption spectroscopy. Moreover, the widespread distribution of LCIP63 across diverse organisms, encompassing marine diatoms, eukaryotic algae, cyanobacteria, bacteria, and archaea, implies its fundamental significance. Further studies are thus necessary to demonstrate or invalidate whether Mn(II) is indeed present in these enzymes.

Investigations of the biochemical properties of the ι-CAs from *B. territorii*, initiated by Capasso and Supuran, endeavor sought to comprehend the enzymatic activity, inhibition, and structure of the bacterial ι-CAs. Heterologous expression of the bacterial ι-CA encoded by the genome of *B. territorii* revealed catalytic efficiency, sensitivity to inhibition by various compounds, and activation by specific amino acids and amines. Notably, the incorporation of Zn^{2+} or Ca^{2+} during enzyme expression yielded active enzymes, with Zn^{2+} likely being the preferred metal ion, as determined by atomic absorption spectroscopy. However, by using that technique it was not possible to understand whether the metal ion was bound within the active site or has a structural, non catalytic role. These findings not only deepened our understanding of BteCAι biochemical properties but also lay the groundwork for future exploration and therapeutic developments targeting bacterial infections.

Recent investigations into ι-CAs have unveiled intriguing facets of these enzymes. Notably, the classification of ι-CAs into COG4875 and COG4337 variants represents a significant advancement, shedding light on the diversity within this enzyme family and suggesting potential functional disparities between the two types. The structural elucidation of COG4337 proteins from *B. natans* and *Anabaena* sp. PCC7120 has revealed that these ι-CAs displayed enzymatic activity even in the absence of metal ion cofactors, challenging our conventional understanding of CAs as metalloenzymes. Structural analyses have provided insights into the substrate and inhibitor binding pockets, highlighting conserved residues crucial for catalysis. Additionally, the proposed alteration in proton-shuttle residues and the hypothesized lack of reversible dehydration reaction in metal-free ι-CAs further underscore their distinctive biochemical properties. While these findings enhance our comprehension of ι-CAs, further experimental validation is imperative to confirm these hypotheses and elucidate the precise mechanisms underlying their catalytic activities.

The recent identification of a novel fusion gene encoding both β-CA and ι-CA in the haptophyte *I. galbana*, adds another layer of complexity to CA research. This unique gene fusion event suggests a fascinating evolutionary trajectory specific to this algal lineage, reflecting dynamic adaptations in response to environmental and physiological factors. Functional assays reveal intriguing characteristics of this fusion enzyme, shedding light on its role in carbon acquisition, photosynthesis, and cellular physiology. Comparative genomic studies across haptophyte species hold promise for elucidating the broader implications of this fusion gene and its evolutionary significance in the context of algal biology.

References

[1] Y. Hirakawa, M. Senda, K. Fukuda, H.Y. Yu, M. Ishida, M. Taira, et al., Characterization of a novel type of carbonic anhydrase that acts without metal cofactors, BMC Biol. 19 (1) (2021) 105–119.

[2] A. Aspatwar, M.E.E. Tolvanen, H. Barker, L. Syrjänen, S. Valanne, S. Purmonen, et al., Carbonic anhydrases in metazoan model organisms: molecules, mechanisms, and physiology, Physiol. Rev. 102 (3) (2022) 1327–1383.

[3] Y. Hirakawa, Y. Hanawa, K. Yoneda, I. Suzuki, Evolution of a chimeric mitochondrial carbonic anhydrase through gene fusion in a haptophyte alga, FEBS Lett. 596 (23) (2022) 3051–3059.

[4] C. Capasso, C.T. Supuran, An overview of the selectivity and efficiency of the bacterial carbonic anhydrase inhibitors, Curr. Med. Chem. 22 (18) (2015) 2130–2139.

[5] C. Capasso, C.T. Supuran, An overview of the alpha-, beta- and gamma-carbonic anhydrases from bacteria: can bacterial carbonic anhydrases shed new light on evolution of bacteria? J. Enzym. Inhib. Med. Ch 30 (2) (2015) 325–332.

[6] C.T. Supuran, C. Capasso, An overview of the bacterial carbonic anhydrases, Metabolites 7 (4) (2017) 56–74.

[7] C.T. Supuran, C. Capasso, Biomedical applications of prokaryotic carbonic anhydrases, Expert. Opin. Ther. Pat. 28 (10) (2018) 745–754.

[8] R.S. Kumar, J.G. Ferry, Prokaryotic carbonic anhydrases of Earth's environment, Subcell. Biochem. 75 (2014) 77–87.

[9] K.S. Smith, J.G. Ferry, Prokaryotic carbonic anhydrases, FEMS Microbiol. Rev. 24 (4) (2000) 335–366.

[10] N. Mangan, M. Brenner, Systems analysis of the CO_2 concentrating mechanism in cyanobacteria, Elife 3 (2014) 02043–02059.

[11] F.Y. Gong, G.X. Liu, X.Y. Zhai, J. Zhou, Z. Cai, Y. Li, Quantitative analysis of an engineered CO_2-fixing Escherichia coli reveals great potential of heterotrophic CO_2 fixation, Biotechnol. Biofuels 8 (2015) 86–96.

[12] Y.J. Wang, D.J. Stessman, M.H. Spalding, The CO_2 concentrating mechanism and photosynthetic carbon assimilation in limiting CO_2: how Chlamydomonas works against the gradient, Plant. J. 82 (3) (2015) 429–448.

[13] S. Del Prete, S. Bua, F.A.S. Alasmary, Z. AlOthman, S. Tambutte, D. Zoccola, et al., Comparison of the sulfonamide inhibition profiles of the alpha-carbonic anhydrase isoforms (SpiCA1, SpiCA2 and SpiCA3) encoded by the genome of the scleractinian coral Stylophora pistillata, Mar. Drugs 17 (3) (2019).

[14] S. Del Prete, S. Bua, D. Zoccola, F.A.S. Alasmary, Z. AlOthman, L.S. Alqahtani, et al., Comparison of the anion inhibition profiles of the alpha-CA isoforms (SpiCA1, SpiCA2 and SpiCA3) from the scleractinian coral Stylophora pistillata, Int. J. Mol. Sci. 19 (7) (2018).

[15] S. Del Prete, D. Vullo, N. Caminiti-Segonds, D. Zoccola, S. Tambutte, C.T. Supuran, et al., Protonography and anion inhibition profile of the alpha-carbonic anhydrase (CruCA4) identified in the Mediterranean red coral Corallium rubrum, Bioorg. Chem. 76 (2018) 281–287.

[16] S. Del Prete, D. Vullo, D. Zoccola, S. Tambutte, C. Capasso, C.T. Supuran, Kinetic properties and affinities for sulfonamide inhibitors of an alpha-carbonic anhydrase (CruCA4) involved in coral biomineralization in the Mediterranean red coral *Corallium rubrum*, Bioorg. Med. Chem. 25 (13) (2017) 3525–3530.

[17] S. Del Prete, D. Vullo, D. Zoccola, S. Tambutte, C.T. Supuran, C. Capasso, Activation profile analysis of CruCA4, an alpha-carbonic anhydrase involved in skeleton formation of the mediterranean red coral, *Corallium rubrum*, Molecules 23 (1) (2017).

[18] C. Capasso, Carbonic anhydrases: a superfamily of ubiquitous enzymes, Int. J. Mol. Sci. 24 (8) (2023) 7014–7017.

[19] C. Capasso, C.T. Supuran, Carbonic anhydrase and bacterial metabolism: a chance for antibacterial drug discovery, Expert. Opin. Ther. Pat. (2024) 1–10.

[20] V. De Luca, V. Carginale, C.T. Supuran, C. Capasso, The gram-negative bacterium *Escherichia coli* as a model for testing the effect of carbonic anhydrase inhibition on bacterial growth, J. Enzym. Inhib. Med. Ch 37 (1) (2022) 2092–2098.

[21] S. Del Prete, R. Merlo, A. Valenti, R. Mattossovich, M. Rossi, V. Carginale, et al., Thermostability enhancement of the alpha-carbonic anhydrase from *Sulfurihydrogenibium yellowstonense* by using the anchoring-and-self-labelling-protein-tag system (ASL(tag), J. Enzyme Inhib. Med. Chem. 34 (1) (2019) 946–954.

[22] E. Kupriyanova, N. Pronina, D. Los, Carbonic anhydrase - a universal enzyme of the carbon-based life, Photosynthetica 55 (1) (2017) 3–19.

[23] K.S. Smith, C. Jakubzick, T.S. Whittam, J.G. Ferry, Carbonic anhydrase is an ancient enzyme widespread in prokaryotes, Proc. Natl Acad. Sci. USA 96 (26) (1999) 15184–15189.

[24] C.S. Gai, J.N. Lu, C.J. Brigham, A.C. Bernardi, A.J. Sinskey, Insights into bacterial CO_2 metabolism revealed by the characterization of four carbonic anhydrases in *Ralstonia eutropha* H16, Amb. Express 4 (2014) 1–12.

[25] P. Piazzetta, T. Marino, N. Russo, Mechanistic explanation of the weak carbonic Anhydrase's Esterase Activity, Molecules 22 (6) (2017).

[26] F. Briganti, S. Mangani, A. Scozzafava, G. Vernaglione, C.T. Supuran, Carbonic anhydrase catalyzes cyanamide hydration to urea: is it mimicking the physiological reaction? J. Biol. Inorg. Chem. 4 (5) (1999) 528–536.

[27] C.T. Supuran, Bacterial carbonic anhydrases as drug targets: toward novel antibiotics? Front. Pharmacol. 2 (2011) 1–6.

[28] C.T. Supuran, How many carbonic anhydrase inhibition mechanisms exist? J. Enzym. Inhib. Med. Ch 31 (3) (2016) 345–360.

[29] C.T. Supuran, Carbonic anhydrase versatility: from pH regulation to CO(2) sensing and metabolism, Front. Mol. Biosci. 10 (2023) 1326633.

[30] S. Del Prete, R. Perfetto, M. Rossi, F.A.S. Alasmary, S.M. Osman, Z. AlOthman, et al., A one-step procedure for immobilising the thermostable carbonic anhydrase (SspCA) on the surface membrane of, J. Enzym. Inhib. Med. Ch 32 (1) (2017) 1120–1128.

[31] F. Migliardini, V. De Luca, V. Carginale, M. Rossi, P. Corbo, C.T. Supuran, et al., Biomimetic CO_2 capture using a highly thermostable bacterial α-carbonic anhydrase immobilized on a polyurethane foam, J. Enzym. Inhib. Med. Ch 29 (1) (2014) 146–150.

[32] D. Vullo, V. De Luca, A. Scozzafava, V. Carginale, M. Rossi, C.T. Supuran, et al., The first activation study of a bacterial carbonic anhydrase (CA). The thermostable α-CA from YO3AOP1 is highly activated by amino acids and amines, Bioorg. Med. Chem. Lett. 22 (20) (2012) 6324–6327.

[33] A.K. Sweetman, C.R. Smith, C.N. Shulse, B. Maillot, M. Lindh, M.J. Church, et al., Key role of bacteria in the short-term cycling of carbon at the abyssal seafloor in a low particulate organic carbon flux region of the eastern Pacific Ocean, Limnol. Oceanogr. 64 (2) (2019) 694–713.

[34] H. Dutta, A. Dutta, The microbial aspect of climate change, Energy Ecol. Env. 1 (4) (2016) 209–232.

[35] S. Talekar, B.H. Jo, J.S. Dordick, J. Kim, Carbonic anhydrase for CO capture, conversion and utilization. Curr. Opin. Biotech. 74 (2022) 230–240.

[36] S. Fabbricino, S. Del Prete, M.E. Russo, C. Capasso, A. Marzocchella, P. Salatino, In vivo immobilized carbonic anhydrase and its effect on the enhancement of CO(2) absorption rate. J. Biotechnol. 336 (2021) 41–49.

[37] M.E. Russo, C. Capasso, A. Marzocchella, P. Salatino, Immobilization of carbonic anhydrase for CO(2) capture and utilization, Appl. Microbiol. Biotechnol. 106 (9-10) (2022) 3419–3430.

[38] M.E. Russo, G. Olivieri, C. Capasso, V. De Luca, A. Marzocchella, P. Salatino, et al., Kinetic study of a novel thermo-stable alpha-carbonic anhydrase for biomimetic CO_2 capture, Enzyme Microb. Technol. 53 (4) (2013) 271–277.

[39 A. Di Fiore, V. Alterio, S.M. Monti, G. De Simone, K. D'Ambrosio, Thermostable carbonic anhydrases in biotechnological applications, Int. J. Mol. Sci. 16 (7) (2015) 15456–15480.

[40] S. Bury-Mone, G.L. Mendz, G.E. Ball, M. Thibonnier, K. Stingl, C. Ecobichon, et al., Roles of alpha and beta carbonic anhydrases of *Helicobacter pylori* in the urease-dependent response to acidity and in colonization of the murine gastric mucosa, Infect. Immun. 76 (2) (2008) 497–509.

[41] J.K. Modak, Y.C. Liu, C.T. Supuran, A. Roujeinikova, Structure-activity relationship for sulfonamide inhibition of *Helicobacter pylori* α-carbonic anhydrase, J. Med. Chem. 59 (24) (2016) 11098–11109.

[42] D.Scott, E.Marcus, G.Sachs, Inventors carbonic anhydrase inhibitors as drugs to eradicate *Helicobacter pylori* in the mammalian, including human stomach patent US20040204493A1, 2004.

[43] A. Amedei, C. Capasso, G. Nannini, C.T. Supuran, Microbiota, bacterial carbonic anhydrases, and modulators of their activity: links to human diseases? Mediators Inflamm. 2021 (2021) 6926082.

[44] C. Campestre, V. De Luca, S. Carradori, R. Grande, V. Carginale, A. Scaloni, et al., Carbonic anhydrases: new perspectives on protein functional role and inhibition in *Helicobacter pylori*, Front. Microbiol. 12 (2021) 629163.

[45] V. De Luca, V. Carginale, C.T. Supuran, C. Capasso, The gram-negative bacterium *Escherichia coli* as a model for testing the effect of carbonic anhydrase inhibition on bacterial growth, J. Enzyme Inhib. Med. Chem. 37 (1) (2022) 2092–2098.

[46] A. Nocentini, C. Capasso, C.T. Supuran, Carbonic anhydrase inhibitors as novel antibacterials in the era of antibiotic resistance: where are we now? Antibiotics (Basel) 12 (1) (2023).

[47] C.T. Supuran, C. Capasso, Antibacterial carbonic anhydrase inhibitors: an update on the recent literature, Expert. Opin. Ther. Pat. 30 (12) (2020) 963–982.

[48] P. Lio, Statistical bioinformatic methods in microbial genome analysis, Bioessays 25 (3) (2003) 266–273.

[49] S.F. Altschul, E.V. Koonin, Iterated profile searches with PSI-BLAST–a tool for discovery in protein databases, Trends Biochem. Sci. 23 (11) (1998) 444–447.

[50] S.F. Altschul, T.L. Madden, A.A. Schaffer, J. Zhang, Z. Zhang, W. Miller, et al., Gapped BLAST and PSI-BLAST: a new generation of protein database search programs, Nucl. Acids Res. 25 (17) (1997) 3389–3402.

[51] A.A. Schaffer, L. Aravind, T.L. Madden, S. Shavirin, J.L. Spouge, Y.I. Wolf, et al., Improving the accuracy of PSI-BLAST protein database searches with composition-based statistics and other refinements, Nucl. Acids Res. 29 (14) (2001) 2994–3005.

[52] R. Clement, S. Lignon, P. Mansuelle, E. Jensen, M. Pophillat, R. Lebrun, et al., Responses of the marine diatom *Thalassiosira pseudonana* to changes in CO(2) concentration: a proteomic approach, Sci. Rep. 7 (2017) 42333.

[53] E.L. Jensen, R. Clement, A. Kosta, S.C. Maberly, B. Gontero, A new widespread subclass of carbonic anhydrase in marine phytoplankton, Isme J. 13 (8) (2019) 2094–2106.

[54] R. Clement, E. Jensen, L. Prioretti, S.C. Maberly, B. Gontero, Diversity of CO_2-concentrating mechanisms and responses to CO_2 concentration in marine and freshwater diatoms, J. Exp. Bot. 68 (14) (2017) 3925–3935.

[55] R. Clement, L. Dimnet, S.C. Maberly, B. Gontero, The nature of the CO_2-concentrating mechanisms in a marine diatom, *Thalassiosira pseudonana*, N. Phytol. 209 (4) (2016) 1417–1427.

[56] E.L. Jensen, S.C. Maberly, B. Gontero, Insights on the functions and ecophysiological relevance of the diverse carbonic anhydrases in microalgae, Int. J. Mol. Sci. 21 (8) (2020).

[57] E.L. Jensen, V. Receveur-Brechot, M. Hachemane, L. Wils, P. Barbier, G. Parsiegla, et al., Structural contour map of the iota carbonic anhydrase from the diatom *Thalassiosira pseudonana* using a multiprong approach, Int. J. Mol. Sci. 22 (16) (2021).

[58] N. Naveenkumar, V.M. Prabantu, S. Vishwanath, R. Sowdhamini, N. Srinivasan, Structures of distantly related interacting protein homologs are less divergent than non-interacting homologs, FEBS Open. Bio. 12 (12) (2022) 2147–2153.

[59] N.R. Uda, V. Seibert, F. Stenner-Liewen, P. Müller, P. Herzig, G. Gondi, et al., Esterase activity of carbonic anhydrases serves as surrogate for selecting antibodies blocking hydratase activity, J. Enzym. Inhib. Med. Ch 30 (6) (2015) 955–960.

[60] J.G. Ferry, The gamma class of carbonic anhydrases, Biochim. Biophys. Acta 1804 (2) (2010) 374–381.

[61] V. Alterio, E. Langella, G. De Simone, S.M. Monti, Cadmium-containing carbonic anhydrase CDCA1 in marine diatom *Thalassiosira weissflogii*, Mar. Drugs 13 (4) (2015) 1688–1697.

[62] V. Alterio, E. Langella, F. Viparelli, D. Vullo, G. Ascione, N.A. Dathan, et al., Structural and inhibition insights into carbonic anhydrase CDCA1 from the marine diatom *Thalassiosira weissflogii*, Biochimie 94 (5) (2012) 1232–1241.

[63] S. Del Prete, A. Nocentini, C.T. Supuran, C. Capasso, Bacterial iota-carbonic anhydrase: a new active class of carbonic anhydrase identified in the genome of the Gram-negative bacterium *Burkholderia territorii*, J. Enzyme Inhib. Med. Chem. 35 (1) (2020) 1060–1068.

[64] A. Nocentini, C.T. Supuran, C. Capasso, An overview on the recently discovered iota-carbonic anhydrases, J. Enzyme Inhib. Med. Chem. 36 (1) (2021) 1988–1995.

[65] B. De Smet, M. Mayo, C. Peeters, J.E.A. Zlosnik, T. Spilker, T.J. Hird, et al., sp nov and sp nov., two novel complex species from environmental and human sources, Int. J. Syst. Evol. Micr. 65 (2015) 2265–2271.

[66] V. De Luca, A. Petreni, A. Nocentini, A. Scaloni, C.T. Supuran, C. Capasso, Effect of sulfonamides and their structurally related derivatives on the activity of iota-carbonic anhydrase from *Burkholderia territorii*, Int. J. Mol. Sci. 22 (2) (2021).

[67] A. Petreni, V. De Luca, A. Scaloni, A. Nocentini, C. Capasso, C.T. Supuran, Anion inhibition studies of the Zn(II)-bound iota-carbonic anhydrase from the Gram-negative bacterium *Burkholderia territorii*, J. Enzyme Inhib. Med. Chem. 36 (1) (2021) 372–376.

[68] V. De Luca, A. Petreni, V. Carginale, A. Scaloni, C.T. Supuran, C. Capasso, Effect of amino acids and amines on the activity of the recombinant iota-carbonic anhydrase from the Gram-negative bacterium *Burkholderia territorii*, J. Enzyme Inhib. Med. Chem. 36 (1) (2021) 1000–1006.

[69] A. Angeli, E. Berrino, S. Carradori, C.T. Supuran, M. Cirri, F. Carta, et al., Amine-and amino acid-based compounds as carbonic anhydrase activators, Molecules 26 (23) (2021).
[70] C.T. Supuran, Carbonic anhydrase activators, Fut. Med. Chem. 10 (5) (2018) 561–573.
[71] C.T. Supuran, Drug interactions of carbonic anhydrase inhibitors and activators, Exp. Opin. Drug. Metab. Toxicol. 20 (3) (2024) 143–155.
[72] E. Licsandru, M. Tanc, I. Kocsis, M. Barboiu, C.T. Supuran, A class of carbonic anhydrase I - selective activators, J. Enzyme Inhib. Med. Chem. 32 (1) (2017) 37–46.
[73] C. Temperini, A. Innocenti, A. Scozzafava, C.T. Supuran, Carbonic anhydrase activators: kinetic and X-ray crystallographic study for the interaction of D- and L-tryptophan with the mammalian isoforms I-XIV, Bioorg. Med. Chem. 16 (18) (2008) 8373–8378.
[74] Y. Hirakawa, M. Senda, K. Fukuda, H.Y. Yu, M. Ishida, M. Taira, et al., Characterization of a novel type of carbonic anhydrase that acts without metal cofactors, BMC Biol. 19 (1) (2021) 105.
[75] Y. Hirakawa, Y. Hanawa, K. Yoneda, I. Suzuki, Evolution of a chimeric mitochondrial carbonic anhydrase through gene fusion in a haptophyte alga, FEBS Lett. 596 (23) (2022) 3051–3059.

CHAPTER SIX

Sulfonamide inhibitors of bacterial carbonic anhydrases

Alessio Nocentini*

Sezione di Scienze Farmaceutiche, NEUROFARBA Department, University of Florence, Sesto Fiorentino, Florence, Italy

*Corresponding author. e-mail address: alessio.nocentini@unifi.it

Contents

1. Introduction   143
2. Mechanism of action of sulfonamide inhibitors of bacterial CA isoforms   145
3. Inhibition of bacterial CA isoforms with sulfonamide compounds   149
4. Conclusions   184
References   185

Abstract

The increasing prevalence of antibiotic-resistant bacteria necessitates the exploration of novel therapeutic targets. Bacterial carbonic anhydrases (CAs) have been known for decades, but only in the past ten years they have garnered significant interest as drug targets to develop antibiotics having a diverse mechanism of action compared to the clinically used drugs. Significant progress has been made in the field in the past three years, with the validation *in vivo* of CAs from *Neisseria gonorrhoeae*, and vancomycin-resistant enterococci as antibiotic targets. This chapter compiles the state-of-the-art research on sulfonamide derivatives described as inhibitors of all known bacterial CAs. A section delves into the mechanisms of action of sulfonamide compounds with the CA classes identified in pathogenic bacteria, specifically α, β, and γ classes. Therefore, the inhibitory profiling of the bacterial CAs with classical and clinically used sulfonamide compounds is reported and analyzed. Another section covers various other series of sulfonamide CA inhibitors studied for the development of new antibiotics. By synthesizing current research findings, this chapter highlights the potential of sulfonamide inhibitors as a novel class of antibacterial agents and paves the way for future drug design strategies.

1. Introduction

Carbonic anhydrases (CAs) are a superfamily of ubiquitous metalloenzymes that catalyze the reversible hydration of carbon dioxide into bicarbonate and a proton. This reaction plays a crucial role in the

The Enzymes, Volume 55
ISSN 1874-6047, https://doi.org/10.1016/bs.enz.2024.06.006

143

physiological processes of organisms all over the phylogenetic tree [1,2]. Evolution interestingly converged to eight distinct CA families, α-, β-, γ-, δ-, ζ-, η, θ- and ι-CAs, encoded by as many gene families [3,4]. The active site of most CAs contains a zinc ion (Zn^{2+}), which is essential for catalysis. The fifteen CA isoforms present in human all belong to the α-class and are involved in many physiological and pathological processes, leading to their validation as important therapeutic targets to treat a range of disorders [5,6]. hCA inhibitors belonging to the sulfonamide/sulfamate type have been long used as diuretics, systemic anti-convulsants, topically acting anti-glaucoma agents or for the treatment of altitude sickness, and more recently studied as anti-obesity, anti-inflammatory, anti-neuropathic pain and anti-tumor agents/diagnostic tools [6–8]. Additionally, CAs belonging to various families were identified in many microorganisms behaving as human pathogens [8,9]. These isozymes were shown to be crucial for the virulence, growth, or acclimatization of the parasite, making the use of CAIs highly promising to fight human infections caused by protozoa, fungi, and bacteria [9,10]. Specifically for the topic of this chapter, α-, β-, γ- and ι-class CAs have been identified to date in bacteria. The inhibition of bacterial CAs presents a promising strategy for the development of novel antimicrobial agents [11–14]. As for hCAs, among the various classes of bacterial CA inhibitors, sulfonamides have garnered significant attention due to their potency and specificity. Primary sulfonamides ($R\text{-}SO_2NH_2$, where R can be an aromatic, heterocyclic, aliphatic, or sugar scaffold), as well as the structurally related isosters sulfamides ($R\text{-}NH\text{-}SO_2NH_2$) and sulfamates ($R\text{-}O\text{-}SO_2NH_2$), function by binding to the zinc ion in the active site of CAs, thereby blocking their catalytic activity [15,16]. This inhibition can disrupt essential physiological processes in bacteria, leading to potential therapeutic applications in treating bacterial infections and mitigating bacterial resistance [13]. The historical context of sulfonamide inhibitors dates back to the discovery of the antibacterial properties of sulfanilamide (Fig. 1) in the early 20th century. Since then, sulfonamides have evolved from simple antibacterial agents to sophisticated inhibitors designed through structure-based drug design and high-throughput screening techniques. These advancements have allowed for the development of inhibitors with enhanced selectivity and potency against bacterial CAs [13]. This chapter delves into the mechanisms of action and development of sulfonamide inhibitors, and their potential applications in the treatment of bacterial infections.

2. Mechanism of action of sulfonamide inhibitors of bacterial CA isoforms

The CA catalytic reaction follows a two-step mechanism (Fig. 1), with the enzyme featuring a metal-bound hydroxide ion serving as the catalytically active species [5]. Initially, in the first step of the reaction, the zinc-bound hydroxide ion acts as a strong nucleophile towards the CO_2 molecule, which is bound in a hydrophobic pocket nearby. This interaction results in the formation of HCO_3^-, which is subsequently displaced from the active site by a water molecule. The second step, which is the rate-determining step, involves the regeneration of the metal hydroxide species through a proton transfer reaction from the Zn^{2+}-bound water to either an exogenous proton acceptor or to an active site residue, represented by B in Fig. 1. Inhibitors bearing a primary sulfonamide group, which becomes deprotonated within the CA active site ($-SO_2NH^-$) through an acid-base reaction with the metal-bound hydroxide ion, bind to the catalytic zinc ion upon the displacement of the water molecule and adopting a tetrahedral geometry (Fig. 1). Additional stabilizations to the adduct are provided by interactions specifically occurring within the different bacterial isozyme classes [16].

Few members of the bacterial α-CA family have been structurally characterized to date [17–20]. Analysis of the available structures reveals that these enzymes retain the typical fold of human α-CAs, featuring a

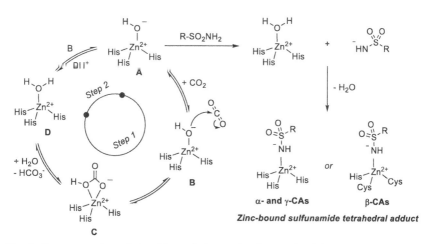

Fig. 1 Representation of the CA catalytic mechanism and inhibition mechanism by sulfonamide derivatives.

central twisted β-sheet. The active site is also located in a conical cavity that extends from the protein surface to the molecule's center. Notably, all bacterial α-isozymes exhibit a more compact structure compared to their human counterparts due to the deletion of three protein segments, one of which is located on the edge of the active site, resulting in a wider entrance to the active site cavity. A few inhibitors, most of which are sulfonamides, were cocrystallized with α-CA isoforms. The crystal structure of the α-CA from *Neisseria gonorrhoeae* described by Huang et al. in 1998 represents the first structural resolution ever of a bacterial α-CA as well as the first evidence of an adduct with a sulfonamide inhibitor, that is acetazolamide (AAZ, Fig. 2) [21]. However, neither structure of the α-NgCA deposited in the PDB (ID 1KOP or 1KOQ) by Huang et al. has the ligand AAZ bound to the enzyme, as it was instead claimed. In 2022, Flaherty and collaborators filled this gap reporting the crystallographic resolution of NgCA in complex with AAZ and a similar derivative [22]. The overall architectures of α-NgCA and hCA II are similar (Fig. 2). The interaction of the sulfonamide group with the zinc ion and with Thr177 (199 in hCA II) is practically identical to that observed in hCA II, and includes two hydrogen bonds. Another similarity is the hydrogen bonds between the ring nitrogen atoms of the inhibitor and the OH group of Thr178 (200 in hCA II), whilst a difference is a weaker H-bond (due to greater distance) between the acetamido N atom and Gln90 (92 in hCA II) in α-NgCA. Moreover, in the complex with hCA II the acetyl group of the acetamido side-chain forms a van der Waals contact with Phe131. This interaction is not observed in the complex with α-NgCA, because of the segment absence in the bacterial enzyme (Fig. 2).

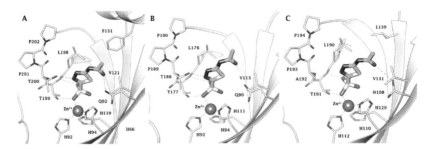

Fig. 2 Active site view of α-CA isoforms in adduct with AAZ: (A) hCA II (PDB 3HS4), (B) NgαCA (PDB 8DYQ), and (C) HpαCA (PDB 4YGF). Amino acids are labeled with one-letter codes. The zinc ion is represented as a gray sphere. H-bonds are shown as black dashed lines.

More lately, Modakh et al. reported the crystal structure of the α-CA from *Helicobacter pylori* complexed with sulfonamide inhibitors such as AAZ (Fig. 2C), methazolamide (MZA), benzolamide (BZA) and ethoxzolamide (EZA) [23,24]. Most of the secondary structure elements present in hCA II are retained in HpαCA, but several amino acid mutations occur between the two isoforms. While the interaction of the sulfonamide group with the zinc ion and with Thr191 (199 in hCA II) is practically identical to that described above for αNgCA, the Thr/Ala192 mutation annuls the H–bond interaction involving the thiadiazole ring (Fig. 2). The latter thus stacks against the hydrophobic side of the cone-shaped surface leading into the active site and makes van der Waals contacts with Val131, Leu190, and Ala192. The binding mode is further stabilized through a hydrogen bond between the acetamide group O atom and the side chain of Asn108.

As well, a few members of the bacterial β-CA family have been structurally characterized to date [25–28]. The active site of β-CAs exists at the interface between the two identic monomers composing the catalytically active unit of the enzyme and contains hydrophobic and hydrophilic areas similar to α-CAs. The monomers consist primarily of α-helices that surround an ordered β-sheet and show α/β fold unique to these enzymes. The residues of the catalytic triad are highly conserved among β-CA isoforms and consist of one His and two Cys residues. Crystal structures have revealed two different zinc environments in β-CA, denoted type-I and type-II, that can switch in a pH-dependent manner. The type-I active site is locked since the carboxyl group of an aspartic acid coordinates the zinc ion as the fourth ligand. For pH values higher than 8.3, the enzyme active site was instead found in its open and active form, due to the salt bridge that the aspartate residue forms with an arginine residue conserved in all the β-CA [25]. A molecule of water/hydroxide ion completes the tetrahedral geometry coordination pattern of the metal ion. The only inhibitor cocrystallized with a bacterial such enzyme is the sulfamide anion bound to β-CA psCA3 from *Pseudomonas aeruginosa* [29]. Sulfamide binds to the active site zinc ion by the deprotonated NH^- moiety with one oxygen forming a hydrogen bond with the main chain nitrogen of Gly103, and the other oxygen forming a hydrogen bond with Gln33 within the dimer interface. In contrast, the only experimental structural information of a β-CA in adduct with a sulfonamide inhibitor comes from crystallographic studies with the CA from the green alga *Coccomyxa*, whose structure was solved in complex with AAZ (Fig. 3) [28]. The deprotonated sulfonamide group binds to the zinc ion in a tetrahedral geometry. Additionally, the

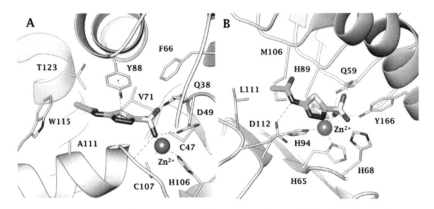

Fig. 3 Active site view of: (A) the crystallographic adduct of the β-CA from the green alga *Coccomyxa* with AAZ (PDB 3UCJ); (B) predicted binding mode of AAZ to the γ-CA from *Vibrio cholerae* built by homology. Amino acids are labeled with one-letter codes. The zinc ion is represented as a gray sphere. H-bonds are shown as black dashed lines.

sulfonamide NH moiety forms a hydrogen bond with the side chain of Asp49, while one oxygen atom accepts a hydrogen bond from the side chain NH₂ of Gln389. The other oxygen atom is in van der Waals contact with Val71 and Ala108. The orientation of the inhibitor is further constrained by its thiadiazole ring, which is positioned in the channel and stacked by the side chain of Tyr88. The acetamido group is embedded in a predominantly hydrophobic region of the active site cleft.

Few members of the γ-CA class have been structurally characterized to date, with CAM, the CA from the archeon *Methanosarcina thermophila*, being the prototype [30,31]. The active enzyme is a homotrimer resulting from the packing of three identical subunits. The active site is localized at the interfaces between two subunits. In the active site, a zin ion is coordinated by two His residues, which extend from equivalent positions of adjacent turn of one monomer, and by another His residue located in a neighboring monomer. Two water molecules complete the zinc coordination sphere forming a distorted trigonal bipyramidal geometry. The crystal structure of YrdA from *Escherichia coli*, solved in two different crystal forms, is the only available crystal structure for a bacterial γ-CA [32]. An active site form was also detected in the crystal structure, where a fourth His residue coordinates the zinc ion in the active site, resulting in almost tetrahedral coordination, which was denominated as the closed conformation. No inhibitor was cocrystallized with a γ-CA to date. Nonetheless, in 2021 the binding mode of aromatic sulfonamides and clinically licensed drugs was studied *in silico* with the three CA isoforms from the human pathogen *Vibrio cholerae*, that are α-VchCA, β-VchCA, and γ-VchCA [33]. Specifically, the study built by homology the structure of β-VchCA and γ-VchCA. The sulfonamide moiety in the deprotonated form was predicted to coordinate the zinc ion at the bottom of the cylindrical active site by tetrahedral geometry. The sulfonamide NH^- and $S{=}O$ groups engage H-bonds with Q59 and H68 side chains, with the NH^- group also acting as an H-bond acceptor by the Y159 side chain OH moiety. The ligand binding is stabilized by VdW contacts of the thiadiazole ring with His94 and Met106 and by an H-bond occurring between the acetamido NH group and Asp112.

3. Inhibition of bacterial CA isoforms with sulfonamide compounds

The following paragraph collects all sulfonamide inhibitors of bacterial CAs reported to date. The K_I values reported in Tables 1–13 are the mean from three different assays by a Stopped Flow technique with errors in the range of 5–10%.

The panel of compounds depicted in Figs. 4 and 5 has been utilized as a standard cohort of sulfonamide derivatives for the inhibitory characterization of a multitude of CA isoforms, among which a lot of bacterial isozymes [34–59]. Among these compounds, several simple aromatic and heteroaromatic sulfonamides **1–34** (Fig. 4) and the clinically licensed agents AAZ-EPA (Fig. 5). The latter include acetazolamide (AAZ), methazolamide (MZA), ethoxzolamide (EZA), and dichlorophenamide (DCP) are classical, systemically working antiglaucoma CAIs. Dorzolamide (DZA) and brinzolamide (BRZ) are topically-acting antiglaucoma agents. Benzolamide (BZA) is an orphan drug belonging to this class of pharmacological agents. Zonisamide (ZNS), sulthiame (SLT), and sulfamic acid ester topiramate (TPM) are widely used antiepileptic drugs. Sulpiride (SLP) and indisulam (IND) were also shown by our group to belong to this class of pharmacological agents, together with the COX2 selective inhibitors celecoxib (CLX) and valdecoxib (VLX). Sac-charin (SAC) and the diuretic hydrochlorothiazide (HCT) are also known to act as CAIs. Famotidine (FAM) and epacadostat (EPA) are CAI sulfamide drugs clinically used respectively as a histamine H2 receptor antagonist and a selective indoleamine-2,3-dioxygenase 1 inhibitor.

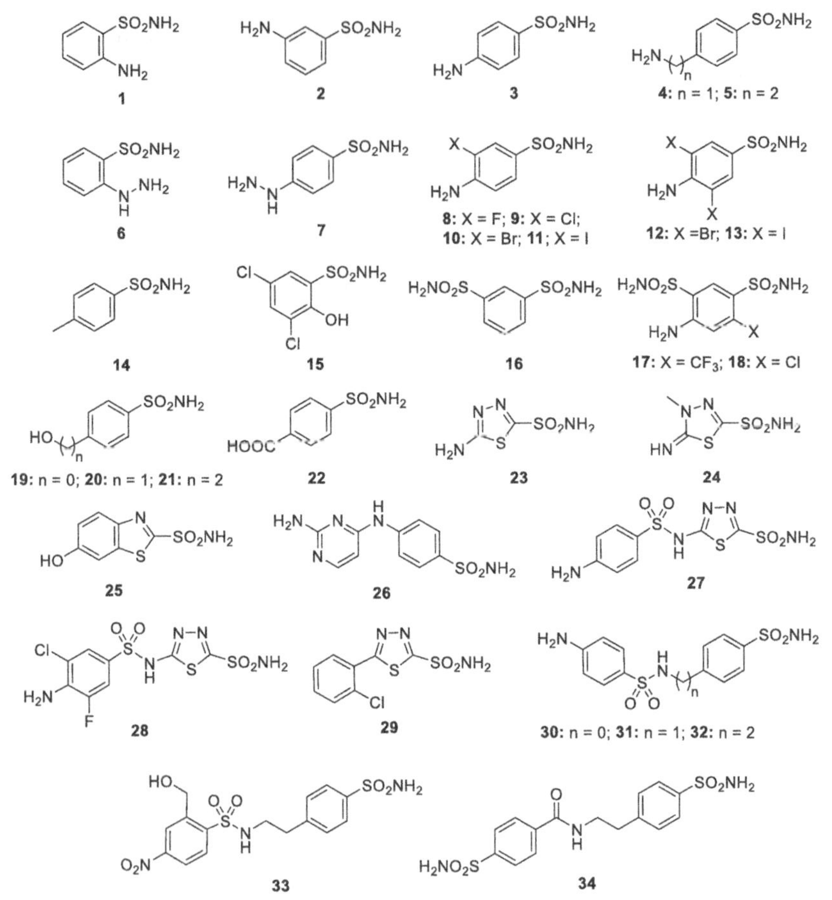

Fig. 4 Structure of sulfonamide derivatives **1–34**.

In detail, Tables 1–4 collect the inhibition profiles of the aforementioned compounds against the human CAs I and II, as off-target standard isozymes and these CAs from pathogenic bacteria: HpαCA and HpβCA from *Helicobacter pylori* [34,35], the β-class isoforms MtCA1, MtCA2 and MtCA3 from *Micobacterium tuberculosis* [36–38], the β-class isoforms BsCA1 and BsCA2 from *Brucella suis* [39,40], the β-class isoforms StCA1 and StCA2 from *Salmonella enterica* [41], VchαCA, VchβCA and VchγCA from *Vibrio cholerae* [42–44], PgβCA and PgγCA from *Porphyromonas gingivalis* [45], the β-class isoforms from LpCA1 and LpCA2 from *Legionella pneumophila* [46], the β-class isoform SmuCA from *Streptococcus mutans* [47], BpsγCA and BpsβCA from *Burkholderia pseudomallei* [48], the β-class

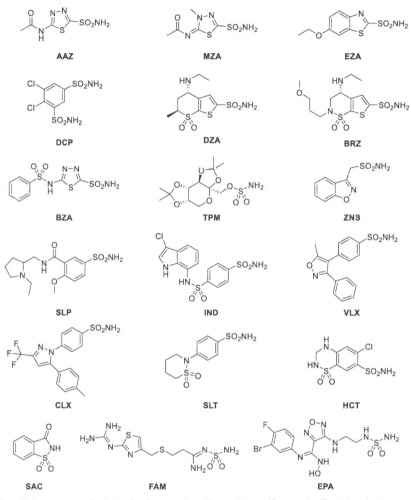

Fig. 5 Structure of clinically licensed sulfonamide/sulfamate/sulfamide derivatives **AAZ-EPA**.

isoform FtuCA from *Francisella tularensis* [49], the β-class isoform CpeCA from *Clostridium perfringens* [50], EcoβCA and EcoγCA from *Escherichia coli* [51,52], MscβCA and MscγCA from *Mammaliicoccus sciuri* [53]. Tables 5 and 6 collects instead the inhibition profiles of **1–34** and AAZ-EPA against CAs from nonpathogenic bacteria, that are: the α-class isoform SspCA from *Sulfurihydrogenibium yellowstonense* [54], the α-class isoform SazCA from *Sulfurihydrogenibium azorense* [55], the γ-class isoforms NcoCA, PhaCA and CpsCA respectively from the Antarctic bacteria *Nostoc*

Table 1 Inhibition profile of sulfonamide derivatives 1–34 against human hCA I and II, and CAs from pathogenic bacteria: HpαCA and HpβCA from *Helicobacter pylori*, the β-class isoforms MtCA1, MtCA2, and MtCA3 from *Micobacterium tuberculosis*; the β-class isoforms BsCA1 and BsCA2 from *Brucella suis*, the β-class isoforms StCA1 and StCA2 from *Salmonella enterica*, VchαCA, VchβCA and VchγCA from *Vibrio cholerae*.

Cmp	K_I (nM)																
	hCA I	hCA II	HpαCA	HpβCA	MtCA1	MtCA2	MtCA3	BsCA1	BsCA2	StCA1	StCA2	VchαCA	VchβCA	VchγCA
1	45,400	295	426	16,400	9230	33,700	6240	5870	5048	5580	7240	440	463	672
2	28,000	300	nt	nt	nt	nt	nt	nt	nt	nt	nt	nt	nt	nt
3	25,000	240	454	1845	9840	29,600	7110	2500	5080	6510	8800	471	447	95
4	25,000	170	873	2360	8690	30,700	7330	768	5910	6450	7750	447	>10,000	81
5	21,000	160	1150	3500	9560	29,100	3420	880	8560	5930	6760	402	>10,000	69
6	24,000	125	541	15,250	nt	nt	nt	1035	5639	12,100	12,900	52	361	73
7	28,000	300	316	8650	nt	nt	nt	2400	4763	6140	6980	125	785	94
8	8300	60	1230	1359	8740	28,900	7900	1070	83	6550	4220	199	>10,000	74
9	9800	110	378	1463	7520	27,700	1510	800	79	9550	9600	139	9120	74
10	6500	40	452	1235	186	31,600	7320	nt	nt	nt	nt	133	>10,000	95
11	6000	70	510	1146	7710	32,400	5810	nt	nt	nt	nt	99	>10,000	544

12	73	9650	nt	nt	nt	nt	nt	243	60	7060	6310	nt	nt	nt
13	124	14,000	nt	nt	nt	nt	nt	345	84	6910	7000	nt	nt	nt
14	320	78,500	430	2470	4920	38,900	7020	1580	5560	9380	6440	219	>10,000	76
15	54	7300	nt	nt	nt	nt	nt	nt	nt	nt	nt	nt	nt	nt
16	8.2	79	nt	nt	nt	nt	nt	nt	nt	nt	nt	nt	nt	nt
17	63	5800	412	973	8100	29,600	2350	4830	87	6790	8160	63	879	87
18	75	8400	49	640	1720	32,500	21700	940	691	7300	7030	45	4450	563
19	94	9500	nt	nt	nt	nt	nt	nt	nt	nt	nt	nt	nt	nt
20	80	5500	876	236	9560	38,300	7820	754	758	6820	7540	54	4430	764
21	125	21,000	1134	218	5510	34,500	2510	865	5130	5750	6090	57	757	902
22	133	23,000	1052	450	8210	39,200	7400	340	697	6010	8190	71	817	271
23	60	8600	323	2590	11,540	2090	7630	1210	847	770	765	23	68	66
24	19	9300	549	768	12,650	2380	7920	1430	11	671	703	12	82	70
25	30	95	207	39	nt	nt	nt	48	83	642	860	4.7	2218	4153
26	33	109	84	38	750	2360	91	21	769	246	631	60	515	5100

(continued)

Table 1 Inhibition profile of sulfonamide derivatives 1–34 against human hCA I and II, and CAs from pathogenic bacteria: HpαCA and HpβCA from *Helicobacter pylori*, the β-class isoforms MtCA1, MtCA2, and MtCA3 from *Micobacterium tuberculosis*, the β-class isoforms BsCA1 and BsCA2 from *Brucella suis*, the β-class isoforms StCA1 and StCA2 from *Salmonella enterica*, VchαCA, VchβCA and VchγCA from *Vibrio cholerae*. (*cont'd*)

Cmp	K_I (nM)													
	hCA I	hCA II	HpαCA	HpβCA	MtCA1	MtCA2	MtCA3	BsCA1	BsCA2	StCA1	StCA2	VchαCA	VchβCA	VchγCA
27	6.1	2.0	268	64	950	978	3100	70	108	68	44	4.2	349	88
28	1.4	0.3	nt	nt	nt	nt	nt	186	663	51	38	nt	nt	nt
29	690	12	105	37	7480	45,200	7600	33	166	424	338	0.59	859	5570
30	42	6	nt	nt	nt	nt	nt	nt	nt	nt	nt	nt	nt	nt
31	164	46	131	87	612	3210	2210	1050	94	91	40	43	304	556
32	185	50	114	71	853	2290	170	745	68	72	31	30	3530	6223
33	69	11	nt	nt	nt	nt	nt	nt	nt	nt	nt	nt	nt	nt
34	40	5.2	nt	nt	nt	nt	nt	27	90	124	96	nt	nt	nt

The best K_I values *per* isoform are underlined. nt: not tested.

Table 2 Inhibition profile of sulfonamide derivatives **1–34** against CAs from pathogenic bacteria: PgβCA and PgγCA from *Porphyromonas gingivalis*, the β-class isoforms from LpCA1 and LpCA2 from *Legionella pneumophila*, the β-class isoform SmuCA from *Streptococcus mutans*, BpsβCA and BpsγCA from *Burkholderia pseudomallei*, the β-class isoform FtuCA from *Francisella tularensis*, the β-class isoform CpeCA from *Clostridium perfringens*, EcoβCA and EcoγCA from *Escherichia coli*, MscβCA and MscγCA from *Mammaliicoccus sciuri*.

Cmd	PgβCA	PgγCA	LpCA1	LpCA2	SmuCA	BpsβCA	BpsγCA	FtuCA	CpeCA	EcoβCA	EcoγCA	MscβCA	MscγCA
								K_I (nM)					
1	nt	nt	nt	nt	nt	nt	nt	nt	nt	nt	nt	nt	nt
2	477	4220	939	455	2750	>50,000	574	2535	451	705	314	355	797
3	715	893	946	277	1975	3895	1720	6076	402	790	193	409	888
4	783	3600	757	516	2200	3900	>50,000	18,600	3690	2840	160	193	354
5	475	3840	734	375	1580	>50,000	>50,000	15,200	2430	3321	622	253	176
6	nt	nt	nt	nt	nt	nt	nt	nt	nt	nt	nt	nt	nt
7	nt	nt	nt	nt	nt	nt	nt	nt	nt	nt	nt	nt	nt
8	818	680	770	592	2155	>50,000	>50,000	8070	443	>10,000	605	93	2010
9	4525	662	866	396	3840	>50,000	12,500	17,950	320	>10,000	671	95	426
10	6620	201	988	181	246	>50,000	>50,000	>50,000	713	2712	718	75	478
11	nt	nt	nt	nt	nt	nt	nt	nt	nt	nt	nt	nt	nt

(continued)

Table 2 Inhibition profile of sulfonamide derivatives **1–34** against CAs from pathogenic bacteria: PgβCA and PgγCA from *Porphyromonas gingivalis*, the β-class isoforms from LpCA1 and LpCA2 from *Legionella pneumophila*, the β-class isoform SmuCA from *Streptococcus mutans*, BpsβCA and BpsγCA from *Burkholderia pseudomallei*, the β-class isoform FtuCA from *Francisella tularensis*, the β-class isoform CpeCA from *Clostridium perfringens*, EcoβCA and EcoγCA from *Escherichia coli*, MscβCA and MscγCA from *Mammaliicoccus sciuri*. (cont'd)

Cmd	PgβCA	PgγCA	LpCA1	LpCA2	SmuCA	BpsβCA	BpsγCA	FtuCA	CpeCA	EcoβCA	EcoγCA	MscβCA	MscγCA
								K_I (nM)					
12	nt	nt	nt	nt	nt	nt	nt	nt	nt	nt	nt	nt	nt
13	nt	nt	nt	nt	nt	nt	nt	nt	nt	nt	nt	nt	nt
14	710	945	556	624	2520	>50,000	>50,000	>50,000	379	3015	221	83	647
15	5040	218	913	622	12,000	4100	>50,000	>50,000	4670	8561	2577	202	2429
16	364	>10,000	1060	933	1710	3170	1550	6550	210	457	246	95	552
17	4765	711	929	593	1870	425	14,000	875	8755	6246	1779	81	33,000
18	3898	1040	642	496	3430	307	23,500	>50,000	7635	4385	1953	79	17,200
19	8955	223	2260	631	4355	4345	9650	6630	145	2340	4238	603	915
20	>10,000	326	969	280	468	2215	10,800	14,500	131	6445	1013	619	2931
21	>10,000	178	3540	721	13,500	417	1825	8110	314	502	1975	232	736
22	>10,000	560	2390	476	>50,000	>50,000	1500	8950	1268	205	2064	555	722

23	7100	510	541	382	414	3065	18,400	655	297	4122	197	417	539
24	>10,000	595	913	391	2050	2500	1810	5120	460	440	712	553	820
25	nt	nt	nt	nt	nt	nt	nt	nt	nt	nt	nt	nt	nt
26	9150	685	472	321	250	>50,000	1838	3595	179	416	1894	909	1893
27	7645	1450	90	45	455	266	1810	13,250	105	726	883	92	1626
28	nt	nt	nt	nt	nt	nt	nt	nt	nt	nt	nt	nt	nt
29	nt	nt	nt	nt	nt	nt	nt	nt	nt	nt	nt	nt	nt
30	3405	4100	319	52	>50,000	5200	1805	10,800	513	93	3501	83	72
31	9240	4650	60	50	438	1500	1700	8950	312	322	4045	92	217
32	7960	3400	40	25	430	>50,000	24,500	>50,000	51	82	4262	96	46
33	6450	3540	101	79	307	1650	1335	3500	697	473	819	85	83
34	nt	nt	nt	nt	nt	nt	nt	nt	nt	nt	nt	nt	nt

The best K_I values *per* isoform are underlined. nt = not tested.

Table 3 Inhibition profile of the clinically licensed sulfonamide derivatives **AAZ-EPA** against human hCA I and II, and CAs from pathogenic bacteria: HpαCA and HpβCA from *Helicobacter pylori*, the β-class isoforms MtCA1, MtCA2, and MtCA3 from *Micobacterium tuberculosis*, the β-class isoforms BsCA1 and BsCA2 from *Brucella suis*, the β-class isoforms StCA1 and StCA2 from *Salmonella enterica*, VchaCA, VchβCA andVchyCA from *Vibrio cholerae*.

Cmp	K_i (nM)													
	hCA I	hCA II	HpαCA	HpβCA	MtCA1	MtCA2	MtCA3	BsCA1	BsCA2	StCA1	StCA2	VchaCA	VchβCA	VchyCA
AAZ	250	12	21	40	481	9	104	63	303	59	84	6.8	4512	473
MZA	50	14	225	176	781	660	562	54	642	134	68	3.6	6260	494
EZA	25	8	193	33	1030	27	594	17	420	528	721	0.69	6450	85
DCP	1200	38	378	105	872	2010	611	58	112	90	95	37	2352	1230
DZA	50,000	9	4360	73	744	99	137	21	923	445	607	6.3	4728	87
BRZ	45,000	3	210	128	839	127	201	26	625	687	412	2.5	845	93
BZA	15	9	315	54	810	467	338	75	117	85	98	4.2	846	78
TPM	250	10	172	32	612	474	3020	57	99	624	697	>1000	874	69
ZNS	56	35	231	254	28,680	876	208	1850	406	5430	5700	982	8570	725
SLP	1200	40	204	35	2300	266	7920	19	84	5640	8730	>1000	6245	78
IND	31	15	413	143	97	717	7840	50	130	6860	6900	8.7	7700	91

VLX	54,000	43	nt	nt	1297	682	7810	<u>19</u>	612	6850	6580	90	8200	817
CLX	50,000	21	nt	nt	10,350	713	7760	<u>18</u>	128	5830	6110	>1000	4165	834
SLT	374	9	nt	nt	5160	664	6720	nt	nt	6800	4670	88	455	464
HCT	328	290	nt	nt	nt	nt	nt	nt	nt	6810	6870	79	87	500
SAC	18,540	5950	nt	nt	7960	792	7150	nt	nt	26,700	14,800	>1000	<u>275</u>	464
FAM	922	58	<u>21</u>	50	nt	nt	nt	nt	nt	nt	nt	nt	nt	nt
EPA	8262	917	nt	nt	nt	nt	nt	nt	nt	nt	nt	nt	nt	nt

The best K_I values *per* isoform are underlined. nt: not tested.

Table 4 Inhibition profile of the clinically licensed sulfonamide derivatives **AAZ-EPA** against and CAs from pathogenic bacteria: PgβCA and PgγCA from *Porphyromonas gingivalis*, the β-class isoforms from LpCA1 and LpCA2 from *Legionella pneumophila*, the β-class isoform SmuCA from *Streptococcus mutans*, BpsβCA and BpsγCA from *Burkholderia pseudomallei*, the β-class isoform FtuCA from *Francisella tularensis*, the β-class isoform CpeCA from *Clostridium perfringens*, EcoβCA and EcoγCA from *Escherichia coli*, MscβCA and MscγCA from *Mammaliicoccus sciuri*.

Cmd	K_I (nM)												
	PgβCA	PgγCA	LpCA1	LpCA2	SmuCA	BpsβCA	BpsγCA	FtuCA	CpeCA	EcoβCA	EcoγCA	MscβCA	MscγCA
AAZ	214	324	77	72	344	745	149	770	49	227	248	628	245
MZA	393	343	201	88	445	186	1595	2920	113	480	921	863	95
EZA	280	613	71	103	2430	3850	1865	5188	72	557	5538	698	97
DCP	>10,000	1032	1670	64	2700	529	>50,000	>50,000	68	>10,000	889	nt	8580
DZA	2415	685	2070	336	4315	3670	2260	>50,000	37	629	2007	909	44
BRZ	408	722	648	467	>50,000	4270	1270	>50,000	121	2048	4842	815	100
BZA	2675	741	159	148	4040	185	653	8400	497	276	94	501	444
TPM	4250	>10,000	665	882	>50,000	>50,000	3010	>50,000	204	3359	648	466	478
ZNS	345	157	831	820	3620	4060	>50,000	>50,000	389	3189	755	4551	923
SLP	1470	418	253	245	2710	>50,000	5600	>50,000	389	97	914	807	2203
IND	1353	131	1090	525	412	4375	1800	>50,000	173	2392	387	588	369

VLX	2395	755	536	879	444	>50,000	>50,000	7810	160	2752	891	<u>509</u>	1540
CLX	415	169	990	421	2425	>50,000	>50,000	7015	312	1894	944	871	3024
SLT	3140	424	485	463	1865	>50,000	>50,000	8900	63	285	446	824	405
HCT	1572	380	15,800	745	2615	3490	>50,000	9030	219	5010	4903	593	18,300
SAC	2244	273	20,500	447	3320	nt	nt	>50,000	>50,000	6693	3643	667	29,700
FAM	nt	nt	nt	nt	nt	nt	nt	nt	nt	2769	274	nt	nt
EPA	nt	nt	nt	nt	nt	nt	nt	nt	nt	2560	744	nt	nt

The best K_I values *per* isoform are underlined. nt= not tested.

Table 5 Inhibition profile of sulfonamide derivatives **1–34** against CAs from non-pathogenic bacteria: the α-class isoform SspCA from *Sulfurihydrogenibium yellowstonense*, the α-class isoform SazCA from *Sulfurihydrogenibium azorense*, the γ-class isoforms NcoCA, PhaCA and CpsCA from the Antarctic bacteria *Nostoc commune*, *Pseudoalteromonas haloplanktis* and *Colwellia psychrerythraea*, respectively and the ι-class BteCA from *Burkholderia territorii*.

Cmp	K_I (nM)					
	SspCA	SazCA	NcoCA	PhaCA	CpsCA	BteCA
1	707	380	nt	nt	nt	nt
2	nt	nt	492	5460	83	325
3	652	361	488	9210	867	477
4	350	274	825	>100,000	562	97
5	73	261	742	6150	767	786
6	695	5630	nt	nt	nt	nt
7	803	659	nt	nt	nt	nt
8	68	208	20,600	>100,000	724	481
9	41	186	23,750	>100,000	460	346
10	76	193	872	7460	625	96
11	45	172	nt	nt	nt	nt
12	nt	nt	nt	nt	nt	nt
13	nt	nt	nt	nt	nt	nt
14	848	308	785	>100,000	646	446
15	nt	nt	2810	8450	684	357
16	nt	nt	683	>100,000	471	568
17	61	78	632	8980	907	239
18	87	84	269	8465	556	329
19	nt	nt	4034	9150	5390	594
20	493	247	7570	8760	6650	540
21	162	208	777	>100,000	6600	404

22	72	96	75	8730	5050	467
23	137	103	85	6310	767	303
24	433	126	480	6100	7660	434
25	66	9.1	nt	nt	nt	nt
26	<u>6.5</u>	10	78	<u>5400</u>	854	<u>93</u>
27	<u>5.5</u>	<u>4.1</u>	67	7690	860	268
28	nt	nt	nt	nt	nt	nt
29	nt	nt	nt	nt	nt	nt
30	nt	nt	553	8465	726	365
31	7.8	20	74	9290	893	408
32	9.7	16	<u>40</u>	8885	834	698
33	nt	nt	402	8090	5020	307
34	nt	nt	nt	nt	nt	nt

The best K_I values *per* isoform are underlined. nt: not tested.

commune, Pseudoalteromonas haloplanktis and *Colwellia psychrerythraea* [56–58] and the ι-class BteCA from *Burkholderia territorii* [59].

The collected inhibition profiles are highly variable and do not allow to work out an exhaustive SAR. The most potent inhibitors *per* isoform within each series are listed below (K_Is are reported in brackets and underlined in the Tables) to gather information for drug design purposes. HpαCA: **18** (49 nM), AAZ and FAM (21 nM); HpβCA: **25**, **26**, **29** (37–39 nM), EZA and TPM (32–33 nM); MtCA1: **10** (186 nM) and IND (97 nM); MtCA2: **27** (978 nM) and AAZ (9 nM); MtCA3: **26** (91 nM) and AAZ (104 nM); BsCA1: **26**, EZA, DZA, SLP, VLX, CLX (17–21 nM); BsCA2: **24** (11 nM) and SLP (84 nM); StCA1: **28** (51 nM) and AAZ (59 nM); StCA2: **28**, **32** (31–38 nM) and MZA (68 nM); VchαCA: **29** (0.59 nM) and EZA (0.69 nM); VchβCA: **23** (68 nM) and SAC (275 nM); VchγCA: **5**, **6**, **8**, **9**, **23**, **24** (67–74 nM) and BZA, TPM and SLP (69–78 nM); PgβCA: **16** (324 nM) and AAZ (214 nM); PgγCA: **10**, **21** (178–201 nM), IND and ZNS (131–157 nM); LpCA1: **32** (40 nM), EZA and AAZ

Table 6 Inhibition profile of the clinically licensed sulfonamide derivatives **AAZ-EPA** against CAs from nonpathogenic bacteria: the α-class isoform SspCA from *Sulfurihydrogenibium yellowstonense*, the α-class isoform SazCA from *Sulfurihydrogenibium azorense*, the γ-class isoforms NcoCA, PhaCA and CpsCA from the Antarctic bacteria *Nostoc commune, Pseudoalteromonas haloplanktis* and *Colwellia psychrerythraea*, respectively and the ι-class BteCA from *Burkholderia territorii*.

Cmd	K_I (nM)					
	SspCA	SazCA	NcoCA	PhaCA	CpsCA	BteCA
AAZ	<u>4.5</u>	<u>0.9</u>	76	403	502	<u>519</u>
MZA	8.2	2.4	191	<u>95</u>	594	8466
EZA	9.3	1.3	264	679	513	5024
DCP	8.5	25	345	2465	639	825
DZA	8.0	2.2	67	831	868	794
BRZ	12.3	2.0	81	735	457	3703
BZA	8.1	3.7	48	24,600	439	724
TPM	6.6	11	<u>36</u>	877	500	787
ZNS	140	35	86	490	771	806
SLP	67	2.5	60	>100,000	505	958
IND	84	4.8	92	<u>87</u>	491	638
VLX	5.3	28	53	735	<u>343</u>	934
CLX	6.9	24	88	761	402	960
SLT	7.8	29	82	898	<u>380</u>	954
HCT	7.2	54	59	790	<u>345</u>	780
SAC	876	79	408	867	601	7081
FAM	nt	nt	nt	nt	nt	943
EPA	nt	nt	nt	nt	nt	955

The best K_I values *per* isoform are underlined. nt: not tested.

(71–77 nM); LpCA2: **32** (25 nM), DCP and AAZ (64–72 nM); SmuCA: **10**, **26** (246–250 nM), AAZ and IND (344–412 nM); BpsβCA: **18**, **27** (266–307 nM), BZA and MZA (185–186 nM); BpsγCA: **2** (574 nM),

AAZ (149 nM); FtuCA: **23** (655 nM) and AAZ (770 nM); CpeCA: **32** (51 nM) and DZA (37 nM); EcoβCA: **30, 32** (82–93 nM) and SLP (97 nM); EcoγCA: **4** (160 nM) and BZA (94 nM); MscβCA: **10, 14, 17, 18, 30** (75–83 nM), TPM, BZA and VLX (466–509 nM); MscγCA: **32** (46 nM) and DZA (44 nM); SspCA: **26, 27** (5.5–6.5 nM) and AAZ (4.5 nM); SazCA: **27** (4.1 nM) and AAZ (0.9 nM); NcoCA: **32** (40 nM) and TPM (36 nM); PhaCA: **26** (5400 nM), MZA and IND (87–95 nM); CpsCA: **2** (83 nM), SLT, VLX and HCT (343–380 nM); BteCA: **4, 10, 26** (93–97 nM) and AAZ (519 nM).

The inhibitory profiling of these bacterial CA isoforms has been utilized over the last two decades to develop a new series of inhibitors in search of effective antibiotics. In 2009, Guzel et al. described a series of pyridinium derivatives of the 2-(hydrazinocarbonyl)-3-aryl-1H-indole-5-sulfonamide scaffold (**32** and **36-38**, Fig. 6 and Table 7) [60]. Many such compounds showed excellent inhibitory activity against MtCA1 and MtCA3, among which **36–38** outperformed the lead **32** and similar derivatives with sub-nanomolar K_Is (0.92–0.97 nM) against both MtCAs.

In the same year, Maresca et al. reported a series of diazenylbenzene-sulfonamides again as inhibitors of MtCA1 and MtCA3 [61]. Most compounds acted as low micromolar inhibitors of the two enzymes with the most efficient ones **39–41** (Fig. 6, Table 7) exhibiting K_Is in the range 2690–4620 nM.

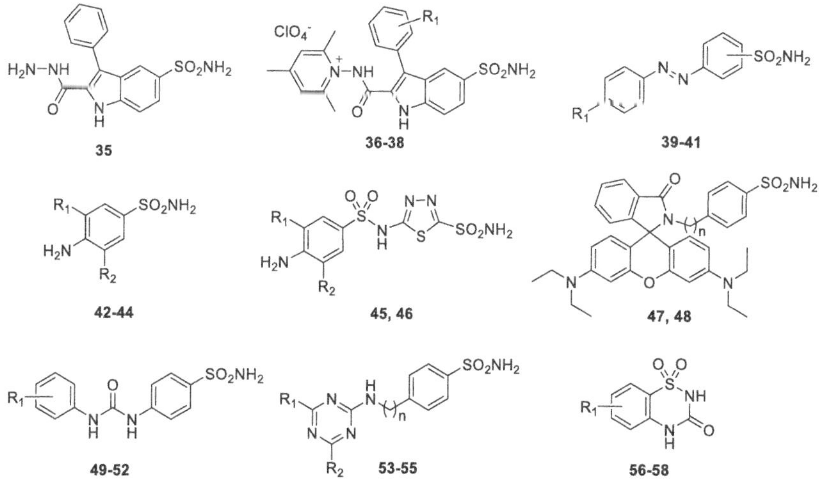

Fig. 6 Structure of sulfonamide derivatives **35–58.**

Table 7 Inhibition profile of sulfonamide compounds **35–58** against the β-class iso-forms MtCA1, MtCA2 and MtCA3 from *M. tuberculosis*.

Cmp					K_I (nM)		
	m/p	n	R_1	R_2	MtCA1	MtCA2	MtCA3
35	—	—	—	—	48	nt	31
36	—	—	3–Br	—	3.2	nt	0.97
37	—	—	3–OCH_3	—	0.92	nt	1.8
38	—	—	4–Cl	—	21	nt	0.96
39	*m*	—	OH	—	8970	nt	9230
40	*m*	—	NH_2	—	7000	nt	8680
41	*p*	—	$NHCH_3$	—	7690	nt	9180
42	—	—	F	Br	940	4310	3190
43	—	—	F	I	1220	4490	2790
44	—	—	Cl	Br	580	4620	2690
45	—	—	F	Br	140	430	270
46	—	—	F	I	120	410	300
47	—	0	—	—	4410	414	342
48	—	2	—	—	3700	409	422
49	—	—	4–Cl	—	4.8	nt	53
50	—	—	pentaF	—	5.0	nt	6.4
51	—	—	4–*n*Bu	—	35	nt	63
52	—	—	3–NO_2	—	67	nt	6.5
53	—	0	$NHCH_3$		580	9.6	22
54	—	2	$NHCH_2COOH$	F	610	8.1	>1000
55	—	2	F	F	42	8.6	2.1
56	—	—	7–CH_3	—	1050	41	18
57	—	—	5,7–diCH_3	—	934	38	15
58	—	—	5–F	—	1290	39	16

In 2013, the same research group reported dihalogenated sulfanilamides and benzolamides as inhibitors of the three MtCAs [62]. The latter were all inhibited by the reported compounds with efficacies between the sub-micromolar to the micromolar range, depending on the substitution pattern at the sulfanilamide moiety/fragment. The best inhibitors within the halogenated benzolamide subset were **45** and **46** (Fig. 6, Table 7) with K_Is in the range of 120–430 nM, whereas the best halogenated sulfanilamides (subset **42–44**) were less effective (K_Is in the range of 580–4620 nM).

The three β-class CAs from *M. tuberculosis* were also included as biological targets in a study by Winum and collaborators describing a series of fluorescent sulfonamide CAIs obtained by attaching rhodamine B to the benzenesulfonamide scaffold [63]. The most performing MtCA inhibitors were **47** and **48** (Fig. 6, Table 7) acting in the medium nanomolar range against MtCA2 and MtCA3 (K_Is in the range of 342–422 nM), whilst a 10-fold less potent action was measured against MtCA1 (K_Is of 4410 and 3700 nM, respectively).

MtCA1 and MtCA3 were potently inhibited by a series of ureido substituted benzenesulfonamides, as reported by Pacchiano et al. in 2011 [64]. Subset **49–52** (Fig. 6, Table 7) collects the most effective MtCA inhibitors of the series with K_Is lying in the low nanomolar range between 4.8 and 67 nM.

A new series of fluorine containing 1,3,5-triazinyl sulfonamide derivatives was described in 2013 and evaluated for the inhibition of MtCA1, MtCA2, and MtCA3 [65]. The three enzymes were efficiently inhibited with K_I values in the nanomolar or submicromolar range, depending on the substitution of one or both fluorine atoms at the 1,3,5-triazine ring. The most efficient of the reported compounds were **53–55** (Fig. 6, Table 7), among which derivatives **55** outperformed for inhibitory potency with K_Is in the range 2.1–42 nM, against the three isoforms.

Lately, a series of 2H-benzo[e][1,2,4]thiadiazin-3(4H)-one 1,1-dioxide (BTD) based CAIs were studied as new anti-mycobacterial agents, considering their chemical features to meet the criteria for a potent inhibition of β-class CA isozymes [66]. The BTDs derivatives effectively inhibited the mycobacterial CAs, especially MtCA2 and MtCA3, with K_I values up to a low nanomolar range (MtCA3, K_Is in the range 15–2250 nM; MtCA2, K_Is in the range 38–4480 nM). Importantly, the most potent MtCA inhibitors, among which **56–58** (Fig. 6, Table 7) demonstrated efficacy in inhibiting the growth of *M. tuberculosis* strains resistant to both rifampicin and isoniazid, standard reference drugs for Tuberculosis treatment.

In the studies describing the inhibitory characterization of HpαCA and HpβCA with sulfonamides **1–38** and **AAZ-EPA** [34,35], another series of compounds featuring a 4-tert-butyl-phenylcarboxamido/sulfonamido tail at the benzenesulfonamide/1,3,4-thiadiazole-2-sulfonamide scaffold (Fig. 7, Table 8) was evaluated for their inhibition effects against the two bacterial CAs. Compounds of this subset emerged as the most effective inhibitors of the studies, outperforming the classical and clinical sulfonamide derivatives. Specifically, compounds **64** and **70** showed the lowest K_Is that are 13 and 12 nM against HpαCA and 28 and 24 nM against HpβCA, respectively.

Importantly in the context of *H. pylori* infections, the sulfonamide compound EZA showed significant activity *in vitro* against various anti-*H. pylori* strains, such as SS1 and 26,695 without leading to the development of significant drug resistance [67].

In 2010, a number of sulfonamides incorporating sugar moieties have been investigated for inhibition of the β-class BsCA1 from *B. suis* [68]. Small structural changes in the sugar moiety led to dramatic differences in enzyme inhibitory activity for this small series of compounds. The most effective BsCA1 inhibitors are found in subset **71–74** (Fig. 8, Table 9). Among these, the galactose and ribose sulfanilamides **71** and **72** stood out, with K_Is of 8.9 nM and 9.2 nM, respectively. Compound **73** and AAZ also significantly inhibited the growth of bacterial cell cultures.

In 2015, N–glycosyl-N–hydroxysulfamides were described by Winum and collaborators as potent inhibitors of BsCA1 and BsCA2 [69]. The most effective derivatives are found in the subset **75–77** (Fig. 8, Table 9). Overall, BsCA II was better inhibited than BsCA I by these compounds with inhibition constants in the range 60–799 nM *vs* 522–958 nM. The

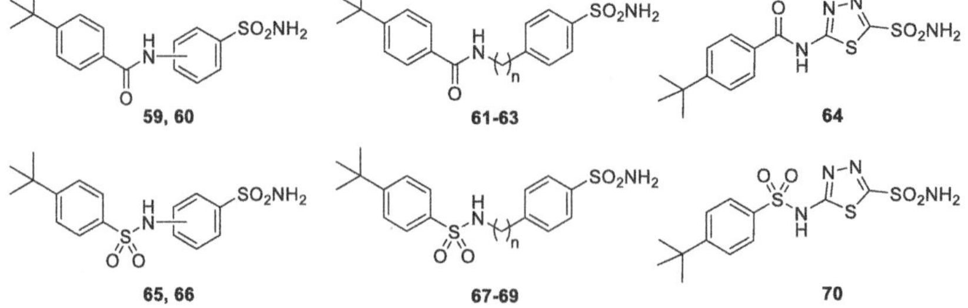

Fig. 7 Structure of sulfonamide derivatives **59–70**.

Table 8 Inhibition profile of sulfonamide compounds **59–70** against HpαCA and HpβCA from *H. pylori*.

Cmp	o/m	n	HpαCA	HpβCA
			K_I (nM)	
59	*o*	—	536	23,500
60	*m*	—	316	241
61	—	0	79	158
62	—	1	539	101
63	—	2	62	44
64	—	—	13	28
65	*o*	—	640	24,800
66	*m*	—	318	213
67	—	0	60	150
68	—	1	31	96
69	—	2	27	45
70	—	—	12	24

acetylated derivatives were generally better BsCA II inhibitors compared to the corresponding deacetylated compounds.

Derivatives **78** and **79** (Fig. 8, Table 9) are the most potent inhibitors of BsCA1 and BsCA2 from a series of 4 substituted pyridine-3-sulfonamides derivatives reported by Monti et al. in 2018 [70]. In this case, BsCA1 was more effectively inhibited by these sulfonamides than BsCA2 with inhibition in the range 34–624 nM *vs* 62 – > 10 mM.

BsCA1 and BsCA2 were also studied as bacterial CA target by Ceruso et al. with a series of Schiff base benzenesulfonamide derivatives, among which **80–82** (Fig. 8, Table 9) showed the best inhibitory performance [71]. Specifically, these compounds showed medium-low nanomolar efficacy against BsCA1, while BsCA2 was strongly inhibited by most compounds of the series, with **80–82** even exhibiting single-digit K_I values (2.3–3.3 nM).

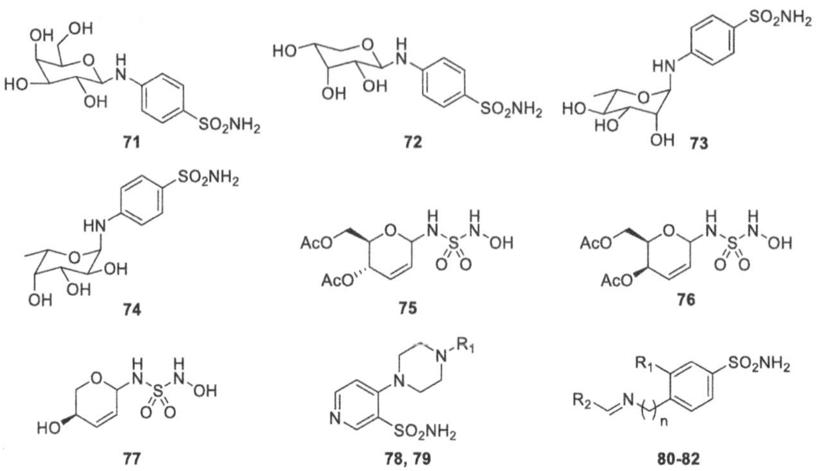

Fig. 8 Structure of sulfonamide derivatives **71–82**.

Table 9 Inhibition profile of sulfonamide compounds **71–82** against the β-class isoforms BsCA1 and BsCA2 from *B. suis*.

Cmp				K_I (nM)	
	n	R₁	R₂	BsCA1	BsCA2
71	—	—	—	9.2	nt
72	—	—	—	8.9	nt
73	—	—	—	19	nt
74	—	—	—	28	nt
75	—	—	—	565	60
76	—	—	—	732	76
77	—	—	—	811	89
78	—	4-Cl-C$_6$H$_4$	—	61	62
79	—	CH$_2$C$_6$H$_5$	—	34	97
80	0	F	C$_6$H$_5$	11	2.3
81	2	H	4-CN-C$_6$H$_4$	85	3.3
82	2	H	5-Br-indol-3-yl	29	2.5

Four generations of poly(amidoamine) dendrimers incorporating benzenesulfonamide moieties (G_0, G_1, G_2 and G_3; G1 is compound **83** shown in Fig. 9) were investigated as inhibitors of PgβCA and PgγCA from *P. gingivalis* and VchαCA from *V. cholerae* [72]. G1–G3, incorporating 8, 16 and 32 sulfonamides moieties respectively, showed a quite similar inhibitory action against VchαCA (K_Is ranging between 8.1 and 9.2 nM). PgβCA was also inhibited by all dendrimers investigated here, with K_Is in the range 77.0–84.9 nM, with compound **83** being the most acting one. PgγCA was poorly inhibited by dendrimers G_0–G_3 with K_Is in the range 283–682 nM.

In 2014 Ceruso et al. described new benzenesulfonamides incorporating GABA or N-α-acetyl-L-lysine scaffolds, both bearing guanidine functionalities or not, as water solubilizing moieties [73]. These compounds were assayed for their inhibition of PgγCA and VchαCA. The latter was highly inhibited by most such compounds, among which **85, 86, 88** and **89** (Fig. 9, Table 10) even showing single-digit K_I values. This series of derivatives inhibited instead PgγCA in a medium-high nanomolar range, with compounds **84, 87,** and **88** being the most effective ones (K_Is in the range 40–54 nM).

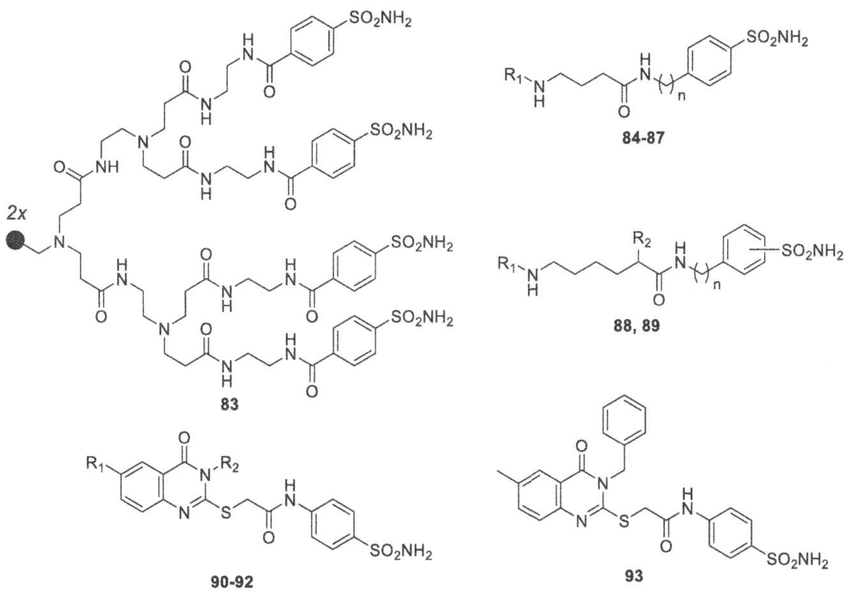

Fig. 9 Structure of sulfonamide derivatives **83–93**.

Table 10 Inhibition profile of sulfonamide compounds **83–93** against PgβCA and PgγCA from *P. gingivalis* and VchαCA from *V. cholerae*.

Cmp	m/p	n	R$_1$	R$_2$	K_I (nM)		
					PgβCA	PgγCA	VchαCA
83	—	—	—	—	77	563	8.1
84	—	1	H	—	nt	43	65
85	—	1	Boc	—	nt	490	8.8
86	—	1	(=NBoc)NHBoc	—	nt	74	6.1
87	—	2	(=NH)NH$_2$	—	nt	40	29
88	*m*	—	NHAc	Boc	nt	54	8.8
89	*p*	1	(=NH)NH$_2$	NHAc	nt	81	5.8
90	—	—	CH$_3$	CH$_2$CH=CH$_2$	nt	19	6.0
91	—	—	CH$_3$	CH$_2$C$_6$H5	nt	3.5	2.7
92	—	—	NO$_2$	CH$_2$CH=CH$_2$	nt	24	6.6
93	—	—	—	—	nt	31	5.5

In 2014 and 2015, Alafeefy et al. descrived a series of quinazolinone benzenesulfonamides as inhibitors of PgγCA and VchαCA [74,75]. Many such compounds, such as those found in the subset **90–93** (Fig. 9, Table 10), greatly inhibited VchαCA with single-digit K_I values (2.7–6.6 nM). The same compounds demonstrated slightly less effective inhibition profiles against PgγCA compared to VchαCA, though maintaining a good to medium efficacy. The best PgγCA inhibitor was compound **91** with a K_I of 3.5 nM.

In 2009 Abuaita and Withney used a sulfonamide CAI, that is EZA, to experimentally demonstrate the implication of the CA-mediated conversion of CO_2 into bicarbonate in the pathogen virulence induction [76]. As a matter of fact, bicarbonate is the first identified positive effector for ToxT activity, a regulatory protein that directly activates transcription of the genes encoding cholera toxin and other virulence genes. In this context, EZA inhibited bicarbonate-mediated virulence induction.

VchαCA and VchβCA were the object of a study published in 2017 by Mohamed et al. implying a series of sulfonamides incorporating a

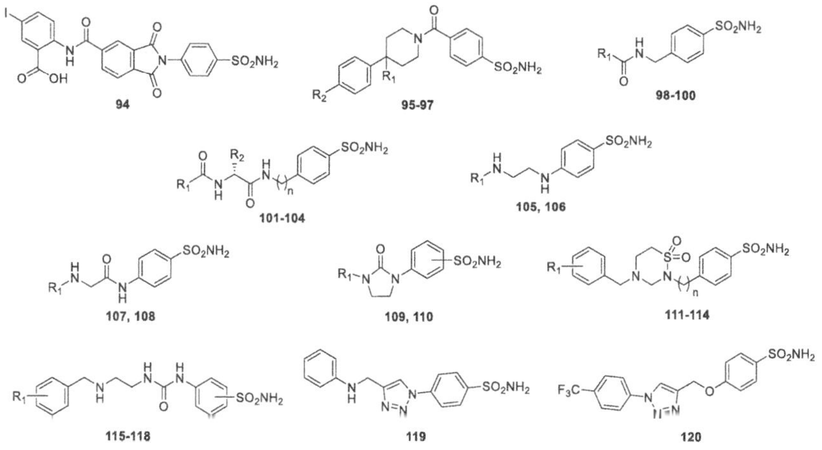

Fig. 10 Structure of sulfonamide derivatives **94–120**.

phthalimido moiety as human and bacterial CA inhibitors [77]. Compound **94** (Fig. 10, Table 11), though showing a limited inhibitory efficacy against both isozymes (K_Is of 281 and 5400 nM, respectively) was the most active of the study.

In 2019, Gitto et al. reported a series of benzenesulfonamides bearing piperidine and piperazine moieties as inhibitors of the three CAs from *V. cholerae* [78]. Compounds **95** and **96** (Fig. 10, Table 11) were the most potent inhibitors against VchαCA with K_Is of 43 and 45 nM, whilst **97** showed the best inhibitory performance against VchβCA (K_I of 90 nM). No inhibition was detected against VchγCA.

The same authors lately described the first ligand-based pharmacophore model to study selective inhibitors of VchβCA [79]. By a virtual screening applied to a collection of sulfonamides, nine compounds were retrieved, synthesized and assayed for inhibition of the three VchCAs. Compounds **98–100** (Fig. 10, Table 11) were among the most effective ones and showed low nanomolar efficacy against VchαCA (K_Is in the range 7.7–18 nM). The biphenyl compound **98** stood out as the most potent VchβCA inhibitor identified (K_I of 96 nM), while the highest potency against VchγCA was measured for derivatives **99** and **100**.

In 2021, the same authors defined VchCAs as drug targets, reporting a series of benzenesulfonamides bearing amino acids as linking groups by applying a structural simplification approach [80]. Many such compounds showed potency and selectivity of action over human CAs, with **101–104** (Fig. 10, Table 11) being the most effective ones. VchαCA was again the

Table 11 Inhibition profile of sulfonamide compounds **94–118** against VchαCA, VchβCA andVchγCA from *V. cholerae* and BpsγCA and BpsβCA from *B. pseudomallei*.

Cmp	m/p	n	R$_1$	R$_2$	K_I (nM)[a]				
					VchαCA	VchβCA	VchγCA	BpsβCA	BpsγCA
94	—	—	—	—	281	5400	nt	nt	nt
95	—	—	H	H	43	3307	>10,000	nt	nt
96	—	—	OH	Cl	45	3945	>10,000	nt	nt
97	—	—	CN	H	90	806	>10,000	nt	nt
98	—	—	biphen-4-yl	—	12	96	177	nt	nt
99	—	—	4-Br-C$_6$H$_4$	—	18	179	98	nt	nt
100	—	—	CH$_2$CH$_2$SCH$_3$	—	7.7	538	80	nt	nt
101	—	0	C$_6$H$_5$	H	0.4	884	563	nt	nt
102	—	0	C$_6$H$_5$	CH(CH$_3$)$_2$	9.2	389	24	nt	nt
103	—	0	4-Cl-C$_6$H$_4$	H	9.6	60	282	nt	nt
104	—	1	furan-2-yl	CH$_3$	8.5	611	93	nt	nt
105	—	—	CH$_2$C$_6$H$_5$	—	23	12,400	>50,000	nt	nt

			R						
106	—	—	$CH(CH_3)_2$	—	7.5	12,500	>50,000	nt	nt
107	—	—	$4\text{-}I\text{-}C_6H_4$	—	23	>50,000	>50,000	nt	nt
108	—	—	$6\text{-}CH_3\text{-pyridin-2-yl}$	—	17	>50,000	>50,000	nt	nt
109	p	—	$CH_2C_6H_5$	—	2.6	>50,000	8100	nt	nt
110	m	—	$4\text{-}Cl\text{-}C_6H_4$	—	4.0	>50,000	>50,000	nt	nt
111	—	0	H	—	76	432	7300	656	37
112	—	0	pentaF	—	8.5	4444	8109	82	824
113	—	2	4-F	—	6.1	211	8235	221	192
114	—	2	$4\text{-}NO_2$	—	94	43	8575	5500	19
115	m	—	4-F	—	631	4476	89	649	2428
116	m	—	4-OH	—	254	4735	89	6931	3621
117	m	—	2-OH, 4-Br	—	41	4966	95	7261	350
118	p	—	2-F	—	478	6858	97	2573	432

most efficiently inhibited isozyme (K_Is in the range 0.4–9.6 nM). Compounds **102** and **103** also acted as the most potent inhibitors of VchγCA (K_I of 24 nM) and VchβCA (K_I of 60 nM).

Lately, Fantacuzzi et al. reported a wide series of *p*- and *m*-benzenesulfonamides featuring moieties with a different flexibility degree to evaluate their impact on VchCAs inhibition [81]. The compounds, among the best inhibiting ones, were found in subset **105–110** (Fig. 10, Table 11) and were potent inhibitors of VchαCA, with K_Is in the range 2.6–23 nM. Unfortunately, they displayed weak or no activity against the other two VchCAs, with only a few measured low micromolar K_I values. Most of these derivatives were investigated in cells for their antibacterial effects against different clinical isolates of the *V. cholerae*, which are SI-Vc22, SI-Vc71, and SI-Vc912. However, the MIC determination showed a weak direct-acting activity of these compounds (MIC values > 64 µg/mL). These results are not entirely surprising, considering that inhibiting VchCAs may not harm the bacterium *in vitro*. Instead, this inhibition could be crucial *in vivo*, particularly concerning virulence.

In 2021, Akgul et al. prepared a series of taurultam benzenesulfonamides, that were tested as inhibitors of VchαCA, VchβCA, VchγCA, BpsβCA and BpsγCA [82]. VchαCA, and BpsγCA to a lesser extent, were the most affected isoforms with K_Is in the low nanomolar and medium nanomolar range, respectively. Specifically, compounds found in subset **111–114** (Fig. 10, Table 11) showed the best inhibitory performance. Among these, derivative **114** exhibited the highest VchβCA and BpsγCA inhibition over other such compounds with K_Is of 43 and 19 nM. In contrast, it was a medium VchαCA inhibitor (K_I of 90 nM) and a weak micromolar modulator of VchγCA and BpsβCA.

The same panel of five bacterial CAs were considered as drug targets in a study by Ali et al. describing a series of benzylaminoethylureido-tailed benzenesulfonamides such as **115–118** (Fig. 10, Table 11) [83]. Most of the compounds presented excellent inhibition against VchγCA compared to VchαCA and especially VchβCA, with K_Is in the range of 82–191 nM. A subset of sulfonamides, such as **115**, exhibited good inhibitory action against BpsβCA with K_I values in the range of 394–743 nM.

Compounds **119** and **120** (Fig. 10, Table 12) belong to a two-subset series of triazole benzenesulfonamide assayed in 2019 for the inhibition of VchαCA, VchβCA, and MtCA3 [84]. The first subset of derivatives (*e.g.* **119**) potently inhibited VchαCA with K_Is between 0.72 and 22.6 nM (Table 10), whilst the remaining compounds (*e.g.* **120**) preferentially

Table 12 Inhibition profile of sulfonamide compounds **119–130** against VchaCA, VchβCA, VchγCA and VchyCA from *V. cholerae*, BpsγCA and BpsβCA from *B. pseudomallei*, HpaCA from *H. pylori*, MtCA3 from *M. tuberculosis* and the β-class StCA2 from *S. enterica*.

Cmp	m/p	n	R₁	Kᵢ (nM)							
				VchaCA	VchβCA	VchγCA	BpsβCA	BpsγCA	HpaCA	MtCA3	StCA2
119	—	—	—	0.72	284	nt	nt	nt	nt	279	nt
120	—	—	—	57	102	nt	nt	nt	nt	28	nt
121	m	0	3-CH₃	0.36	7127	nt	nt	nt	nt	nt	nt
122	p	0	4-heptyl	0.30	8057	nt	nt	nt	nt	nt	nt
123	p	2	[1,3]dioxole	0.30	6777	nt	nt	nt	nt	nt	nt
124	—	—	C₆H₅	220	2610	8230	580	820	950	nt	nt
125	—	—	COOH	210	7750	3640	360	7000	640	nt	nt
126	—	—	—	870	8840	4700	90	4700	2430	nt	nt
127	—	—	—	33	865	nt	>10,000	405	nt	270	84
128	—	—	—	45	6651	nt	>10,000	321	nt	214	78
129	—	0	H	2.2	96	nt	4094	304	nt	202	815
130	—	2	4-OCH₃	3.2	85	nt	>10,000	>10,000	nt	9.3	>10,000

inhibited VchβCA (K_Is in the range 54.8–102.4 nM) and MtCA3 (K_Is in the range 28.2–192.5 nM).

From 2017 to 2020, Angeli et al. described several series of selenium (or occasionally other chalcogen atoms) as bacterial CA inhibitors [85–87]. Initially, a series of acyl selenoureido benzensulfonamides was reported to inhibit the sole VchαCA and VchβCA. Excellent inhibitory activity in the subnanomolar range was observed against VchαCA, with the most active compounds **121–123** (Fig. 11, Table 12) having K_Is in the range 0.30–0.36 nM (Table 10). The β-class isoform was much less inhibited with K_Is in the micromolar range. The same author therefore described a series of benzenesulfonamides incorporating selenazoles with diverse substitution patterns, which were evaluated against a wider panel of bacterial CAs, that are VchαCA, VchβCA, VchγCA, BpsβCA, BpsγCA and HpαCA. VchαCA and BpsβCA were the most inhibited isoforms (K_Is in the range 0.30–0.36 nM and 90–6200 nM respectively, Table 12), especially the best-performing derivatives that are found in the set **124–126** (Fig. 11, Table 12). The β- and γ-class enzymes from *V. cholerae* showed a weaker susceptibility to this class of inhibitors, with K_Is ranging between 2610–9120 nM and 3640–9290 nM respectively. Finally, compounds from this set showed a low micromolar to high nanomolar inhibitory action against BpsγCA and HpαCA, exemplified by **124–126** having K_Is between 820 and 7000 nM.

Thus, Angeli et al. reported benzenesulfonamides bearing thio- and seleno-acetamides linkers as inhibitors of VchαCA, VchβCA, BpsβCA, BpsγCA, MtCA3 and StCA2. Most compounds had not effects against BpsβCA, whilst VchαCA was the most affected isozyme of the evaluated panel. In detail, compounds of subset **127–130** (Fig. 11, Table 12) showed K_Is in the range 2.2–45 nM against VchαCA and medium nanomolar efficacy against BpsγCA and MtCA3 (K_Is in the range 202–405 nM).

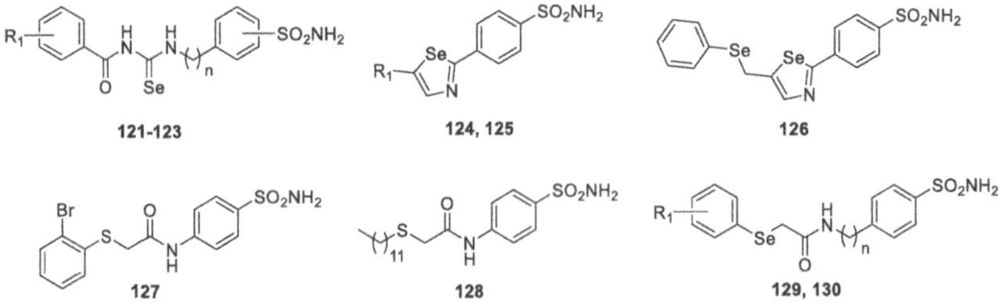

Fig. 11 Structure of chalcogenide-incorporating benzenesulfonamides **121–130**.

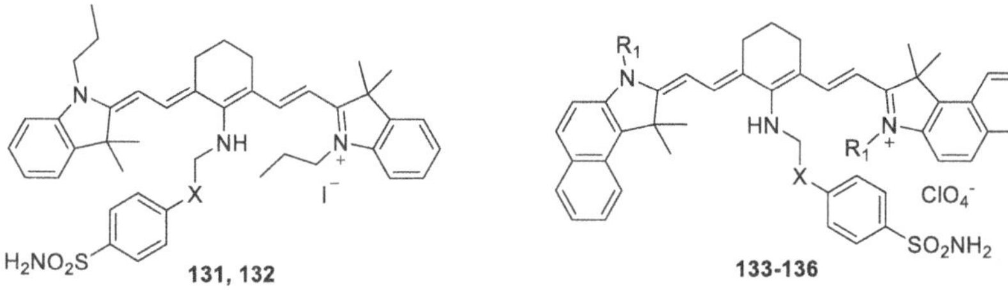

Fig. 12 Structure of heptamethine-incorporating benzenesulfonamides **133–136.**

Variable inhibition profiles, from low nanomolar to low micromolar ranges, were measured against VchβCA and StCA2.

Lately, heptamethine-based compounds decorated with benzenesulfonamide moieties were proposed to inhibit bacterial CAs and to be photoactivated by specific wavelengths setting a photodynamic therapy for treating hard-to-treat infections [88]. Compounds **131–136** (Fig. 12, Table 13) were assayed for their inhibition of a panel of isoforms displaying a generally good, medium nanomolar efficacy. **131**, **132** and **136** showed moderate to good antibacterial activity against Gram-positive bacteria such as *E. faecalis*, *S. aureus*, and *S. epidermidis*. When these compounds were irradiated with a 652 nm laser, a 10- to 100-fold increase was measured in the compound antibacterial action.

Twenty years after the discovery of NgαCA and its structural characterization by X-ray crystallography Flaherty and coworkers designed a series of AAZ-based compounds with the acetamido group of the lead replaced by aliphatic (acyclic and cyclic), aromatic and heterocyclic moieties, not all incorporating the amide functionality [89]. This study established both structure-activity relationships that contribute to both inhibitory action against NgαCA and antimicrobial activity. Many such derivatives, among which **137–144** (Fig. 13, Table 14) are the most effective ones, showed improved NgαCA inhibitory action than AAZ (K_I of 74 nM) up to a subnanomolar range (K_Is in the range 0.7–37 nM), though with limited increase in selectivity of action over hCA II. Compounds such as **140** and **143** displayed minimum inhibitory concentration values as low as 0.25 μg/mL equating to an 8- to 16-fold improvement in antigonococcal activity compared to AAZ (MIC of 0.25 μg/mL). MIC values up to 0.5 μg/mL were thus measured for **140** against a panel of 30 clinical isolates, and it outperformed azithromycin for antimicrobial activity. In a subsequent study, the efficacy of AAZ and EZA was compared against 22 clinical isolates of *N. gonorrhoeae*, including some drug-

Table 13 Inhibition profile of sulfonamide compounds **125–130** against VchaCA, VchβCA and VchγCA from *V. cholerae*, BpsγCA and BpsβCA from *B. pseudomallei*, HpaCA from *H. pylori*, MtCA3 from *M. tuberculosis* and the β-class StCA2 from *S. enterica*.

Cmp	X	R$_1$	K_I (nM)					
			MscCA	EcoβCA	EcoγCA	PsCA1	PsCA3	SmuCA
131	CH$_2$	—	62	55	120	60	89	102
132	CH$_2$NHCONH	—	56	69	93	47	48	73
133	CH$_2$	(CH$_2$)$_4$SO$_3$Na	59	60	117	65	77	99
134	CH$_2$NHCONH	(CH$_2$)$_4$SO$_3$Na	68	70	111	89	81	93
135	CH$_2$	CH$_3$	67	74	110	71	78	103
136	CH$_2$NHCONH	CH$_3$	74	65	124	58	94	85

resistant strains [90]. Despite EZA was found to be a weaker *in vitro* NgαCA inhibitor (K_I of 94 nM) than AAZ (K_I of 74 nM), it exhibited superior antimicrobial properties in cell-based assays, with MIC values ranging from 0.015 to 0.25 μg/mL, compared to AAZ with a MIC range of 0.125 to 16 μg/mL. No resistance to the two sulfonamides was detected, and postantibiotic effects were observed up to 10 h after treatment.

Following this, Seleem and colleagues demonstrated the *in vivo* efficacy of AAZ in a mouse model of *N. gonorrhoeae* infection [91]. Their study showed that AAZ reduced the gonococcal burden in the vaginas of infected mice by 90% after three days of treatment with 50 mg/kg AAZ.

Flaherty and coworkers also described the α- and γ-class enzymes encoded by *Enterococcus faecium* (EfαCA and EfγCA) as alternative antimicrobial drug targets to treat vancomycin-resistant enterococci (VRE) infections [92,93]. Initially, AAZ was used again as an investigational inhibitor, detecting a rather effective MIC against *E. faecium* strain HM-965 (MIC of 2 μg/mL) as well as against several other multidrug-resistant strains. Thus, the series of AAZ-based compounds, with the acetamido group of the lead replaced by aliphatic (acyclic and cyclic), aromatic and heterocyclic moieties, showed valuable inhibition profiles against EfαCA and EfγCA and MIC values for the inhibition of *E. faecium* HM-965 strain. The best antibiotic action against this strain was observed for compounds **138–141** and **143** (Fig. 13, Table 14) with MIC values in the range of 0.007–0.25 μg/mL. MIC values up to 0.007 μg/mL were thus measured for a subset of derivatives, among which **138–144**, against a panel of 11 clinical VRE isolates, with many of them outperforming linezolid in antimicrobial activity. Most amide compounds, especially **138–144** were also effective EfCAα inhibitors (K_Is in the range of 6.4–69.6 nM) and rather effective EfγCA inhibitors (K_Is in the range of 56.4–390 nM), but

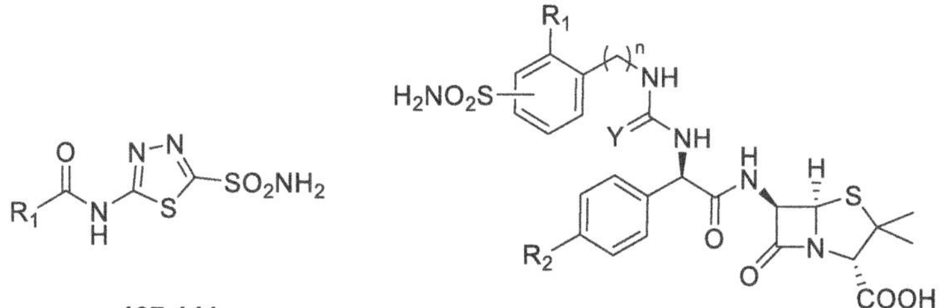

Fig. 13 Structure of thiadiazolesulfonamide derivatives **137–144** and penicillin-incorporating benzenesulfonamide compounds **145–150**.

Table 14 Inhibition profile of sulfonamide compounds **137–144** against NgCA from *Neisseria gonorrhoeae* and EfaCA and EfyCA from *Enterococcus faecium* and MIC values against *N. gonorrhoeae* CDC 181 and *E. faecium* HM-965 strains.

Cmp	R_1	K_I (nM)			MIC (µg/mL)	
		NgaCA	EfaCA	EfyCA	*N. gonorrhoeae* CDC 181	*E. faecium* HM-965
AAZ	CH_3	74	57	323	4	2
137	$CH_2CH_2CH_3$	27	50	325	8	0.5
138	$CH_2C(CH_3)_3$	9.1	70	390	16	0.015
139	cyclohexyl	9.8	49	131	2	0.25
140	CH_2–cyclohexyl	37	14	305	0.5	0.06
141	CH_2CH_2–cyclohexyl	0.7	67	346	1	0.007
142	C_6H_5	79	12	310	4	2
143	$CH_2CH_2C_6H_5$	8.3	6.4	148	0.25	0.06
144	CH_2–morpholin–1–yl	30	20	56	>64	1

Table 15 Inhibition profile of sulfonamide compounds **145–150** against NgCA from *Neisseria gonorrhoeae* and EcoβCA and EcoγCA from *E. coli*, and MIC values against *N. gonorrhoeae* FA1090 and CDC 181 strains.

Cmp	n	m/p	X	R_1	R_2	K_I (nM)			MIC (μg/mL)	
						NgaCA	EcoβCA	EcoγCA	*N. gonorrhoeae* FA1090	*N. gonorrhoeae* CDC 181
145	0	*m*	O	H	H	84	43	513	0.03	1
146	0	*p*	O	H	H	22	76	221	0.25	1
147	0	*p*	S	H	H	46	123	199	0.5	1
148	0	*p*	O	H	OH	7.1	71	216	0.25	0.5
149	0	*p*	S	Cl	OH	60	191	132	0.03	0.5
150	1	*p*	S	H	OH	114	464	185	0.125	1

there was not a straightforward correlation between EfCA inhibition and the measured MIC.

Recently, the same authors demonstrated the potential of using bacterial CAIs for gut VRE decolonization in an animal model [94]. AAZ, **138**, **141** and **144** were tested with experiments in mice infected with the *E. faecium* strain HM-952, and their gut bioburden was quantified in fecal pellets after 8 days of treatment with 20 mg/kg inhibitor. The treated animals showed significantly reduced *E. faecium* fecal bioburden compared to those treated with the vehicle and linezolid as a control drug. After 3 days of CAI treatment, the *E. faecium* fecal bioburden was reduced by 90%, with the greatest reduction observed after 5 days. AAZ was the most effective among the tested compounds, with this being attributed to its reduced gut absorption compared to the other more lipophilic inhibitors. Linezolid achieved only an 86.5% reduction after 8 days, indicating that sulfonamide CAIs might be significantly more effective decolonization agents against VRE than the standard clinically used drug.

Lately, a multitarget drug design approach was applied for the first time to inhibitors of bacterial CA combining a benzenesulfonamide scaffold with the β-lactam antibiotics amoxicillin and ampicillin [95]. The hybrid compounds were assayed for their inhibition of NgαCA, EcoβCA and EcoγCA. Derivatives of subset **145–150** (Fig. 13, Table 15) stood out as the most performing ones both in terms of inhibitory action (K_Is *versus* NgαCA ranging between 7.1 and 114 nM) and antibiotic efficacy against an *N. gonorrhoeae* FA1090 strain (MIC values in the range 0.03–0.5 μg/mL) outperforming ampicillin and amoxicillin 4- to 8-fold. A selection of derivatives was also tested against three multidrug-resistant *N. gonorrhoeae* strains such as CDC 181, with **148** and **149** exhibiting highly promising MIC values of 0.5 μg/mL. Importantly, an *in silico* simulation was used to show that these hybrid derivatives maintained the mechanism of action of the lead β-lactams covalently binding the penicillin binding protein 2 of *N. gonorrhoeae*.

4. Conclusions

This chapter provides an overview of the state-of-the-art research on sulfonamide inhibitors targeting bacterial CA isoforms. The emergence of antibiotic-resistant bacteria has underscored the need for new therapeutic targets. Consequently, bacterial CAs have attracted significant attention in drug design, offering a potential pathway for developing antibiotics with mechanisms of action distinct from those of currently used drugs. Over the past three years, the *in vivo* validation of CAs from *N. gonorrhoeae* and vancomycin-resistant enterococci as antibiotic targets has significantly renewed interest in this field. A paragraph was dedicated to the structural annotation of bacterial CAs, of the α, β, and γ classes, in complex with sulfonamide inhibitors using acetazolamide AAZ as a standard inhibitor. Thus, Tables 1–6 collected all inhibitory profiles of bacterial CAs with classical (Fig. 1) and clinically used (Fig. 2) sulfonamide compounds reported to date to provide information for drug design purposes. The subsequent section detailed all other series of sulfonamide CA inhibitors (Figs. 3–13) evaluated *in vitro* against bacterial CAs and, in some cases, against bacterial strains (Tables 7–15). This comprehensive collection underscores the potential of sulfonamide inhibitors as a novel class of antibacterial agents and paves the way for new and effective drug design strategies.

References

[1] K.S. Smith, C. Jakubzick, T.S. Whittam, J.G. Ferry, Carbonic anhydrase is an ancient enzyme widespread in prokaryotes, Proc. Natl. Acad. Sci. U. S. A. 96 (26) (1999) 15184–15189.

[2] T.H. Maren, Carbonic anhydrase: chemistry, physiology, and inhibition, Physiol. Rev. 47 (1967) 595–781.

[3] C.T. Supuran, Structure and function of carbonic anhydrases, Biochem. J. 473 (14) (2016) 2023–2032.

[4] Y. Hirakawa, M. Senda, K. Fukuda, et al., Characterization of a novel type of carbonic anhydrase that acts without metal cofactors, BMC Biol. 19 (1) (2021) 105.

[5] V. Alterio, A. Di Fiore, K. D'Ambrosio, C.T. Supuran, G. De Simone, Multiple binding modes of inhibitors to carbonic anhydrases: how to design specific drugs targeting 15 different isoforms? Chem. Rev. 112 (2012) 4421–4468.

[6] A. Nocentini, W.A. Donald, C.T. Supuran, Human carbonic anhydrases: tissue distribution, physiological role, and druggability, in: A. Nocentini, C.T. Supuran (Eds.), Carbonic Anhydrases, Academic Press, 2019, pp. 151–185.

[7] C.T. Supuran, Carbonic anhydrases: novel therapeutic applications for inhibitors and activators, Nat. Rev. Drug Discov. 7 (2008) 168–181.

[8] C.T. Supuran, Emerging role of carbonic anhydrase inhibitors, Clin. Sci. 135 (10) (2021) 1233–1249.

[9] C. Capasso, C.T. Supuran, Bacterial, fungal and protozoan carbonic anhydrases as drug targets, Expert. Opin. Ther. Targets 19 (2015) 1689–1704.

[10] C.T. Supuran, An overview of novel antimicrobial carbonic anhydrase inhibitors, Expert. Opin. Ther. Targets 27 (10) (2023) 897–910.

[11] D.P. Flaherty, M.N. Seleem, C.T. Supuran, Bacterial carbonic anhydrases: underexploited antibacterial therapeutic targets, Future Med. Chem. 13 (19) (2021) 1619–1622.

[12] C. Capasso, C.T. Supuran, An overview of the alpha-, beta- and gamma-carbonic anhydrases from bacteria: can bacterial carbonic anhydrases shed new light on evolution of bacteria? J. Enzyme Inhib. Med. Chem. 30 (2) (2015) 325–332.

[13] C.T. Supuran, C. Capasso, Antibacterial carbonic anhydrase inhibitors: an update on the recent literature, Expert. Opin. Ther. Pat. 30 (12) (2020) 963–982.

[14] A. Nocentini, C. Capasso, C.T. Supuran, Carbonic anhydrase inhibitors as novel antibacterials in the era of antibiotic resistance: where are we now? *Antibiotics* 12 (1) (2023) 142.

[15] C.T. Supuran, How many carbonic anhydrase inhibition mechanisms exist? J. Enzyme Inhib. Med. Chem. 31 (2016) 345–360.

[16] A. Nocentini, C.T. Supuran, Advances in the structural annotation of human carbonic anhydrases and impact on future drug discovery, Expert. Opin. Drug Discov. 14 (11) (2019) 1175–1197.

[17] A. Di Fiore, C. Capasso, V. De Luca, S.M. Monti, V. Carginale, C.T. Supuran, et al., X-ray structure of the first 'extremo-alpha-carbonic anhydrase', a dimeric enzyme from the thermophilic bacterium Sulfurihydrogenibium yellowstonense YO3AOP1, Acta Crystallogr. D. Biol. Crystallogr. 69 (Pt 6) (2013) 1150–1159.

[18] G. De Simone, S.M. Monti, V. Alterio, M. Buonanno, V. De Luca, M. Rossi, et al., Crystal structure of the most catalytically effective carbonic anhydrase enzyme known, SazCA from the thermophilic bacterium Sulfurihydrogenibium azorense, Bioorg. Med. Chem. Lett. 25 (9) (2015) 2002–2006.

[19] M.E. Compostella, P. Berto, F. Vallese, G. Zanotti, Structure of alpha-carbonic anhydrase from the human pathogen Helicobacter pylori, Acta Crystallogr. Sect. F Struct. Biol. Commun. 71 (Pt 8) (2015) 1005–1011.

[20] A. Di Fiore, K. D'Ambrosio, J. Ayoub, V. Alterio, G. De Simone, α-Carbonic anhydrases, in: A. Nocentini, C.T. Supuran (Eds.), Carbonic Anhydrases, Academic Press, 2019, pp. 19–54.

[21] S. Huang, Y. Xue, E. Sauer-Eriksson, L. Chirica, S. Lindskog, B.H. Jonsson, Crystal structure of carbonic anhydrase from *Neisseria gonorrhoeae* and its complex with the inhibitor acetazolamide, J. Mol. Biol. 283 (1) (1998) 301–310.

[22.] A.K. Marapaka, A. Nocentini, M.S. Youse, W. An, K.J. Holly, C. Das, et al., Structural characterization of thiadiazolesulfonamide inhibitors bound to *Neisseria gonorrhoeae* α-carbonic anhydrase, ACS Med. Chem. Lett. 14 (1) (2022) 103–109 6.

[23] J.K. Modak, Y.C. Liu, M.A. Machuca, C.T. Supuran, A. Roujeinikova, Structural basis for the inhibition of helicobacter pylori alpha-carbonic anhydrase by sulfonamides, PLoS One 10 (5) (2015) e0127149.

[24] J.K. Modak, Y.C. Liu, C.T. Supuran, A. Roujeinikova, Structure-activity relationship for sulfonamide inhibition of Helicobacter pylori alpha-carbonic anhydrase, J. Med. Chem. 59 (24) (2016) 11098–11109.

[25] A.B. Murray, R. McKenna, β-Carbonic anhydrases, in: A. Nocentini, C.T. Supuran (Eds.), Carbonic Anhydrases, Academic Press, 2019, pp. 55–78.

[26] M.A. Pinard, S.R. Lotlikar, C.D. Boone, D. Vullo, C.T. Supuran, M.A. Patrauchan, et al., Structure and inhibition studies of a type II beta-carbonic anhydrase psCA3 from Pseudomonas aeruginosa, Bioorg Med. Chem. 23 (15) (2015) 4831–4838.

[27] A. Suarez Covarrubias, A.M. Larsson, M. Högbom, J. Lindberg, T. Bergfors, C. Björkelid, et al., Structure and function of carbonic anhydrases from *Mycobacterium tuberculosis*, J. Biol. Chem. 280 (19) (2005) 18782–18789.

[28] S. Huang, T. Hainzl, C. Grundström, C. Forsman, G. Samuelsson, A.E. Sauer-Eriksson, Structural studies of β-carbonic anhydrase from the green alga Coccomyxa: inhibitor complexes with anions and acetazolamide, PLoS One 6 (12) (2011) e28458.

[29] A.B. Murray, M. Aggarwal, M. Pinard, D. Vullo, M. Patrauchan, C.T. Supuran, et al., Structural mapping of anion inhibitors to β-carbonic anhydrase psCA3 from *Pseudomonas aeruginosa*, ChemMedChem 13 (19) (2018) 2024–2029.

[30] M. Ferraroni, γ-Carbonic anhydrases, in: A. Nocentini, C.T. Supuran (Eds.), Carbonic Anhydrases, Academic Press, 2019, pp. 79–106.

[31] C. Kisker, H. Schindelin, B.E. Alber, J.G. Ferry, D.C. Rees, A left-handed β-helix revealed by the crystal structure of a carbonic anhydrase from the archaeon *Methanosarcina thermophila*, EMBO J. 15 (1996) 2323–2330.

[32] H.M. Park, J.H. Park, J.W. Choi, J. Lee, B.Y. Kim, C.H. Jung, et al., Structures of the γ-class carbonic anhydrase homologue YrdA suggest a possible allosteric switch, Acta Crystallogr. D. Biol. Crystallogr. 68 (Pt 8) (2012) 920–926.

[33] A. Bonardi, A. Nocentini, S.M. Osman, F.A. Alasmary, T.M. Almutairi, D.S. Abdullah, et al., Inhibition of α-, β- and γ-carbonic anhydrases from the pathogenic bacterium *Vibrio cholerae* with aromatic sulphonamides and clinically licenced drugs—a joint docking/molecular dynamics study, J. Enzyme Inhib. Med. Chem. 36 (1) (2021) 469–479.

[34] I. Nishimori, T. Minakuchi, K. Morimoto, S. Sano, S. Onishi, H. Takeuchi, et al., Carbonic anhydrase inhibitors: DNA cloning and inhibition studies of the alpha-carbonic anhydrase from *Helicobacter pylori*, a new target for developing sulfonamide and sulfamate gastric drugs, J. Med. Chem. 49 (6) (2006) 2117–2126.

[35] I. Nishimori, T. Minakuchi, T. Kohsaki, S. Onishi, H. Takeuchi, D. Vullo, et al., Carbonic anhydrase inhibitors: the beta-carbonic anhydrase from *Helicobacter pylori* is a new target for sulfonamide and sulfamate inhibitors, Bioorg Med. Chem. Lett. 17 (13) (2007) 3585–3594.

[36] T. Minakuchi, I. Nishimori, D. Vullo, A. Scozzafava, C.T. Supuran, Molecular cloning, characterization, and inhibition studies of the Rv1284 beta-carbonic anhy-drase from *Mycobacterium tuberculosis* with sulfonamides and a sulfamate, J. Med. Chem. 52 (8) (2009) 2226–2232.

[37] F. Carta, A. Maresca, A.S. Covarrubias, S.L. Mowbray, T.A. Jones, C.T. Supuran, Carbonic anhydrase inhibitors. Characterization and inhibition studies of the most active beta-carbonic anhydrase from *Mycobacterium tuberculosis*, Rv3588c, Bioorg Med. Chem. Lett. 19 (23) (2009) 6649–6654.

[38] I. Nishimori, T. Minakuchi, D. Vullo, A. Scozzafava, A. Innocenti, C.T. Supuran, Carbonic anhydrase inhibitors. Cloning, characterization, and inhibition studies of a new beta-carbonic anhydrase from *Mycobacterium tuberculosis*, J. Med. Chem. 52 (9) (2009) 3116–3120.

[39] P. Joseph, F. Turtaut, S. Ouahrani-Bettache, J.L. Montero, I. Nishimori, T. Minakuchi, et al., Cloning, characterization, and inhibition studies of a beta-car-bonic anhydrase from *Brucella suis*, J. Med. Chem. 53 (5) (2010) 2277–2285.

[40] P. Joseph, S. Ouahrani-Bettache, J.L. Montero, I. Nishimori, T. Minakuchi, D. Vullo, et al., A new β-carbonic anhydrase from *Brucella suis*, its cloning, characterization, and inhibition with sulfonamides and sulfamates, leading to impaired pathogen growth, Bioorg Med. Chem. 19 (3) (2011) 1172–1178.

[41] I. Nishimori, T. Minakuchi, D. Vullo, A. Scozzafava, C.T. Supuran, Inhibition studies of the β-carbonic anhydrases from the bacterial pathogen *Salmonella enterica* serovar Typhimurium with sulfonamides and sulfamates, Bioorg Med. Chem. 19 (16) (2011) 5023–5030.

[42] S. Del Prete, S. Isik, D. Vullo, V. De Luca, V. Carginale, A. Scozzafava, et al., DNA cloning, characterization, and inhibition studies of an α-carbonic anhydrase from the pathogenic bacterium *Vibrio cholerae*, J. Med. Chem. 55 (23) (2012) 10742–10748.

[43] S. Del Prete, D. Vullo, V. De Luca, V. Carginale, M. Ferraroni, S.M. Osman, et al., Sulfonamide inhibition studies of the β-carbonic anhydrase from the pathogenic bacterium *Vibrio cholerae*, Bioorg Med. Chem. 24 (5) (2016) 1115–1120.

[44] S. Del Prete, D. Vullo, V. De Luca, V. Carginale, S.M. Osman, Z. AlOthman, et al., Comparison of the sulfonamide inhibition profiles of the α-, β- and γ-carbonic anhydrases from the pathogenic bacterium *Vibrio cholerae*, Bioorg Med. Chem. Lett. 26 (8) (2016) 1941–1946.

[45] S. Del Prete, D. Vullo, S.M. Osman, A. Scozzafava, Z. AlOthman, C. Capasso, et al., Sulfonamide inhibition study of the carbonic anhydrases from the bacterial pathogen *Porphyromonas gingivalis*: the β-class (PgiCAb) versus the γ-class (PgiCA) enzymes, Bioorg Med. Chem. 22 (17) (2014) 4537–4543.

[46] I. Nishimori, D. Vullo, T. Minakuchi, A. Scozzafava, C. Capasso, C.T. Supuran, Sulfonamide inhibition studies of two β-carbonic anhydrases from the bacterial pathogen *Legionella pneumophila*, Bioorg Med. Chem. 22 (11) (2014) 2939–2946.

[47] N. Dedeoglu, V. DeLuca, S. Isik, H. Yildirim, F. Kockar, C. Capasso, et al., Sulfonamide inhibition study of the β-class carbonic anhydrase from the caries producing pathogen *Streptococcus mutans*, Bioorg Med. Chem. Lett. 25 (11) (2015) 2291–2297.

[48] D. Vullo, S. Del Prete, P. Di Fonzo, V. Carginale, W.A. Donald, C.T. Supuran, et al., Comparison of the sulfonamide inhibition profiles of the β- and γ-carbonic anhydrases from the pathogenic bacterium *Burkholderia pseudomallei*, Molecules 22 (3) (2017) 421.

[49] S. Del Prete, D. Vullo, S.M. Osman, Z. AlOthman, C.T. Supuran, C. Capasso, Sulfonamide inhibition profiles of the β-carbonic anhydrase from the pathogenic bacterium Francisella tularensis responsible of the febrile illness tularemia, Bioorg Med. Chem. 25 (13) (2017) 3555–3561.

[50] D. Vullo, R.S.S. Kumar, A. Scozzafava, J.G. Ferry, C.T. Supuran, Sulphonamide inhibition studies of the β-carbonic anhydrase from the bacterial pathogen *Clostridium perfringens*, J. Enzyme Inhib. Med. Chem. 33 (1) (2018) 31–36.

[51] S. Del Prete, V. De Luca, S. Bua, A. Nocentini, V. Carginale, C.T. Supuran, et al., The effect of substituted benzene-sulfonamides and clinically licensed drugs on the catalytic activity of CynT2, a carbonic anhydrase crucial for *Escherichia coli* life cycle, Int. J. Mol. Sci. 21 (11) (2020) 4175.

[52] S. Del Prete, S. Bua, C.T. Supuran, C. Capasso, *Escherichia coli* γ-carbonic anhydrase: characterisation and effects of simple aromatic/heterocyclic sulphonamide inhibitors, J. Enzyme Inhib. Med. Chem. 35 (1) (2020) 1545–1554.

[53] V. De Luca, S. Giovannuzzi, C.T. Supuran, C. Capasso, May sulfonamide inhibitors of carbonic anhydrases from *Mammaliicoccus sciuri* prevent antimicrobial resistance due to gene transfer to other harmful Staphylococci? Int. J. Mol. Sci. 23 (22) (2022) 13827.

[54] D. Vullo, V.D. Luca, A. Scozzafava, V. Carginale, M. Rossi, C.T. Supuran, et al., The alpha-carbonic anhydrase from the thermophilic bacterium *Sulfurihydrogenibium yellowstonense* YO3AOP1 is highly susceptible to inhibition by sulfonamides, Bioorg Med. Chem. 21 (6) (2013) 1534–1538.

[55] D. Vullo, V. De Luca, A. Scozzafava, V. Carginale, M. Rossi, C.T. Supuran, et al., The extremo-α-carbonic anhydrase from the thermophilic bacterium *Sulfurihydrogenibium azorense* is highly inhibited by sulfonamides, Bioorg. Med. Chem. 21 (15) (2013) 4521–4525.

[56] D. Vullo, V. De Luca, S. Del Prete, V. Carginale, A. Scozzafava, S.M. Osman, et al., Sulfonamide inhibition studies of the γ-carbonic anhydrase from the Antarctic bacterium *Colwellia psychrerythraea*, Bioorg. Med. Chem. Lett. 26 (4) (2016) 1253–1259.

[57] D. Vullo, V. De Luca, S. Del Prete, V. Carginale, A. Scozzafava, C. Capasso, et al., Sulfonamide inhibition studies of the γ-carbonic anhydrase from the Antarctic cyanobacterium Nostoc commune, Bioorg. Med. Chem. 23 (8) (2015) 1728–1734.

[58] D. Vullo, V. De Luca, S. Del Prete, V. Carginale, A. Scozzafava, C. Capasso, et al., Sulfonamide inhibition studies of the γ-carbonic anhydrase from the Antarctic bacterium *Pseudoalteromonas haloplanktis*, Bioorg. Med. Chem. Lett. 25 (17) (2015) 3550–3555.

[59] V. De Luca, A. Petreni, A. Nocentini, A. Scaloni, C.T. Supuran, C. Capasso, Effect of sulfonamides and their structurally related derivatives on the activity of ι-carbonic anhydrase from *Burkholderia territorii*, Int. J. Mol. Sci. 22 (2) (2021) 571.

[60] O. Güzel, A. Maresca, A. Scozzafava, A. Salman, A.T. Balaban, C.T. Supuran, Discovery of low nanomolar and subnanomolar inhibitors of the mycobacterial beta-carbonic anhydrases Rv1284 and Rv3273, J. Med. Chem. 52 (13) (2009) 4063–4067.

[61] A. Maresca, F. Carta, D. Vullo, A. Scozzafava, C.T. Supuran, Carbonic anhydrase inhibitors. Inhibition of the Rv1284 and Rv3273 beta-carbonic anhydrases from Mycobacterium tuberculosis with diazenylbenzenesulfonamides, Bioorg. Med. Chem. Lett. 19 (17) (2009) 4929–4932.

[62] A. Maresca, A. Scozzafava, D. Vullo, C.T. Supuran, Dihalogenated sulfanilamides and benzolamides are effective inhibitors of the three β-class carbonic anhydrases from *Mycobacterium tuberculosis*, J. Enzyme Inhib. Med. Chem. 28 (2) (2013) 384–387.

[63] M. Rami, A. Innocenti, J.L. Montero, A. Scozzafava, J.Y. Winum, C.T. Supuran, Synthesis of rhodamine B-benzenesulfonamide conjugates and their inhibitory activity against human α- and bacterial/fungal β-carbonic anhydrases, Bioorg. Med. Chem. Lett. 21 (18) (2011) 5210–5213.

[64] F. Pacchiano, F. Carta, D. Vullo, A. Scozzafava, C.T. Supuran, Inhibition of β-carbonic anhydrases with ureido-substituted benzenesulfonamides, Bioorg. Med. Chem. Lett. 21 (1) (2011) 102–105.

[65] M. Ceruso, D. Vullo, A. Scozzafava, C.T. Supuran, Sulfonamides incorporating fluorine and 1,3,5-triazine moieties are effective inhibitors of three β-class carbonic anhydrases from *Mycobacterium tuberculosis*, J. Enzyme Inhib. Med. Chem. 29 (5) (2014) 686–689.

[66] S. Bua, A. Bonardi, G.R. Mük, A. Nocentini, P. Gratteri, C.T. Supuran, Benzothiadiazinone-1,1-dioxide carbonic anhydrase inhibitors suppress the growth of drug-resistant *Mycobacterium tuberculosis* strains, Int. J. Mol. Sci. 25 (5) (2024) 2584.

[67] J.K. Modak, A. Tikhomirova, R.J. Gorrell, et al., Anti-Helicobacter pylori activity of ethoxzolamide, J. Enzyme Inhib. Med. Chem. 34 (1) (2019) 1660–1667.

[68] D. Vullo, I. Nishimori, A. Scozzafava, S. Köhler, J.Y. Winum, C.T. Supuran, Inhibition studies of a beta-carbonic anhydrase from Brucella suis with a series of water soluble glycosyl sulfanilamides, Bioorg. Med. Chem. Lett. 20 (7) (2010) 2178–2182.

[69] J. Ombouma, D. Vullo, S. Köhler, P. Dumy, C.T. Supuran, J.Y. Winum, N-glycosyl-N-hydroxysulfamides as potent inhibitors of Brucella suis carbonic anhydrases, J. Enzyme Inhib. Med. Chem. 30 (6) (2015) 1010–1012.

[70] S.M. Monti, A. Meccariello, M. Ceruso, K. Szafrański, J. Sławiński, C.T. Supuran, Inhibition studies of Brucella suis β-carbonic anhydrases with a series of 4-substituted pyridine-3-sulphonamides, J. Enzyme Inhib. Med. Chem. 33 (1) (2018) 255–259.

[71] M. Ceruso, F. Carta, S.M. Osman, Z. Alothman, S.M. Monti, C.T. Supuran, Inhibition studies of bacterial, fungal and protozoan β class carbonic anhydrases with Schiff bases incorporating sulfonamide moieties, Bioorg. Med. Chem. 23 (15) (2015) 4181–4187.

[72] F. Carta, S.M. Osman, D. Vullo, Z. AlOthman, S. Del Prete, C. Capasso, et al., Poly (amidoamine) dendrimers show carbonic anhydrase inhibitory activity against α-, β-, γ- and η-class enzymes, Bioorg. Med. Chem. 23 (21) (2015) 6794–6798.

[73] M. Ceruso, S. Del Prete, Z. AlOthman, S.M. Osman, A. Scozzafava, C. Capasso, et al., Synthesis of sulfonamides with effective inhibitory action against Porphyromonas gingivalis γ-carbonic anhydrase, Bioorg. Med. Chem. Lett. 24 (16) (2014) 4006–4010.

[74] A.M. Alafeefy, M. Ceruso, A.M. Al-Tamimi, S. Del Prete, C.T. Supuran, C. Capasso, Inhibition studies of quinazoline-sulfonamide derivatives against the γ-CA (PgiCA) from the pathogenic bacterium, Porphyromonas gingivalis, J. Enzyme Inhib. Med. Chem. 30 (4) (2015) 592–596.

[75] A.M. Alafeefy, M. Ceruso, A.M. Al-Tamimi, S. Del Prete, C. Capasso, C.T. Supuran, Quinazoline-sulfonamides with potent inhibitory activity against the α-carbonic anhydrase from Vibrio cholerae, Bioorg. Med. Chem. 22 (19) (2014) 5133–5140.

[76] B.H. Abuaita, J.H. Withey, Bicarbonate induces *Vibrio cholerae* virulence gene expression by enhancing ToxT activity, Infect. Immun. 77 (11) (2009) 5202 –5202.

[77] M.A. Mohamed, A.A. Abdel-Aziz, H.M. Sakr, A.S. El-Azab, S. Bua, C.T. Supuran, Synthesis and human/bacterial carbonic anhydrase inhibition with a series of sulfonamides incorporating phthalimido moieties, Bioorg. Med. Chem. 25 (8) (2017) 2524–2529.

[78] R. Gitto, L. De Luca, F. Mancuso, S. Del Prete, D. Vullo, C.T. Supuran, et al., Seeking new approach for therapeutic treatment of cholera disease via inhibition of bacterial carbonic anhydrases: experimental and theoretical studies for sixteen benzenesulfonamide derivatives, J. Enzyme Inhib. Med. Chem. 34 (1) (2019) 1186–1192.

[79] F. Mancuso, L. De Luca, A. Angeli, E. Berrino, S. Del Prete, C. Capasso, et al., In silico-guided identification of new potent inhibitors of carbonic anhydrases expressed in *Vibrio cholerae*, ACS Med. Chem. Lett. 11 (11) (2020) 2294–2299.

[80] F. Mancuso, L. De Luca, F. Bucolo, M. Vrabel, A. Angeli, C. Capasso, et al., 4-Sulfamoylphenylalkylamides as inhibitors of carbonic anhydrases expressed in *Vibrio cholerae*, ChemMedChem 16 (24) (2021) 3787–3794.

[81] M. Fantacuzzi, I. D'Agostino, S. Carradori, et al., Benzenesulfonamide derivatives as *Vibrio cholerae* carbonic anhydrases inhibitors: a computational-aided insight in the structural rigidity-activity relationships, J. Enzyme Inhib. Med. Chem. 38 (1) (2023) 2201402.

[82] O. Akgul, A. Angeli, S. Selleri, C. Capasso, C.T. Supuran, F. Carta, Taurultams incorporating arylsulfonamide: first in vitro inhibition studies of α-, β- and γ-class Carbonic Anhydrases from *Vibrio cholerae* and *Burkholderia pseudomallei*, Eur. J. Med. Chem. 219 (2021) 113444.

[83] M. Ali, A. Angeli, M. Bozdag, F. Carta, C. Capasso, U. Farooq, et al., Benzylaminoethylureido-tailed benzenesulfonamides show potent inhibitory activity against bacterial carbonic anhydrases, ChemMedChem 15 (24) (2020) 2444–2447.

[84] S. Bua, S.M. Osman, S. Del Prete, C. Capasso, Z. AlOthman, A. Nocentini, et al., Click-tailed benzenesulfonamides as potent bacterial carbonic anhydrase inhibitors for targeting *Mycobacterium tuberculosis* and *Vibrio cholerae*, Bioorg. Chem. 86 (2019) 183–186.

[85] A. Angeli, G. Abbas, S. Del Prete, F. Carta, C. Capasso, C.T. Supuran, Acyl selenoureido benzensulfonamides show potent inhibitory activity against carbonic anhydrases from the pathogenic bacterium *Vibrio cholerae*, Bioorg. Chem. 75 (2017) 170–172.

[86] A. Angeli, M. Pinteala, S.S. Maier, S. Del Prete, C. Capasso, B.C. Simionescu, et al., Inhibition of bacterial α-, β- and γ-class carbonic anhydrases with selenazoles incorporating benzenesulfonamide moieties, J. Enzyme Inhib. Med. Chem. 34 (1) (2019) 244–249.

[87] A. Angeli, M. Pinteala, S.S. Maier, B.C. Simionescu, A. Milaneschi, G. Abbas, et al., Evaluation of thio- and seleno-acetamides bearing benzenesulfonamide as inhibitor of carbonic anhydrases from different pathogenic bacteria, Int. J. Mol. Sci. 21 (2) (2020) 598.

[88] S. Carradori, A. Angeli, P.S. Sfragano, X. Yzeiri, M. Calamante, D. Tanini, et al., Photoactivatable heptamethine-based carbonic anhydrase inhibitors leading to new anti-antibacterial agents, Int. J. Mol. Sci. 24 (11) (2023) 9610.

[89] C.S. Ewitt, N.S. Abutaleb, A.E.M. Elhassanny, A. Nocentini, X. Cao, D.P. Amos, et al., Structure-activity relationship studies of acetazolamide-based carbonic anhydrase inhibitors with activity against *Neisseria gonorrhoeae*, ACS Infect. Dis. 7 (7) (2021) 1969–1984.

[90] N.S. Abutaleb, A.E.M. Elhassanny, A. Nocentini, et al., Repurposing FDA-approved sulphonamide carbonic anhydrase inhibitors for treatment of *Neisseria gonorrhoeae*, J. Enzyme Inhib. Med. Chem. 37 (1) (2022) 51–61.

[91] N.S. Abutaleb, A.E.M. Elhassanny, M.N. Seleem, In vivo efficacy of acetazolamide in a mouse model of *Neisseria gonorrhoeae* infection, Microb. Pathog. 164 (2022) 105454.

[92] J. Kaur, X. Cao, N.S. Abutaleb, A. Elkashif, A.L. Graboski, A.D. Krabill, et al., Optimization of acetazolamide-based scaffold as potent inhibitors of vancomycin-resistant *Enterococcus*, J. Med. Chem. 63 (17) (2020) 9540–9562.

[93] W. An, K.J. Holly, A. Nocentini, R.D. Imhoff, C.S. Hewitt, N.S. Abutaleb, et al., Structure-activity relationship studies for inhibitors for vancomycin-resistant *Enterococcus* and human carbonic anhydrases, J. Enzyme Inhib. Med. Chem. 37 (1) (2022) 1838–1844.

[94] N.S. Abutaleb, A. Shrinidhi, A.B. Bandara, M.N. Seleem, D.P. Flaherty, Evaluation of 1,3,4-thiadiazole carbonic anhydrase inhibitors for gut decolonization of vancomycin-resistant *Enterococci*, ACS Med. Chem. Lett. 14 (4) (2023) 487–492.

[95] A. Bonardi, A. Nocentini, S. Giovannuzzi, N. Paoletti, A. Ammara, S. Bua, et al., Development of penicillin-based carbonic anhydrase inhibitors targeting multidrug-resistant *Neisseria gonorrhoeae* (in press.), J. Med. Chem. (2024), https://doi.org/10. 1021/acs.jmedchem.4c00740

CHAPTER SEVEN

Non-sulfonamide bacterial CA inhibitors

Fabrizio Carta[*]
NEUROFARBA Department, Sezione di Scienze Farmaceutiche e Nutraceutiche, University of Florence, Sesto Fiorentino, Florence, Italy
*Corresponding author. e-mail address: fabrizio.carta@unifi.it

Contents

1. Introduction   193
2. Anionic species   194
3. Coumarins   201
4. Phenols   202
5. *N*-Methyl thiosemicarbazones   203
6. *N*-Hydroxy ureas   205
7. Miscellaneous   207
8. Conclusions   207
References   208

Abstract

Non-sulfonamide chemical moieties able to inhibit the bacterial (b) expressed Carbonic Anhydrases (CAs; EC 4.2.1.1) constitute an important alternative to the prototypic modulators discussed in Chapter 6, as give access to large and variegate chemical classes, also of the natural origin. This contribution reports the main classes of compounds profiled *in vitro* on the bCAs and thus may be worth developing for the validation process of this class of enzymes.

1. Introduction

Bacterial genomes encode for the α-, β-, γ- and ι-CAs, and all of them do contain the catalytic Zn (II) ion buried within their active site coordinated by means of different aminoacid patterns [1–4]. Despite such differences in the structural organization of the catalytic core, the challenge to selectively disrupt the bacterial (b) expressed CAs over the human (h) ones remains [5]. An immediate approach to tackle such an issue is represented by the development of non-sulfonamidic inhibitors. The rationale behind this approach is to make use of the chemical/physical features of CA targeting new moieties in order to discriminate between host and non-host expressed isozymes [5,6]. It is intended that such a discriminative approach accounts both for the selective interaction of the ligand with the target enzyme as well as for its ability to reach it. The latter consideration is high relevant since bacteria do contain their enzymes into membrane confined spaces. A very interesting example of CAs compartmentalization is for the α-family which in bacteria are exclusively expressed within the periplasmic space of Gram-negative strains and thus not present in Gram-positives [7].

Herein are reported the main chemical moieties of the non-sulfonamide type which are currently considered at different stages for the inhibition of bCAs.

2. Anionic species

Anionic species are widely present in biological environments and do have significant regulating activities on the eukaryotic and prokaryotic enzymatic pools by means of variegate mechanisms which include competitive inhibition, allosteric regulation, metal co-factoring and pH buffering [8]. Such species have been largely profiled *in vitro* for their potential modulating features on bCAs, and usually show K_I values in the sub- or millimolar ranges [9].

The rational in profiling the anionic species lays on the complexing features of some of them for transition metal ions in solution or within the active sites of metalloenzymes such as the CAs [10–15]. Specifically, the panel of anions usually considered are reported in Table 1.

Table 1 List of ions profiled as bCAIs.

F^-	N_3^-	SnO_3^{2-}	$B_4O_7^{2-}$	SO_4^{2-}
Cl^-	HCO_3^-	SeO_4^{2-}	ReO_4^-	ClO_4^-
Br^-	CO_3^{2-}	TeO_4^{2-}	RuO_4^-	BF_4^-
I^-	NO_3^-	OsO_5^{2-}	$S_2O_8^{2-}$	FSO_3^-
CNO^-	NO_2^-	$S_2O_7^{2-}$	$SeCN^-$	SO_4^{2-}
SCN^-	HS^-	$P_2O_7^{4-}$	CS_3^{2-}	PF_6^-
CN^-	HSO_3^-	$V_2O_7^{4-}$	Et_2NCS^{2-}	$CF_3SO_3^-$

The list of scientific contributions reporting the *in vitro* evaluation of ionic species on the bCAs is quite vast, also in consideration that almost each identified CA isoform reported in the literature was initially assessed for the ability of such species to modulate the enzymatic activity [16–24]. None the anionic species listed in Table 1 has been reported in adduct with any bCA and its binding mode assessed by means of X-ray crystallography or any of the biophysical techniques, such as circular dichroism (CD), electron para-magnetic spectroscopy (EPR), electronic spectroscopy (UV/Vis), and NMR techniques. The most representative ions such as bromine [25], hydrogen sulfide [26], azide [27], bicarbonate [28], thiocyanate [29] and cyanamide [30] ions were experimentally assessed by means of X-ray crystallography for their interactions within the hCA II enzyme assumed as model (Fig. 1).

It is reasonable to assume that the binding modes of such ions, when endowed of appreciable enzymatic activity, are preserved in all bCAs of the α-type. For the remaining genetic families any indication of the ion binding may be presumed from non-direct measurements of the biophysical type and/or from *in silico* investigations.

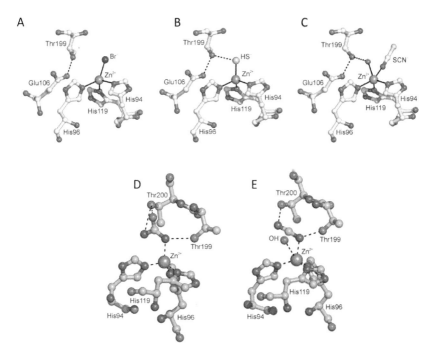

Fig. 1 X-Ray crystallographic adducts of (A) bromine [25], (B) hydrogen sulphide [26], (C) thiocyanate [29], (D) bicarbonate [28] and (E) cyanamide [30] with hCA II.

To date, neither *in vitro* cellular, *in vivo* experiments or clinical studies are reported with the intent to afford such ions a biomedical application based on their CA modulating features. More importantly, such ions do represent valuable models to develop highly efficient CA modulating species of the organic/metallorganic type. Examples of this kind include the dithio- (DTC) [31], monothio- (MTC) and xanthates (XTC) [32] moieties which are all chemically derived from the progenitor trithio-carbonate ion (CS_3^{2-}) [33] (Fig. 2).

The first scientific contribution which reported the *in vitro* kinetic evaluation of DTCs on bCAs dates in 2013 [34]. Specifically, a series of *N*-mono- and *N,N*-disubstituted DTCs was profiled for their activity on the *Mycobacterium tuberculosis* expressed β-CAs mtCA 1 (Rv1284) and mtCA 3 (Rv3273) and compared to the hCAs I and II (Table 2).

Both mtCAs considered in the study were effectively inhibited by this DTC series as they reported K_I values spanning between sub- to medium-nanomolar values, depending on the substitution patterns at the metal binding group. Among the compound series reported, the DTC **FC14–584B** in Fig. 3 resulted particularly effective in inhibiting plated *M. marinum* colonies as well in infected zebrafish larvae. In the last case marginal cytotoxicity was reported [35].

An expanded DTC chemical series along with the isosteric MTCs were explored for their *in vitro* modulating activities on the mtCA3 isoform, which is considered as particularly relevant for the *Mycobacterium tuberculosis* metabolisms [36] (Fig. 4).

The kinetic results in Table 3A showed this enzyme susceptible of inhibition by some DTCs with K_I values comprised between 2.4 and 812 nM. Among the series, derivatives **1**, **3**, **5**, **7** and **10** exhibited minimal toxicity profiles (Table 3B) and were considered worth considering for further evaluation *in vivo* [36].

A selected contribution dealing with the evaluation of DTCs on bCAs is the work of Giovannuzzi et al. on *Neisseria gonorrhoeae* α-CA (NgCA) [37]. A series of 31 DTCs reported K_Is spanning between 83.7 nM and 5.1 μM for such isoform. Furthermore, the series was investigated *in vitro*

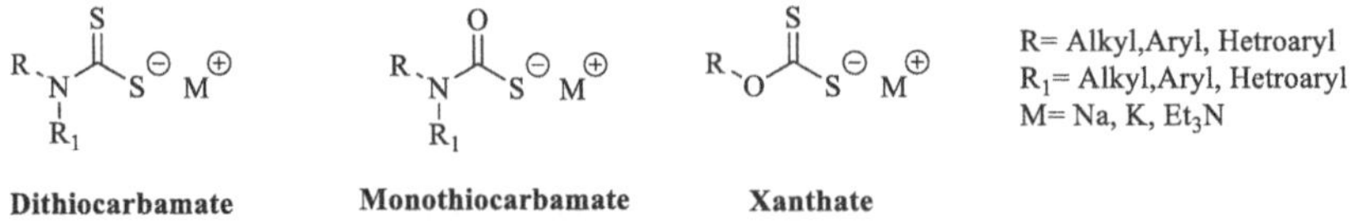

R= Alkyl,Aryl, Hetroaryl
R_1= Alkyl,Aryl, Hetroaryl
M= Na, K, Et_3N

Dithiocarbamate   Monothiocarbamate   Xanthate

Fig. 2 General structures of **DTC**, **MTC** and **TTCs**.

Table 2 Inhibition data of the human (h) α-CA isoforms hCA I and II, and the mycobacterial β-CA isoforms mtCAs 1 and 3 with DTCs [34].

Cmpnd	R[1]	R[2]	K_I (nM)*				M
			hCA I	hCA II	mtCA 1	MtCA 3	
1	H	Ph	4.8	4.5	5.6	2.5	Et$_3$NH
2	H	O[(CH$_2$CH$_2$)]$_2$N	4.8	3.6	6.1	2.4	K
3	H	MeN[(CH$_2$CH$_2$)]$_2$N	33.5	33.0	4.7	2.6	K
4	H	2-buyl	21.1	29.4	6.0	3.6	K
5	H	O[(CH$_2$CH$_2$)]$_2$N(CH$_2$)$_2$	31.8	36.3	7.1	2.8	K
6**	H	N[(CH$_2$CH$_2$)N]$_3$	31.9	13.5	4.2	4.0	K
7	H	PhC-I$_2$	4.1	0.7	7.1	87.3	Na
8	h	4-Pyridyl CH$_2$	3.5	16.6	5.4	5.7	Et$_3$NH
9	H	[(CH$_2$)$_5$N]CH$_2$CH$_2$	4.5	20.3	9.1	8.8	k
10	H	2-thiazolyl	3.9	4.6	89.4	9.5	Et$_3$NH
11	H	KOOCCH$_2$	13.1	325	7.8	8.3	K
12	H	imidazole-1-yl-(CH$_2$)$_3$	8.6	24.7	5.3	8.7	A
13	Me	Me	699	6910	893	659	Na
14	Et	Et	790	3100	615	431	Na

(continued)

Table 2 Inhibition data of the human (h) α-CA isoforms hCA I and II, and the mycobacterial β-CA isoforms mtCAs 1 and 3 with DTCs [34]. (cont'd)

Cmpnd	R^1	R^2	K_i (nM)*				M
			hCA I	hCA II	mtCA 1	MtCA 3	
15		(CH$_2$)$_5$	0.96	27.5	90.5	4.1	Na
16	n-Pr	n-Pr	1838	55.5	74.8	80.0	Na
17	n-Bu	n-Bu	43.1	50.9	81.7	72.8	Na
18	iso-Bu	iso-Bu	0.97	0.95	86.2	43.0	Na
19	n-Hex	n-Hex	48.0	51.3	95.4	51.7	Na
20	Et	n-Bu	157	27.8	91.6	63.5	Na
21	HOCH$_2$CH$_2$	HOCH$_2$CH$_2$	9.2	4.0	7.5	6.0	Na
22	Me	Ph	39.6	21.5	25.2	46.8	Na
23	Me	Ph CH$_2$	69.9	25.4	72.0	62.5	Na
24		O[(CH$_2$CH$_2$)]$_2$	0.88	0.95	0.94	0.91	Na
25		NaS$_2$CN[(CH$_2$CH$_2$)]$_2$	12.6	0.92	7.7	8.0	Na
26		(NC)(Ph)C(CH$_2$CH$_2$)$_2$	48.4	40.8	93.0	61.2	N
27*		(s)-[CH$_2$ CH$_2$ CH$_2$ CH(COONa)]	2.5	17.3	7.1	6.4	Na
—		AZA (acetazolamide)	250	12	481	104	—

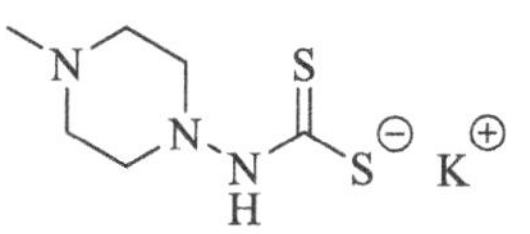

FC14-584B

Fig. 3 Chemical structure of the DTC FC14-584B.

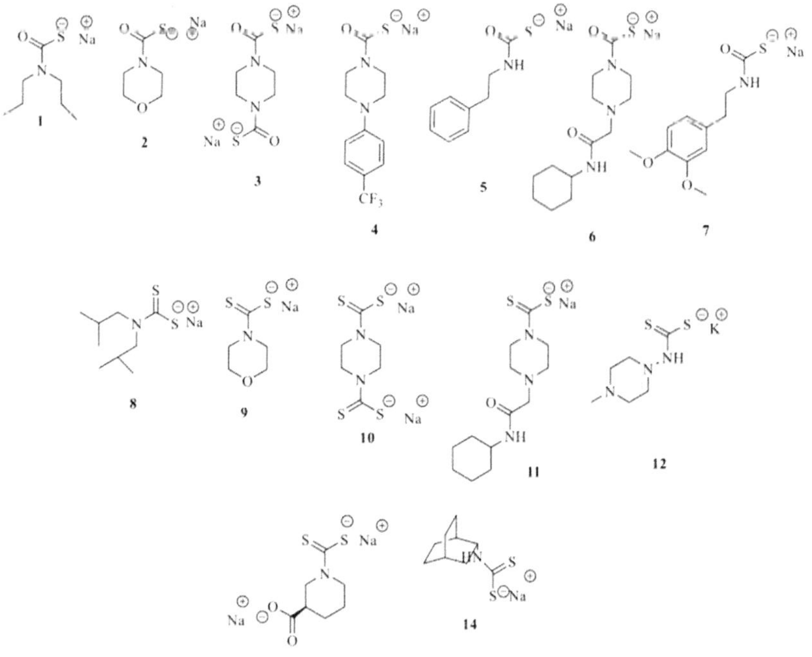

Fig. 4 Chemical structures of MTCs **1–7** and DTCs **8–14** [36].

for their antibacterial activity on plated *N. gonorrhoeae* strains. Only DTCs featured with remarkable hydrophobic scaffolds did exhibit moderate activity although the associated NgCA inhibition was rather limited [37].

Similar trend was reported for a series of MTC containing moieties which were profiled on the α-NgCA and the α- and γ-CAs (*i.e.* α-EfCA and γ-EfCA) from the Gram–positive *Enterococcus faecium* [38]. The *in vitro* K_I values were in the medium–high nanomolar range and again the most potent inhibitors resulted poorly effective in reducing the growth of such bacterial strains *in vitro*.

Table 3 (A) Inhibition data of Mtb β-CA3 and human CA isoforms hCA I and II for compounds **1–14** and compared to the acetazolamide (**AAZ**) as a standard drug; (B) Compounds **1, 3, 5, 7** and **10** that inhibit Mtb β-CA3 efficiently and show minimal toxicity [37].

Compounds	K_I (nM)[a]		
	Mtb β-CA3	hCA I	hCA II
1	558.6	>2000[26]	46.7[26]
2	747.26	569[26]	>2000[26]
3	812.0	876[26]	22.4[26]
4	83.3	>2000[26]	43.6[26]
5	93.0	>2000[26]	43.7[26]
6	780.7	949[26]	45.9[26]
7	97.9	891[26]	26.7[26]
8	43.0	0.97[25]	0.95[25]
9	2.4	0.88[25]	0.95[25]
10	8.0	12.6[25]	0.92[25]
11	>2000	415[27]	67.2[27]
12	2.6	33.5[25]	33.0[25]
13	>2000	496[27]	80.5[27]
14	>2000	494[27]	48.7[27]
AAZ	104	250.0	12.0

Compounds	K_I (nM)[a]	Safe concentration (μM)[b]
1	558.6	400
3	812.0	500
5	93.0	400
7	97.9	500
10	8.0	500

3. Coumarins

Validation of the coumarins as CAIs, greatly extended the variety of structural diversity beneficial for the development of CA modulators [39–45]. The inhibition mechanism of such a moiety was investigated in detail on the hCA II/6-(1S-Hydroxy-3-methylbutyl)-7-methoxy-2H-chromen-2-one adduct, and revealed a suicide-based mechanism which relies on the esterase activity [43] which is exclusively possessed by this class (Fig. 5). Later, evidences for the thioesterase activity were also reported [46].

Structural differences among human (h) and bacterial (b) expressed α-CAs, although minimal, are present and usually do affect the proteins multimeric organization and thus have effects on the catalytic activities and ligand binding affinities [43].

An alternative inhibition mechanism was proved for the 2-thioxocoumarin scaffold, which make use of the *exo*-sulfur atom to establish hydrogen bonds with the water coordinating Zn (II) ion (Fig. 6) [47].

Coumarins as inhibitors of the bCAs belonging to the α-class are likely to share the same mechanisms with the mammals, unless alternative mechanisms yet to be discovered, take place. Although no adducts of the type in Fig. 5 with the bCAs are currently available, robust evidence for the esterase intrinsic activities of such isoforms are reported in the literature, whereas the β-, γ- and ι-CAs are clearly devoid [48]. Such a consideration

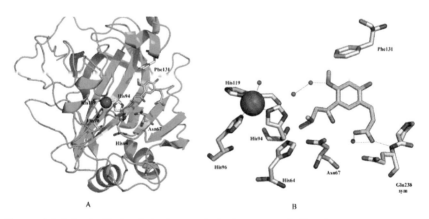

Fig. 5 (A) hCA II-/(S,Z)-3-(2-hydroxy-5-(1-hydroxy-3-methylbutyl)-4-methoxyphenyl) acrylic acid adduct; (B) Detailed interactions of the hCA II/ (S,Z)-3-(2-hydroxy-5-(1-hydroxy-3-methylbutyl)-4-methoxyphenyl)acrylic acid adduct. The protein is depicted as a green ribbon, the catalytic Zn (II) ion as the violet sphere, and the three protein ligands (His94, 96, and 119) are in CPK colors as stick models [43].

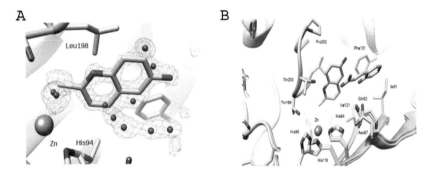

Fig. 6 (A) $F_o - F_c$ omit map of the hCA II-thioxocoumarin adduct; (B) Superposition of the hCA II-thioxocoumarin adduct (sky blue, PDB: 4WL4) with the hCA II-hydrolyzed coumarin adduct (PDB: 5BNL) (silver) [47].

narrows the application of coumarins and/or their bioisosteric derivatives only to Gram-negative bacteria which exclusively encode for and confine α-CAs into their periplasmic space [7].

The first and relevant contribution on this topic is from Giovannuzzi et al. which reported a small panel of simple coumarins able to effectively inhibit the α-CAs from *Vibrio cholerae* (VchCAα) and *Neisseria gonorrheae* (NgCAα) [49]. Of relevance, is the consistent selectivity of such compounds for the bCAs over the hCAs I, II and that is reasonably ascribed to the slightly larger and thus better accommodating catalytic cleft of the former [49].

The impact of this study in the field is wider than expected [50] as naturally occurring compounds of the coumarin type pave the way for a plethora of synthetically related structures such as the sulfocoumarins [51,52], benzoxaphosphepine 2-oxides [53], those containing chalcogen elements or their biologically active derivatives [54–57] which are not currently reported as bCA inhibitors.

4. Phenols

The phenolic moiety is abundantly present in a plethora of natural substances, and specifically in those of vegetarian origin [58–62]. For the purposes of this contribution it is sufficient to report that such a moiety interacts to the Zn (II) coordinated water molecule at the bottom of the hCA II catalytic cleft, which eventually is involved in additional stabilizing interactions with the exposed amino acid residues [59] (Fig. 7).

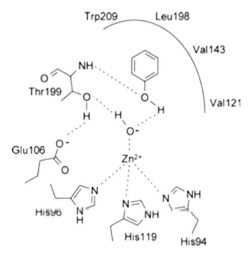

Fig. 7 Schematic representation of the binding mode of phenol to the hCA II active site [59].

The most complete investigation on phenolic compounds relates the natural products carvacrol and thymol, which have been investigated as inhibitors of the *H. pylori* CAs HpCAα and HpCAβ [63] (Table 4).

Carvacrol resulted a medium potency inhibitor of both HpCAs and ineffective against the hCAs I and II (K_Is > 100 μM), whereas for the thymol a micromolar inhibition potency was observed only for the HpCAβ [63]. Quite interestingly both compounds were assessed for the anti-*H. pylori* activity in terms of growth impairment, biofilm production and release of outer membrane vesicles [64,65].

Most of the scientific contributions in the field dealing with phenols relate hCAs and thus such a moiety remains largely under-investigated for development as a new class of bCA inhibitors. Similarly, the literature lacks deeper consideration for phenol related moieties such as the selenols [66], the 2-mercaptobenzoxazoles [67] and the cathecols [68].

5. *N*-Methyl thiosemicarbazones

The only study on *N*-methyl thiosemicarbazones (TSC) on bCAs was reported in 2022 by D'Agostino et al. [69]. *In silico* target predictions accounted for such a moiety as a candidate worth developing for inhibition of CAs [69]. Compounds **1–9** in Fig. 8 were assessed for the inhibitory

Table 4 Inhibition data of hCAs I, II, VI, HpCAα, HpCAβ, PgiCAβ, SmuCA and MgCA with Carvacrol and Thymol [63].

Compound	Structure	K_i (μM)[a]							
		hCA I	hCA II	hCA IV	HpCA α	HpCA β	PgiCA β	SmuCA	MgCA
Carvacrol		>100	>100	>100	8.4	13.3	28.0	86.6	6.8
Thymol		>100	>100	>100	>100	3.4	48.9	84.9	>100
AAZ		0.25	0.012	0.011	0.021	0.040	0.214	0.344	40.0

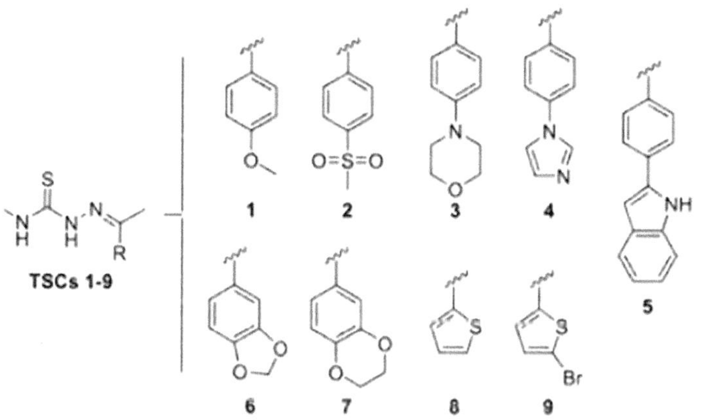

Fig. 8 Chemical structures of TSCs 1–9.

activity of a panel of 6 bCAs from bacterial strains of biomedical interest such as *S. aureus* (SauβCA), *E. coli* (EcoβCA and EcoγCA), *P. aeruginosa* (psCA3-β), *H. pylori* (HpαCA and HpβCA) and were compared to the off-target hCAs I and II (Table 5).

The data above showed TSCs **1**, **2** and **9** as highly selective for the isoforms EcoγCA, followed by the psCA3-β, whereas the derivatives **3** and **7** showed good inhibition for the EcoβCA. The compound **5** had interesting affinity for both the HpαCA and HpβCA isoforms [69]. The promising *in vitro* affinity data for the considered bCAs were not sustained from the *in vitro* antibacterial experiments. Nevertheless, TSCs are an interesting moiety for the high capacity to interact with metal ions largely diffused in metalloenzymes such as Cu (II), Zn (II) and Mg (II) among others.

6. *N*-Hydroxy ureas

This class of compounds was reported as effective inhibitor of the hCAs, and more importantly they were found to coordinate in a bidentate fashion the catalytic Zn (II) ion by means of the NH and OH groups of the CONHOH fragment [70,71]. To date, the only scientific report accounting for this moiety to be profiled *in vitro* on bCAs is from Berrino et al. [72]. Specifically, a series of *N'*-aryl-*N*-hydroxyureas was investigated *in vitro* for the ability to inhibit the *V. Cholerae* VchCAα, VchCAβ, VchCAγ and the β- and γ-CAs from *Porphyromonas gingivalis* (*i.e.* PgiCAβ and PgiCAγ). The study comprised also

Table 5 Inhibition data of bCAs and hCAsI and II for TSCs 1–9.

CAS	1	2	3	4	5	6	7	8	9	AAZ
SauβCA	688.1	609	609	720.7	>1000	>1000	530.9	603.2	394.2	0.63
EcoβCA	415.9	307.1	91.5	>1000	>1000	682.1	96.5	890.3	472.2	0.23
EcoγCA	86.2	80	807.1	875	688.9	779.9	388.3	220.5	83.2	0.25
psCA3-β	92.1	93.4	>1000	>1000	698.3	632.4	>1000	285.8	86.7	0.0076
HpαCA	714.9	718.8	97.7	522.9	95.2	824.5	574.5	310.9	727	0.0021
HpβCA	689.6	>1000	788.9	692.1	348.3	767.4	860.5	700	586	0.004
hCA I	919.3	944.5	841.3	>1000	>1000	>1000	>1000	903.6	967	0.25
hCA II	910.4	896.6	828.8	>1000	>1000	>1000	>1000	859.3	879.4	0.012

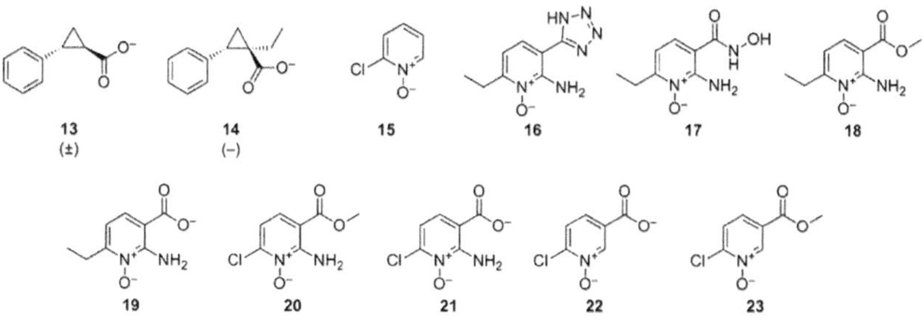

Fig. 9 Structures of compounds selected in the study [73].

the δ-CA from the marine diatom *Thalassiosira weissflogii* (TweCAδ), herein not considered in detail as outside the scope of the contribution. Notably, most of the reported *N*-hydroxyureas were ineffective inhibitors for the hCAs I and II (*i.e.* K_Is > 10 mM), whereas both the VchCAα and VchCAβ were effectively inhibited with K_I values spanning between 97.5 nM–7.26 μM for the former and 52.5 nM–1.81 μM for the latter respectively. Conversely the VchCAγ was far less inhibited from the series of compounds considered (*i.e.*, K_Is between 4.75 and 8.87 μM). A good discrimination was reported among the PgiCAβ and the PgiCAγ being the latter exclusively inhibited with K_I values in the range of 59.8 nM–6.42 μM [72].

7. Miscellaneous

A very interesting contribution on the search of new moieties potentially useful for validation of the bCAs as antiinfectives was reported by Annunziato et al. [73]. In the study the maximum common substructure (MCS) approach was integrated with *in silico* modeling and *in vitro* assays, allowed to identify new chemotypes, which are characterized by high selectivity for the bCAs over the off-target ones. The structures of the compounds considered are listed in Fig. 9.

8. Conclusions

The discovery of non-coumarin inhibitors of the bCAs gave access to many chemical structures, mainly of natural origin inspiration. Such results are a considerable reinforcement to the currently available tools in drug design and development aiming to tackle the multi-drug resistance phenomena which heavily affects most of the antibiotics efficacy [74]. Despite the variety of chemical moieties identified and briefly reported in this contribution, the validation of the bCAs as drug targets remains a challenge. Almost all the selected compounds fail when investigated for their bacterial growth inhibition ability and none is currently under consideration for development by means of *in vitro* models of infection.

References

[1] D.P. Flaherty, M.N. Seleem, C.T. Supuran, Bacterial carbonic anhydrases: underexploited antibacterial therapeutic targets, Future Med. Chem. 19 (2021) 1619–1622, https://doi.org/10.4155/fmc-2021-0207

[2] C. Capasso, C.T. Supuran, Biomedical applications of prokaryotic carbonic anhydrases: an update, Expert Opin. Ther. Pat. (2024) 1–13, https://doi.org/10.1080/13543776.2024.2365407

[3] C. Capasso, C.T. Supuran, Antibacterial carbonic anhydrase inhibitors: an update on the recent literature, Expert. Opin. Ther. Pat. (12) (2020) 963–982, https://doi.org/10.1080/13543776.2020.1811853

[4] A. Aspatwar, J.-Y. Winum, F. Carta, C.T. Supuran, M. Hammaren, M. Parikka, et al., Carbonic anhydrase inhibitors as novel drugs against mycobacterial β-carbonic anhydrases: an update on in vitro and in vivo studies, Molecules (11) (2018) 2911, https://doi.org/10.3390/molecules23112911

[5] J.-Y. Winum, F. Carta, C. Ward, P. Mullen, D. Harrison, S.P. Langdon, et al., Ureido-substituted sulfamates show potent carbonic anhydrase IX inhibitory and antiproliferative activities against breast cancer cell lines, Bioorg. Med. Chem. Lett. 14 (2012) 4681–4685, https://doi.org/10.1016/j.bmcl.2012.05.083

[6] C. Capasso, C.T. Supuran, Carbonic anhydrase and bacterial metabolism: a chance for antibacterial drug discovery, Expert. Opin. Ther. Pat. (2024) 1–10, https://doi.org/10.1080/13543776.2024.2332663

[7] V. De Luca, V. Carginale, C.T. Supuran, C. Capasso, The gram-negative bacterium Escherichia coli as a model for testing the effect of carbonic anhydrase inhibition on bacterial growth, J. Enzyme Inhib. Med. Chem. (1) (2022) 2092–2098, https://doi.org/10.1080/14756366.2022.2101644

[8] M. Moustakas, The role of metal ions in biology, biochemistry and medicine, Mater. (Basel) (3) (2021) 549, https://doi.org/10.3390/ma14030549

[9] A. Nocentini, A. Angeli, F. Carta, J.-Y. Winum, R. Zalubovskis, S. Carradori, et al., Reconsidering anion inhibitors in the general context of drug design studies of modulators of activity of the classical enzyme carbonic anhydrase, J. Enzyme Inhib. Med. Chem. (1) (2021) 561–580, https://doi.org/10.1080/14756366.2021.1882453

[10] G. De Simone, C.T. Supuran, (In)organic anions as carbonic anhydrase inhibitors, J. Inorg. Biochem. 111 (2012) 117–129, https://doi.org/10.1016/j.jinorgbio.2011.11.017

[11] I. Bertini, C. Luchinat, A. Scozzafava, Binding affinity of bicarboxylate ions for cobalt (II) bovine carbonic anhydrase, Bioinorg. Chem. 9 (1978) 93–100, https://doi.org/10.1016/s0006-3061(00)80283-5

[12] I. Bertini, C. Luchinat, A. Scozzafava, Interaction of cobalt-bovine carbonic anhydrase with the acetate ion, Biochim. Biophys. Acta 452 (1976) 239–244, https://doi.org/10.1016/0005-2744(76)90076-0

[13] I. Bertini, C. Luchinat, R. Pierattelli, A.J. Vila, The interaction of acetate and formate with cobalt carbonic anhydrase. An NMR study, Eur. J. Biochem. 208 (1992) 607–615, https://doi.org/10.1111/j.1432-1033.1992.tb17225.x

[14] I. Bertini, C. Luchinat, A. Scozzafava, A 31 P NMR study of phosphate in presence of cobalt(II)- and copper(II)- substituted bovine carbonic anhydrase B, FEBS Lett. 93 (1978) 251–254, https://doi.org/10.1016/0014-5793(78)81115-6

[15] J.S. Taylor, J.E. Coleman, Electron spin resonance of 63Cu and 65Cu carbonic anhydrases. Resolution of nitrogen ligand superhyperfine structure, J. Biol. Chem. 248 (1973) 749–755.

[16] L.J. Urbanski, D. Vullo, S. Parkkila, C.T. Supuran, An anion and small molecule inhibition study of the β-carbonic anhydrase from *Staphylococcus aureus*, J. Enzyme Inhib. Med. Chem. 1 (2021) 1088–1092, https://doi.org/10.1080/14756366.2021.1931863

[17] A. Nocentini, C.S. Hewitt, M.D. Mastrolorenzo, D.P. Flaherty, C.T. Supuran, Anion inhibition studies of the α carbonic anhydrases from *Neisseria gonorrhoeae*, J. Enzyme Inhib. Med. Chem. 1 (2021) 1061–1066, https://doi.org/10.1080/14756366.2021.1929202

[18] A. Petreni, V. De Luca, A. Scaloni, A. Nocentini, C. Capasso, C.T. Supuran, Anion inhibition studies of the Zn(II)-bound ι-carbonic anhydrase from the Gram-negative bacterium *Burkholderia territorii*, J. Enzyme Inhib. Med. Chem. 1 (2021) 372–376, https://doi.org/10.1080/14756366.2020.1867122

[19] S. Del Prete, V. De Luca, A. Nocentini, A. Scaloni, M.D. Mastrolorenzo, C.T. Supuran, et al., Anion inhibition studies of the beta-carbonic anhydrase from *Escherichia coli*, Molecules 11 (2020) 2564, https://doi.org/10.3390/molecules25112564

[20] V. De Luca, D. Vullo, S. Del Prete, V. Carginale, S.M. Osman, Z. AlOthman, et al., Cloning, characterization and anion inhibition studies of a γ-carbonic anhydrase from the Antarctic bacterium *Colwellia psychrerythraea*, Bioorg. Med. Chem. 4 (2016) 835–840, https://doi.org/10.1016/j.bmc.2016.01.005

[21] M. Ferraroni, S. Del Prete, D. Vullo, C. Capasso, C.T. Supuran, Crystal structure and kinetic studies of a tetrameric type II β-carbonic anhydrase from the pathogenic bacterium *Vibrio cholerae*, Acta Pt 12 (2015) 2449–2456, https://doi.org/10.1107/S1399004715018635

[22] N. Dedeoglu, V. De Luca, S. Isik, H. Yildirim, F. Kockar, C. Capasso, et al., Cloning, characterization and anion inhibition study of a β-class carbonic anhydrase from the caries producing pathogen *Streptococcus mutans*, Bioorg. Med. Chem. 13 (2015) 2995–3001, https://doi.org/10.1016/j.bmc.2015.05.007

[23] I. Nishimori, D. Vullo, T. Minakuchi, A. Scozzafava, S.M. Osman, Z. AlOthman, et al., Anion inhibition studies of two new β-carbonic anhydrases from the bacterial pathogen *Legionella pneumophila*, Bioorg. Med. Chem. Lett. 4 (2014) 1127–1132, https://doi.org/10.1016/j.bmcl.2013.12.124

[24] D. Vullo, R.S.S. Kumar, A. Scozzafava, C. Capasso, J.G. Ferry, C.T. Supuran, Anion inhibition studies of a β-carbonic anhydrase from *Clostridium perfringens*, Bioorg. Med. Chem. Lett. 24 (2013) 6706–6710, https://doi.org/10.1016/j.bmcl.2013.10.037

[25] B.M. Jönsson, K. Håkansson, A. Liljas, The structure of human carbonic anhydrase II in complex with bromide and azide, FEBS Lett. 322 (1993) 186–190, https://doi.org/10.1016/0014-5793(93)81565-h

[26] S. Mangani, K. Håkansson, Crystallographic studies of the binding of protonated and unprotonated inhibitors to carbonic anhydrase using hydrogen sulphide and nitrate anions, Eur. J. Biochem. 3 (1992) 867–871, https://doi.org/10.1111/j.1432-1033.1992.tb17490.x

[27] S.K. Nair, D.W. Christianson, Crystallographic studies of azide binding to human carbonic anhydrase II, Eur. J. Biochem. 1 (1993) 507–515, https://doi.org/10.1111/j.1432-1033.1993.tb17788.x

[28] B. Sjöblom, M. Polentarutti, K. Djinovic-Carugo, Structural study of X-ray induced activation of carbonic anhydrase, Proc. Natl. Acad. Sci. U. S. A. 106 (2009) 10609–106113, https://doi.org/10.1073/pnas.0904184106

[29] A.E. Eriksson, T.A. Jones, A. Liljas, Refined structure of human carbonic anhydrase II at 2.0 A resolution, Proteins 4 (1988) 274–282, https://doi.org/10.1002/prot. 340040406

[30] A. Guerri, F. Briganti, A. Scozzafava, C.T. Supuran, S. Mangani, Mechanism of cyanamide hydration catalyzed by carbonic anhydrase II suggested by cryogenic X-ray diffraction, Biochemistry 40 (2000) 12391–12397, https://doi.org/10.1021/bi000937c

[31] F. Carta, M. Aggarwal, A. Maresca, A. Scozzafava, R. McKenna, C.T. Supuran, Dithiocarbamates: a new class of carbonic anhydrase inhibitors. Crystallographic and kinetic investigations, Chem. Commun. (Camb.) 13 (2012) 1868–1870, https://doi.org/10.1039/c2cc16395k

[32] D. Vullo, M. Durante, F.S. Di Leva, S. Cosconati, E. Masini, A. Scozzafava, et al., Monothiocarbamates strongly inhibit carbonic anhydrases in vitro and possess intraocular pressure lowering activity in an animal model of glaucoma, J. Med. Chem. 12 (2016) 5857–5867, https://doi.org/10.1021/acs.jmedchem.6b00462

[33] C. Temperini, A. Scozzafava, C.T. Supuran, Carbonic anhydrase inhibitors. X-ray crystal studies of the carbonic anhydrase II–trithiocarbonate adduct—an inhibitor mimicking the sulfonamide and urea binding to the enzyme, Bioorg. Med. Chem. Lett. 2 (2010) 474–478, https://doi.org/10.1016/j.bmcl.2009.11.124

[34] A. Maresca, F. Carta, D. Vullo, C.T. Supuran, Dithiocarbamates strongly inhibit the β-class carbonic anhydrases from *Mycobacterium tuberculosis*, J. Enzyme Inhib. Med. Chem. 2 (2013) 407–411, https://doi.org/10.3109/14756366.2011.641015

[35] A. Aspatwar, M. Hammarén, S. Koskinen, B. Luukinen, H. Barker, F. Carta, et al., β-CA-specific inhibitor dithiocarbamate Fc14-584B: a novel antimycobacterial agent with potential to treat drug-resistant tuberculosis, J. Enzyme Inhib. Med. Chem. 1 (2017) 832–840, https://doi.org/10.1080/14756366.2017.1332056

[36] A. Aspatwar, M. Hammaren, M. Parikka, S. Parkkila, F. Carta, M. Bozdag, et al., In vitro inhibition of *Mycobacterium tuberculosis* β-carbonic anhydrase 3 with Mono- and dithiocarbamates and evaluation of their toxicity using zebrafish developing embryos, J. Enzyme Inhib. Med. Chem. 1 (2020) 65–71, https://doi.org/10.1080/14756366. 2019.1683007

[37] S. Giovannuzzi, N.S. Abutaleb, C.S. Hewitt, F. Carta, A. Nocentini, M.N. Seleem, et al., Dithiocarbamates effectively inhibit the α-carbonic anhydrase from *Neisseria gonorrhoeae*, J. Enzyme Inhib. Med. Chem. 1 (2022) 1–8, https://doi.org/10.1080/14756366.2021.1988945

[38] S. Giovannuzzi, A.K. Marapaka, N.S. Abutaleb, F. Carta, H.-W. Liang, A. Nocentini, et al., Inhibition of pathogenic bacterial carbonic anhydrases by monothiocarbamates, J. Enzyme Inhib. Med. Chem. 1 (2023) 2284119, https://doi.org/10.1080/14756366. 2023.2284119

[39] A.G. Atanasov, S.B. Zotchev, V.M. Dirsch, C.T. Supuran, Natural products in drug discovery: advances and opportunities, Nat. Rev. Drug. Discov. 20 (2021) 200–216, https://doi.org/10.1038/s41573-020-00114-z

[40] A. Stefanachi, F. Leonetti, L. Pisani, M. Catto, A. Carotti, Coumarin: a natural, privileged and versatile scaffold for bioactive compounds, Molecules 23 (2018) 250, https://doi.org/10.3390/molecules23020250

[41] C.T. Supuran, Coumarin carbonic anhydrase inhibitors from natural sources, J. Enzyme Inhib. Med. Chem. 35 (2020) 1462–1470, https://doi.org/10.1080/14756366.2020.1788009

[42] A. Maresca, C. Temperini, H. Vu, N.B. Pham, S.A. Poulsen, A. Scozzafava, et al., Non-zinc mediated inhibition of carbonic anhydrases: coumarins are a new class of suicide inhibitors, J. Am. Chem. Soc. 131 (2009) 3057–3062, https://doi.org/10.1021/ja809683v

[43] A. Maresca, C. Temperini, L. Pochet, B. Masereel, A. Scozzafava, C.T. Supuran, Deciphering the mechanism of carbonic anhydrase inhibition with coumarins and thiocoumarins, J. Med. Chem. 53 (2010) 335–344, https://doi.org/10.1021/acs.jmedchem.5b00969

[44] R.A. Davis, D. Vullo, A. Maresca, C.T. Supuran, S.A. Poulsen, Natural product coumarins that inhibit human carbonic anhydrases, Bioorg. Med. Chem. 21 (2013) 1539–1543, https://doi.org/10.1016/j.bmc.2012.07.021

[45] C.T. Supuran, Carbonic anhydrase inhibitors from marine natural products, Mar. Drugs 20 (2022) 721, https://doi.org/10.3390/md20110721

[46] M. Tanc, F. Carta, A. Scozzafava, C.T. Supuran, α-Carbonic anhydrases possess thioesterase activity, ACS Med. Chem. Lett. 3 (2015) 292–295, https://doi.org/10.1021/ml500470b

[47] M. Ferraroni, F. Carta, A. Scozzafava, C.T. Supuran, Thioxocoumarins show an alternative carbonic anhydrase inhibition mechanism compared to coumarins, J. Med. Chem. 1 (2016) 462–473, https://doi.org/10.1021/acs.jmedchem.5b01720

[48] A. Innocenti, C.T. Supuran, Paraoxon, 4-nitrophenyl phosphate and acetate are substrates of α- but not of β-, γ- and ζ-carbonic anhydrases, Bioorg. Med. Chem. Lett. 20 (2010) 6208–6212, https://doi.org/10.1016/j.bmcl.2010.08.110

[49] S. Giovannuzzi, C.S. Hewitt, A. Nocentini, C. Capasso, D.P. Flaherty, C.T. Supuran, Coumarins effectively inhibit bacterial α-carbonic anhydrases, J. Enz. Inh. Med. Chem. 37 (2022) 333–338, https://doi.org/10.1080/14756366.2021.2012174

[50] C.T. Supuran, Coumarin carbonic anhydrase inhibitors from natural sources, J. Enz. Inhib. Med. Chem. 1 (2020) 1462–1470, https://doi.org/10.1080/14756366.2020.1788009

[51] K. Tars, D. Vullo, A. Kazaks, J. Leitans, A. Lends, A. Grandane, et al., J. Med. Chem. 1 (2013) 293–300, https://doi.org/10.1021/jm301625s

[52] A. Grandane, M. Tanc, L. Di Cesare Mannelli, F. Carta, C. Ghelardini, R. Žalubovskis, et al., 6-Substituted sulfocoumarins are selective carbonic anhdydrase IX and XII inhibitors with significant cytotoxicity against colorectal cancer cells, J. Med. Chem. 9 (2015) 3975–3983, https://doi.org/10.1021/acs.jmedchem.5b00523

[53] A. Pustenko, A. Balašova, A. Nocentini, C.T. Supuran, R. Žalubovskis, 3H-1,2-Benzoxaphosphepine 2-oxides as selective inhibitors of carbonic anhydrase IX and XII, J. Enzyme Inhib. Med. Chem. 1 (2023) 216–224, https://doi.org/10.1080/14756366.2022.2143496

[54] D. Tanini, A. Capperucci, M. Locuoco, M. Ferraroni, G. Costantino, A. Angeli, et al., Benzoselenoates: a novel class of carbonic anhydrase inhibitors, Bioorg Chem. 122 (2022) 105751, https://doi.org/10.1016/j.bioorg.2022.105751

[55] A. Angeli, M. Ferraroni, A. Capperucci, D. Tanini, G. Costantino, C.T. Supuran, Selenocarbamates as a prodrug-based approach to carbonic anhydrase inhibition, ChemMedChem 11 (2022) e202200085.

[56] A.S. Al-Tamimi, M. Etxebeste-Mitxeltorena, C. Sanmartín, A. Jiménez-Ruiz, L. Syrjänen, S. Parkkila, et al., Discovery of new organoselenium compounds as antileishmanial agents, Bioorg. Chem. 86 (2019) 339–345, https://doi.org/10.1016/j.bioorg.2019.01.069

[57] D. Tanini, S. Carradori, A. Capperucci, L. Lupori, S. Zara, M. Ferraroni, et al., Chalcogenides-incorporating carbonic anhydrase inhibitors concomitantly reverted oxaliplatin-induced neuropathy and enhanced antiproliferative action, Eur. J. Med. Chem. 225 (2021) 113793, https://doi.org/10.1016/j.ejmech.2021.113793

[58] C. Paloukopoulou, S. Govari, A. Soulioti, I. Stefanis, A. Angeli, A. Matheeussen, et al., Phenols from Origanum dictamnus L. and Thymus vulgaris L. and their activity against Malassezia globosa carbonic anhydrase, Nat. Prod. Res. 6 (2022) 1558–1564, https://doi.org/10.1080/14786419.2021.1880406

[59] V. Alterio, A. Di Fiore, K. D'Ambrosio, et al., Multiple binding modes of inhibitors to carbonic anhydrases: how to design specific drugs targeting 15 different isoforms? Chem. Rev. 8 (2012) 4421–4468, https://doi.org/10.1021/cr200176r

[60] A. Karioti, F. Carta, C.T. Supuran, Phenols and polyphenols as carbonic anhydrase inhibitors, Molecules 12 (2016) 1649, https://doi.org/10.3390/molecules2112164

[61] A. Karioti, M. Ceruso, F. Carta, A.R. Bilia, C.T. Supuran, New natural product carbonic anhydrase inhibitors incorporating phenol moieties, Bioorg. Med. Chem. 22 (2015) 7219–7225, https://doi.org/10.1016/j.bmc.2015.10.018

[62] A. Nocentini, S. Bua, S. Del Prete, Y.E. Heravi, A.A. Saboury, A. Karioti, et al., Natural polyphenols selectively inhibit beta-carbonic anhydrase from the dandruff-producing fungus *Malassezia globosa*: Activity and modeling studies, ChemMedChem 8 (2018) 816–823, https://doi.org/10.1002/cmdc.201800015

[63] R. Grande, S. Carradori, V. Puca, et al., Selective inhibition of *Helicobacter pylori* carbonic anhydrases by Carvacrol and Thymol could impair biofilm production and the release of outer membrane vesicles, Int. J. Mol. Sci. 21 (2021) 11583, https://doi.org/10.3390/ijms222111583

[64] V. Puca, G. Turacchio, B. Marinacci, et al., Antimicrobial and antibiofilm activities of carvacrol, amoxicillin and salicylhydroxamic acid alone and in combination vs. *Helicobacter pylori*: towards a new multi-targeted therapy, Int. J. Mol. Sci. 5 (2023) 4455, https://doi.org/10.3390/ijms24054455

[65] M. Ronci, S. Del Prete, V. Puca, et al., Identification and characterization of the α-CA in the outer membrane vesicles produced by *Helicobacter pylori*, J. Enzyme Inhib. Med. Chem. 1 (2019) 189–195, https://doi.org/10.1080/14756366.2018.1539716

[66] D. Tanini, A. Capperucci, M. Ferraroni, F. Carta, A. Angeli, C.T. Supuran, Direct and straightforward access to substituted alkyl selenols as novel carbonic anhydrase inhibitors, Eur. J. Med. Chem. 185 (2020) 111811, https://doi.org/10.1016/j.ejmech.2019.111811

[67] M. Bozdag, C.T. Supuran, D. Esposito, A. Angeli, F. Carta, S.M. Monti, et al., 2-Mercaptobenzoxazoles: a class of carbonic anhydrase inhibitors with a novel binding mode to the enzyme active site, Chem. Commun. (Camb.) 59 (2020) 8297–8300, https://doi.org/10.1039/D0CC02857F

[68] K. D'Ambrosio, S. Carradori, S. Cesa, A. Angeli, S.M. Monti, C.T. Supuran, et al., Catechols: a new class of carbonic anhydrase inhibitors, Chem. Commun. (Camb.) 85 (2020) 13033–13036, https://doi.org/10.1039/d0cc05172a

[69] I. D'Agostino, G.E. Mathew, P. Angelini, R. Venanzoni, G. Angeles Flores, A. Angeli, et al., Biological investigation of N-methyl thiosemicarbazones as anti-microbial agents and bacterial carbonic anhydrases inhibitors, J. Enzyme Inhib. Med. Chem. 1 (2022) 986–993, https://doi.org/10.1080/14756366.2022.2055009

[70] A. Scozzafava, C.T. Supuran, Hydroxyurea is a carbonicanhydrase inhibitor, Bioorg. Med. Chem. 11 (2003) 2241–2246, https://doi.org/10.1016/s0968-0896(03)00112-3

[71] C. Temperini, A. Innocenti, A. Scozzafava, C.T. Supuran, N-hydroxyurea—a versatile zinc binding function in thedesign of metalloenzyme inhibitors, Bioorg. Med. Chem. Lett. 16 (2006) 4316–4320, https://doi.org/10.1016/j.bmcl.2006.05.068

[72] E. Berrino, M. Bozdag, S. Del Prete, F.A.S. Alasmary, L.S. Alqahtani, Z. AlOthman, et al., Inhibition of α-, β-, γ-, and δ-carbonic anhydrases from bacteria and diatoms with N'-aryl-N-hydroxy-ureas, J. Enzyme Inhib. Med. Chem. 1 (2018) 1194–1198, https://doi.org/10.1080/14756366.2018.1490733

[73] G. Annunziato, A. Angeli, F. D'Alba, A. Bruno, M. Pieroni, D. Vullo, et al., Discovery of new potential anti-infective compounds based on carbonic anhydrase inhibitors by rational target-focused repurposing approaches, ChemMedChem 17 (2016) 1904–1914, https://doi.org/10.1002/cmdc.201600180

[74] https://www.who.int/health-topics/antimicrobial-resistance (Last access 10/06/2024).

CHAPTER EIGHT

Helicobacter pylori CAs inhibition

Bianca Laura Bernardoni[a], Concettina La Motta[a], Simone Carradori[b,*], and Ilaria D'Agostino[a]

[a]Department of Pharmacy, University of Pisa, Pisa, Italy
[b]Department of Pharmacy, G. d'Annunzio University of Chieti-Pescara, Chieti, Italy
*Corresponding author. e-mail address: simone.carradori@unich.it

Contents

1. *Helicobacter pylori*, a bacterium threatening public health: an overview of the pathogen, its related diseases, and current treatments   213
2. HpCAs as promising targets for anti-Hp agents: functional role and kinetics   215
3. Structural insight into HpCAs   218
4. Modulation of HpCAs for therapeutic purposes: CAI libraries from the literature   219
5. Sulfonamido-containing CAIs   221
6. Anion and acid CAIs   228
7. Phenolic CAIs   230
8. Conclusions: how far HpCAs came and future perspectives   232
Acknowledgments   232
References   232

Abstract

Infections from *Helicobacter pylori* (Hp) are endangering Public Health safety worldwide, due to the associated high risk of developing severe diseases, such as peptic ulcer, gastric cancer, diabetes, and cardiovascular diseases. Current therapies are becoming less effective due to the rise of (multi)drug-resistant phenotypes and an urgent need for new antibacterial agents with innovative mechanisms of action is pressing. Among the most promising pharmacological targets, Carbonic Anhydrases (EC: 4.2.1.1) from Hp, namely HpαCA and HpβCA, emerged for their high druggability and crucial role in the survival of the pathogen in the host. Thereby, in the last decades, the two isoenzymes were isolated and characterized offering the opportunity to profile their kinetics and test different series of inhibitors.

1. *Helicobacter pylori*, a bacterium threatening public health: an overview of the pathogen, its related diseases, and current treatments

Helicobacter pylori (Hp) is a spiral-shaped Gram-negative bacterium whose name is recurrent in hospital corridors and, also, in non–medical conversations,

The Enzymes, Volume 55
ISSN 1874-6047, https://doi.org/10.1016/bs.enz.2024.05.013

213

due to the high rate of infections that the pathogen causes per year [1,2], with half the population worldwide suffering from the disease [3,4]. Although large, the number of diagnoses is highly underestimated due to the difficulty in collecting clinical data in many countries, especially underdeveloped ones. Also, Hp infection is missing in the routine health screenings which could quickly identify the pathogen when no specific disease symptom has already been noticed. However, in general, existing diagnostic methods, *e.g.,* fecal antigen testing, urea breath testing, and immunohistochemical staining from tissue biopsies [5], have different sensitivity and specificity or cannot be able to detect the infection in older patients with decreased levels of anti-Hp IgG correlated to the progression of gastric lesions [3]. Since 2017, an alert regarding Hp infections was launched by the World Health Organization which included the pathogen in the Global Priority Pathogens List, and, in particular, in the "high priority category" [6]. Albeit Hp is not monitored by the European Antimicrobial Resistance Surveillance [7], various organizations aimed at making people aware of its infectious disease and manage the related data/guidelines, such as The Helicobacter Foundation [8] and the European Registry on Helicobacter pylori management (Hp-EuReg) [9]. Hp is also reported as endemic worldwide in the CDC Yellow Book, a resource for international travelers provided by the US government that lists all the pathogens and/or diseases related to different countries with the corresponding health care recommendations, such as vaccines and advice [10,11].

The reasons why Hp is in the spotlight of Health organizations and Medicinal Chemistry research lines are evident and validated during the last century. In particular, apart from the low health and social costs of the infectious disease itself, Hp is highly correlated with a large panel of pathological conditions. First of all, Hp is a recognized risk factor for gastritis, long-lasting inflammation, and peptic ulcers. Also, it is a well-established carcinogen agent, causing gastric cancers [12], particularly gastric adenocarcinoma and gastric mucosa-associated lymphoid tissue (MALT) lymphoma [13,14]. Moreover, Hp infection has recently been associated with other extra-gastric diseases, such as overweighting or obesity [15,16], hepatic [17,18], cardiovascular [19,20], and autoimmune [21] diseases, anemia [22], Alzheimer's disease [23], metabolic syndrome, diabetes mellitus [24,25], and many others [26].

The bacterium has a peculiar mechanism of host colonization: its attachment to the gastric mucosa is mediated by chemotaxis and Transducer-like proteins (Tlps), while Hp ability to navigate through the mucous layer is due to the flagellum. However, survival in the stomach is allowed by a dual enzymatic system composed of Urease (EC: 3.5.1.5) and Carbonic Anhydrases

(CAs, EC: 4.2.1.1) [27], which will be discussed below. Interestingly, several virulence factors are encoded by Hp, and, among them, some are essential to manipulate the host cell signaling, *e.g.,* cytotoxin-associated gene A (CagA) and vacuolating cytotoxin gene A (VacA). Moreover, Hp lipopolysaccharide (LPS) and peptidoglycan components generate a strong chronic inflammatory response in the gastric mucosa. Mucin enables the binding of Hp to the host cells but its turnover is impaired by Hp-mediated expression alteration [28]. As most microorganisms, Hp can form biofilm, in which the extracellular matrix embeds a specific architecture of bacterial communities, that improve Hp adherence performance and protect the bacterium from antibacterial treatments [29].

Apart from its innate resistance to antibiotics, Hp strains can also acquire resistomes thanks to the production of extracellular outer membrane vesicles (OMVs) containing nucleic acids, *e.g.,* eDNA, lipids, sugars, and proteins, which permit cell-cell communication and interactions, also resulting in horizontal gene transfer [30,31]. Most interestingly, Hp OMVs are implicated in biofilm formation [32,33], also involved in drug resistance.

To date, standard pharmacological therapy for Hp infection is based on a triple approach, comprising at least two antibiotics (generally clarithromycin and amoxicillin) and one proton pump (H^+/K^+-ATPase) inhibitor (PPI), aimed at affecting bacterial cell wall, protein synthesis, and genome [28]. This was in agreement with the last Maastricht/Florence consensus report released in 2022 which gave guidelines for Hp infection management and, in particular, recommended the administration of the triple therapy or the so-called BQT approach (bismuth, tetracycline, metronidazole, and a PPI) [34]. Unfortunately, several drug-resistant phenotypes of Hp were isolated, being not more susceptible to some of the most employed antibiotics currently in clinics, *e.g.,* metronidazole, amoxicillin, clarithromycin, tetracycline, and levofloxacin [35,36], with different molecular mechanisms [37]. Thus, new therapeutical options and anti-Hp agents acting through innovative modes of action have been and continue to be developed, such as new generations of old antibiotics, capsules, probiotics, and CA inhibitors (CAIs), which will be discussed in detail [38].

2. HpCAs as promising targets for anti-Hp agents: functional role and kinetics

As widely discussed elsewhere in this book, CAs are ubiquitous metalloenzymes involved in several physiopathological pathways by catalyzing simple reactions, such as the hydration of carbon dioxide (CO_2) into bicarbonate ion (HCO_3^-) and proton (H^+). The superfamily of these ligases is composed of different families or classes, not phylogenetically correlated and named with the letters of the Greek alphabet: α, β, γ, δ, ζ, η, θ, and ι. However, while humans encode only for α-CAs, bacteria express α, β, γ, and ι isoenzymes, and Hp synthesizes two isoenzymes: HpβCA, localized in the cytoplasm, and HpαCA released in the periplasmatic space.

Interestingly, Hp is both a microaerophile, thereby growing in environments with low O_2 partial pressure, and, a capnophile, thus, it requires an atmosphere with 5–10% CO_2 levels [39]. Hence, the role of CO_2 and the maintenance of optimal pH in Hp lifecycle are clear [40,41]. However, the specific functions of the two CA enzymes in the bacterium have been debated for a long time.

First, Acetazolamide (**AAZ**, Fig. 4) resulted to show moderate efficacy in fighting the Hp colonization of the host in *in vivo* studies [42], although being a pan-CAI and showing affinity towards HpCAs. However, HpCAs inhibition was found to be beneficial in the treatment of peptic ulcers [43] and was employed to reduce gastric acid secretion years before the discovery of Hp as the causative agent of gastric infection. In fact, preliminary studies seemed to attribute a decrease in acid secretion to **AAZ** [44] but higher doses and/or several days were required to gain the effect [42,45]. Then, traces of evidence highlighted **AAZ** effect on chronic gastritis patients [46] and the compound was found to relieve typical ulcer symptoms [46]. Conversely, the involvement of HpCAs in the acclimation of the bacterium in the host stomach was demonstrated [47,48]. Indeed, Hp is a neutrophile and grows preferentially at neutral pH; the intracellular pH of Hp is nearly neutral whereas the stomach lumen reaches values close to 2. Thus, the bacterium has developed a dual enzyme system composed of the Ni-metalloenzyme Urease and the Zn-metalloenzyme CAs as an adaptive mechanism to survive in this environment [49–52]. Briefly, urea is toxic for the bacterium at neutral pH: the exposure of urease-positive strains to high urea levels at neutral pH does not allow their survival since the degradation of urea into ammonia (NH_3) and CO_2 leads to an increase in the cytoplasmatic pH. Otherwise, urease-negative mutans were found to survive in the presence/absence of urea at low pH values (4.5–7.2) [53]. In general, at high H^+ concentrations, the transmembrane urease channel UreI is activated and allows the diffusion of urea into the cell. The urease-catalyzed production of NH_3 and CO_2, both gaseous and able to rapidly diffuse into the periplasm, contributes to neutralizing the acid entered from the extracellular environment. Also, the released CO_2 is a substrate of HpCAs and undergoes fixation and carbonatation, thereby, being stored in the cell (Fig. 1) [54].

Interestingly, urease activity was found to decrease in Hp CA mutants with individual or concomitant inactivation of HpCAs, and high levels of urea (50–150 mM) as a medium supplement involved in a significant increase in the bacterium growth. Also, the addition of 50 mM of bicarbonate in the medium resulted in enhanced growth of the bacterium whereas alkalization by sodium hydroxide did not significantly affect Hp survival. In wild-type strains, HpαCA was found to be upregulated after

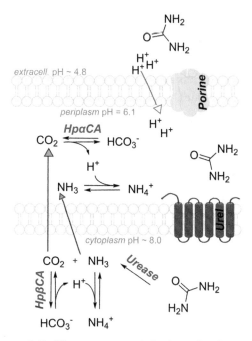

Fig. 1 Involvement of HpCA enzymes and dual mechanism with Urease in *H. pylori*. A molecule of urea enters the bacterial cell *via* specific porines located in the outer membrane, crosses the periplasmatic space thanks to the urea channel (Urel), activated and open at acid pH, and reaches the cytoplasm. Herein, it is converted by Urease into CO_2 and NH_3, gaseous molecules that fast diffuse into the periplasm. CO_2 is hydrated to H_2CO_3 by β- and α-CA isoforms, first in the cytosol and then in the periplasm, respectively, and dissociates to HCO_3^- and H^+. The latter protonates NH_3 into NH_4^+. The NH_3/NH_4^+ system buffers the periplasmatic pH to 6.1. Violet arrows indicate passive diffusion of gaseous CO_2 and NH_3, while yellow arrows indicate active absorption of urea through porin and Urel. Proteins structures are for illustrative purposes and not comprehensive or related to the actual 3D structure of the proteins. *The figure was prepared by means of ChemDraw 22.0.0 (PerkinElmer).*

exposition to low pH in the absence of urea whereas HpβCA mutant strains showed a reduced growth rate in the acid environment (pH 6.25). On the other hand, the dual HpCAs mutant strain resulted to be not sensitive to the pH lowering and its growth was unaffected by the acid environment [53].

Hence, due to the relevant function of HpCAs and, in general, bacterial CAs in the vitality or virulence of the pathogen encoding for them, their inhibition is believed to be a valuable strategy in the development of new antibacterial agents, especially in the frame of urgent needs for innovative treatment options [55–60]. In the case of Hp, targeting at least one of its two isoenzymes could result in the inhibition of the growth of the bacterium and, thereby, a resolution of the infection [61], as already proved for other pathogens [62,63]. In particular, besides **AAZ**, another CAI, ethoxzolamide (**EZA**, Fig. 4) was found to exert an antibacterial activity on different Hp strains, including SS1 and 26695, and minimal inhibitory concentration (MIC) values, were assessed to range from 300 to 500 μM, with a concentration-dependent bactericidal effect [64,65].

Additionally, mass spectrometry analyses revealed the presence of HpαCA in Hp OMVs generated *in vitro* from both the bacterium planktonic and biofilm phenotypes, with the former being in higher amounts, as assessed by protonography [56]. Thus, the protein can be easily transferred to other strains which do not encode for it. As regards the kinetics, the two HpCAs are catalytically efficient with almost identical activity to that of the hCA I for the CO_2 hydration reaction (Table 1).

3. Structural insight into HpCAs

The 3D structure of HpαCA was solved in 2015 by Compostella et al. [69,70], while no experimental data are available for the β isoform. By

Table 1 Kinetic parameters for CO_2 hydration reaction catalyzed by HpCAs and hCA I and inhibition constants (K_I) for AAZ [66-68].

Isoenzyme		Activity level	$K_{cat}(10^5 s^{-1})$	K_M (mM)	K_{cat}/K_M $(10^7 M^{-1}s^{-1})$	K_I AAZ (nM)
hCA I		Medium	2.0	4.0	5.0 at pH 7.5, 20 °C	250
HpαCA	Truncated enzyme [66,67]	Low	2.56	17.0	1.47 at pH 8.9, 25 °C	-
	Full-length enzyme [68]		2.35	16.6	1.56 at pH 8.9, 25 °C	21
HpβCA [68]		Medium	7.1	14.7	4.8 at pH 8.3, 20 °C	40

observing the primary sequences of the isoenzymes, HpαCA is located beside the cytosol, in particular, in the periplasmic environment, contains a signal peptide, and has no transmembrane domains, whereas HpβCA is located in the cytoplasm, as analyzed by the Protter tool [71] (Fig. 2).

The α-isoform of Hp consists of a single polypeptide chain of 228 residues organized in 5 α-helices and 14 β-strands, generating a compact globular domain ($43 \times 48 \times 50 \text{ Å}^3$) (Fig. 2). The protein is very similar to human (h) CAs and shares a 28% sequence identity with hCA II, with 258 Cα atoms superimposed with a low RMSD value. However, the length of some surface loops is longer in the pathogen enzyme with respect to the human isoform [75]. All α-CA enzymes, including that from Hp, are characterized by a high degree of conservation of the catalytic site. In particular, the zinc ion (the enzyme cofactor) is located at the bottom of a deep cleft (cone-shape cavity) and is tetrahedrally coordinated by three histidine residues, which are His110, His112, and His129 for HpαCA (Fig. 2), and a hydroxide ion (OH^-) or a water molecule. The latter is also involved in a hydrogen bond with the hydroxyl moiety of a threonine (Thr191) that also interacts with the carboxylate function of Glu116. Also, the binding of the substrate is performed by Leu190, and the proton shuttling is guaranteed by His85. Conversely, as peculiar for bacterial αCAs, the entrance portion of the protein is wider than that of hCA II [76].

As regards HpβCA, Alphafold released the predicted structure of the isoenzyme with a very high confidence level, as ensured by the per-residue confidence score (pLDDT) > 90 on most of the sequence [73]. However, the protein is composed of 221 amino acids and according to the QuatIdent tool is part of a homohexamer [77]. PSIPred analysis considers the monomer to consist of high-confidence predicted 7 α-helices and 5 β-stands [78]. As reported for several β-CAs, the zinc ion should be coordinated by two cysteines (Cys102 and Cys38), one histidine (His99) residues and one aspartate (Asp40) in case of low pH values and closed configuration, whereas the latter is involved in a salt bridge with the arginine (Arg42) of the catalytic diad in the active form (Fig. 2).

4. Modulation of HpCAs for therapeutical purposes: CAI libraries from the literature

The intricate role of CA in bacterial metabolism and its significance in pathogenicity, by furnishing essential components (CO_2 and HCO_3^-),

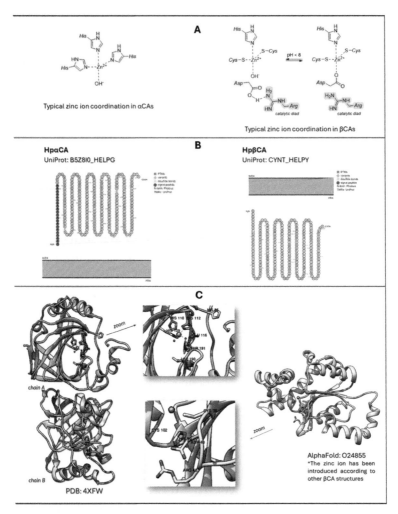

Fig. 2 Structural insight into HpCAs. (A) Typical ion coordination in the active site of α (left) and β (right) CAs; the latter in both the closed and open configuration [72]. (B) Protein visualization of HpαCA (left) and HpβCA (right) obtained through the Protter online tool; accession numbers of UniProt Database are B5Z8I0_HELPG and CYNT_HELPY, respectively. (C) 3D Structures of HpαCA (left) and HpβCA (right) with a zoom of the active site. The former is from X-ray crystal (PDB ID: 4XFW [69,70], corresponding to UniProt ID: B5Z8I0_HELPG), the latter from AlphaFold prediction (AlphaFold ID: O24855 [73], corresponding to UniProt ID: CYNT_HELPY). Protein chains (A and B for HpαCA in rose and yellow, respectively, whereas in cyan for HpβCA) are represented as ribbons, the amino acid residues in the catalytic site are represented in sticks and labeled. The zinc ion is represented as a green sphere and for HpβCA has been added manually. *Panel A was prepared through ChemDraw 22.0.0 (PerkinElmer); panel B was obtained by inserting the UniProt IDs on the query on https://wlab.ethz.ch/protter [71]; panel C was prepared through the UCSF Chimera tool [74].*

ensuring pH homeostasis, and influencing biofilm stability, underline its potential as a novel therapeutic target for infection treatment [79]. Moreover, the growing phenomenon of (multi-)drug resistance across various antibiotic classes emphasizes the urgent need for novel antibacterial agents. In this context, novel targets have been investigated aiming to develop alternative strategies to treat Hp infections, including urease, respiratory complex I, menaquinone biosynthesis pathway [80], flagellar motility [81], and HpCAs [48,82]. As widely described above, Hp relies on HpαCA and HpβCA to ensure its survival in the highly acidic gastric environment, characterized by pH levels as low as 1.5–2.0 [40,47,50,54], often in collaboration with other proteins such as urease [41,83]. Therefore, inhibiting bacterial CAs emerges as an unconventional yet promising approach to curb bacterial growth and treat Hp infections [57]. To date, CAIs can be classified into four groups based on their binding mode within the catalytic site: whether they bind to the catalytic metal ion with the so-called Zinc-Binding Group (ZBG), establish an interaction with the metal-coordinated water molecule, obstruct the active site entrance, or interact with pockets in the enzyme distinct from the active site (outside the active site) (Fig. 3) [84–86].

However, reported CAIs targeting HpCAs are limited to date due to their very recently exploited significance, falling into only two of these classes and can be easily grouped based on their chemical structure: sulfonamides and anions, which belong to the zinc-binding inhibitors, and phenols, which belong to the water-coordinating CAIs [57,87].

5. Sulfonamido-containing CAIs

The most representative group includes CAIs that directly bind to the Zn^{2+} ion such as sulfonamides, sulfamates, and sulfamides [85,86]. As before mentioned, in the early 70s, **AAZ** along with **EZA** (Fig. 4), were effectively employed by Puscas and Buzas in treating gastric and duodenal ulcers [88–90]. First, their therapeutic efficacy was attributed to the inhibition of hCA I and II present in the gastric mucosa, which were considered to be involved in gastric acid secretion due to H^+ ions generated by the hydration of CO_2. However, it wasn't until later that their antiulcer effect was attributed to their inhibitory activity against HpCAs [48,61,66–68]. In 2006, Nishimori and coworkers tested a Traditional Test Panel (TTP) composed of more than 45 CAIs containing the sulfonamide moiety and numerous sulfonamides and sulfamates used in clinics on

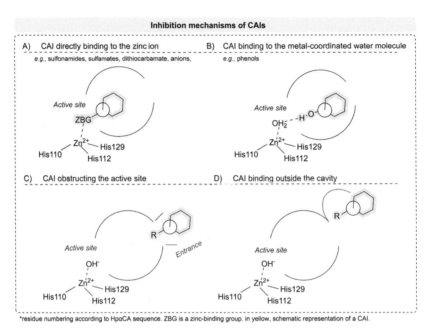

Fig. 3 Classification of CAIs based on their binding mechanisms and some representative chemical classes.

HpαCA by using the stopped-flow CO_2 hydrase assay [66,67]. Actually, the authors used a C-terminal truncated form of HpαCA, consisting of 202 amino acid residues [67], and later the recombinant Hp enzyme, which lacked the first N-terminal signal sequence of 18 amino acid residues [66]. Even though their catalytic activity was similar to that of the full-length HpαCA, in 2008 the same group finally reported the *in vitro* enzymatic data of the complete TTP on both HpαCA and HpβCA [61,68].

The RCB Protein Data Bank database [91] provides valuable insights for medicinal chemists regarding the binding site of HpαCA. Indeed, six X-ray crystal structures have been reported for the isoenzyme in complex with sulfonamide-containing compounds (in yellow, Fig. 4), including **AAZ**, **EZA**, benzolamide (**BZA**), and derivatives **1–3** (Fig. 4) [76,92].

With the exception of compound **2**, which was modeled into the active site of HpαCA using coordinates for compound **1**, they reported and analyzed new crystal structures of HpαCA adducts of the sulfonamides **BZA**, **EZA**, **1**, and **3** [92]. All these compounds exhibited a potent inhibition of hCA II (K_I = 8–60 nM, Table 2), with **AAZ** and compound **3** also showing K_Is in the low nanomolar range on HpαCA (K_I = 21 and 8 nM, respectively, Table 2).

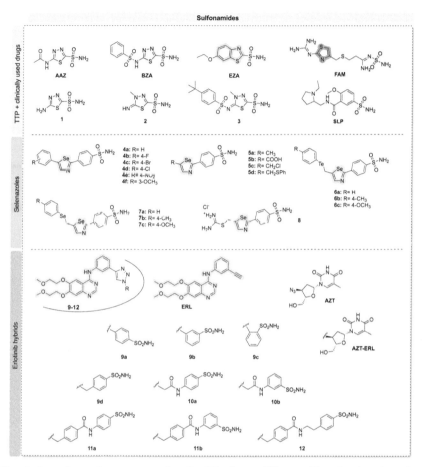

Fig. 4 Overview of representative HpCAIs from different libraries bearing the sulfonamide group. Compounds contain sulfonamide moiety, highlighted in yellow, and are divided into Traditional Test Panel (TTP) and clinically used drugs, selenazoles, and Erlotinib hybrids. Heteroaromatic cores, *e.g.*, **AAZ**-derived 1,3,4-thiadiazole, benzo [d]thiazole, thiazole, and 1,3-selenazole are highlighted in cyan, salmon, violet, and lime, respectively, whereas Erlotinib portion is highlighted in pink.

Additionally, except for compounds **1** and **2**, they all showed good inhibitory activity against HpβCA [90]. Structurally, 5-amino-1,3,4-thiadiazole-2-sulfonamide is contained in **AAZ**, benzolamide (**BZA**) and compound **1**, whereas 5-imino-4,5-dihydro-1,3,4-thiadiazole-2-sulfonamide core is peculiar of **2** and **3** and, finally, **EZA** features a benzothiazole scaffold, thereby, possessing a quite broad spectrum of heterocyclic chemotypes (Fig. 4). Interestingly, compounds **1** and **2** are simplified or *truncated*

Table 2 Inhibition data of sulfonamide-containing CAIs on HpCAs and hCAs I and II by means of stopped-flow CO_2 hydrase assay [61,92-97].

| | Compound | K_I (nM) | | | |
		HpαCA	HpβCA	hCA I	hCA II
TTP + clinically used drugs [61,92,93]	AAZ	21	40	250	12
	BZA	315	54	15	9
	EZA	193	33	25	8
	1	323	2590	8600	60
	2	225	176	50	14
	3	8	40	338	24
	SLP	204	35	1200	40
	FAM	21	50	922	58
Selenazoles [94,95]	4a	1950	n.d.	33	30
	4b	2110	n.d.	638	86
	4c	790	n.d.	206	89
	4d	2030	n.d.	414	68
	4e	2210	n.d.	540	245
	4f	1730	n.d.	42	71
	5a	2110	n.d.	16	47
	5b	1640	n.d.	42	77
	5c	1630	n.d.	7.3	73
	5d	2710	n.d.	935	75
	6a	2720	n.d.	135	80
	6b	2420	n.d.	8.2	187
	6c	2840	n.d.	48	9.5
	7a	2430	n.d.	289	9.8
	7b	2430	n.d.	30	9.6
	7c	2800	n.d.	7.1	7.7
	8	2040	n.d.	15	80
Erlotinib hybrids [96]	9a	63	n.d.	70	2.0
	9b	70.1	n.d.	85	22
	9c	252	n.d.	3502	5.0
	9d	19	n.d.	9.3	4.5
	10a	56	n.d.	3.3	10
	10b	72	n.d.	382	7.4
	11a	48	n.d.	35	5.4
	11b	44	n.d.	31	3.0
	12	805	n.d.	2.6	5.0
	ERL	n.a.	n.d.	n.a.	3976
	AZT	n.a.	n.d.	n.a.	684
	ERL-AZT	19	n.d.	7.1	3.0

K_I values are reported as means of three independent experiments by a stopped-flow technique.[97] Errors are in the range of ± 5-10% of the reported values. n.a.: not active at the highest concentration tested (100 μM). n.d.: not determined.

derivatives of **AAZ** and **3**, respectively. Compound **1** exhibited weaker inhibition of HpαCA (K_I = 323 nM, Table 2) compared to **AAZ**, which showed a K_I equal to 21 nM. Moreover, the acetamido group of **AAZ** establishes beneficial interactions with the enzyme, including a hydrogen bond with the Nδ atom of Asn108 and a van der Waals contact with the side chain of Lys88 [92]. The common 2-sulfonamide moiety and thiadiazole ring of **AAZ**, **BZA**, and **3** lay in the hydrophobic part of the enzyme active site, whereas the phenyl and the *tert*-butyl-phenyl tails of sulfonamides **BZA** and **3**, respectively, extend towards the rim of the active site [76,92,98]. **BZA** showed a better inhibition against hCA II (K_I = 9 nM, Table 2) compared to HpαCA (K_I = 315 nM, Table 2), a trend similarly observed with **EZA**. This can be explained using crystallographic data. Upon superimposing the bacterial and human proteins, it became evident that the bacterial enzyme possesses a more open hydrophilic active site pocket relative to hCA II, due to the presence of a short α-helix in hCA II. This helix, named helix130–136, interacts hydrophobically with bulkier inhibitors like **BZA**, hence enhancing their hCA II inhibition. On the other hand, **3**, which differs from the **BZA** structure by featuring a CH_3 substituent at position 4 of the thiadiazole ring and a *tert*-butyl substituent in the *para*-position of the phenyl ring, demonstrates a stronger HpαCA inhibition compared to **BZA** and a slight selectivity also over hCA II. In this case, the *tert*-butyl-phenyl fragment clashes with the helix130–133 of hCA II, but accommodates well in the small loop present in the bacterial isoenzyme. Moreover, in the HpαCA, the aliphatic *tert*-butyl substituent fosters hydrophobic interactions with the side chain of Leu139 and the aliphatic portion of the side chain of Lys133, contributing to improving the affinity towards the enzyme. Like **BZA**, the oxygen atoms of the sulfonamido group in **3** form hydrogen bonds with Lys88 and Asn108. Additionally, the CH_3 substituent at the thiadiazole ring, absent in **BZA**, establishes a van der Waals contact with Pro194. These interactions counteract an unfavorable contact with the polar side chain of Asp107, resulting in tighter binding of **3** to HpαCA (K_I = 8.2 nM) compared to **BZA**. The presence of polar residues in the region where the helix130–136 is absent in HpαCA, including Asp107 (corresponding to Ile91 in hCA II), Lys133, and Thr196 (corresponding to Leu204 in hCA II) suggests the importance of polar interactions. Therefore, these results revealed that sulfonamides with polar or amphipathic groups in position 5 of the five-membered heterocyclic ring are likely to have a preference toward HpαCA over hCA II [92].

Later, the anti-Hp activity of **EZA** was assessed on various Hp strains in order to understand the potential emergence of drug-resistance phenomena to the compound. It demonstrated efficacy against clinical isolates resistant to metronidazole, clarithromycin, and/or amoxicillin, indicating a distinct mechanism from that of traditional antibiotics [64]. Moreover, in-depth biological evaluations were performed on Hp strains SS1 and 26695 and the rate of single-step spontaneous resistance acquisition was determined to be less than 10^{-8}, demonstrating that resistance does not develop easily [65]. Drug-resistant clinical isolates contained mutations in genes encoding proteins involved in cell wall synthesis and in control of gene expression. However, the bacterial CA genes were not among them, remaining susceptible to being inhibited by **EZA** [64,65].

Notably, the antipsychotic sulpiride (**SLP**, Fig. 2) proved to be an effective inhibitor of both HpαCA ($K_I = 204$ nM, Table 2) and HpβCA ($K_I = 35$ nM, Table 2) [61]. Antibacterial susceptibility testing was conducted on three Hp strains, including NCTC11637, SS1, and 26695, demonstrating that **SLP** effectively inhibits the growth at a concentration of 200 μg/mL [99]. Interestingly, one of the most explored clinical strategies to fight drug resistance involves the co-administration of antibacterial drugs with different mechanisms of action. This trend is mirrored in medicinal chemistry through multitargeting and hybridization approaches, which aim to combine dual or multiple activities in a single molecule [100–102]. In 2018, Angeli et al. unveiled the potential dual activity of famotidine (**FAM**, Fig. 4), a well-used third-generation antiulcer histamine H_2 receptor antagonist in clinical use for decades for the treatment of peptic ulcer and Zollinger-Ellison syndrome [103,104]. The compound contains a sulfamidate moiety, a bioisostere of the sulfonamide group and a recognized ZBG, which resulted to be fundamental in inhibiting a panel of hCA isoforms (hCA I–XIV) and the two HpCAs (Table 2). In fact, **FAM** emerged as a nanomolar CAI of human isoenzymes II, VI, IX, and XII (with K_Is ranging from 3 to 125 nM) and the X-ray study of its complex with hCAs I and II shed light on the binding mode within the active sites of the enzymes [93]. Moreover, **FAM** was found to strongly inhibit HpCAs, with a K_I of 21 and 50 nM for the α and β isoforms (Table 2), respectively, suggesting that its antiulcer effects may stem not only from H_2 receptor antagonism but also from compromising Hp survival within the gastric niche by inhibiting the pathogen CAs, as previously demonstrated for **AAZ** [54,93]. Compared to **AAZ**, **FAM** exhibited a more potent isoform selectivity as an HpCAI rather than hCAI. These findings support its potential as a lead compound for the development of multi-target anti-infective agents.

In 2019, a series of benzenesulfonamide-bearing selenazoles (**4–8**, Fig. 4) first reported as inhibitors of tumor-associated hCAs IX and XII [94] were evaluated for their inhibitory activity targeting six different bacterial CAs, including HpαCA (Table 2) [95]. The design of selenium-containing compounds was based on previously reported promising results on the scaffold as antibacterials [105–107]. However, derivatives **4–8** were found to inhibit HpαCA with K_Is ranging between 0.79 and 2.84 µM (Table 2), resulting in a relatively flat profile that left little space for structure-activity relationships (SARs) discussion. Nevertheless, all these sulfonamides demonstrated inhibition levels at least two orders of magnitude lower than **AAZ** [95]. The rationale of incorporating known CA-inhibiting chemotypes with known active compounds is intriguing to the scientific community and enriches research pipelines [100–102,108]. In this context, Benito et al. applied the hybridization approach to developing new anti-Hp agents incorporating the Erlotinib (**ERL**, Fig. 4) scaffold, an antitumoral drug currently used in clinics to treat non-small-cell lung and pancreatic cancers [109–111], and the sulfonamide function, leading to potent inhibition of their catalytic activity [96]. Despite lacking inherent antibacterial activity, **ERL** was selected considering the ability of Hp to induce the overexpression of the epidermal growth factor receptor (EGFR) in host cells and the role of **ERL** as a first-generation EGFR inhibitor [112,113] in various cellular processes such as survival, proliferation, differentiation, and migration [114]. Moreover, **ERL** chemical structure allows for the derivatization of the terminal acetylene function through an efficient and versatile click chemistry reaction [115]. Ten **ERL** derivatives (**9–12**, Fig. 4) were synthesized, nine of which, compounds **9–12**, contained the terminal sulfonamide moiety with varying spacers to explore the optimal distance for effective interaction with CAs. Additionally, one derivative, compound **AZT–ERL** (Fig. 4), was designed by combining the **ERL** portion with the Zidovudine (**AZT**) core [96]. **AZT** is a known antiviral agent, acting as a competitive inhibitor of the viral reverse transcriptase, with demonstrated antimicrobial activity against Gram-negative bacteria such as *Escherichia coli* and *Klebsiella pneumoniae* [116]. The hybrid compounds were expected to exhibit interaction with human EGFR through the **ERL** portion and target HpαCA *via* the benzenesulfonamide tails, thereby, displaying antibacterial activity [96]. All the ERL-clicked derivatives were tested on a panel of CAs from bacterial species also involved in cancer development [116–119], including HpαCA, EcoβCA from *E. coli*, SmuβCA from *Streptococcus mutans* and PgiβCA and PgiγCA from *Porphyromonas gingivalis*. In terms of inhibitory activity (Table 2), all derivatives exhibited potent inhibition of hCAs I and II and HpαCA at low nanomolar concentrations, demonstrating significant selectivity against other bacterial CAs and suggesting their potential as narrow-spectrum antibacterial agents. **AZT** and **ERL** were also tested and found inactive against all tested CA isoforms except for hCA II (K_I of 684 nM and 3.97 µM, respectively). Regarding HpαCA, **9–11a-b**, endowed with the sulfonamide function in the *para-* and *meta*-position on the phenyl ring, respectively, showed a potent inhibition with comparable K_I values ranging between 43.6 and 72 nM. Conversely, the *ortho*-substituted derivative **9c** lost more than 3.5-time activity with respect to the other regioisomers **9a-b**. These data align with the reported molecular studies revealing a good fitting of the sulfonamide in the *para*-position on the phenyl group, compared to the other positions [120]. Interestingly, the benzyl derivative **9d** emerged as the best-in-class HpαCA inhibitor with a K_I of 19 nM. On the other hand, compound **12**, with the longest spacer, exhibited the highest activity against hCA II, but conversely, it displayed the weakest inhibition on HpαCA with a K_I of 805 nM. Additionally, antibacterial activity against Hp ATCC 43504 was evaluated. Compounds **9c** and **10b** resulted in being the most potent in the series with MICs corresponding to 8 and 16 µg/mL, respectively. Determination of minimum bactericidal concentrations (MBCs) unveiled a bactericidal effect (MBC/MIC ratio ≤ 4) for both compounds. However, microbiological results may underestimate the effective antibacterial activity of ERL-clicked derivatives due to their apparent low solubility [96]. Notably, the ERL-clicked derivatives exhibited greater potency against bacterial CAs compared to previously reported small molecules and drugs bearing sulfonamide moieties (moderate-high nanomolar CAIs) [66] or selenium (low micromolar CAIs) [95]. Their significant achievement lies in their nanomolar inhibition of HpαCA while showing no activity against CAs encoded by *E. coli*, *S. mutans*, and *P. gingivalis* [96]. This advancement represents a step forward in developing narrow-spectrum antibacterial agents, crucial in combating the rise of drug-resistant phenotypes in the post–antibiotic era.

6. Anion and acid CAIs

In 2013, to explore and identify new possible HpαCAIs, Maresca et al. tested a large series of inorganic anions, as well as sulfamides, sulfamic acid, boronic acids, and others. They share the same binding mechanism of sulfonamides as all these compounds are known to target the metal ion in metalloenzymes, like CAs [121,122]. Representative compounds with inhibition data are detailed in Table 3.

Table 3 Inhibition data of anion and acid CAIs on HpCAs and hCAs I and II by means of stopped-flow CO_2 hydrase assay [97,121,123].

		K_I (μM)			
	Compound	**HpαCA**	**HpβCA**	**hCA I**	**hCA II**
Anions and Acids[121,123]	SO_4^{2-} (**13**)	820	570	630	>200000
	$H_2NSO_2NH_2$ (**14**)	73	72	310	1130
	H_2NSO_3H (**15**)	80	94	21	390
	Ph-$B(OH)_2$ (**16**)	97	73	58600	23100
	Ph-AsO_3H_2 (**17**)	440	92	31700	49200
	CS_3^{2-} (**18**)	380	210	87	8.8
	Et_2NCS^{2-} (**19**)	4.9	7.4	790	3100

K_I values are reported as means of three independent experiments by a stopped-flow technique.[97] Errors are in the range of ± 5-10% of the reported values. Anions were tested as sodium salts.

In general, α-CA isoforms are generally more sensitive to this class of inhibitors and, also in this case, compounds **14–16** and **18** showed K_I values in the range of 4.9–97 μM for HpαCA (Table 3). Interestingly, the substitution of one or two O atoms of the sulfate **13** with NH_2 moieties, as in sulfamic acid **14** or sulfonamide **15**, led to a significant increase of the inhibitory activity on both the HpCA isoforms (K_I = 72–94 μM, Table 3). Phenyl boronic acid **16** and phenylarsonic acid **17** exhibited good HpβCA in the millimolar range, being surprisingly very selective over the human CAs as reported in Table 3. Therefore, together with compound **19**, they represent the most selective anions for HpCAs over human isoforms in the current literature. However, while compound **16** maintained a similar K_I (97 μM, Table 3) on HpαCA, derivative **17** showed a decrease in inhibitory potency (K_I = 440 μM, Table 3). On the other hand, we can observe a difference in activity between trithiocarbonate **18** and trithiocarbamate **19**, with the latter showing an enhanced inhibition of HpCAs with two orders of magnitude compared to **18**, from 280 and 210 to 4.9 and 7.4 μM (Table 3), for β- and α-CA isoforms, respectively. In addition, **19** showed a more selective profile on the bacterial isoforms over hCAs. Conversely, most of these anions generally showed an even stronger inhibition of

HpβCA compared to the α-class isoenzyme. Moreover, dithiocarbamate **19** is the only anion in the tested library with slightly lower K_I on HpαCA. These findings unveil a new chemical class capable of targeting HpCAs with a distinct ZBG from sulfonamides, suggesting promising avenues for the development of CAIs targeting Hp [121].

7. Phenolic CAIs

Phenols fit into another class of CAIs since their mechanism involves indirect binding to the zinc ion. In particular, the phenolic OH moiety establishes hydrogen bonds with the metal-coordinated water molecule securing the inhibitor–enzyme binding [84,85]. Despite generally being less potent CAIs compared to sulfonamides, their prevalence in the natural world renders them an easily accessible class for investigation [124]. Two regioisomer phenols, carvacrol (**CAR**) and thymol (**THY**), have been recently investigated as HpCAs inhibitors (Fig. 5) [125,126]. These compounds are natural phenolic monoterpenoids abundant in the essential oils of *Lamiaceae* and *Apiaceae* families. Among the terpenes, they are proven to be clinically relevant as antimicrobial, antioxidant, and anti-inflammatory agents [127,128]. In addition, both **CAR** and **THY** have been approved by the European Union and Food and Drug Administration as a safe food additive as well as in cosmetics and agriculture fields [129,130].

As regards their anti-HpCA activity, **CAR** showed a similar micromolar activity profile on both the HpCAs, with a slightly higher affinity for the α isoform (K_I = 8.4 and 13.3 μM on HpαCA and HpβCA, respectively, Table 4).

On the other hand, **THY** displayed focused and selective inhibition of the β-isoenzyme of Hp, with a K_I value of 3.4 μM. Both the phenols did

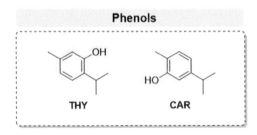

Fig. 5 Chemical structures of CAR and THY.

Table 4 Inhibition data of CAR, THY, and the reference AAZ on HpCAs and hCAs I and II by means of stopped-flow CO_2 hydrase assay [97,125].

	Compound	K_I (µM)			
		HpαCA	**HpβCA**	**hCA I**	**hCA II**
Phenols [125]	**CAR**	8.4	13.3	>100	>100
	THY	>100	3.4	>100	>100

K_I values are reported as means of three independent experiments by a stopped-flow technique.[97] Errors are in the range of ±5-10% of the reported values.

not show activity against hCAs I and II. Interestingly, they were also evaluated against other bacterial β-CAs from *P. gingivalis* and *S. mutans*, resulting in being weaker inhibitors (K_Is = 28–49 µM for PgiβCA, 85–87 µM for SmuCA), which are crucial targets for assessing selectivity as they are major components of the oral microbiome [125]. Moreover, **CAR** and **THY** were tested on Hp ATCC 43504, showing MIC and MBC values of 128 and 256 µg/mL, respectively [125,131,132]. They also inhibited the development of mature Hp biofilms at sub-MIC values, as assessed by the alamarBlue reduction (80% for **CAR** and 85% for **THY**), colony-forming unit counts, and crystal violet staining, and were able to prevent the release of OMVs [125]. Furthermore, the selective antimicrobial potential of both **CAR** and **THY** against three probiotic bacteria encoding for CAs was investigated. Interestingly, the MIC values of the natural phenols against CA-encoding probiotic bacteria were surprisingly high (4–8 µg/mL for *L. reuteri* DSM 17938 and > 16 µg/mL for *L. rhamnosus* GG ATCC 53103 and *L. acidophilus* ATCC SD5214), indicating selective toxicity towards Hp. These findings could suggest a potential synergistic effect for an anti-Hp therapy through the possible association of the CAI with those probiotic species [125]. **CAR** and **THY** also demonstrated minimal toxicity against AGS cells, with IC_{50} values of 300 and 200 µM, respectively, thus ensuring a safe administration by the oral route and after contact with the stomach environment [131,132]. In 2023, **CAR** was additionally explored in combination with either amoxicillin or a urease inhibitor (salicylhydroxamic acid), revealing enhanced antimicrobial and antibiofilm efficacy when combined, thereby representing a novel and promising strategy to combat Hp infection [126].

8. Conclusions: how far HpCAs came and future perspectives

In light of the escalating challenge of multi-drug resistance, the search for antibacterial agents with novel mechanisms of action is imperative. Recent studies have identified and validated HpCAs as promising targets for bacterial drug development, with sulfonamides emerging as nanomolar inhibitors. Beyond them, only a limited number of phenols, sulfamides, dithiocarbamates, and boron-containing compounds have been explored for their inhibitory potential against Hp enzymes. Furthermore, HpβCA remains underexplored due to the lack of an X-ray structure that unable a computer-aided drug design of selective inhibitors. The necessity for inhibiting both HpCAs for effective antibacterial activity or targeting only one remains unclear and due to their function in the survival of the bacterium in the acid environment, a specific cell-based assay is required to better investigate their antibacterial activity. Therefore, continuous efforts are needed to address these issues and advance the development of innovative anti-Hp treatments.

Acknowledgments

This work was partially supported by a grant from the Italian Ministry of University and Research for financial support under the FISR program, project FISR_04819 BacCAD. This work is based upon work from COST Action EURESTOP, CA21145, supported by COST (European Cooperation in Science and Technology).

References

[1] W.M. Duck, J. Sobel, J.M. Pruckler, Q. Song, D.L. Swerdlow, C.R. Friedman, et al., Antimicrobial resistance incidence and risk factors among Helicobacter pylori–infected persons, United States, Emerg. Infect. Dis. J. – CDC 10 (6) (2004), https://doi.org/10.3201/eid1006.030744

[2] R. Borka Balas, L.E. Meliț, C.O. Mărginean, Worldwide prevalence and risk factors of Helicobacter pylori infection in children, Children (Basel) 9 (2022) 1359, https://doi.org/10.3390/children9091359

[3] J.K.Y. Hooi, W.Y. Lai, W.K. Ng, M.M.Y. Suen, F.E. Underwood, D. Tanyingoh, et al., Global prevalence of Helicobacter pylori infection: systematic review and meta-analysis, Gastroenterology 153 (2017) 420–429, https://doi.org/10.1053/j.gastro.2017.04.022

[4] Y. Li, H. Choi, K. Leung, F. Jiang, D.Y. Graham, W.K. Leung, Global prevalence of Helicobacter pylori infection between 1980 and 2022: a systematic review and meta-analysis, Lancet Gastroenterol. Hepatol. 8 (2023) 553–564, https://doi.org/10.1016/S2468-1253(23)00070-5

[5] S.M. Shakir, F.A. Shakir, M.R. Couturier, Updates to the diagnosis and clinical management of Helicobacter pylori infections, Clin. Chem. 69 (2023) 869–880, https://doi.org/10.1093/clinchem/hvad081

[6] G.V. Asokan, T. Ramadhan, E. Ahmed, H. Sanad, WHO global priority pathogens list: a bibliometric analysis of medline-pubmed for knowledge mobilization to infection prevention and control practices in bahrain, Oman Med. J. 34 (2019) 184–193, https://doi.org/10.5001/omj.2019.37

[7] European Centre for Disease Prevention, World Health Organization, editors. Antimicrobial Resistance Surveillance in Europe 2023–2021 Data. Stockholm, 2023.

[8] The Helicobacter Found. https://helico.com/.

[9] O.P. Nyssen, L. Moreira, N. García-Morales, A. Cano-Català, I. Puig, F. Mégraud, et al., European Registry on Helicobacter pylori Management (Hp-EuReg): Most relevant results for clinical practice, Front. Gastroenterol. 1 (2022), https://doi.org/10.3389/fgstr.2022.965982

[10] Helicobacter pylori | CDC Yellow Book, 2024. https://wwwnc.cdc.gov/travel/yellowbook/2024/infectious-diseases/helicobacter-pylori (accessed April 10, 2024).

[11] J.B. Nemhauser, CDC Yellow Book. 2024: Health Information for International Travel. OUP USA, 2023.

[12] M. Xiong, C. Yu, B. Ren, M. Zhong, Q. Peng, M. Zeng, et al., Global knowledge mapping and emerging trends in Helicobacter pylori-related precancerous lesions of gastric cancer research: a bibliometric analysis from 2013 to 2023, Medicine (Baltim.) 102 (2023) e36445, https://doi.org/10.1097/MD.0000000000036445

[13] National Cancer Institute. Helicobacter pylori (H. pylori) and Cancer. https://www.cancer.gov/about-cancer/causes-prevention/risk/infectious-agents/h-pylori-fact-sheet (accessed April 11, 2024).

[14] A. Elbehiry, E. Marzouk, M. Aldubaib, A. Abalkhail, S. Anagreyyah, N. Anajirih, et al., Helicobacter pylori infection: current status and future prospects on diagnostic, therapeutic and control challenges, Antibiotics 12 (2023) 191, https://doi.org/10.3390/antibiotics12020191

[15] F. Kamarehei, Y. Mohammadi, The effect of Helicobacter pylori infection on overweight: a systematic review and meta-analysis, Iran. J. Public Health 51 (2022) 2417–2424, https://doi.org/10.18502/ijph.v51i11.11159

[16] J. Chen, J. Ma, X. Liu, S. Duan, N. Liang, S. Yao, The association between Helicobacter pylori infection with overweight/obesity: a protocol for a systematic review and meta-analysis of observational studies, Medicine (Baltim.) 99 (2020) e18703, https://doi.org/10.1097/MD.0000000000018703

[17] K. Okushin, T. Tsutsumi, K. Ikeuchi, A. Kado, K. Enooku, H. Fujinaga, et al., Helicobacter pylori infection and liver diseases: epidemiology and insights into pathogenesis, World J. Gastroenterol. 24 (2018) 3617–3625, https://doi.org/10.3748/wjg.v24.i32.3617

[18] M. Mohammadi, A. Attar, M. Mohammadbeigi, A. Peymani, S. Bolori, F. Fardsanei, The possible role of Helicobacter pylori in liver diseases, Arch. Microbiol. 205 (2023) 281, https://doi.org/10.1007/s00203-023-03602-z

[19] L.D.N. Tali, G.F.N. Faujo, J.L.N. Konang, J.P. Dzoyem, L.B.M. Kouitcheu, Relationship between active Helicobacter pylori infection and risk factors of cardiovascular diseases, a cross-sectional hospital-based study in a Sub-Saharan setting, BMC Infect. Dis. 22 (2022) 731, https://doi.org/10.1186/s12879-022-07718-3

[20] P.G. Jamkhande, S.G. Gattani, S.A. Farhat, Helicobacter pylori and cardiovascular complications: a mechanism based review on role of Helicobacter pylori in cardiovascular diseases, Integr. Med. Res. 5 (2016) 244–249, https://doi.org/10.1016/j.imr.2016.05.005

[21] L. Wang, Z.-M. Cao, L.-L. Zhang, X. Dai, Z. Liu, Y. Zeng, et al., Helicobacter pylori and autoimmune diseases: involving multiple systems, Front. Immunol. 13 (2022), https://doi.org/10.3389/fimmu.2022.833424

[22] M. Ortiz, B. Rosado-Carrión, R. Bredy, Role of Helicobacter pylori infection in Hispanic patients with anemia, Bol. Asoc. Med. P. R. 106 (2014) 13–18.

[23] A. Douros, Z. Ante, C.A. Fallone, L. Azoulay, C. Renoux, S. Suissa, et al., Clinically apparent Helicobacter pylori infection and the risk of incident Alzheimer's disease: a population-based nested case-control study, Alzheimers Dement. 20 (2024) 1716–1724, https://doi.org/10.1002/alz.13561

[24] A. Eshraghian, S.A. Hashemi, A. Hamidian Jahromi, H. Eshraghian, S.M. Masoompour, M.A. Davarpanah, et al., Helicobacter pylori infection as a risk factor for insulin resistance, Dig. Dis. Sci. 54 (2009) 1966–1970, https://doi.org/10.1007/s10620-008-0557-7

[25] K. Aldossari, Association of Helicobacter pylori infection with metabolic and hormonal disorders: 2448, Off. J. Am. Coll. Gastroenterology | ACG | Gastroenterology | ACG 110 (2015) S1016.

[26] F.-W. Tsay, P.-I. Hsu, H. pylori infection and extra-gastroduodenal diseases, J. Biomed. Sci. 25 (2018) 65, https://doi.org/10.1186/s12929-018-0469-6

[27] I. D'Agostino, S. Carradori. Urease. In Metalloenzymes From Bench to Bedside, chapter 5.1, pp. 393-410, 1st Edition - Elsevier, London, UK, September 15, 2023, Editors: Claudiu T. Supuran, William Alexander Donald, ISBN: 9780128239742. https://doi.org/10.1016/B978-0-12-823974-2.00035-8.

[28] A. Ali, K.I. AlHussaini, Helicobacter pylori: a contemporary perspective on pathogenesis, diagnosis and treatment strategies, Microorganisms 12 (2024) 222, https://doi.org/10.3390/microorganisms12010222

[29] H. Yonezawa, T. Osaki, S. Kamiya, Biofilm formation by Helicobacter pylori and its involvement for antibiotic resistance, Biomed. Res. Int. 2015 (2015) 914791, https://doi.org/10.1155/2015/914791

[30] M.F. González, P. Díaz, A. Sandoval-Bórquez, D. Herrera, A.F.G. Quest, Helicobacter pylori outer membrane vesicles and extracellular vesicles from Helicobacter pylori-infected cells in gastric disease development, Int. J. Mol. Sci. 22 (2021) 4823, https://doi.org/10.3390/ijms22094823

[31] A. Kulp, M.J. Kuehn, Biological functions and biogenesis of secreted bacterial outer membrane vesicles, Annu. Rev. Microbiol. 64 (2010) 163–184, https://doi.org/10.1146/annurev.micro.091208.073413

[32] R. Grande, E. Di Campli, S. Di Bartolomeo, F. Verginelli, M. Di Giulio, M. Baffoni, et al., Helicobacter pylori biofilm: a protective environment for bacterial recombination, J. Appl. Microbiol. 113 (2012) 669–676, https://doi.org/10.1111/j.1365-2672.2012.05351.x

[33] R. Grande, M.C. Di Marcantonio, I. Robuffo, A. Pompilio, C. Celia, L. Di Marzio, et al., Helicobacter pylori ATCC 43629/NCTC 11639 outer membrane vesicles (OMVs) from biofilm and planktonic phase associated with extracellular DNA (eDNA), Front. Microbiol. 6 (2015) 1369, https://doi.org/10.3389/fmicb.2015.01369

[34] P. Malfertheiner, F. Megraud, T. Rokkas, J.P. Gisbert, J.-M. Liou, C. Schulz, et al., Management of Helicobacter pylori infection: the Maastricht VI/Florence consensus report, Gut (2022) gutjnl-2022-327745, https://doi.org/10.1136/gutjnl-2022-327745

[35] L. Boyanova, P. Hadzhiyski, R. Gergova, R. Markovska, Evolution of Helicobacter pylori resistance to antibiotics: a topic of increasing concern, Antibiotics (Basel) 12 (2023) 332, https://doi.org/10.3390/antibiotics12020332

[36] L. Bujanda, O.P. Nyssen, J. Ramos, D.S. Bordin, B. Tepes, A. Perez-Aisa, et al., Effectiveness of Helicobacter pylori treatments according to antibiotic resistance, Am. J. Gastroenterol. 119 (2024) 646–654, https://doi.org/10.14309/ajg.0000000000002600

[37] E. Tshibangu-Kabamba, Y. Yamaoka, Helicobacter pylori infection and antibiotic resistance—from biology to clinical implications, Nat. Rev. Gastroenterol. Hepatol. 18 (2021) 613–629, https://doi.org/10.1038/s41575-021-00449-x

[38] G.M. Buzás, P. Birinyi, Newer, older, and alternative agents for the eradication of Helicobacter pylori infection: a narrative review, Antibiotics 12 (2023) 946, https://doi.org/10.3390/antibiotics12060946

[39] S.A. Park, A. Ko, N.G. Lee, Stimulation of growth of the human gastric pathogen Helicobacter pylori by atmospheric level of oxygen under high carbon dioxide tension, BMC Microbiol. 11 (2011) 96, https://doi.org/10.1186/1471-2180-11-96

[40] Y. Wen, J. Feng, D.R. Scott, E.A. Marcus, G. Sachs, Involvement of the HP0165-HP0166 two-component system in expression of some acidic-pH-upregulated genes of Helicobacter pylori, J. Bacteriol. 188 (2006) 1750–1761, https://doi.org/10.1128/JB.188.5.1750-1761.2006

[41] D.R. Scott, E.A. Marcus, Y. Wen, S. Singh, J. Feng, G. Sachs, Cytoplasmic histidine kinase (HP0244)-regulated assembly of urease with urei, a channel for urea and its metabolites, CO_2, NH3, and NI14+, is necessary for acid survival of Helicobacter pylori, J. Bacteriol. 192 (2010) 94–103, https://doi.org/10.1128/JB.00848-09

[42] R. Shahidzadeh, A. Opekun, A. Shiotani, D.Y. Graham, Effect of the carbonic anhydrase inhibitor, acetazolamide, on Helicobacter pylori infection in vivo: a pilot study, Helicobacter 10 (2005) 136–138, https://doi.org/10.1111/j.1523-5378.2005.00306.x

[43] D. Puscas, I. Paun, R. Ursea, N. Lenghel, C.A. Dinga, Carbonic anhydrase inhibitors in the treatment of gastro-duodenal ulcer, G. E. N. 30 (1975) 93–100.

[44] A.E. Lindner, N. Cohen, J. Berkowitz, H.D. Janowitz, A note on the oral dose of acetazolamide required to inhibit acid secretion in man, Gastroenterology 46 (1964) 273–275.

[45] L. Poller, The effect of acetazoleamide, an inhibitor of carbonic anhydrase, on gastric secretion, Br. J. Pharmacol. Chemother. 11 (1956) 263–266.

[46] R. Cheli, M. Dodero, The effect of acetazolamide on gastric secretion in normal subjects and in patients of chronic gastritis confirmed by biopsy, Minerva Gastroenterol. 2 (1956) 120–122.

[47] Y. Wen, J. Feng, D.R. Scott, E.A. Marcus, G. Sachs, The HP0165-HP0166 two-component system (ArsRS) regulates acid-induced expression of HP1186 alpha-carbonic anhydrase in Helicobacter pylori by activating the pH-dependent promoter, J. Bacteriol. 189 (2007) 2426–2434, https://doi.org/10.1128/JB.01492-06

[48] G.M. Buzás, C.T. Supuran, The history and rationale of using carbonic anhydrase inhibitors in the treatment of peptic ulcers. In memoriam Ioan Puşcaş (1932–2015), J. Enzyme Inhib. Med. Chem. 31 (2016) 527–533, https://doi.org/10.3109/14756366.2015.1051042

[49] D.R. Scott, D. Weeks, C. Hong, S. Postius, K. Melchers, G. Sachs, The role of internal urease in acid resistance of Helicobacter pylori, Gastroenterology 114 (1998) 58–70, https://doi.org/10.1016/s0016-5085(98)70633-x

[50] G. Sachs, J.A. Kraut, Y. Wen, J. Feng, D.R. Scott, Urea transport in bacteria: acid acclimation by gastric Helicobacter spp, J. Membr. Biol. 212 (2006) 71–82, https://doi.org/10.1007/s00232-006-0867-7

[51] E.A. Marcus, G. Sachs, Y. Wen, J. Feng, D.R. Scott, Role of the Helicobacter pylori sensor kinase ArsS in protein trafficking and acid acclimation, J. Bacteriol. 194 (2012) 5545–5551, https://doi.org/10.1128/JB.01263-12

[52] D. Tsikas, E. Hanff, G. Brunner, Helicobacter pylori, its urease and carbonic anhydrases, and macrophage nitric oxide synthase, Trends Microbiol. 25 (2017) 601–602, https://doi.org/10.1016/j.tim.2017.05.002

[53] F. Nils Stähler, L. Ganter, K. Lederer, M. Kist, S. Bereswill, Mutational analysis of the Helicobacter pylori carbonic anhydrases, FEMS Immunol. Med. Microbiol. 44 (2005) 183–189, https://doi.org/10.1016/j.femsim.2004.10.021

[54] E.A. Marcus, A.P. Moshfegh, G. Sachs, D.R. Scott, The periplasmic alpha-carbonic anhydrase activity of Helicobacter pylori is essential for acid acclimation, J. Bacteriol. 187 (2005) 729–738, https://doi.org/10.1128/JB.187.2.729-738.2005

[55] C.T. Supuran, Inhibition of bacterial carbonic anhydrases and zinc proteases: from orphan targets to innovative new antibiotic drugs, Curr. Med. Chem. 19 (2012) 831–844, https://doi.org/10.2174/092986712799034824

[56] C. Campestre, V. De Luca, S. Carradori, R. Grande, V. Carginale, A. Scaloni, et al., Carbonic anhydrases: new perspectives on protein functional role and inhibition in Helicobacter pylori, Front. Microbiol. 12 (2021) 629163, https://doi.org/10.3389/fmicb.2021.629163

[57] C.T. Supuran, An overview of novel antimicrobial carbonic anhydrase inhibitors, Expert. Opin. Ther. Targets (2023) 1–14, https://doi.org/10.1080/14728222.2023.2263914

[58] C. Capasso, C.T. Supuran, An overview of the selectivity and efficiency of the bacterial carbonic anhydrase inhibitors, Curr. Med. Chem. 22 (2015) 2130–2139.

[59] I. D'Agostino, G.E. Mathew, P. Angelini, R. Venanzoni, G. Angeles Flores, A. Angeli, et al., Biological investigation of N-methyl thiosemicarbazones as antimicrobial agents and bacterial carbonic anhydrases inhibitors, J. Enzyme Inhib. Med. Chem. 37 (2022) 986–993, https://doi.org/10.1080/14756366.2022.2055009

[60] M. Fantacuzzi, I. D'Agostino, S. Carradori, F. Liguori, F. Carta, M. Agamennone, et al., Benzenesulfonamide derivatives as Vibrio cholerae carbonic anhydrases inhibitors: a computational-aided insight in the structural rigidity-activity relationships, J. Enzyme Inhib. Med. Chem. 38 (2023) 2201402, https://doi.org/10.1080/14756366.2023.2201402

[61] I. Nishimori, S. Onishi, H. Takeuchi, C.T. Supuran, The alpha and beta classes carbonic anhydrases from Helicobacter pylori as novel drug targets, Curr. Pharm. Des. 14 (2008) 622–630, https://doi.org/10.2174/138161208783877875

[62] C.S. Hewitt, N.S. Abutaleb, A.E.M. Elhassanny, A. Nocentini, X. Cao, D.P. Amos, et al., Structure–activity relationship studies of acetazolamide-based carbonic anhydrase inhibitors with activity against Neisseria gonorrhoeae, ACS Infect. Dis. 7 (2021) 1969–1984, https://doi.org/10.1021/acsinfecdis.1c00055

[63] N.S. Abutaleb, A.E.M. Elhassanny, A. Nocentini, C.S. Hewitt, A. Elkashif, B.R. Cooper, et al., Repurposing FDA-approved sulphonamide carbonic anhydrase inhibitors for treatment of Neisseria gonorrhoeae, J. Enzyme Inhib. Med. Chem. 37 (2022) 51–61, https://doi.org/10.1080/14756366.2021.1991336

[64] J.K. Modak, A. Tikhomirova, R.J. Gorrell, M.M. Rahman, D. Kotsanas, T.M. Korman, et al., Anti-Helicobacter pylori activity of ethoxzolamide, J. Enzyme Inhibition Medicinal Chem. 34 (2019) 1660–1667, https://doi.org/10.1080/14756366.2019.1663416

[65] M.M. Rahman, A. Tikhomirova, J.K. Modak, M.L. Hutton, C.T. Supuran, A. Roujeinikova, Antibacterial activity of ethoxzolamide against Helicobacter pylori strains SS1 and 26695, Gut Pathog. 12 (2020) 20, https://doi.org/10.1186/s13099-020-00358-5

[66] I. Nishimori, T. Minakuchi, K. Morimoto, S. Sano, S. Onishi, H. Takeuchi, et al., Carbonic anhydrase inhibitors: DNA cloning and inhibition studies of the alpha-carbonic anhydrase from Helicobacter pylori, a new target for developing sulfonamide and sulfamate gastric drugs, J. Med. Chem. 49 (2006) 2117–2126, https://doi.org/10.1021/jm0512600

[67] I. Nishimori, D. Vullo, T. Minakuchi, K. Morimoto, S. Onishi, A. Scozzafava, et al., Carbonic anhydrase inhibitors: cloning and sulfonamide inhibition studies of a carboxyterminal truncated α-carbonic anhydrase from Helicobacter pylori, Bioorg. Med. Chem. Lett. 16 (8) (2006) 2182, https://doi.org/10.1016/j.bmcl.2006.01.044

[68] I. Nishimori, T. Minakuchi, T. Kohsaki, S. Onishi, H. Takeuchi, D. Vullo, et al., Carbonic anhydrase inhibitors: the beta-carbonic anhydrase from Helicobacter pylori is a new target for sulfonamide and sulfamate inhibitors, Bioorg. Med. Chem. Lett. 17 (2007) 3585–3594, https://doi.org/10.1016/j.bmcl.2007.04.063

[69] M.E. Compostella, P. Berto, F. Vallese, G. Zanotti, Structure of α-carbonic anhydrase from the human pathogen Helicobacter pylori, Acta Crystallogr. F. Struct. Biol. Commun. 71 (2015) 1005–1011, https://doi.org/10.1107/S2053230X15010407

[70] RCSB Protein Data Bank. RCSB PDB - 4XFW. https://www.rcsb.org/structure/4XFW.

[71] U. Omasits, C.H. Ahrens, S. Müller, B. Wollscheid, Protter: interactive protein feature visualization and integration with experimental proteomic data, Bioinformatics 30 (6) (2014) 884, https://doi.org/10.1093/bioinformatics/btt607

[72] C. Capasso, C.T. Supuran, Targeting carbonic anhydrases, Targeting Carbonic Anhydrases, Future Science Ltd, 2014, pp. 2–4, https://doi.org/10.4155/fseb2013.13.45

[73] AlphaFold Protein Structure Database. AlphaFold—O24855. https://alphafold.ebi.ac.uk/entry/O24855.

[74] E.F. Pettersen, T.D. Goddard, C.C. Huang, G.S. Couch, D.M. Greenblatt, E.C. Meng, et al., UCSF Chimera—a visualization system for exploratory research and analysis, J. Comput. Chem. 25 (2004) 1605–1612, https://doi.org/10.1002/jcc.20084

[75] L.C. Chirica, B. Elleby, S. Lindskog, Cloning, expression and some properties of alpha-carbonic anhydrase from Helicobacter pylori, Biochim. Biophys. Acta 1544 (2001) 55–63, https://doi.org/10.1016/s0167-4838(00)00204-1

[76] J.K. Modakh, Y.C. Liu, M.A. Machuca, C.T. Supuran, A. Roujeinikova, Structural basis for the inhibition of Helicobacter pylori α-carbonic anhydrase by sulfonamides, PLoS One 10 (2015) e0127149, https://doi.org/10.1371/journal.pone.0127149

[77] H.-B. Shen, K.-C. Chou, QuatIdent: a web server for identifying protein quaternary structural attribute by fusing functional domain and sequential evolution information, J. Proteome Res. 8 (2009) 1577–1584, https://doi.org/10.1021/pr800957q

[78] L.J. McGuffin, K. Bryson, D.T. Jones, The PSIPRED protein structure prediction server, Bioinformatics 16 (2000) 404–405, https://doi.org/10.1093/bioinformatics/16.4.404

[79] C. Capasso, C.T. Supuran, Carbonic anhydrase and bacterial metabolism: a chance for antibacterial drug discovery, Expert. Opin. Ther. Pat. (2024) 1–10, https://doi.org/10.1080/13543776.2024.2332663

[80] N.A. Vita, S.M. Anderson, M.D. LaFleur, R.E. Lee, Targeting Helicobacter pylori for antibacterial drug discovery with novel therapeutics, Curr. Opin. Microbiol. 70 (2022) 102203, https://doi.org/10.1016/j.mib.2022.102203

[81] S. Suerbaum, N. Coombs, L. Patel, D. Pscheniza, K. Rox, C. Falk, et al., Identification of antimotilins, novel inhibitors of Helicobacter pylori flagellar motility that inhibit stomach colonization in a mouse model, mBio 13 (2022) e0375521, https://doi.org/10.1128/mbio.03755-21

[82] C.T. Supuran, Bacterial carbonic anhydrases as drug targets: toward novel antibiotics? Front. Pharmacol. 2 (2011) 34, https://doi.org/10.3389/fphar.2011.00034

[83] S. Bury-Moné, G.L. Mendz, G.E. Ball, M. Thibonnier, K. Stingl, C. Ecobichon, et al., Roles of alpha and beta carbonic anhydrases of Helicobacter pylori in the urease-dependent response to acidity and in colonization of the murine gastric mucosa, Infect. Immun. 76 (2008) 497–509, https://doi.org/10.1128/IAI.00993-07

[84] V. Alterio, A. Di Fiore, K. D'Ambrosio, C.T. Supuran, G. De Simone, Multiple binding modes of inhibitors to carbonic anhydrases: how to design specific drugs targeting 15 different isoforms? Chem. Rev. 112 (2012) 4421–4468, https://doi.org/10.1021/cr200176r

[85] C.T. Supuran, How many carbonic anhydrase inhibition mechanisms exist? J. Enzyme Inhib. Med. Chem. 31 (2016) 345–360, https://doi.org/10.3109/14756366.2015.1122001

[86] C.T. Supuran, Advances in structure-based drug discovery of carbonic anhydrase inhibitors, Expert. Opin. Drug. Discov. 12 (2017) 61–88, https://doi.org/10.1080/17460441.2017.1253677

[87] C.T. Supuran, C. Capasso, Antibacterial carbonic anhydrase inhibitors: an update on the recent literature, Expert. Opin. Ther. Pat. 30 (2020) 963–982, https://doi.org/10.1080/13543776.2020.1811853

[88] I. Puşcaş, Treatment of gastroduodenal ulcers with carbonic anhydrase inhibitors, Ann. N. Y. Acad. Sci. 429 (1984) 587–591, https://doi.org/10.1111/j.1749-6632.1984.tb12392.x

[89] I. Puscas, G. Buzas, Treatment of duodenal ulcers with ethoxzolamide, an inhibitor of gastric mucosa carbonic anhydrase, Int. J. Clin. Pharmacol. Ther. Toxicol. 24 (1986) 97–99.

[90] C.T. Supuran, Novel carbonic anhydrase inhibitors for the treatment of Helicobacter pylori infection, Expert. Opin. Investig. Drugs (2024) 1–10, https://doi.org/10.1080/13543784.2024.2334714

[91] H.M. Berman, T. Battistuz, T.N. Bhat, W.F. Bluhm, P.E. Bourne, K. Burkhardt, et al., The protein data bank, Acta Crystallogr. D. Biol. Crystallogr. 58 (2002) 899–907, https://doi.org/10.1107/s0907444902003451

[92] J.K. Modak, Y.C. Liu, C.T. Supuran, A. Roujeinikova, Structure-activity relationship for sulfonamide inhibition of Helicobacter pylori α-carbonic anhydrase, J. Med. Chem. 59 (2016) 11098–11109, https://doi.org/10.1021/acs.jmedchem.6b01333

[93] A. Angeli, M. Ferraroni, C.T. Supuran, Famotidine, an antiulcer agent, strongly inhibits Helicobacter pylori and human carbonic anhydrases, ACS Med. Chem. Lett. 9 (2018) 1035–1038, https://doi.org/10.1021/acsmedchemlett.8b00334

[94] A. Angeli, E. Trallori, M. Ferraroni, L. Di Cesare Mannelli, C. Ghelardini, C.T. Supuran, Discovery of new 2, 5-disubstituted 1,3-selenazoles as selective human carbonic anhydrase IX inhibitors with potent anti-tumor activity, Eur. J. Med. Chem. 157 (2018) 1214–1222, https://doi.org/10.1016/j.ejmech.2018.08.096

[95] A. Angeli, M. Pinteala, S.S. Maier, S. Del Prete, C. Capasso, B.C. Simionescu, et al., Inhibition of bacterial α-, β- and γ-class carbonic anhydrases with selenazoles incorporating benzenesulfonamide moieties, J. Enzyme Inhib. Med. Chem. 34 (2019) 244–249, https://doi.org/10.1080/14756366.2018.1547287

[96] G. Benito, I. D'Agostino, S. Carradori, M. Fantacuzzi, M. Agamennone, V. Puca, et al., Erlotinib-containing benzenesulfonamides as anti-Helicobacter pylori agents through carbonic anhydrase inhibition, Future Med. Chem. (2023), https://doi.org/10.4155/fmc-2023-0208

[97] R.G. Khalifah, The carbon dioxide hydration activity of carbonic anhydrase. I. Stop-flow kinetic studies on the native human isoenzymes B and C, J. Biol. Chem. 246 (1971) 2561–2573.

[98] PLOS ONE Staff. Correction: structural basis for the inhibition of Helicobacter pylori α-carbonic anhydrase by sulfonamides. PLoS One 10 (2015) e0132763. https://doi.org/10.1371/journal.pone.0132763.

[99] S. Morishita, I. Nishimori, T. Minakuchi, S. Onishi, H. Takeuchi, T. Sugiura, et al., Cloning, polymorphism, and inhibition of beta-carbonic anhydrase of Helicobacter pylori, J. Gastroenterol. 43 (2008) 849–857, https://doi.org/10.1007/s00535-008-2240-3

[100] A. Redij, S. Carradori, A. Petreni, C.T. Supuran, M.P. Toraskar, Coumarin-pyrazoline hybrids as selective inhibitors of the tumor-associated carbonic anhydrase IX and XII, Anticancer. Agents Med. Chem. 23 (2023) 1217–1223, https://doi.org/10.2174/1871520623666230220162506

[101] C.T. Supuran, Multitargeting approaches involving carbonic anhydrase inhibitors: hybrid drugs against a variety of disorders, J. Enzyme Inhib. Med. Chem. 36 (2021) 1702–1714, https://doi.org/10.1080/14756366.2021.1945049

[102] E. Berrino, S. Carradori, A. Angeli, F. Carta, C.T. Supuran, P. Guglielmi, et al., Dual carbonic anhydrase IX/XII inhibitors and carbon monoxide releasing molecules modulate LPS-mediated inflammation in mouse macrophages, Antioxidants (Basel) 10 (2021) 56, https://doi.org/10.3390/antiox10010056

[103] D.M. Campoli-Richards, S.P. Clissold, Famotidine. Pharmacodynamic and pharmacokinetic properties and a preliminary review of its therapeutic use in peptic ulcer disease and Zollinger-Ellison syndrome, Drugs 32 (1986) 197–221, https://doi.org/10.2165/00003495-198632030-00001

[104] C.W. Howden, G.N. Tytgat, The tolerability and safety profile of famotidine, discussion 35, Clin. Ther. 18 (1996) 36–54, https://doi.org/10.1016/s0149-2918(96)80177-9

[105] M. Koketsu, H. Ishihara, M. Hatsu, Novel compounds, 1,3-selenazine derivatives, as antibacterial agents against Escherichia coli and Staphylococcus aureus, Res. Commun. Mol. Pathol. Pharmacol. 101 (1998) 179–186.

[106] K.Z. Łączkowski, A. Biernasiuk, A. Baranowska-Łączkowska, S. Zielińska, K. Sałat, A. Furgała, et al., Synthesis, antimicrobial and anticonvulsant screening of small library of tetrahydro-2H-thiopyran-4-yl based thiazoles and selenazoles, J. Enzyme Inhib. Med. Chem. 31 (2016) 24–39, https://doi.org/10.1080/14756366.2016.1186020

[107] A.T. Mbaveng, A.G. Ignat, B. Ngameni, V. Zaharia, B.T. Ngadjui, V. Kuete, In vitro antibacterial activities of p-toluenesulfonyl-hydrazinothiazoles and hydrazinoselenazoles against multi-drug resistant Gram-negative phenotypes, BMC Pharmacol. Toxicol. 17 (2016) 3, https://doi.org/10.1186/s40360-016-0046-0

[108] I. D'Agostino, S. Zara, S. Carradori, V. DeLuca, C. Clemente, C.H.M. Kocken, et al., Antimalarial agents targeting plasmodium falciparum carbonic anhydrase: towards dual acting artesunate hybrid compounds, ChemMedChem (2023) e202300267, https://doi.org/10.1002/cmdc.202300267

[109] N. Starling, J. Neoptolemos, D. Cunningham, Role of erlotinib in the management of pancreatic cancer, Ther. Clin. Risk Manag. 2 (2006) 435–445, https://doi.org/10.2147/tcrm.2006.2.4.435

[110] B. Piperdi, R. Perez-Soler, Role of erlotinib in the treatment of non-small cell lung cancer: clinical outcomes in wild-type epidermal growth factor receptor patients, Drugs 72 (Suppl 1) (2012) 11–19, https://doi.org/10.2165/1163018-S0-000000000-00000

[111] C. Arienti, S. Pignatta, A. Tesei, Epidermal growth factor receptor family and its role in gastric cancer, Front. Oncol. 9 (2019) 1308, https://doi.org/10.3389/fonc.2019.01308

[112] J. Kim, N. Kim, J.H. Park, H. Chang, J.Y. Kim, D.H. Lee, et al., The effect of Helicobacter pylori on epidermal growth factor receptor-induced signal transduction and the preventive effect of celecoxib in gastric cancer cells, Gut Liver 7 (2013) 552–559, https://doi.org/10.5009/gnl.2013.7.5.552

[113] B.E. Chichirau, S. Diechler, G. Posselt, S. Wessler, Tyrosine kinases in Helicobacter pylori infections and gastric cancer, Toxins (Basel) 11 (2019) 591, https://doi.org/10.3390/toxins11100591

[114] S. Sigismund, D. Avanzato, L. Lanzetti, Emerging functions of the EGFR in cancer, Mol. Oncol. 12 (2018) 3–20, https://doi.org/10.1002/1878-0261.12155

[115] A. Angeli, C.T. Supuran, Click chemistry approaches for developing carbonic anhydrase inhibitors and their applications, J. Enzyme Inhib. Med. Chem. 38 (2023) 2166503, https://doi.org/10.1080/14756366.2023.2166503

[116] A. Doléans-Jordheim, E. Bergeron, F. Bereyziat, S. Ben-Larbi, O. Dumitrescu, M.-A. Mazoyer, et al., Zidovudine (AZT) has a bactericidal effect on enterobacteria and induces genetic modifications in resistant strains, Eur. J. Clin. Microbiol. Infect. Dis. 30 (2011) 1249–1256, https://doi.org/10.1007/s10096-011-1220-3

[117] S.A.S. Alanazi, K.T.A. Alduaiji, B. Shetty, H.A. Alrashedi, B.L.G. Acharya, S. Vellappally, et al., Pathogenic features of Streptococcus mutans isolated from dental prosthesis patients and diagnosed cancer patients with dental prosthesis, Microb. Pathog. 116 (2018) 356–361, https://doi.org/10.1016/j.micpath.2018.01.037

[118] A. Basu, R. Singh, S. Gupta, Bacterial infections in cancer: a bilateral relationship, Wiley Interdiscip. Rev. Nanomed. Nanobiotechnol 14 (2022) e1771, https://doi.org/10.1002/wnan.1771

[119] R.L. Emanuele Liardo, A.M. Borzì, C. Spatola, B. Martino, G. Privitera, F. Basile, et al., Effects of infections on the pathogenesis of cancer, Indian. J. Med. Res. 153 (2021) 431–445, https://doi.org/10.4103/ijmr.IJMR_339_19

[120] A. Innocenti, R.A. Hall, C. Schlicker, A. Scozzafava, C. Steegborn, F.A. Mühlschlegel, et al., Carbonic anhydrase inhibitors. Inhibition and homology modeling studies of the fungal beta-carbonic anhydrase from Candida albicans with sulfonamides, Bioorg. Med. Chem. 17 (2009) 4503–4509, https://doi.org/10.1016/j.bmc.2009.05.002

[121] A. Maresca, D. Vullo, A. Scozzafava, C.T. Supuran, Inhibition of the alpha- and beta-carbonic anhydrases from the gastric pathogen Helycobacter pylori with anions, J. Enzyme Inhib. Med. Chem. 28 (2013) 388–391, https://doi.org/10.3109/14756366.2011.649268

[122] C. Temperini, A. Scozzafava, C.T. Supuran, Carbonic anhydrase inhibitors. X-ray crystal studies of the carbonic anhydrase II–trithiocarbonate adduct—an inhibitor mimicking the sulfonamide and urea binding to the enzyme, Bioorg. Med. Chem. Lett. 20 (2010) 474–478, https://doi.org/10.1016/j.bmcl.2009.11.124

[123] A. Innocenti, A. Scozzafava, C.T. Supuran, Carbonic anhydrase inhibitors. Inhibition of cytosolic isoforms I, II, III, VII and XIII with less investigated inorganic anions, Bioorg. Med. Chem. Lett. 19 (7) (2009) 1855, https://doi.org/10.1016/j.bmcl.2009.02.088

[124] C.T. Supuran, Carbonic anhydrase inhibitors from marine natural products, Mar. Drugs 20 (2022) 721, https://doi.org/10.3390/md20110721

[125] R. Grande, S. Carradori, V. Puca, I. Vitale, A. Angeli, A. Nocentini, et al., Selective inhibition of Helicobacter pylori carbonic anhydrases by carvacrol and thymol could impair biofilm production and the release of outer membrane vesicles, Int. J. Mol. Sci. 22 (2021) 11583, https://doi.org/10.3390/ijms222111583

[126] V. Puca, G. Turacchio, B. Marinacci, C.T. Supuran, C. Capasso, P. Di Giovanni, et al., Antimicrobial and antibiofilm activities of carvacrol, amoxicillin and sali-cylhydroxamic acid alone and in combination vs. Helicobacter pylori: towards a new multi-targeted therapy, Int. J. Mol. Sci. 24 (2023) 4455, https://doi.org/10.3390/ijms24054455

[127] K. Kachur, Z. Suntres, The antibacterial properties of phenolic isomers, carvacrol and thymol, Crit. Rev. Food Sci. Nutr. 60 (2020) 3042–3053, https://doi.org/10.1080/10408398.2019.1675585

[128] L. Marinelli, A. Di Stefano, I. Cacciatore, Carvacrol and its derivatives as antibacterial agents, Phytochem. Rev. 17 (2018) 903–921, https://doi.org/10.1007/s11101-018-9569-x

[129] A. Kowalczyk, M. Przychodna, S. Sopata, A. Bodalska, I. Fecka, Thymol and thyme essential oil-new insights into selected therapeutic applications, Molecules 25 (2020) 4125, https://doi.org/10.3390/molecules25184125

[130] W. Mączka, M. Twardawska, M. Grabarczyk, K. Wińska, Carvacrol—a natural phenolic compound with antimicrobial properties, Antibiotics (Basel) 12 (2023) 824, https://doi.org/10.3390/antibiotics12050824

[131] F. Sisto, S. Carradori, P. Guglielmi, C.B. Traversi, M. Spano, A.P. Sobolev, et al., Synthesis and biological evaluation of carvacrol-based derivatives as dual inhibitors of H. pylori strains and AGS cell proliferation, Pharmaceuticals (Basel) 13 (2020) 405, https://doi.org/10.3390/ph13110405

[132] F. Sisto, S. Carradori, P. Guglielmi, M. Spano, D. Secci, A. Granese, et al., Synthesis and evaluation of thymol-based synthetic derivatives as dual-action inhibitors against different strains of H. pylori and AGS cell line, Molecules 26 (2021) 1829, https://doi.org/10.3390/molecules26071829

CHAPTER NINE

Neisseria gonorrhoeae carbonic anhydrase inhibition

Molly S. Youse, Katrina J. Holly, and Daniel P. Flaherty*

Borch Department of Medicinal Chemistry and Molecular Pharmacology, College of Pharmacy, Purdue University, West Lafayette, IN, United States
*Corresponding author. e-mail address: dflaher@purdue.edu

Contents

1. Introduction	243
2. Enzyme genetic families encoded in *Neisseria gonorrhoeae*	244
2.1 α-Carbonic anhydrase	244
2.2 β-Carbonic anhydrase	249
2.3 γ-Carbonic anhydrase	257
3. Enzyme inhibition	258
3.1 FDA-approved CAIs	258
3.2 Reported *N. gonorrhoeae* carbonic anhydrase inhibitors	260
4. Conclusion	272
References	272

Abstract

Carbonic anhydrases (CAs) are ubiquitous enzymes that are found in all kingdoms of life. Though different classes of CAs vary in their roles and structures, their primary function is to catalyze the reaction between carbon dioxide and water to produce bicarbonate and a proton. *Neisseria gonorrhoeae* encodes for three distinct CAs (NgCAs) from three different families: an α-, a β-, and a γ-isoform. This chapter details the differences between the three NgCAs, summarizing their subcellular locations, roles, essentiality, structures, and enzyme kinetics. These bacterial enzymes have the potential to be drug targets; thus, previous studies have investigated the inhibition of NgCAs—primarily the α-isoform. Therefore, the classes of inhibitors that have been shown to bind to the NgCAs will be discussed as well. These classes include traditional CA inhibitors, such as sulfonamides, phenols, and coumarins, as well as non-traditional inhibitors including anions and thiocarbamates.

1. Introduction

Carbonic anhydrases (CAs) are a class of metalloenzymes that facilitate the hydration of carbon dioxide to bicarbonate and a proton. This

The Enzymes, Volume 55
ISSN 1874-6047, https://doi.org/10.1016/bs.enz.2024.05.008

reversible reaction is required for a variety of essential physiological processes including bicarbonate transport, pH homeostasis, cellular respiration, metabolism, and fatty acid biosynthesis [1–4]. Eight distinct families of CAs—α-, β-, γ, δ, ζ, η, θ, and ι—have been identified throughout all kingdoms of life and are characterized based on their structures, catalytic metal ions, enzymatic activities, and organisms in which they are found [5–10]. *Neisseria gonorrhoeae*, the Gram-negative pathogen that causes the venereal disease gonorrhea, expresses an α-, β-, and γ-CA [11–13]. The α-CA class was originally discovered in human erythrocytes in 1933 [14,15] and since then, sixteen total α-CA subfamilies have been identified in humans [16]. As the most widely-distributed and thoroughly studied of the eight CA families, α-CAs have also been identified in other eukaryotes [17], plants [18], algae [19] fungi [20], and bacteria [1,21]. Since the α-CA class was shown to be clinically relevant early on, most of the research dedicated to developing drugs for CAs have focused on α-CA inhibition [22]. The β-CA class was discovered shortly after α-CA. Originally identified in plant chloroplasts in 1939 [23], β-CAs are also found in algae [24], bacteria [25], archaea [26], and fungi [27,28]. Similarly, γ-CAs are present in archaea [29], bacteria [30,31], plants [32], and algae [33].

The three NgCAs have been studied to various degrees. Research focused on the α-NgCA began nearly three decades ago [11,34,35] and the enzyme has since been studied as a drug target, with multiple classes of molecules being evaluated as α-NgCA inhibitors [36–40]. The physiological role of β-NgCA has only recently been published [12], and very little is known about γ-NgCA, though inferences on its structure and enzyme kinetics can be made based on other bacterial γ-CAs.

2. Enzyme genetic families encoded in *Neisseria gonorrhoeae*

2.1 α-Carbonic anhydrase

2.1.1 Subcellular location

The α-CA isoform from *N. gonorrhoeae* (α-NgCA) is the most-studied of the three CA isoforms in *N. gonorrhoeae*, as the gene was first cloned and purified in 1996 and was structurally characterized two years later [11,34]. When originally expressed in *E. coli*, an N-terminal 26-residue signal peptide in α-NgCA was cleaved at Ala26 by *E. coli* machinery and localized to the periplasm. This was the first evidence suggesting the enzyme's

subcellular location, though at that time it was not known if this localization mechanism observed in *E. coli* translated to *N. gonorrhoeae* [11,35].
Since then, other Gram-negative bacterial α-CAs have been identified and
are known to have N-terminal signal sequences as well, resulting in periplasmic localization. For example, an α-CA in *Ralstonia eutropha* was
confirmed to localize in the periplasm by fusing the enzyme to Red
Fluorescent Protein (RFP) and imaged using fluorescent microscopy to
visualize concentrated fluorescence near the outer perimeter of the cell
[41]. Furthermore, *R. eutropha* containing a mutant α-CA without its
signaling peptide had lower growth than the wild type strain, suggesting
the lack of signaling peptide may have prevented α-CA from localizing in
the periplasm and thus preventing the enzyme from fulfilling its physiological role in the cell [41]. Localization of α-CA has also been studied in
Helicobacter pylori, where its α-CA signal peptide is present in whole cell
extract but absent in outer membrane vesicle lysate [42]. Additionally,
hydratase activity in the outer membrane vesicles of *H. pylori* has been
attributed to the presence of α-CA in the periplasmic space [42]. Taken
together, these two examples suggest that although experiments have not
been conducted to confirm where α-NgCA resides, it likely is localized in
the periplasm. Furthermore, DeepLocPro [43], a tool that uses deep
learning to predict subcellular localization of prokaryotic proteins, predicts
with high probability (>99%) that the enzyme is periplasmic.

2.1.2 Hypothesized role in bacterial physiology

The exact function of α-NgCA in *N. gonorrhoeae* has yet to be experimentally determined, although the localization of the enzyme to the periplasm may provide some insight on its physiological roles [1,21]. It is presumed that bacterial α-CAs, including α-NgCA, convert diffused CO_2 to
bicarbonate to facilitate metabolism, as both CO_2 and bicarbonate are
essential substrates for enzymes that participate in the TCA cycle such as
pyruvate carboxylase and acetyl-CoA carboxylase [1,21,41,44–46]. Though
this has not been demonstrated in α-NgCA, specific roles of α-CAs in other
pathogens have previously been investigated. For example, the α-CA in
R. eutropha, Caa, was found to be responsible for sequestering CO_2 and
converting it to bicarbonate to provide the cell with necessary carbon sources
[41]. In the absence of Caa, CO_2 traverses freely back and forth across the
outer membrane and cell growth is therefore negatively affected, likely
because metabolic pathways can no longer proceed without Caa generating
bicarbonate [41]. Similarly, associations between periplasmic α-CA activity

and CO_2/bicarbonate concentrations have been observed in *R. palustris* [47]. Interestingly, the α-CA in *H. pylori* is essential for buffering the periplasm when exposed to low pH, allowing the bacteria to maintain its inner membrane integrity [48,49]. Since *H. pylori* colonizes the stomach, the α-CA role in periplasmic homeostasis is an important mechanism to aid cell survival. *N. gonorrhoeae* colonizes in the vagina, an acidic environment as well, so α-NgCA may also play a part in buffering the periplasm, though there are no published results to support this.

2.1.3 Essentiality
It is presumed that α-NgCA is essential, as a study using RNA-seq and Tn-Seq to investigate the gonococcal transcriptome identified the enzyme as essential for gonococcal survival and growth [13]. Essentiality of α-NgCA has also been confirmed through the use of gene silencing, in which an antisense peptide nucleic acid (PNA) was used to reduce the mRNA transcription of α-NgCA and resulted in a complete loss of viability (unpublished results). Since α-NgCA presumably produces bicarbonate to supply the bacteria with an essential carbon source, it follows that α-NgCA is essential for survival. Although CO_2 can spontaneously react with water to produce bicarbonate, it is proposed that the reaction does not occur readily enough for Gram-negative bacteria without the catalysis facilitated by CAs [45]. Since CO_2 can diffuse over the outer membrane of the cell unlike the negatively charged bicarbonate anion, CO_2 may traverse back and forth across the membrane more quickly than the hydration reaction can naturally occur [50–52]. While some bacteria have a membrane-localized bicarbonate transporter (MpsAB) to supply intracellular bicarbonate, there is no evidence of such a transporter in *N. gonorrhoeae* [12,52]. Without a mechanism to uptake extracellular bicarbonate, α-NgCA may be essential to maintain an adequate bicarbonate concentration in the periplasm.

2.1.4 Structure
The original crystal structures for α-NgCA were reported to include a ligand-bound structure with the CA inhibitor acetazolamide, though only two apo structures (PDB:1KOP, 1KOQ) were added to the Protein Data Bank [34]. The structure of the enzyme (Fig. 1A and B), consisting of 252 residues and a molecular weight of 28,085 Da, resembles that of human α-carbonic anhydrases I and II (hCAI and hCAII), with which α-NgCA shares a sequence similarity of 35% and 34%, respectively (Fig. 2) [11,34,53,54]. Both α-NgCA and hCAII have twisted central β-sheet

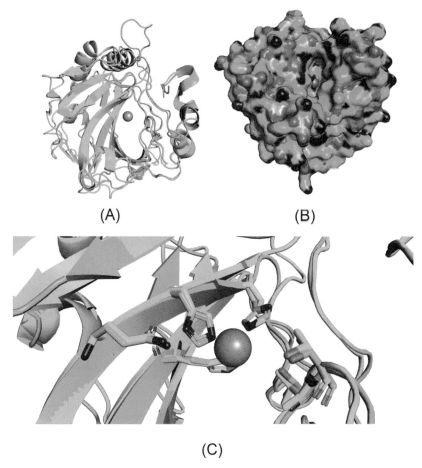

Fig. 1 (A) Crystal structure of α-NgCA (PDB: 8DPC, cyan) [57] overlaid with hCAII (PDB: 1CA2, green) [55]. (B) Surface representation of α-NgCA. (C) Overlay of α-NgCA and hCAII active sites. Relevant residues represented as sticks.

made up of ten β-strands, though the β-strands and loops are shorter than that of the human isoform (Fig. 1A) [11,34,55]. The active site is comparable to all other α-CA active sites, which are in a deep cavity and are comprised of a Zn^{2+} ion coordinated to three histidines (His92, His94, and His111 in α-NgCA) and a water molecule (Fig. 1C) [1,21,34,44,53,56]. The α-NgCA active site also contains two threonines (Thr177 and Thr178), and although these threonines are not conserved across all α-CAs, they are conserved in the hCAII binding site, adding to the similarities between the enzymes (Fig. 1C) [34,57]. Despite these similarities, α–NgCA

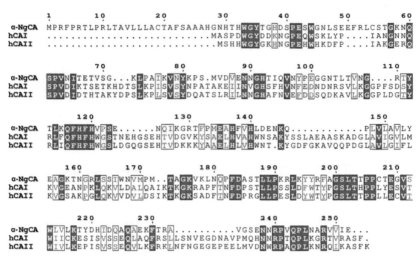

Fig. 2 Sequence alignment of α-NgCA (UniProt ID: Q50940) with hCAI (UniProt ID: P00915) and hCAII (UniProt ID: P00918). Characters highlighted red indicate strict identity; characters with blue box reflect similarity within the group. Figure generated with ESPript 3.0 [58].

is missing three segments of protein that are present in hCAII, contributing to a loss of two 3_{10}-helices [21,56]. These helices form a hydrophobic cleft in hCAII, while the absence of them in α-NgCA gives rise to a wider hydrophobic channel below the entrance of the active site [57]. This difference may prove useful in the design of inhibitors for α-NgCA, as exploiting the subtle differences in active site shape and residues can be used as a strategy to gain selectivity for the bacterial enzyme over the human enzyme.

2.1.5 Enzyme kinetics

The conversion of CO_2 to bicarbonate by α-CAs is well-established and is described as a two-step ping-pong mechanism (Fig. 3) [59–62]. In the first step of catalysis, a Zn^{2+}-bound nucleophilic hydroxide ion attacks the CO_2 bound in the active site, which forms a Zn^{2+}-bound bicarbonate ion. A water molecule in the active site then coordinates to the Zn^{2+} ion, displacing the bicarbonate ion. A proton is transferred from a water molecule via a His residue (His66 in α-NgCA, His64 in hCAII) that acts as a proton shuttle, which then releases the proton to bulk solvent [60,62,63]. This step is the rate-limiting step of the reaction and regenerates the hydroxide ion on the Zn^{2+} ion, allowing for the catalytic process to repeat [62,64]. Initial studies of

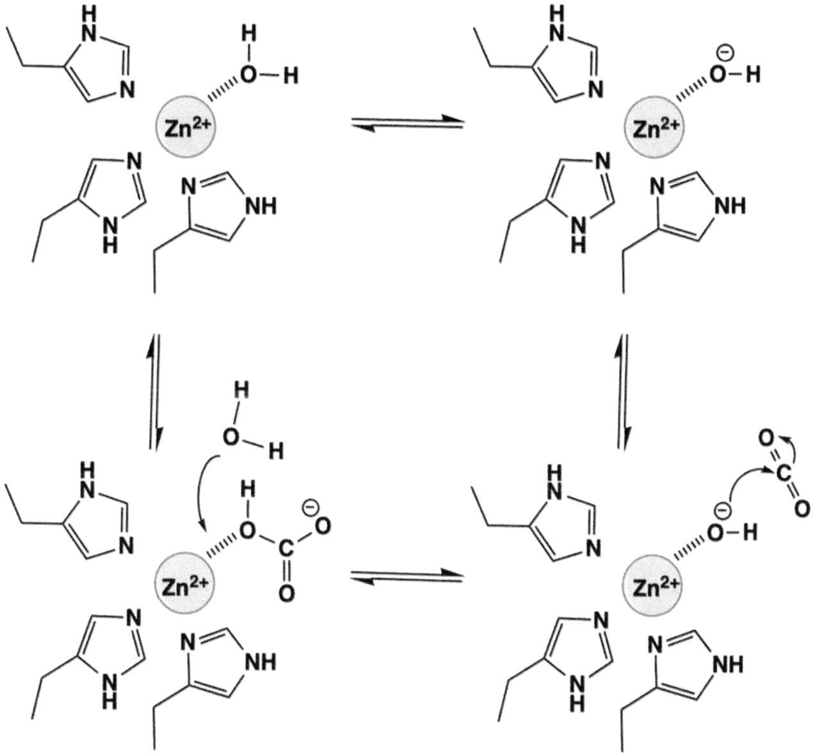

Fig. 3 The catalytic CO_2 hydration reaction facilitated by α-CA through a ping-pong mechanism.

α-NgCA investigated the enzyme kinetics of CO_2 hydration via a stopped-flow apparatus and demonstrated the enzyme follows Michaelis-Menten kinetics and has a high catalytic efficiency, with a k_{cat} of 1.0×10^6 s^{-1} at pH 9.0 and a k_{cat}/K_m of 5.4×10^7 M^{-1} s^{-1} [11]. This high enzymatic activity is comparable to other bacterial α-CAs (Table 1) [65–67]. Though all classes of bacterial CAs are efficient catalysts, α-CAs generally have higher catalytic turnovers for the CO_2 hydration reaction compared to that of β- and γ-CAs, which may be due to the larger catalytic pockets featured in α- isoforms [1].

2.2 β-Carbonic anhydrase

2.2.1 Subcellular location

Despite the β–family being the most widespread CA class in bacteria [68], the β-NgCA isoform, encoded by the *ngo2079* gene [12], is less-studied than α-NgCA. It is presumed that β-NgCA is localized in the cytoplasm, as other

Table 1 Catalytic activity of bacterial α-CAs compared to hCAI and hCAII.

Enzyme	k_{cat} (s^{-1})	k_{cat}/K_m (M^{-1}s^{-1})
hCAI	2.0×10^5	5.0×10^7
hCAII	1.40×10^6	1.5×10^8
α-NgCA	1.0×10^6	7.4×10^7
α-EfCA	8.1×10^5	4.9×10^7
α-hpCA[a]	2.4×10^5	1.4×10^7
α-VchCA[b]	8.2×10^5	7.0×10^7
SazCA[c]	4.4×10^6	3.5×10^8
SspCA[c]	9.4×10^5	1.1×10^8

[a]Measured at pH 8.9, reported by Chirica et al. [65].
[b]Measured at pH 7.5, reported by Del Prete et al. [66].
[c]Reported by De Luca et al. [67].

bacterial β-CAs including those in *E. coli* [51] and *H. pylori* [69] are confirmed to be cytosolic. Additionally, localization prediction by DeepLocPro [43] suggests with 98% probability that β-NgCA is cytoplasmic. Interestingly, a small set of Gram-negative β-CAs have putative signal peptides similar to the signal peptides seen in α-NgCA and other bacterial α-CAs [44]. SignalP 6.0, a machine-learning tool that can be utilized to predict the presence and cleavage site of five types of signal peptides [70], identifies that β-CAs in *Haliscomeno-bacter hydrossis* (Uniprot code: F4KZC7), *Gillisia limnaea*, (Uniprot code: H2BTS7), and *P. gingivalis* (Uniprot code: A0A0K2J5Z9) have signal peptides predicted to be cleaved after residues 18, 13, and 16, respectively. N-terminal signal peptides in bacteria allow for proteins to be translocated from the cytoplasm to the periplasm [71]. Thus, it is hypothesized that these cleaved β-CAs may reach the periplasm and participate in roles similar to that of α-CAs, though no experimental evidence has confirmed this [44]. When submitted to SignalP 6.0, there was no signal peptide identified in β-NgCA, which may further suggest that the enzyme is confined to the cytoplasm.

2.2.2 Hypothesized role in bacterial physiology
Evidence suggests that β-CAs in bacteria are necessary for growing in ambient air conditions where atmospheric levels of CO_2 are low. One study reported a β-CA-deficient mutant of *Clostridium perfringens* was unable to grow on

semi-defined media in the absence of CO_2, despite showing normal growth in 10% CO_2. When the wild-type copy of the β-CA gene was inserted into the mutant strain, growth was restored [72]. In similar studies with *Corynebacterium glutamicum*, *R. eutropha*, and *S. pneumoniae*, deletions of their respective β-CA genes resulted in severely diminished growth in ambient air conditions [41,73,74]. In both cases, growing the β-CA deletion mutants in the presence of supplemented CO_2 restored growth to normal levels compared to the wild-type strains [41,73]. This association between β-CA and CO_2 dependency for growth was recently evaluated in *N. gonorrhoeae*, as approximately 40–60% of clinical gonococcal isolates require supplemental CO_2 to initiate growth in the lag phase [12,75]. Unpublished data using gene silencing has shown that inhibiting β-NgCA expression via an anti-βCA PNA has a bactericidal effect in ambient air, suggesting that β-NgCA is essential for gonococcal survival, at least under normal conditions. The effect of β-NgCA on growth was further elucidated by Rubin et al. [12]. First, a genome-wide associated study (GWAS) identified a missense mutation in *ngo2079* (the gene that encodes β-NgCA) to be associated with the ability for *N. gonorrhoeae* to grow in absence of CO_2. This was further validated in another experiment in which genomic DNA from FA6140, a strain that grows in ambient air, was used to transform FA19, a strain that requires supplemental CO_2 for growth. Through this analysis, it was found that a single nucleotide polymorphism in *ngo2079* was associated with transformants that could now grow without supplemental CO_2 [12]. This hypomorphic mutation was identified as G19E in β-NgCA, wherein β-NgCA^{19E} is not only necessary, but also is the only mutation needed, for *N. gonorrhoeae* to grow in ambient air, while variants with β-NgCA^{19G} require supplemental CO_2 [12].

Even though β-NgCA^{19E} allows for *N. gonorrhoeae* growth in the absence of CO_2, the strains with β-NgCA^{19E} have a longer lag phase when grown in ambient air conditions than when grown with supplemental CO_2 [12]. This suggests the higher concentration of CO_2 allows for the generation of more intracellular bicarbonate, allowing for a shorter lag phase. Importantly, the longer lag phase period observed in the ambient air growth of β-NgCA^{19E} strains is not shortened when grown in media supplemented with either sodium bicarbonate or oxaloacetate, a downstream product of bicarbonate. Likewise, β-NgCA^{19G} strains still exhibit no growth in these ambient air, media-supplemented conditions [12]. Since growth is not affected by the supplementation of bicarbonate nor oxaloacetate, it is presumed that *N. gonorrhoeae* does not have a system to transport extracellular bicarbonate into the cell. Thus, β-NgCA appears to play an important role in regulating CO_2 and bicarbonate in the cytoplasm, as the enzyme's catalytic role may be largely responsible in controlling bicarbonate concentrations to produce downstream metabolites. A similar role has been associated with cynT, a β-CA in *E. coli*, where it was proposed that the absence of cynT would decrease the activity of bicarbonate-dependent carboxylation reactions that are associated with metabolism such as those catalyzed by acetyl CoA-carboxylase and carbamoyl-phosphate synthetase [51]. Notably, however, *E. coli* does not express an α-CA while *N. gonorrhoeae* does, and it remains unclear how α-NgCA may also participate in bicarbonate production for metabolic purposes as well.

Additionally, β-NgCA appears to have a role in buffering the pH intracellularly. The 19E and 19G variant strains have the same survival rate at pH 5.5; however, at pH 4.5, β-NgCA^{19G} variant dies more quickly than β-NgCA^{19E} [12]. This is likely an important aspect for *N. gonorrhoeae* survival in a physiological context, as a healthy vaginal pH is moderately acidic (approximately 3.8 to 5) [12,76]. Interestingly, the two variants also generate different metabolic landscapes, which in turn results in *N. gonorrhoeae* β-NgCA^{19G} being less susceptible to ciprofloxacin than strains with β-NgCA^{19E}. Though the mechanism for this acquired drug resistance facilitated by β-NgCA^{19G} is unknown, it is likely due to its association with the ciprofloxacin resistance mutation GyrA^{91F} rather than the β-NgCA mutation having a direct influence on ciprofloxacin resistance [12].

2.2.3 Essentiality

As explained above, β-NgCAs in general appear to be essential for growth in ambient air. Based on the finding that hypomorphic β-NgCA confers CO_2 dependence, Rubin et al. knocked out the β-NgCA gene in FA1090, an *N. gonorrhoeae* strain that does not require supplemental CO_2 to grow (and therefore must express the β-NgCA^{19E} hypomorph). It was found that the knockout strain depends on supplemental CO_2 to grow just as strains with β-NgCA^{19G} do, demonstrating that β-NgCA^{19E} itself is not strictly essential for in vitro growth [12]. Furthermore, when NgCA^{19E} and NgCA^{19G} strains are both grown in supplemental CO_2, they grow at the same rate regardless of starting inoculum size, suggesting that the mutation is only associated with growth advantages and not with the general fitness of the bacteria [12].

2.2.4 Structure

The β-NgCA sequence is highly conserved across *Neisseria* spp., as β-NgCA shares 96% sequence identity with the β-CA in *N. meningitidis* and greater than 75% sequence identity with β-CAs in *N. flavescens*, *N. animalis*, *N. weaveri*, and *N. arctica* (Fig. 4) [54]. Though a crystal structure for the 199-residue, 22,086 Da enzyme (UniProt accession code: Q5F557) has not yet been solved, prior evidence indicates β-CAs are dimeric structures made up of two identical monomers that each feature a Zn^{2+} ion in the active site. Structures from AlphaFold [77] predict β-NgCA to be a homodimeric structure that features a central beta sheet consisting of four parallel β-strands and a fifth anti-parallel β-strand arranged in a 2-1-3-4-5 pattern, a motif that is characteristic of β-CAs (Fig. 5A) [78]. Its quaternary structure is most likely a tetramer based on most other structurally characterized β-CAs in Gram-negative bacteria [25,79,80], however, it may exist as a dimer [81], hexamer [82], or octamer [83]. Within the β CA class are two subclasses, type I and type II, which differ in their active site architecture. In both types, the active site is located at the dimer interface and is a narrow hydrophobic tunnel [78]. However, type I enzymes coordinate with the active site Zn^{2+} ion via a $Cys_2His(X)$ mechanism, where X is an exchangeable ligand such as water or acetic acid, whereas type II coordination occurs through a $Cys_2HisArg$ motif [78]. Although type I β-CAs also have an Arg residue close to the active site,

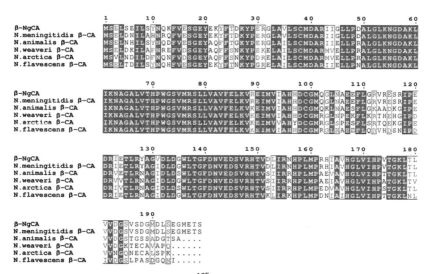

Fig. 4 Sequence alignment for β-NgCA[19E] with β-CAs from other *Neisseria* species that share the highest sequence similarity. Uniprot accession codes for CAs: *N. gonorrhoeae* – Q5F55; *N. meningitidis* – Q7DD5; *N. animalis* – A0A5P3MTN0; *N. weaveri* – A0A3S4Z8Q7; *N. arctica* – A0A0JOYV11; C0ERG8 – *N. flavescens*. Characters highlighted red indicate strict identity; characters with blue box reflect similarity within the group. Figure generated with ESPript 3.0 [58].

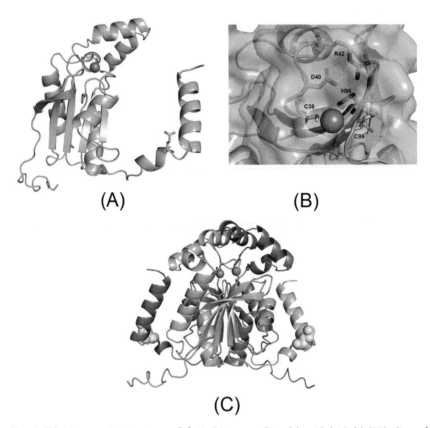

Fig. 5 (A) Monomeric structure of β-NgCA as predicted by AlphaFold [77]. Site of hypomorphic mutation G19E is shown with glutamate represented as sticks. (B) Active site of β-NgCA with key active site residues Cys38, Cys99, and His96 represented as sticks. Asp40 forms a dyad with Arg42 rather than coordinating with Zn, suggesting β-NgCA is a type I CA. (C) Predicted dimer of β-NgCA. Green spheres highlight the site of the hypomorphic mutation G19E, which is far from the active site and dimerization interface.

it forms a dyad with an Asp residue rather than coordinating Zn^{2+} [83–85]. In the β-NgCA active site, it appears that Arg40 is positioned away from the Zn^{2+} ion and is instead interacting with Asp42 (Fig. 5B), suggesting that β-NgCA is type I. Additionally, at least two type II β-CAs—ECCA and HICA—have non-catalytic bicarbonate binding sites [79]. The bicarbonate binds through a Trp-Arg-Tyr triad [79], and although this binding has only been observed in these two type II enzymes thus far, other structurally characterized type II enzymes still have this same binding triad [86,87]. This Trp-Arg-Tyr triad is absent from β-NgCA, further supporting that it is a type I enzyme. Based on the already-characterized enzymes, bacterial β-CAs can be either type I or type II, so there is no reason to believe that β-NgCA is type II solely based on the premise that ECCA and HICA are type II.

2.2.5 Enzyme kinetics

As with all kinetically characterized CAs, β-CAs are highly efficient enzymes: one of the fastest β-CAs investigated to date is the CA expressed in *Pisum sativum* (pea plant), which has a k_{cat} of 4.0×10^5 s^{-1} and a k_{cat}/K_m of 1.8×10^8 M^{-1} s^{-1} at pH 9 [88], a rate comparable to some of the fastest α-CAs like hCAII (k_{cat}/K_m of 1.5×10^8 M^{-1} s^{-1}) [60]. Comparing the active site of the *Pisum sativum* β-CA to the active site of hCAII reveals that they highly resemble each other, and upon superimposition, it appears that the *Pisum sativum* β-CA is a mirror image of hCAII [83]. This piece of evidence, along with other kinetic studies, suggests that the catalytic mechanism of β-CAs mimics that of the α-CAs. Analysis of the CO_2 hydration rates by type I β-CAs from *Methanobacterium thermoautotrophicum* [89], and *Clostridium perfringens* [72] shows that the k_{cat} depends on buffer concentration in a saturable manner, indicating that a ping-pong mechanism is at play with the buffer acting as the second substrate. Thus, it is believed that the catalysis facilitated by β-CA occurs by CO_2 associating with hydrophobic residues in the active site, allowing for a nearby Zn^{2+}-bound hydroxide ion to perform a nucleophilic attack and convert CO_2 to a bicarbonate ion. Subsequently, water and bicarbonate are exchanged at the Zn^{2+} ion, and the water molecule then loses a hydrogen ion to the bulk solvent via proton shuttle to regenerate the hydroxide ion [78].

One of the defining aspects of kinetics in β-CA is the influence of pH, as the CO_2 hydration reaction facilitated by the enzymes is found to be heavily pH-dependent. The k_{cat} and k_{cat}/K_m decrease when the pH decreases, and often the enzymatic activity is reduced or entirely lost at a pH below 8.0 [78,79,88–91]. However, the kinetic profile of these enzymes can be quite complex, as the relationship between pH and catalytic rate seems to be influenced by multiple ionizations [79,89,90]. For example, enzyme kinetics studies with the type I β-CA in *Arabidiposis thaliana* reveal that the pH dependence associated with k_{cat} is dictated by two ionizable proton shuttle groups—one with a pKa of approximately 6.0, and the other with a pKa of approximately 8.7 [90]. The enzymatic activity of type II β-CAs can be even further complicated due to possible conformational changes in these enzymes [78,87,92], though it is not expected that β-NgCA would exhibit these allosteric effects as there is no evidence of a secondary bicarbonate binding site.

Interestingly, one effect on kinetic properties that has been observed in β-NgCA is the influence of the hypomorphic mutation G19E. A stopped flow kinetic inhibition assay was used to measure the exchange between CO_2 and HCO_3^- and revealed β-NgCA^{19E} to have a higher catalytic activity than β-NgCA^{19G}, with a k_{cat} for the CO_2 hydration reaction of 4.5×10^5 s^{-1} and a k_{cat}/K_m of 3.8×10^7. The k_{cat} and k_{cat}/K_m of β-NgCA^{19G} are 1.1×10^4 s^{-1} and 9.1×10^5, respectively. Albeit slower than α-NgCA, β-NgCA^{19E} is nevertheless considered to be a fast-acting enzyme and is in agreement with other enzymatically characterized bacterial β-CAs (Table 2) [80,93–97]. The difference in enzymatic activity between the two variants has also been demonstrated by complementing IPTG-inducible β-NgCA variants in E. coli and measuring the growth of each in ambient air. Complementation for β-NgCA^{19E} required 50 μM IPTG, while 100 μM IPTG was necessary for complementation with β-NgCA^{19G}, providing additional evidence that both variants can restore growth, but at different enzymatic efficiencies [12]. It is still unclear why

Table 2 Catalytic activity of bacterial β-CAs.

Enzyme	k_{cat} (s^{-1})	k_{cat}/K_m (s^{-1}M^{-1})
β-NgCA	4.5×10^5	3.8×10^7
β-NgCA^{E19G}	1.1×10^4	9.1×10^5
β-hpCA[a]	7.1×10^5	4.8×10^7
β-VchCA[b]	3.3×10^5	4.1×10^7
bsCA 1[c]	6.4×10^5	3.9×10^7
bsCA 2[c]	1.1×10^6	8.9×10^7
PCA[d]	7.4×10^5	6.5×10^7
stCA 1[e]	1.0×10^6	8.3×10^7
stCA 2[e]	7.9×10^5	5.2×10^7

[a]Measured at pH 8.3, reported by Nishimori et al. [93].
[b]Measured at pH 8.3, reported by Ferraroni et al. [80].
[c]Measured at pH 8.3, reported by Joseph et al. [94,95].
[d]Measured at pH 8.4, reported by Burghout et al. [96].
[e]Measured at pH 8.3, reported by Vullo et al. [97].

the G19E mutation would influence enzyme kinetics, as the residue does not appear to be located near the active site, nor is it located at the dimerization interface, according to AlphaFold's predicted structure (Fig. 5A and C) [12,77]. However, this difference in activity may explain the aforementioned CO_2 dependence. Since the β-NgCA^{19G} variant is the less efficient enzyme of the two and requires CO_2 to grow, it is hypothesized that the supplemental CO_2 is needed to compensate for the slower activity. The different catalytic efficiencies also provide an explanation as to why the β-NgCA^{19E} variant survives longer at pH 4.5 than the β-NgCA^{19G}, as faster enzyme kinetics would allow for N. gonorrhoeae to maintain intracellular pH for longer.

2.3 γ-Carbonic anhydrase

2.3.1 Essentiality

Little information is known about γ-NgCA. The enzyme's kinetics and its role in bacterial physiology have not yet been studied; however, it is presumed to be non-essential for the pathogen based on a study investigating the transcriptome of N. gonorrhoeae to identify essential gonococcal genes [13]. Similarly, there are few studies that have explored the essentiality of other bacterial γ-CAs. In Enterococcus faecium, compounds that inhibit the γ-CA have been shown to have an antimicrobial effect on the bacteria, suggesting that the enzyme may be essential for growth [98,99]. However, it is unclear whether the bacteriostatic effect in E. faecium is due to the direct inhibition of γ-CA, as target engagement inside the bacterial cell has not been confirmed and as knockout studies have not been implemented to assess how the absence of the enzyme affects bacterial growth [98,99]. Thus, though γ-NgCA is putatively a non-essential enzyme, there still remain many questions to further understand its role in N. gonorrhoeae and the role of bacterial γ-CAs as a whole.

2.3.2 Structure

Structural data for γ -CAs in general pales in comparison to the available data for α-CAs [22]. To date, a structure for the γ-CA in N. gonorrhoeae has not yet been solved, though crystal structures of yrdA, a γ-CA in E. coli [100], along with crystal structures of γ-CAs from the cyanobacteria Thermosynechococcus elongatus [101] and the archaeon Methanosarcina thermophila [102] and Pyrococcus horikoshii [103] have been published. These structures are homotrimers, with each monomer being comprised of parallel β-sheets that together form a left-handed β-helix fold. They also feature a Zn^{2+} ion coordinated by three His residues at each of the three catalytic sites, which are buried at the interfaces between each monomer [100–103]. Though these structures all contain Zn^{2+} as their physiological metal ion, evidence has shown that γ-CAs can tolerate—and sometimes have improved catalytic activity—when other metals are incorporated. For example, Cam, the γ-CA in *M. thermophila*, contains Fe^{2+} and subsequently exhibits increased activity when purified anaerobically rather than aerobically [104,105]. The metal ion in Cam can also be replaced with Co^{2+}, Cu^{2+}, Mn^{2+}, Cd^{2+}, and Ni^{2+}, resulting in varying degrees of activity [105], while Ca^{2+} has been observed in addition to the Zn^{2+} ions in the γ-CA structure of *Pyrococcus horikoshii* [103]. Based on the available information from existing structures and studies, it is therefore reasonable to hypothesize that γ-NgCA, like other γ-CAs, is a homotrimer containing a Zn^{2+} ion at its active site.

3. Enzyme inhibition

3.1 FDA-approved CAIs

Inhibiting carbonic anhydrases in *N. gonorrhoeae* was first suggested in 1965 by Forkman and Laurell [106], who reported that the growth of *N. gonorrhoeae* was inhibited when dosed with acetazolamide (AZM), a known carbonic anhydrase inhibitor (CAI) traditionally prescribed as a diuretic [107] and used to treat a variety of ailments including glaucoma [108], congestive heart failure [109], and altitude sickness [110]. Sanders and Maren furthered this research validating that AZM, along with two other FDA-approved CAIs methazolamide and ethoxzolamide (EZM), exhibit antimicrobial effects against *N. gonorrhoeae* [111]. It was presumed that these drugs were effective against *N. gonorrhoeae* by inhibiting NgCA, as EZM had a lower minimum inhibitory concentration (MIC) against strains grown in ambient air than those grown in elevated levels of CO_2 [111]. However, direct inhibition of NgCA was not measured until decades later by the Lindskog group, who reported EZM to have a K_i of 84 nM via isotope exchange methodology [35]. More recent efforts have employed CO_2 hydration assays to characterize the inhibition of AZM [40] and EZM [112] to the α- and β-NgCAs and compare them to hCAI and II inhibition (Table 3). AZM is slightly more potent against α-NgCA and is approximately 1.5-fold more potent against β-NgCA in comparison to EZM. Though EZM exhibits moderate inhibition against α-NgCA ($K_i = 94$ nM),

Table 3 Inhibitory constants for FDA-approved CAIs against NgCAs and hCAs.

CAI		CO_2 hydration K_i (nM)[a]			
		α-NgCA	β-NgCA	hCAI	hCAII
AZM		74[b]	285[c]	250[b]	12.5[b]
EZM		94[d]	438[c]	24[d]	8[d]

[a]Catalytic CO_2 hydration assay K_i determined from the mean of one experiment performed in triplicate and IC$_{50}$ values entered into Cheng-Prusoff equation. [b]Reported by Hewitt et al. [40]. [c]Unpublished data. [d]Reported by Abutaleb et al. [112].

it is still more potent against both the human enzymes [112]. In general, published studies exploring NgCA inhibition thus far have only assessed α-NgCA, as it was the first of the three enzymes to be discovered and characterized.

The crystal structure of AZM bound to α-NgCA (PDB: 8DYQ) elucidates the binding mode to the bacterial enzyme and reveals how it differs from the human enzymes, particularly from hCAII, as its active site and the residues that participate in ligand binding are very similar to that of α-NgCA [57]. AZM binds to α-NgCA in the same orientation as it binds to other α-CAs [113], with the nitrogens in the thiadiazole ring pointing toward two threonines (Thr177 and Thr178 in α-NgCA; Thr199 and Thr200 in hCAII) in the active site (Fig. 6) [57,114]. Comparing intermolecular interactions, the sulfonamide in AZM forms two interactions with either Thr177 or Thr199 in each of the enzymes, and the two nitrogens in the thiadiazole portion interact with Thr178 and Thr200, respectively. One of the most significant differences between the binding poses of AZM is the lack of interaction with Gln90 in α-NgCA, as the hydrogen bond between the carbonyl and Gln92 is present in the AZM-hCAII complex [114]. This lack of interaction in α-NgCA is caused by the slightly lower angle at which AZM lies in the cavity, causing the distance between the carbonyl and the residue to be longer, which may be a direct result of the aforementioned missing 3_{10} helices present in human CAs. These helices create a hydrophobic cleft in hCAII and likely pushes AZM up within range to facilitate the interaction, while the wider channel in the bacterial enzyme does not have the same effect [57]. The loss of this hydrogen bond was corroborated by biochemical data, in which analogs removing the carbonyl from AZM had no change in inhibition properties

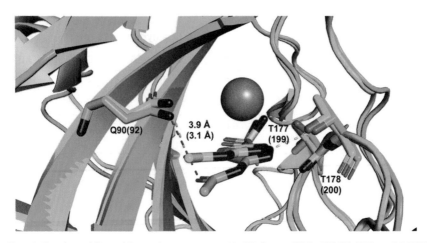

Fig. 6 Overlay of ligand-bound structures α-NgCA (cyan, PDB: 8DYQ) [57] and hCAII (green, PDB: 3HS4) [114] with AZM. Relevant α-NgCA active site residues are labeled with corresponding hCAII residues in parentheses. Purple and orange dashed lines highlight the distances from the carbonyl in AZM to the Gln90/Gln92 residues. Yellow dashed line highlights the essential interaction between the thiadiazole core and Thr178/Thr200.

against α-NgCA while they displayed reduced activity against hCA II (AZM vs **1**, Table 4) [40]. This slight difference in active site shape and its effect on a ligand's pose may be an effective strategy in a drug discovery campaign to gain selectivity over the human enzyme.

3.2 Reported *N. gonorrhoeae* carbonic anhydrase inhibitors

3.2.1 Sulfonamide-based inhibitors

Based on the potency of AZM and its promising therapeutic potential against *N. gonorrhoeae*, the structure–activity relationship (SAR) for a series of AZM-based analogs, derivatized off the amine, was explored and resulted in nanomolar efficacy in the CO_2 hydration assay [40]. Several SAR trends were observed and will be summarized below. First, deletion of the carbonyl in AZM resulted in compound **1** and displayed no effect in inhibition to α-NgCA, though it does result in a 1.5-fold decrease in β-NgCA activity. These analogs confirm that the carbonyl is not necessary for α-NgCA inhibition as alluded to previously. Another trend showed that increasing alkyl branching off the amide (**2** vs. **3**) increases potency against α-NgCA, as does increasing the alkyl chain length (**4** vs. **5**); however, both trends exhibit the opposite effect against β-NgCA. Heterocycles are also well-tolerated against the bacterial enzymes, as exhibited by **6** (K_is = 29.9 nM and 130 nM).

Table 4 Inhibition constants[a] for a subset of AZM derivatives tested against α-NgCA, hCAI, and hCAII by Hewitt et al. [40].

Cmpd		K_i (nM)			
		α-NgCA	β-NgCA[b]	hCAI	hCAII
AZM		74.1	290	250	12.5
1		73.9	430	701	47.2
2		59.8	250	235	37.2
3		30.5	340	167	9.5
4		27.5	300	213	22.3
5		8.5	550	328	22.7
6		29.9	130	14	0.9
7		34.6	190	86.3	19.4
8		36.6	750	58.7	24.5
9		0.70	150	945	0.32
10		8.3	60	855	8.1
11		9.8	130	63.3	20.2
12		>10,000	>10,000	>10,000	>10,000
13		632	1300	465	97.2
14		439	400	1331	67.7

[a]Mean from three separate assays, determined using a stopped flow CO_2 assay.
[b]Unpublished data.

Cyclic alkyl ring size affects only inhibition to β-NgCA, as the cyclopentyl and cyclohexyl analogs are similar in potency against α-NgCA (**7** vs. **8**). Furthermore, there is a clear preference comparing cyclohexane and aromatic phenyl pendant groups in that the cyclic alkane is more active against α-NgCA and less active against β-NgCA (**9** vs. **10**). Comparing alkyl chain length of cyclohexyl-substituted analogs (8 vs. 9 vs. 11) reveals that a two-methylene linker is the optimal length for activity. Of all the AZM analogs tested, **9** is the most potent against α-NgCA (K_i = 0.70 nM) and **10** is the most potent against β-NgCA (K_i = 60 nM) [40].

In addition to derivatizing the amide on the AZM scaffold, modifications to the sulfonamide and the scaffold highlight the importance of specific moieties in AZM that contribute to its potency against α- and β- NgCA. Unsurprisingly, replacing the sulfonamide with a sulfone (**12**) results in complete abolishment of activity against the bacterial and human enzymes, as the inhibition of CAs requires binding to either the Zn^{2+} ion or a solvent molecule in the binding site, neither of which the sulfone moiety is capable of doing. Additionally, replacing the thiadiazole with a thiazole (**13**) or benzene (**14**) greatly reduced inhibition.

Some SAR observations from the AZM analog inhibition study can be explained through the ligand-bound structures of α-NgCA with the CAIs. For example, AZM and **1** are equipotent against the bacterial enzyme likely because the amide in AZM does not participate in any binding in the active site, thus, there are no lost interactions. This relationship between ligand-protein interactions and K_i values is corroborated by the inhibition differences between AZM and **1** against hCAII: **1** is 3.8-fold less potent, which can be attributed to the loss of the carbonyl-Gln92 side chain interaction observed in the crystal structure of AZM bound to hCAII. The importance of the interaction between Thr177 in α-NgCA and the nitrogen adjacent to the sulfonamide, as shown in Fig. 6, can be observed when comparing AZM to **13** and **14**. Binding of the thiazolesulfonamide and the benzenesulfonamide analogs would both result in a loss of this protein-ligand interaction, which may explain the significant decrease in potency compared to AZM.

Crystal structures of α-NgCA bound to three derivatives—**8** (PDB: 8DR2), **10** (PDB: 8DRB), and **11** (PDB: 8DQF)—provide further understanding for AZM analog inhibition [57]. Interestingly, these three analogs have flipped binding poses compared to AZM: the sulfur of the thiadiazole core, rather than the nitrogens, is directed toward the two threonine residues in the active site (Fig. 7). Conversely, Andring et al. [115] showed that **11** binds to hCAII in the traditional binding pose, oriented the same direction as AZM.

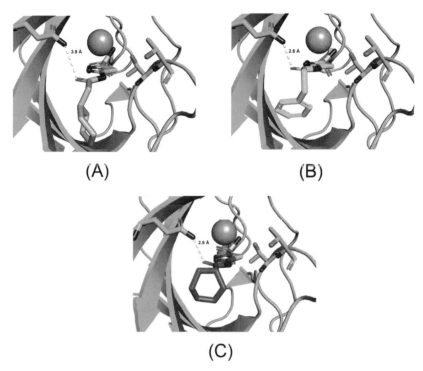

Fig. 7 Ligand-bound structures of α-NgCA with AZM analogs **8** (A, PDB: 8DR2), **10** (B, PDB: 8DRB), and **11** (C, PDB: 8DQF) [57]. Yellow dashed lines highlight the distance between the carbonyl in the analogs and Gln90.

Another difference between AZM and its analogs lies in the interactions with residues in the binding site—compounds **10** and **11**, by virtue of the hydrophobic interactions with Val113 and Leu115 and flipped thiadiazole core, are pulled closer to the side chain Gln90 and form hydrogen bonds. The methylene linker in **8** results in the cyclohexyl ring being pulled down the hydrophobic channel away from Gln90, increasing the distance from the carbonyl to the Gln90 side chain to almost 4.0 Å. This distance is just outside the range for a hydrogen bond indicating either no hydrogen bond or a weak, transient interaction [57]. Compounds **10** and **11** have K_i values approximately 4-fold lower than **8**, perhaps because the two-methylene linker in **10** and the lack of a methylene linker in **11** afford the molecules the ability to engage Gln90 and the hydrophobic residues in the channel at the same time. Taking an already well-established CAI such as AZM and modifying its structure to exploit possible interactions in the active site has thus proven a viable strategy in developing novel, potent inhibitors against NgCA.

3.2.2 Phenol-based inhibitors

Phenol compounds are another group of documented CAIs, first identified as inhibitors against bovine CA [116] and later being shown as competitive inhibitors against hCAII [117] before being investigated as inhibitors for α-NgCA [36]. These compounds are generally significantly less potent in their CA inhibition properties compared to the sulfonamide class. Unsubstituted phenol (**15**), along with 4-(methoxy)phenol (**16**), and 3-amino-4-chlorophenol (**17**) have nanomolar activity against α-NgCA (Table 5) [36]. Adding a second hydroxyl group (**18–20**) to the phenol decreases activity, however, the hydroxyl at either the 2-position (**18**) or 3-position (**19**) results in a gain in selectivity over hCAI, while adding a third hydroxyl (**21**) results in significant selectivity over both hCAI and hCAII. The addition of an amine also results in a loss of activity compared to phenol, and though the meta-substitution (**22**) is favored over the para-substitution (**23**), the latter compound provides significant selectivity over the human enzymes as well. These examples demonstrate how a gain in selectivity may come at a cost, sacrificing more potent inhibition. Overall, the cohort of phenol analogs highlight how subtle changes to the substitution pattern may result in large changes of activity, especially at the 6-position. Similar to the more than 6-fold decrease in activity against α-NgCA between **15** and **16**, the addition of a hydroxyl group to **28** to generate the molecule **29** results in a 7.6-fold decrease in activity. Additionally, incorporating a chlorine at the 6-position to generate **24** from **23** results in a nearly 12-fold decrease in activity and diminishes the selectivity over hCAI and hCAII. Di- or tri-fluorinated phenols (**25–27**) also range in activity against α-NgCA from 3.7 nM to 63.1 nM depending on the substitution pattern, further illuminating the fine-tuning of a chemical moiety that is often required to obtain strong inhibitors against the bacterial enzyme [36].

Although there are no ligand-bound structures for α-NgCA with a phenol, crystal structures of hCAII in complex with phenol [118], salicylic acid (PDB: 6UX1) [119], and caffeic acid (PDB: 6YRI) [120] have been published, which may provide insight to how phenol and phenol analogs bind to α-NgCA, as the binding sites of the enzymes are so similar. Unlike sulfonamides, phenols do not coordinate directly with the Zn^{2+} ion, perhaps accounting for the decrease in activity. Instead, phenol displaces the deep water molecule in the hCAII active site and forms one hydrogen bond with the Zn^{2+}-bound water molecule and one hydrogen bond with Thr199 [118]. The hydroxyl groups in caffeic acid interact with the Zn^{2+}-bound

Table 5 Inhibition constants for a subset of phenols tested against α-NgCA, hCAI, and hCAII by Giovannuzzi et al. [36].

Cmpd	R	K_i (μM)[a]		
		α-NgCA	hCAI	hCAII
15	H	0.9	10.2	5.5
16	4-MeOH	0.6	68.9	95.3
17	3-NH$_2$, 4-Cl	0.8	6.3	4.9
18	2-OH	5.3	>100	5.5
19	3-OH	2.2	>100	9.4
20	4-OH	4.7	10.7	0.1
21	3-OH, 6-OH	3.9	>100	>100
22	3-NH$_2$	1.7	4.9	4.7
23	4-NH$_2$	6.0	>100	>100
24	4-NH$_2$, 6-Cl	71.5	57.8	57.5
25	4-F, 6-F	35.9	>100	>100
26	3-F, 5-F	3.7	38.8	33.9
27	2-F, 4-F, 6-F	63.1	>100	>100
28	4-CHCHCOOH	10.0	1.1	1.3
29	4-CHCHCOOH, 6-OH	76.0	2.4	1.6

[a]Mean from three separate assays, determined using a stopped flow CO_2 assay after a 6 h incubation time.

water, but also hydrogen bond with the deep water molecule rather than displacing it (Fig. 8A) [120], while salicylic acid binds through its carboxylic acid to the Zn^{2+}-bound solvent and forms three bonds with Thr199 and Thr200 (Fig. 8B). From these structures, it can be deduced that the derivatized phenols bind to α-NgCA through the Zn^{2+}-bound water molecule and perhaps form additional hydrogen bonds with the deep-water molecule, Thr177, and Thr178 depending on the shape and size of each molecule.

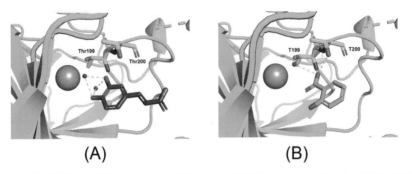

Fig. 8 (A) Caffeic acid (magenta sticks) bound to hCAII by water molecules (PDB: 6YRI) [120]. (B) Salicylic acid (cyan sticks) bound to hCAII through water, Thr199, and Thr200 (PDB: 6UX1) [119].

3.2.3 Coumarins

Coumarins, which have previously been studied as therapeutics for a variety of treatments including tumors [121], HIV [122], and fungi [123], have also been assessed as CAIs after a screening of natural product extracts identified a coumarin derivative to have activity against bovine carbonic anhydrase [124]. The Supuran group later identified coumarins as suicide inhibitors for CAs after observing that co-crystal structures with hCAII (PDB: 3F8E, 5BNL) featured the hydrolyzed products of coumarins, not the coumarins themselves, as the bound ligands (Fig. 9) [125,126]. The hydrolysis of the lactone ring to subsequently form the CA inhibitor can be attributed to the Zn^{2+}-bound hydroxide ion of the CA, which is known to exhibit esterase activity [126,127]. Additionally, coumarin (**30**) and its hydrolyzed product were both tested for inhibition against hCAI and hCAII and were found to have the same K_is, further providing evidence that the true CAIs are the hydrolyzed products rather than the coumarins [126]. In fact, coumarins are weak inhibitors (K_i values in the high micromolar to millimolar range) against hCAs after the 10–15 min-long incubation period typically used to assess CAIs in a CO_2 hydration assay. However, K_i values decrease as incubation time increases up to six hours, suggesting that the esterase reaction of the coumarin and the subsequent formation of the protein–ligand complex is a slow process [126].

The coumarins investigated against α-NgCA thus far, which are all mono- or di-substituted with simple functional groups, exhibited K_i values in the mid-micromolar range (Table 6) [37]. Though the coumarins only have modest potency compared to other classes of CAIs such as the

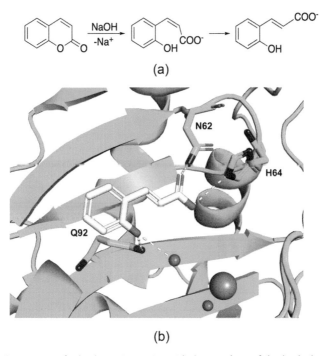

(a)

(b)

Fig. 9 (A) Formation of 2-hydroxycinnamic acid, the product of the hydrolysis reaction between coumarin (**30**) and the Zn^{2+} hydroxide ion, which subsequently binds to CAs. (B) 2-hydroxycinnamic acid (yellow sticks) bound to hCAII (PDB: 5BNL) [125], highlighting its interactions with residues and a water molecule (red sphere).

aforementioned phenols and sulfonamides, this initial structure–activity relationship allows for expansion upon the scaffold to improve activity against NgCA. It was found that compounds **31** and **32** were not only the most potent coumarins against α–NgCA (K_i = 28.6 μM and 42.5 μM, respectively), but were also the most selective over the human isoforms [37]. Bulkier groups at position 3 and 4 led to a decrease in activity (compounds **33–35**), while smaller substitutions such as a hydroxy or methyl at these positions were tolerated (**36–38**). On the other hand, the bulkier NEt_2 group at position 8 on **31** does not affect binding to α–NgCA as 3- and 4- substitutions do, likely because the 8-position is further from the site of hydrolysis on the lactone ring. Whether the functional groups at the 3- and 4-positions affect hydrolysis of the coumarin due to steric hindrance or by modulating the electrophilicity of the carbonyl remains unclear; they also may play a role in enzyme binding interactions rather than affecting the hydrolysis reaction [37].

Table 6 Inhibition constants for a subset of coumarins tested against α-NgCA, hCAI, and hCAII by Giovannuzzi et al. [37].

Cmpd		K_i (µM)[a]		
		α-NgCA	hCAI	hCAII
30	unsub.	81.6	160.0	600.0
31	4-Me, 8-NEt$_2$	28.6	516.5	558.9
32	8-OH, 9-COCH$_3$	42.5	948.9	646.2
33	3-COOEt, 8-OH	394.5	900.1	961.2
34	3-COCH$_3$, 8-OH	394.5	469.7	786.2
35	4-CH$_2$COOEt, 8-OH	469.5	748.9	875.6
36	4-OH	94.7	393.5	513.1
37	3-OH	110.0	498.8	625.2
38	4-Me, 8-NH$_2$	70.9	646.3	485.7

[a]Mean from three separate assays, determined using a stopped flow CO_2 assay after a 6 h incubation time.

The ligand-bound structures of hydrolyzed coumarins with hCAII reveal a unique binding mechanism that was not previously seen with other known CAIs, which bind to either the Zn^{2+} ion or to the Zn^{2+}-bound water molecule. Coumarin products, on the other hand, interact only with water molecules closer to the opening of the active site and participate in hydrogen bonding and pi-pi interactions with nearby residues (Fig. 9B), likely because the shape of the hydrolyzed inhibitors prevents them from reaching the Zn^{2+} ion [128]. Understanding the detailed binding mechanism for each class of NgCA inhibitors is therefore an essential consideration when interpreting results from SARs, as coumarins, phenols, and sulfonamides each have distinct binding modes to the active sites of CAs.

3.2.4 Non-traditional inhibitors

Extensive studies have been conducted to elucidate structure-activity relationships between small drug-like molecules and NgCAs; however, anions and other non-traditional small molecules have been assessed for their inhibition against α-NgCA as well. Monovalent anion inhibition of α-NgCA was first examined by the Lindskog lab [35] a few years after their original work expressing and purifying the enzyme for the first time [11].

Using 4-nitrophenyl acetate as the substrate, the α-NgCA esterase activity was measured to assess inhibition by cyanide, thiocyanate, azide, and cyanate ions (Table 7). The K_i values for these ions against α-NgCA range from .027 mM to 1.16 mM at pH 6.7 [35]. Later, the Flaherty and Supuran labs published a study wherein a stopped flow CO_2 hydratase assay measured the inhibition of the enzyme with a larger set of simple and complex metal-complexing anions, which were also tested against hCAI and hCAII (Table 7) [129]. Through this study, it was found that the halides inhibit α-NgCA with K_i values less than 10 mM, and, notably, F^- ions selectively inhibit the bacterial enzyme by greater than 30-fold over the human isozymes. Selenate and triflate (albeit to a lesser extent) are also NgCA-selective, inhibiting α-NgCA by over 100-fold and 17 fold, respectively. Among the most potent anions are trithiocarbonate and N,N,-diethyldithiocarbamate, which exhibit K_i values of 88 µM and 5.1 µM, respectively, while several other anions including nitrite, bisulfite, sulfate, hydrogensulfide, selenate, and tellurate also inhibit at sub-millimolar levels [129]. The least potent anions perchlorate and hexafluorophosphate show poor inhibition at levels above 100 mM, though this is not surprising as they are known to have poorer affinities to metal ions.

Based on the results from the anion inhibition study revealing trithio-carbonate and N,N-diethyldithiocarbamate to be the most potent of the cohort, dithiocarbamates (DTCs) were investigated in a larger study (Table 8) [38]. This class of inhibitors was first studied against an array of hCAs [130,131] and CAs in *Drosophila melanogaster* [132] and were found to coordinate to the Zn^{2+} ion in CAs through a sulfur in the CS_2^- group [131]. The study against α-NgCA revealed that the originally-tested DTC **39** was actually the least potent of the DTCs, with all of the subsequently tested molecules inhibiting α-NgCA at K_i values at sub-micromolar levels [38]. Compounds **40** and **41** are the most potent inhibitors, each binding at approximately 84 nM, while **42–44** were among the least potent with K_is upward of 700 nM. Comparing **40** to **45** demonstrates the impact of the sulfonamide moiety. Since sulfonamides are known to be a Zn^{2+}-binding group as well, the 3-fold difference in K_i against α-NgCA may be attributed to **40** binding via the sulfonamide instead of the dithiocarbamate as **45** does [38]. In general, the studied DTCs are most effective against hCAII, with most K_i values below 100 nM, while the compounds against hCAI are not as potent—but still exhibit higher activity in comparison to α-NgCA. As evidenced by a ligand-bound crystal structure of hCAII and **46** (PDB: 3P5A), the CS_2^- in the scaffold coordinates directly to the Zn^{2+} ion and interacts with a nearby threonine residue in the CA active site, similarly to how sulfonamides act as a Zn^{2+}-binding group [131].

Table 7 Inhibition constants of anions tested against α-NgCA, hCAI, and hCAII.

Anion	Ki (mM)		
	α-NgCA	hCAI	hCAII
NCO⁻	0.027[a], 0.43[b]	0.03[b]	0.0007[b]
SCN⁻	1.16[a], 0.92[b]	0.2[b]	1.6[b]
CN⁻	0.092[a], 1.0[b]	0.0005[b]	0.02[b]
N_3^-	0.56[a], 2.1[b]	0.0012[b]	1.51[b]
F⁻	8.3[b]	>300[b]	>300[b]
Cl⁻	4.8[b]	6[b]	200[b]
Br⁻	4.0[b]	4[b]	63[b]
I⁻	9.6[b]	0.3[b]	26[b]
NO_2^-	0.59[b]	8.4[b]	63[b]
HSO_3^-	0.66[b]	18[b]	89[b]
SO_4^{2-}	0.83[b]	63[b]	>200[b]
HS⁻	0.55[b]	0.0006[b]	0.04[b]
ClO_4^-	>100[b]	>200[b]	>200[b]
SeO_4^-	0.87[b]	118[b]	112[b]
TeO_4^-	0.76[b]	0.66[b]	0.92[b]
$NH(SO_3)_2^{2-}$	0.25[b]	0.31[b]	0.76[b]
FSO_3^-	0.61[b]	0.79[b]	0.46[b]
CS_3^{2-}	0.088[b]	0.0087[b]	0.0088[b]
$EtNCS_2^-$	0.0051[b]	0.00079[b]	0.0031[b]
PF_6^-	>100[b]	>100[b]	>100[b]
$CF_3SO_3^-$	5.7[b]	>100[b]	>100[b]

[a]Esterase activity using 4-nitrophenyl acetate as the substrate at pH 6.7, from Elleby et al. [35].
[b]Inhibition measured via stopped-flow CO_2 hydrase assay, from Nocentini et al. [129].

Table 8 Inhibition constants for a subset of DTCs and MTCs tested against α-NgCA, hCAI, and hCAII by Giovannuzzi et al. [38,39].

Cmpd	Structure	K_i (nM)[a]		
		α-NgCA	hCAI	hCAII
39		5100	790	3100
40		83.7	97.5	48.1
41		84.4	415	67.2
42		827	485	80.1
43		731	252	30.1
44		723	434	60.2
45		259	425	107
46		483	0.88	0.95
47		395	949	45
48		394	>2000	43
49		417	895	46
50		367	686	>2000

[a]Mean from three separate assays, determined using a stopped flow CO_2 assay.

One of the most recent classes of CAIs investigated against NgCA are the monothiocarbamates (MTCs), published in 2023 [39]. As with DTCs, MTCs were first tested against hCAs prior to being tested against bacterial CAs and exhibited activity against hCAI and hCAII ranging from single-digit nanomolar to greater than 2000 nM [133]. MTCs demonstrate similar activity against α-NgCA as the DTCs, with most of the K_i values being between 500 and 1000 nM, while the most potent MTCs are piperazines with either an additional cyclohexyl (**47**) or phenyl (**48–50**) (Table 8). Though these are still considerably weaker inhibitors than the most potent DTCs, they provide another series of analogs that may be developed as better inhibitors.

4. Conclusion

Neisseria gonorrhoeae expresses three distinct carbonic anhydrases that are unique in their structures and physiological roles. The α-isoform is the most-studied of the three enzymes and a large proportion of NgCA research thus far has focused on α-NgCA inhibition. Sulfonamides are the most potent of the studied inhibitors, with AZM analogs exhibiting low nanomolar inhibition against α-NgCA. Phenols, coumarins, and anions have shown modest inhibition. Recent efforts revealed the role of β-NgCA, as a single-point mutation has been found to confer *N. gonorrhoeae* CO_2 dependence. Additionally, though a crystal structure of β-NgCA has not yet been published, AlphaFold predictions and enzyme inhibition studies have been used to further our understanding of the enzyme. Though previous research has already established bacterial CA roles, essentiality, and subcellular locations, there still remain questions in regard to the specific functions of α-NgCA and β-NgCA and how they may differ or overlap. Furthermore, γ-NgCA is a severely underexplored enzyme in *N. gonorrhoeae* that requires more studying. Lastly, additional studies are still necessary to better understand how the NgCAs and bacterial CAs as a whole may be exploited as potential drug targets for treatment of the pathogen [22].

References

[1] C. Capasso, C.T. Supuran, An overview of the alpha-, beta- and gamma-carbonic anhydrases from Bacteria: can bacterial carbonic anhydrases shed new light on evolution of bacteria? J. Enzyme Inhib. Med. Chem. 30 (2015) 325–332, https://doi.org/10.3109/14756366.2014.910202

[2] D. Lee, J.H. Hong, The fundamental role of bicarbonate transporters and associated carbonic anhydrase enzymes in maintaining ion and pH homeostasis in non-secretory organs, Int. J. Mol. Sci. 21 (2020), https://doi.org/10.3390/ijms21010339

[3] J.R. Krycer, K.H. Fisher-Wellman, D.J. Fazakerley, D.M. Muoio, D.E. James, Bicarbonate alters cellular responses in respiration assays, Biochem. Biophys. Res. Commun. 489 (2017) 399–403, https://doi.org/10.1016/j.bbrc.2017.05.151

[4] D.M. Gibson, E.B. Titchener, S.J. Wakil, Studies on the mechanism of fatty acid synthesis V. Bicarbonate requirement for the synthesis of long-chain fatty acids, Biochim. Biophys. Acta 30 (1958) 376–383, https://doi.org/10.1016/0006-3002(58)90063-5

[5] S. Del Prete, D. Vullo, G.M. Fisher, K.T. Andrews, S.-A. Poulsen, C. Capasso, et al., Discovery of a new family of carbonic anhydrases in the malaria pathogen Plasmodium falciparum—the η-carbonic anhydrases, Bioorg Med. Chem. Lett. 24 (2014) 4389–4396, https://doi.org/10.1016/j.bmcl.2014.08.015

[6] D. Hewett-Emmett, R.E. Tashian, Functional diversity, conservation, and convergence in the evolution of the α-, β-, and γ-carbonic anhydrase gene families, Mol. Phylogenet Evol. 5 (1996) 50–77, https://doi.org/10.1006/mpev.1996.0006

[7] V. Alterio, E. Langella, F. Viparelli, D. Vullo, G. Ascione, N.A. Dathan, et al., Structural and inhibition insights into carbonic anhydrase CDCA1 from the marine diatom Thalassiosira weissflogii, Biochimie 94 (2012) 1232–1241, https://doi.org/10.1016/j.biochi.2012.02.013

[8] M. Lapointe, T.D.B. MacKenzie, D. Morse, An external δ-carbonic anhydrase in a free-living marine dinoflagellate may circumvent diffusion-limited carbon acquisition, Plant. Physiol. 147 (2008) 1427–1436, https://doi.org/10.1104/pp.108.117077

[9] E.L. Jensen, R. Clement, A. Kosta, S.C. Maberly, B. Gontero, A new widespread subclass of carbonic anhydrase in marine phytoplankton, ISME J. 13 (2019) 2094–2106, https://doi.org/10.1038/s41396-019-0426-8

[10] S. Kikutani, K. Nakajima, C. Nagasato, Y. Tsuji, A. Miyatake, Y. Matsuda, Thylakoid luminal θ-carbonic anhydrase critical for growth and photosynthesis in the marine diatom Phaeodactylum tricornutum, Proc. Natl. Acad. Sci. 113 (2016) 9828–9833.

[11] L.C. Chirică, B. Elleby, B.-H. Jonsson, S. Lindskog, The complete sequence, expression in Escherichia coli, purification and some properties of carbonic anhydrase from Neisseria gonorrhoeae, Eur. J. Biochem. 244 (1997) 755–760, https://doi.org/10.1111/j.1432-1033.1997.00755.x

[12] D.H.F. Rubin, K.C. Ma, K.A. Westervelt, K. Hullahalli, M.K. Waldor, Y.H. Grad, CanB is a metabolic mediator of antibiotic resistance in Neisseria gonorrhoeae, Nat. Microbiol. 8 (2023) 28–39, https://doi.org/10.1038/s41564-022-01282-x

[13] C.W. Remmele, Y. Xian, M. Albrecht, M. Faulstich, M. Fraunholz, E. Heinrichs, et al., Transcriptional landscape and essential genes of Neisseria gonorrhoeae, Nucleic Acids Res. 42 (2014) 10579–10595, https://doi.org/10.1093/nar/gku762

[14] N.U. Meldrum, F.J.W. Roughton, Carbonic anhydrase. Its preparation and properties, J. Physiol. 80 (1933) 113–142, https://doi.org/10.1113/jphysiol.1933.sp003077

[15] N.U. Meldrum, F.J.W. Roughton, The state of carbon dioxide in blood, J. Physiol. 80 (1933) 143–170, https://doi.org/10.1113/jphysiol.1933.sp003078

[16] A. Aspatwar, M.E.E. Tolvanen, S. Parkkila, Phylogeny and expression of carbonic anhydrase-related proteins, BMC Mol. Biol. 11 (2010) 25, https://doi.org/10.1186/1471-2199-11-25

[17] R.S. Holmes, Mammalian carbonic anhydrase isozymes: evidence for A third locus, J. Exp. Zool. 197 (1976) 289–295, https://doi.org/10.1002/jez.1401970210

[18] A. Villarejo, S. Burén, S. Larsson, A. Déjardin, M. Monné, C. Rudhe, et al., Evidence for a protein transported through the secretory pathway en route to the higher plant chloroplast, Nat. Cell Biol. 7 (2005) 1224–1231, https://doi.org/10.1038/ncb1330

[19] J. Karlsson, A.K. Clarke, Z. Chen, S.Y. Hugghins, Y. Park, H.D. Husic, et al., A novel α-type carbonic anhydrase associated with the thylakoid membrane in *Chlamydomonas reinhardtii* is required for growth at ambient CO_2, EMBO J. 17 (1998) 1208–1216, https://doi.org/10.1093/emboj/17.5.1208

[20] J.A. Cuesta-Seijo, M.S. Borchert, J.-C. Navarro-Poulsen, K.M. Schnorr, S.B. Mortensen, L. Lo Leggio, Structure of a dimeric fungal α-type carbonic anhydrase, FEBS Lett. 585 (2011) 1042–1048, https://doi.org/10.1016/j.febslet.2011.03.001

[21] C.T. Supuran, C. Capasso, An overview of the bacterial carbonic anhydrases, Metabolites 7 (2017), https://doi.org/10.3390/metabo7040056

[22] D. P Flaherty, M. N Seleem, C. T Supuran, Bacterial carbonic anhydrases: under-exploited antibacterial therapeutic targets, Future Med. Chem. 13 (2021) 1619–1622, https://doi.org/10.4155/fmc-2021-0207

[23] A.C. Neish, Studies on chloroplasts: their chemical composition and the distribution of certain metabolites between the chloroplasts and the remainder of the leaf1, Biochem. J. 33 (1939) 300–308, https://doi.org/10.1042/bj0330300

[24] M. Eriksson, J. Karlsson, Z. Ramazanov, P. Gardeström, G. Samuelsson, Discovery of an algal mitochondrial carbonic anhydrase: molecular cloning and characterization of a low-CO_2-induced polypeptide in Chlamydomonas reinhardtii, Proc. Natl. Acad. Sci. 93 (1996) 12031–12034, https://doi.org/10.1073/pnas.93.21.12031

[25] J.D. Cronk, J.A. Endrizzi, M.R. Cronk, J.W. O'neill, K.Y.J. Zhang, Crystal structure of E. coli β–carbonic anhydrase, an enzyme with an unusual pH–dependent activity, Protein Sci. 10 (2001) 911–922, https://doi.org/10.1110/ps.46301

[26] S. SK, G. FJ, A plant-type (β-class) carbonic anhydrase in the thermophilic metha-noarchaeon Methanobacterium thermoautotrophicum, J. Bacteriol. 181 (1999) 6247–6253, https://doi.org/10.1128/jb.181.20.6247-6253.1999

[27] S. Elleuche, S. Pöggeler, β-carbonic anhydrases play a role in fruiting body devel-opment and ascospore germination in the filamentous fungus Sordaria macrospora, PLoS One 4 (2009) e5177.

[28] R. Götz, A. Gnann, F.K. Zimmermann, Deletion of the carbonic anhydrase-like gene NCE103 of the yeast Saccharomyces cerevisiae causes an oxygen-sensitive growth defect, Yeast 15 (1999) 855–864, https://doi.org/10.1002/(SICI)1097-0061(199907)15:10A<855::AID-YEA425>3.0.CO;2-C

[29] S.A. Zimmerman, J.-F Tomb, J.G. Ferry, Characterization of CamH from Methanosarcina thermophila, founding member of a subclass of the γ class of carbonic anhydrases, J. Bacteriol. 192 (2010) 1353–1360, https://doi.org/10.1128/jb.01164-09

[30] A. Angeli, S. Del Prete, W.A. Donald, C. Capasso, C.T. Supuran, The γ-carbonic anhydrase from the pathogenic bacterium Vibrio cholerae is potently activated by amines and amino acids, Bioorg Chem. 77 (2018) 1–5, https://doi.org/10.1016/j.bioorg.2018.01.003

[31] A. Di Fiore, V. De Luca, E. Langella, A. Nocentini, M. Buonanno, S. Maria Monti, et al., Biochemical, structural, and computational studies of a γ-carbonic anhydrase from the pathogenic bacterium Burkholderia pseudomallei, Comput. Struct. Biotechnol. J. 20 (2022) 4185–4194, https://doi.org/10.1016/j.csbj.2022.07.033

[32] S. Fromm, H.-P. Braun, C. Peterhansel, Mitochondrial gamma carbonic anhydrases are required for complex I assembly and plant reproductive development, N. Phytol. 211 (2016) 194–207, https://doi.org/10.1111/nph.13886

[33] R. Sa, Z. Hong, S. Dorinda, R.A. Jason, R.B.M. Wt, Identification and preliminary characterization of two cDNAs encoding unique carbonic anhydrases from the marine alga Emiliania huxleyi, Appl. Environ. Microbiol. 72 (2006) 5500–5511, https://doi.org/10.1128/AEM.00237-06

[34] S. Huang, Y. Xue, E. Sauer-Eriksson, L. Chirica, S. Lindskog, B.-H. Jonsson, Crystal structure of carbonic anhydrase from Neisseria gonorrhoeae and its complex with the inhibitor acetazolamide, J. Mol. Biol. 283 (1998) 301–310, https://doi.org/10.1006/jmbi.1998.2077

[35] B.È. Elleby, L.C. Chirica, C. Tu, M. Zeppezauer, S. Lindskog, Characterization of carbonic anhydrase from Neisseria gonorrhoeae, Rn vol. 268, (2001).

[36] S. Giovannuzzi, C.S. Hewitt, A. Nocentini, C. Capasso, G. Costantino, D.P. Flaherty, et al., Inhibition studies of bacterial α-carbonic anhydrases with phenols, J. Enzyme Inhib. Med. Chem. 37 (2022) 666–671, https://doi.org/10.1080/14756366.2022.2038592

[37] S. Giovannuzzi, C.S. Hewitt, A. Nocentini, C. Capasso, D.P. Flaherty, C.T. Supuran, Coumarins effectively inhibit bacterial α-carbonic anhydrases, J. Enzyme Inhib. Med. Chem. 37 (2022) 333–338, https://doi.org/10.1080/14756366.2021.2012174

[38] S. Giovannuzzi, N.S. Abutaleb, C.S. Hewitt, F. Carta, A. Nocentini, M.N. Seleem, et al., Dithiocarbamates effectively inhibit the α-carbonic anhydrase from Neisseria gonorrhoeae, J. Enzyme Inhib. Med. Chem. 37 (2022) 1–8, https://doi.org/10.1080/14756366.2021.1988945

[39] S. Giovannuzzi, A.K. Marapaka, N.S. Abutaleb, F. Carta, H.-W. Liang, A. Nocentini, et al., Inhibition of pathogenic bacterial carbonic anhydrases by monothiocarbamates, J. Enzyme Inhib. Med. Chem. 38 (2023) 2284119, https://doi.org/10.1080/14756366.2023.2284119

[40] C.S. Hewitt, N.S. Abutaleb, A.E.M. Elhassanny, A. Nocentini, X. Cao, D.P. Amos, et al., Structure-activity relationship studies of acetazolamide-based carbonic anhydrase inhibitors with activity against Neisseria gonorrhoeae, ACS Infect. Dis. 7 (2021) 1969–1984, https://doi.org/10.1021/acsinfecdis.1c00055

[41] C.S. Gai, J. Lu, C.J. Brigham, A.C. Bernardi, A.J. Sinskey, Insights into bacterial CO_2 metabolism revealed by the characterization of four carbonic anhydrases in Ralstonia eutropha H16, AMB. Express 4 (2014) 2, https://doi.org/10.1186/2191-0855-4-2

[42] M. Ronci, S. Del Prete, V. Puca, S. Carradori, V. Carginale, R. Muraro, et al., Identification and characterization of the α-CA in the outer membrane vesicles produced by Helicobacter pylori, J. Enzyme Inhib. Med. Chem. 34 (2019) 189–195, https://doi.org/10.1080/14756366.2018.1539716

[43] J. Moreno, H. Nielsen, O. Winther, F. Teufel, Predicting the subcellular location of prokaryotic proteins with DeepLocPro, BioRxiv 2024 (2024) 01.04.574157, https://doi.org/10.1101/2024.01.04.574157

[44] C.T. Supuran, C. Capasso, New light on bacterial carbonic anhydrases phylogeny based on the analysis of signal peptide sequences, J. Enzyme Inhib. Med. Chem. 31 (2016) 1254–1260, https://doi.org/10.1080/14756366.2016.1201479

[45] C. Merlin, M. Masters, S. McAteer, A. Coulson, Why is carbonic anhydrase essential to Escherichia coli? J. Bacteriol. 185 (2003) 6415–6424, https://doi.org/10.1128/JB.185.21.6415-6424.2003

[46] K.S. Smith, J.G. Ferry, Prokaryotic carbonic anhydrases, FEMS Microbiol. Rev. 24 (2000) 335–366, https://doi.org/10.1111/j.1574-6976.2000.tb00546.x

[47] L.G. Puskás, M. Inui, K. Zahn, H. Yukawa, A periplasmic, α-type carbonic anhydrase from Rhodopseudomonas palustris is essential for bicarbonate uptake, Microbiology 146 (2000) 2957–2966, https://doi.org/10.1099/00221287-146-11-2957

[48] A. ME, P. MA, S. George, R. SD, The periplasmic α-carbonic anhydrase activity of Helicobacter pylori is essential for acid acclimation, J. Bacteriol. 187 (2005) 729–738, https://doi.org/10.1128/jb.187.2.729-738.2005

[49] B.-M. Stéphanie, L. Mg, E. Bg, T. Marie, S. Kerstin, E. Chantal, et al., Roles of α and β carbonic anhydrases of Helicobacter pylori in the urease-dependent response to acidity and in colonization of the murine gastric mucosa, Infect. Immun. 76 (2008) 497–509, https://doi.org/10.1128/iai.00993-07

[50] J.R. Casey, Why bicarbonate? Biochem. Cell Biol. 84 (2006) 930–939, https://doi.org/10.1139/o06-184

[51] E.I. Kozliak, M.B. Guilloton, M. Gerami-Nejad, J.A. Fuchs, P.M. Anderson, Expression of proteins encoded by the Escherichia coli cyn operon: carbon dioxide-enhanced degradation of carbonic anhydrase, J. Bacteriol. 176 (1994) 5711–5717, https://doi.org/10.1128/jb.176.18.5711-5717.1994

[52] F. Sook-Ha, M. Miki, H. Li, M. Tp, G. Friedrich, The MpsAB bicarbonate transporter is superior to carbonic anhydrase in biofilm-forming bacteria with limited CO_2 diffusion, Microbiol. Spectr. 9 (2021), https://doi.org/10.1128/spectrum.00305-21

[53] C. Supuran, Bacterial carbonic anhydrases as drug targets: Toward novel antibiotics? Front. Pharmacol. 2 (2011) 34.

[54] UniProt: the universal protein knowledgebase in 2023, Nucleic Acids Res. 51 (2023) D523–D531, https://doi.org/10.1093/nar/gkac1052

[55] A.E. Eriksson, T.A. Jones, A. Liljas, Refined structure of human carbonic anhydrase II at 2.0 Å resolution, Proteins Struct. Funct. Bioinforma. 4 (1988) 274–282, https://doi.org/10.1002/prot.340040406

[56] A. Di Fiore, K. D'Ambrosio, J. Ayoub, V. Alterio, G. De Simone, Chapter 2—α-Carbonic anhydrases, in: C.T. Supuran (Ed.), Nocentini ABT-CA, Academic Press, 2019, pp. 19–54, https://doi.org/10.1016/B978-0-12-816476-1.00002-2.

[57] A.K. Marapaka, A. Nocentini, M.S. Youse, W. An, K.J. Holly, C. Das, et al., Structural characterization of thiadiazolesulfonamide inhibitors bound to Neisseria gonorrhoeae α-carbonic anhydrase, ACS Med. Chem. Lett. (2022), https://doi.org/10.1021/acsmedchemlett.2c00471

[58] X. Robert, P. Gouet, Deciphering key features in protein structures with the new ENDscript server, Nucleic Acids Res. 42 (2014) W320–W324, https://doi.org/10.1093/nar/gku316

[59] S. Lindskog, Structure and mechanism of carbonic anhydrase, Pharmacol. Ther. 74 (1997) 1–20, https://doi.org/10.1016/S0163-7258(96)00198-2

[60] D.N. Silverman, S. Lindskog, The catalytic mechanism of carbonic anhydrase: implications of a rate-limiting protolysis of water, Acc. Chem. Res. 21 (1988) 30–36, https://doi.org/10.1021/ar00145a005

[61] C.D. Boone, M. Pinard, R. McKenna, D. Silverman, Catalytic mechanism of α-class carbonic anhydrases: CO_2 hydration and proton transfer, in: S.C. Frost, R. McKenna (Eds.), Carbonic anhydrase: mechanism, regulation, links to disease, and industrial applications, Springer Netherlands, Dordrecht, 2014, pp. 31–52, https://doi.org/10.1007/978-94-007-7359-2_3.

[62] H. Steiner, B.-H. Jonsson, S. Lindskog, The catalytic mechanism of carbonic anhydrase. Eur. J. Biochem. 59 (1975) 253–259, https://doi.org/10.1111/j.1432-1033.1975.tb02449.x

[63] C. Tu, D.N. Silverman, C. Forsman, B.H. Jonsson, S. Lindskog, Role of histidine 64 in the catalytic mechanism of human carbonic anhydrase II studied with a site-specific mutant, Biochemistry 28 (1989) 7913–7918, https://doi.org/10.1021/bi00445a054

[64] D.N. Silverman, R. McKenna, Solvent-mediated proton transfer in catalysis by carbonic anhydrase, Acc. Chem. Res. 40 (2007) 669–675, https://doi.org/10.1021/ar7000588

[65] L.C. Chirica, B. Elleby, S. Lindskog, Cloning, expression and some properties of α-carbonic anhydrase from Helicobacter pylori, Biochim. Biophys. Acta—Protein Struct. Mol. Enzymol. 1544 (2001) 55–63, https://doi.org/10.1016/S0167-4838(00)00204-1

[66] S. Del Prete, S. Isik, D. Vullo, V. De Luca, V. Carginale, A. Scozzafava, et al., DNA cloning, characterization, and inhibition studies of an α-carbonic anhydrase from the pathogenic bacterium Vibrio cholerae, J. Med. Chem. 55 (2012) 10742–10748, https://doi.org/10.1021/jm301611m

[67] V.De Luca, D. Vullo, A. Scozzafava, V. Carginale, M. Rossi, C.T. Supuran, et al., An α-carbonic anhydrase from the thermophilic bacterium Sulphurihydrogenibium azorense is the fastest enzyme known for the CO_2 hydration reaction, Bioorg Med. Chem. 21 (2013) 1465–1469, https://doi.org/10.1016/j.bmc.2012.09.017

[68] K.S. Smith, C. Jakubzick, T.S. Whittam, J.G. Ferry, Carbonic anhydrase is an ancient enzyme widespread in prokaryotes, Proc. Natl. Acad. Sci. 96 (1999) 15184–15189, https://doi.org/10.1073/pnas.96.26.15184

[69] L.C. Chirica, C. Petersson, M. Hurtig, B.-H. Jonsson, T. Borén, S. Lindskog, Expression and localization of α- and β-carbonic anhydrase in Helicobacter pylori, Biochim. Biophys. Acta—Proteins Proteom. 1601 (2002) 192–199, https://doi.org/10.1016/S1570-9639(02)00467-3

[70] F. Teufel, J.J. Almagro Armenteros, A.R. Johansen, M.H. Gíslason, S.I. Pihl, K.D. Tsirigos, et al., SignalP 6.0 predicts all five types of signal peptides using protein language models, Nat. Biotechnol. 40 (2022) 1023–1025, https://doi.org/10.1038/s41587-021-01156-3

[71] A. Juibari, S. Ramezani, M. Rezadoust, Bioinformatics analysis of various signal peptides for periplasmic expression of parathyroid hormone in E. coli, J. Med. Life 12 (2019) 184–191, https://doi.org/10.25122/jml-2018-0049

[72] K.R.S. Sai, H. William, B. CJ, D. PA, B. MS, G. FJ, Biochemistry and physiology of the β class carbonic anhydrase (Cpb) from Clostridium perfringens strain 13, J. Bacteriol. 195 (2013) 2262–2269, https://doi.org/10.1128/jb.02288-12

[73] S. Mitsuhashi, J. Ohnishi, M. Hayashi, M. Ikeda, A gene homologous to β-type carbonic anhydrase is essential for the growth of Corynebacterium glutamicum under atmospheric conditions, Appl. Microbiol. Biotechnol. 63 (2004) 592–601, https://doi.org/10.1007/s00253-003-1402-8

[74] B. Peter, E. Cl, G. Henrik, Q. Beatriz, S. Elles, E. Bjj, et al., Carbonic anhydrase is essential for Streptococcus pneumoniae growth in environmental ambient air, J. Bacteriol. 192 (2010) 4054–4062, https://doi.org/10.1128/jb.00151-10

[75] D.J. Platt, Carbon dioxide requirement of Neisseria gonorrhoeae growing on a solid medium, J. Clin. Microbiol. 4 (1976) 129–132, https://doi.org/10.1128/jcm.4.2.129-132.1976

[76] Y.-P. Lin, W.-C. Chen, C.-M. Cheng, C.-J. Shen, Vaginal pH value for clinical diagnosis and treatment of common vaginitis, Diagnostics 11 (2021), https://doi.org/10.3390/diagnostics11111996

[77] J. Jumper, R. Evans, A. Pritzel, T. Green, M. Figurnov, O. Ronneberger, et al., Highly accurate protein structure prediction with AlphaFold, Nature 596 (2021) 583–589, https://doi.org/10.1038/s41586-021-03819-2

[78] R.S. Rowlett, Structure and catalytic mechanism of the β-carbonic anhydrases, Biochim. Biophys. Acta—Proteins Proteom. 1804 (2010) 362–373, https://doi.org/10.1016/j.bbapap.2009.08.002

[79] J.D. Cronk, R.S. Rowlett, K.Y.J. Zhang, C. Tu, J.A. Endrizzi, J. Lee, et al., Identification of a novel noncatalytic bicarbonate binding site in eubacterial β-carbonic anhydrase, Biochemistry 45 (2006) 4351–4361, https://doi.org/10.1021/bi052272q

[80] M. Ferraroni, S. Del Prete, D. Vullo, C. Capasso, C.T. Supuran, Crystal structure and kinetic studies of a tetrameric type II β-carbonic anhydrase from the pathogenic bacterium *Vibrio cholerae*, Acta Crystallogr. Sect. D. 71 (2015) 2449–2456, https://doi.org/10.1107/S1399004715018635

[81] M. Pinard, S. Lotlikar, M.A. Patrauchan, R. McKenna, Preliminary X-ray crystallographic analysis of β-carbonic anhydrase psCA3 from Pseudomonas aeruginosa, Acta Crystallogr. Sect. F. 69 (2013) 891–894, https://doi.org/10.1107/S1744309113017594

[82] L.D. McGurn, M. Moazami-Goudarzi, S.A. White, T. Suwal, B. Brar, J.Q. Tang, et al., The structure, kinetics and interactions of the β-carboxysomal β-carbonic anhydrase, CcaA, Biochem. J. 473 (2016) 4559–4572, https://doi.org/10.1042/BCJ20160773

[83] M.S. Kimber, E.F. Pai, The active site architecture of *Pisum sativum* β-carbonic anhydrase is a mirror image of that of α-carbonic anhydrases, EMBO J. 19 (2000) 1407–1418, https://doi.org/10.1093/emboj/19.7.1407

[84] L.J. Urbański, A. Di Fiore, L. Azizi, V.P. Hytönen, M. Kuuslahti, M. Buonanno, et al., Biochemical and structural characterisation of a protozoan beta-carbonic anhydrase from Trichomonas vaginalis, J. Enzyme Inhib. Med. Chem. 35 (2020) 1292–1299, https://doi.org/10.1080/14756366.2020.1774572

[85] A.S. Covarrubias, A.M. Larsson, M. Högbom, J. Lindberg, T. Bergfors, C. Björkelid, et al., Structure and function of carbonic anhydrases from *Mycobacterium tuberculosis**, J. Biol. Chem. 280 (2005) 18782–18789, https://doi.org/10.1074/jbc.M414348200

[86] S. Mitsuhashi, T. Mizushima, E. Yamashita, M. Yamamoto, T. Kumasaka, H. Moriyama, et al., X-ray structure of β-carbonic anhydrase from the red alga, *Porphyridium purpureum*, reveals a novel catalytic site for CO_2 hydration*, J. Biol. Chem. 275 (2000) 5521–5526, https://doi.org/10.1074/jbc.275.8.5521

[87] A.S. Covarrubias, T. Bergfors, T.A. Jones, M. Högbom, Structural mechanics of the pH-dependent activity of β-carbonic anhydrase from *Mycobacterium tuberculosis**, J. Biol. Chem. 281 (2006) 4993–4999, https://doi.org/10.1074/jbc.M510756200

[88] I.-M. Johansson, C. Forsman, Kinetic studies of pea carbonic anhydrase, Eur. J. Biochem. 218 (1993) 439–446, https://doi.org/10.1111/j.1432-1033.1993.tb18394.x

[89] S. SK, J. CN, S. Christina, A. SR, G. FJ, Structural and kinetic characterization of an archaeal β-class carbonic anhydrase, J. Bacteriol. 182 (2000) 6605–6613, https://doi.org/10.1128/jb.182.23.6605-6613.2000

[90] R.S. Rowlett, C. Tu, M.M. McKay, J.R. Preiss, R.J. Loomis, K.A. Hicks, et al., Kinetic characterization of wild-type and proton transfer-impaired variants of β-carbonic anhydrase from Arabidopsis thaliana, Arch. Biochem. Biophys. 404 (2002) 197–209, https://doi.org/10.1016/S0003-9861(02)00243-6

[91] R.S. Rowlett, M.R. Chance, M.D. Wirt, D.E. Sidelinger, J.R. Royal, M. Woodroffe, et al., Kinetic and structural characterization of spinach carbonic anhydrase, Biochemistry 33 (1994) 13967–13976, https://doi.org/10.1021/bi00251a003

[92] R.S. Rowlett, C. Tu, J. Lee, A.G. Herman, D.A. Chapnick, S.H. Shah, et al., Allosteric site variants of Haemophilus influenzae β-carbonic anhydrase, Biochemistry 48 (2009) 6146–6156, https://doi.org/10.1021/bi900663h

[93] I. Nishimori, T. Minakuchi, T. Kohsaki, S. Onishi, H. Takeuchi, D. Vullo, et al., Carbonic anhydrase inhibitors: The β-carbonic anhydrase from Helicobacter pylori is a new target for sulfonamide and sulfamate inhibitors, Bioorg Med. Chem. Lett. 17 (2007) 3585–3594, https://doi.org/10.1016/j.bmcl.2007.04.063

[94] P. Joseph, F. Turtaut, S. Ouahrani-Bettache, J.-L. Montero, I. Nishimori, T. Minakuchi, et al., Cloning, characterization, and inhibition studies of a β-carbonic anhydrase from Brucella suis, J. Med. Chem. 53 (2010) 2277–2285, https://doi.org/10.1021/jm901855h

[95] P. Joseph, S. Ouahrani-Bettache, J.-L. Montero, I. Nishimori, T. Minakuchi, D. Vullo, et al., A new β-carbonic anhydrase from Brucella suis, its cloning, characterization, and inhibition with sulfonamides and sulfamates, leading to impaired pathogen growth, Bioorg Med. Chem. 19 (2011) 1172–1178, https://doi.org/10.1016/j.bmc.2010.12.048

[96] P. Burghout, D. Vullo, A. Scozzafava, P.W.M. Hermans, C.T. Supuran, Inhibition of the β-carbonic anhydrase from Streptococcus pneumoniae by inorganic anions and small molecules: toward innovative drug design of antiinfectives? Bioorg Med. Chem. 19 (2011) 243–248, https://doi.org/10.1016/j.bmc.2010.11.031

[97] D. Vullo, I. Nishimori, T. Minakuchi, A. Scozzafava, C.T. Supuran, Inhibition studies with anions and small molecules of two novel β-carbonic anhydrases from the bacterial pathogen Salmonella enterica serovar Typhimurium, Bioorg. Med. Chem. Lett. 21 (2011) 3591–3595, https://doi.org/10.1016/j.bmcl.2011.04.105

[98] J. Kaur, X. Cao, N.S. Abutaleb, A. Elkashif, A.L. Graboski, A.D. Krabill, et al., Optimization of acetazolamide-based scaffold as potent inhibitors of vancomycin-resistant Enterococcus, J Med Chem. 63 (2020) 9540–9562, https://doi.org/10.1021/acs.jmedchem.0c00734

[99] W. An, K.J. Holly, A. Nocentini, R.D. Imhoff, C.S. Hewitt, N.S. Abutaleb, et al., Structure-activity relationship studies for inhibitors for vancomycin-resistant Enterococcus and human carbonic anhydrases, J. Enzyme Inhib. Med. Chem. 37 (2022) 1838–1844, https://doi.org/10.1080/14756366.2022.2092729

[100] H.-M. Park, J.-H. Park, J.-W. Choi, J. Lee, B.Y. Kim, C.-H. Jung, et al., Structures of the γ-class carbonic anhydrase homologue YrdA suggest a possible allosteric switch, Acta Crystallogr. Sect. D. 68 (2012) 920–926, https://doi.org/10.1107/S0907444912017210

[101] K.L. Peña, S.E. Castel, C. De Araujo, G.S. Espie, M.S. Kimber, Structural basis of the oxidative activation of the carboxysomal γ-carbonic anhydrase, CcmM, Proc. Natl. Acad. Sci. U. S. A. 107 (2010) 2455–2460, https://doi.org/10.1073/pnas.0910866107

[102] T.M. Iverson, B.E. Alber, C. Kisker, J.G. Ferry, D.C. Rees, A closer look at the active site of γ-class carbonic anhydrases: high-resolution crystallographic studies of the carbonic anhydrase from Methanosarcina thermophila, Biochemistry 39 (31) (2000) 9222, https://doi.org/10.1021/bi000204s

[103] J. Jeyakanthan, S. Rangarajan, P. Mridula, S.P. Kanaujia, Y. Shiro, S. Kuramitsu, et al., Observation of a calcium-binding site in the γ-class carbonic anhydrase from Pyrococcus horikoshii, Acta Crystallogr. Sect. D. 64 (2008) 1012–1019, https://doi.org/10.1107/S0907444908024323

[104] S.R. MacAuley, S.A. Zimmerman, E.E. Apolinario, C. Evilia, Y.-M. Hou, J.G. Ferry, et al., The archetype γ-class carbonic anhydrase (CAM) contains iron when synthesized in vivo, Biochemistry 48 (2009) 817–819, https://doi.org/10.1021/bi802246s

[105] B.C. Tripp, C.B. Bell III, F. Cruz, C. Krebs, J.G. Ferry, A role for iron in an ancient carbonic anhydrase, J. Biol. Chem. 279 (2004) 6683–6687, https://doi.org/10.1074/jbc.M311648200

[106] A. FORKMAN, A.-B. LAURELL, The effect of carbonic anhydrase inhibitor on the growth of Neisserrae, Acta Pathol. Microbiol. Scand. 65 (1965) 450–456, https://doi.org/10.1111/apm.1965.65.3.450

[107] R. Kassamali, D.A. Sica, Acetazolamide: a forgotten diuretic agent, Cardiol. Rev. 19 (2011).

[108] A.R. Loiselle, E. de Kleine, P. van Dijk, N.M. Jansonius, Intraocular and intracranial pressure in glaucoma patients taking acetazolamide, PLoS One 15 (2020) e0234690, https://doi.org/10.1371/journal.pone.0234690

[109] J. Wongboonsin, C. Thongprayoon, T. Bathini, P. Ungprasert, N.R. Aeddula, M.A. Mao, et al., Acetazolamide therapy in patients with heart failure: a meta-analysis, J. Clin. Med. 8 (2019) 349, https://doi.org/10.3390/jcm8030349

[110] C.M. Toussaint, R.W. Kenefick, F.A. Petrassi, S.R. Muza, N. Charkoudian, Altitude, acute mountain sickness, and acetazolamide: recommendations for rapid ascent, High. Alt. Med. Biol. 22 (2020) 5–13, https://doi.org/10.1089/ham.2019.0123

[111] E. Sanders, T.H. Maren, Inhibition of carbonic anhydrase in Neisseria: effects on enzyme activity and growth, Mol. Pharmacol. 3 (1967) 204–215.

[112] N.S. Abutaleb, A.E.M. Elhassanny, A. Nocentini, C.S. Hewitt, A. Elkashif, B.R. Cooper, et al., Repurposing FDA-approved sulphonamide carbonic anhydrase inhibitors for treatment of Neisseria gonorrhoeae, J. Enzyme Inhib. Med. Chem. 37 (2022) 51–61, https://doi.org/10.1080/14756366.2021.1991336

[113] A. Casini, J. Antel, F. Abbate, A. Scozzafava, S. David, H. Waldeck, et al., Carbonic anhydrase inhibitors: SAR and X-ray crystallographic study for the interaction of sugar sulfamates/sulfamides with isozymes I, II and IV, Bioorg. Med. Chem. Lett. 13 (5) (2003) 841, https://doi.org/10.1016/S0960-894X(03)00029-5

[114] K.H. Sippel, A.H. Robbins, J. Domsic, C. Genis, M. Agbandje-Mckenna, R. McKenna, High-resolution structure of human carbonic anhydrase II complexed with acetazolamide reveals insights into inhibitor drug design, Acta Crystallogr. Sect. F. Struct. Biol. Cryst. Commun. 65 (2009) 992–995, https://doi.org/10.1107/S1744309109036665

[115] J.T. Andring, M. Fouch, S. Akocak, A. Angeli, C.T. Supuran, M.A. Ilies, et al., Structural basis of nanomolar inhibition of tumor-associated carbonic anhydrase IX: X-ray crystallographic and inhibition study of lipophilic inhibitors with acetazolamide backbone, J. Med. Chem. 63 (2020) 13064–13075, https://doi.org/10.1021/acs.jmedchem.0c01390

[116] G.S. Jacob, R.D. Brown III, S.H. Koenig, Interaction of bovine carbonic anhydrase with (neutral) aniline, phenol, and methanol, Biochemistry 19 (1980) 3754–3765, https://doi.org/10.1021/bi00557a017

[117] I. Simonsson, B.-H. Jonsson, S. Lindskog, Phenol, a competitive inhibitor of CO_2 hydration catalyzed by carbonic anhydrase, Biocehm. Biophys. Res. Commun. 108 (1982) 1406–1412, https://doi.org/10.1016/S0006-291X(82)80063-6

[118] S.K. Nair, P.A. Ludwig, D.W. Christianson, Two-site binding of phenol in the active site of human carbonic anhydrase II: structural implications for substrate association, J. Am. Chem. Soc. 116 (1994) 3659–3660, https://doi.org/10.1021/ja00087a086

[119] J. Andring, J. Combs, R. McKenna, Aspirin: a suicide inhibitor of carbonic anhydrase II, Biomolecules 10 (2020), https://doi.org/10.3390/biom10040527

[120] K. D'Ambrosio, S. Carradori, S. Cesa, A. Angeli, S.M. Monti, C.T. Supuran, et al., Catechols: a new class of carbonic anhydrase inhibitors, Chem. Commun. 56 (2020) 13033–13036, https://doi.org/10.1039/D0CC05172A

[121] D.R. Vianna, L. Hamerski, F. Figueiró, A. Bernardi, L.C. Visentin, E.N.S. Pires, et al., Selective cytotoxicity and apoptosis induction in glioma cell lines by 5-oxygenated-6,7-methylenedioxycoumarins from Pterocaulon species, Eur. J. Med. Chem. 57 (2012) 268–274, https://doi.org/10.1016/j.ejmech.2012.09.007

[122] Z. Xu, Q. Chen, Y. Zhang, C. Liang, Coumarin-based derivatives with potential anti-HIV activity, Fitoterapia 150 (2021) 104863, https://doi.org/10.1016/j.fitote.2021.104863

[123] M.A. Al-Haiza, M.S. Mostafa, M.Y. El-Kady, Synthesis and biological evaluation of some new coumarin derivatives, Molecules 8 (2003) 275–286, https://doi.org/10.3390/80200275

[124] H. Vu, N.B. Pham, R.J. Quinn, Direct screening of natural product extracts using mass spectrometry, J. Biomol. Screen. 13 (2008) 265–275, https://doi.org/10.1177/1087057108315739

[125] A. Maresca, C. Temperini, L. Pochet, B. Masereel, A. Scozzafava, C.T. Supuran, Deciphering the mechanism of carbonic anhydrase inhibition with coumarins and thiocoumarins, J. Med. Chem. 53 (2010) 335–344, https://doi.org/10.1021/jm901287j

[127] A. Maresca, C. Temperini, H. Vu, N.B. Pham, S.-A. Poulsen, A. Scozzafava, et al., Non-zinc mediated inhibition of carbonic anhydrases: coumarins are a new class of suicide inhibitors, J. Am. Chem. Soc. 131 (2009) 3057–3062, https://doi.org/10.1021/ja809683v

[127] A. Innocenti, A. Scozzafava, S. Parkkila, L. Puccetti, G. De Simone, C.T. Supuran, Investigations of the esterase, phosphatase, and sulfatase activities of the cytosolic mammalian carbonic anhydrase isoforms I, II, and XIII with 4-nitrophenyl esters as substrates, Bioorg. Med. Chem. Lett. 18 (2008) 2267–2271, https://doi.org/10.1016/j.bmcl.2008.03.012

[129] A. Nocentini, C.S. Hewitt, M.D. Mastrolorenzo, D.P. Flaherty, C.T. Supuran, Anion inhibition studies of the α-carbonic anhydrases from Neisseria gonorrhoeae, J. Enzyme Inhib. Med. Chem. 36 (2021) 1061–1066, https://doi.org/10.1080/14756366.2021.1929202

[130] F. Carta, M. Aggarwal, A. Maresca, A. Scozzafava, R. McKenna, C.T. Supuran, Dithiocarbamates: a new class of carbonic anhydrase inhibitors. Crystallographic and kinetic investigations, Chem. Commun. 48 (2012) 1868–1870, https://doi.org/10.1039/C2CC16395K

[131] F. Carta, M. Aggarwal, A. Maresca, A. Scozzafava, R. McKenna, E. Masini, et al., Dithiocarbamates strongly inhibit carbonic anhydrases and show antiglaucoma action in vivo, J. Med. Chem. 55 (2012) 1721–1730, https://doi.org/10.1021/jm300031j

[132] L. Syrjänen, M.E.E. Tolvanen, M. Hilvo, D. Vullo, F. Carta, C.T. Supuran, et al., Characterization, bioinformatic analysis and dithiocarbamate inhibition studies of two new α-carbonic anhydrases, CAH1 and CAH2, from the fruit fly Drosophila melanogaster, Bioorg. Med. Chem. 21 (2013) 1516–1521, https://doi.org/10.1016/j.bmc.2012.08.046

[133] D. Vullo, M. Durante, F.S. Di Leva, S. Cosconati, E. Masini, A. Scozzafava, et al., Monothiocarbamates strongly inhibit carbonic anhydrases in vitro and possess intraocular pressure lowering activity in an animal model of glaucoma, J. Med. Chem. 59 (2016) 5857–5867, https://doi.org/10.1021/acs.jmedchem.6b00462

CHAPTER TEN

Enterococci carbonic anhydrase inhibition

Katrina J. Holly, Molly S. Youse, and Daniel P. Flaherty*

Borch Department of Medicinal Chemistry and Molecular Pharmacology, College of Pharmacy, Purdue University, West Lafayette, IN, United States
*Corresponding author. e-mail address: dflaher@purdue.edu

Contents

1. Introduction	283
2. Carbonic anhydrase genetic families encoded in *Enterococci*	284
2.1 α-Carbonic anhydrase	284
2.2 γ-Carbonic anhydrase	292
3. Enzyme inhibition	295
3.1 FDA-approved carbonic anhydrase inhibitors	295
3.2 Published carbonic anhydrase inhibitors	298
3.3 Future of carbonic anhydrase inhibitors as anti-enterococcal agents	305
4. Conclusion	307
References	308

Abstract

Carbonic anhydrase metalloenzymes are encoded in genomes throughout all kingdoms of life with a conserved function catalyzing the reversible conversion of CO_2 to bicarbonate. Carbonic anhydrases have been well-investigated in humans, but are still relatively understudied in bacterial organisms, including *Enterococci*. Studies over the past decade have presented bacterial carbonic anhydrases as potential drug targets, with some chemical scaffolds potently inhibiting the *Enterococcus* carbonic anhydrases *in vitro* and displaying antimicrobial efficacy against *Enterococcus* organisms. While carbonic anhydrases in *Enterococci* still have much to be explored, hypotheses may be drawn from similar Gram-positive organisms for which known information exists about carbonic anhydrase function and relevance. Within this chapter is reported information and rational hypotheses regarding the subcellar locations, potential physiological roles, essentiality, structures, and kinetics of carbonic anhydrases in *Enterococci*.

1. Introduction

The presence and activity of carbonic anhydrase (CA) in *Enterococci* is a new frontier that has only begun to be explored within the past decade.

ISSN 1874-6047, https://doi.org/10.1016/bs.enz.2024.05.011

Carbonic anhydrases are a vast family of metalloenzymes present in all kingdoms of life that can play critical roles in pH regulation and metabolism, depending on their context. In the human context, they have been proven to be excellent druggable targets, as many inhibitor scaffolds have been synthesized and successfully cleared clinical development. But as antibiotic resistance continues to emerge as a global health concern and as multi-drug resistant pathogens continue to deplete the antibiotic arsenals available, novel antibacterial drug targets that expose new and unexplored liabilities are of great interest in antibiotic drug discovery. Based on general CA knowledge, known roles of CAs in other pathogens, and preliminary work exploring the function and antimicrobial relevance of both the α- and γ-CA isoforms in *Enterococci* CA, the field of *Enterococci* CA research is primed for new and fascinating discoveries. Herein is a detailed report of both the known and unknown characteristics of *Enterococci* CAs as it relates to their locations and functional roles in *Enterococci* growth, their potential essentiality to the organism, and their structural and kinetic aspects, which taken together may prompt further curiosity into whether these enzymes hold promise as unique and relevant antibiotic targets.

2. Carbonic anhydrase genetic families encoded in *Enterococci*

2.1 α-Carbonic anhydrase

2.1.1 Subcellular location

The subcellular location of α-CA in *Enterococci* has not been explored. Hypotheses regarding the subcellular location of bacterial α-CAs in Gram-negative species propose that it is trafficked to the periplasmic space *via* an N-terminal signal sequence where it acts upon passively diffused CO_2 to convert it to bicarbonate for metabolic processes [1,2]. However, Gram-positive species lack an outer membrane and periplasmic space, leaving its exact niche in *Enterococci* unknown. Assuming it also acts upon passively diffused CO_2 in Gram-positive species, it is likely that it exists within the cytosol of *Enterococci,* if not also within other compartments.

2.1.2 Hypothesized role in bacterial physiology

The topic of distribution and roles of different CA isoforms across bacterial species has proven to be a complex subject area that has been developing over the past three decades [3]. The first reports of CAs encoded in

Enterococci were reported in 1999 when Smith et al. examined genome and protein databases to assemble phylogenies for the α-, β-, and γ-CA isoforms based on isoform-specific open reading frames (ORFs) identified [4]. At that time, *E. faecalis* was the only *Enterococcus* species identified to contain CA-encoding genes, which included α- and γ-CA. In 2015, Capasso and Supuran reported BLAST searches which revealed that α-CAs were only present in Gram-negative bacterial species and that *E. faecalis* only encoded one γ-CA [1]. In 2020, Kaur et al. reported that another member of the *Enterococcus* genus, *E. faecium*, was indeed found to encode both an α- and γ-CA [5]. The α-CAs are more promiscuous in substrate scope than other CA isoforms which are restricted to CO_2/bicarbonate only [1]. This promiscuity includes not only hydrolase activity regarding CO_2 and CS_2 but also esterase activity capable of hydrolyzing a range of ester substrates [1,6]. The exact substrate scope of *Enterococci* α-CAs is still undefined, but it is not unreasonable to postulate that their function extends beyond that of maintenance of CO_2-bicarbonate homeostasis.

Some evidence of physiological role can be gleaned from natural isolates that have modified CA genes. In 2023, Chilambi et al. reported the isolation of an *E. faecalis* clinical strain from a patient with infective endocarditis [7]. This strain was remarkable in that it contained a point mutation in the α-CA gene, introducing a premature stop codon at position 209, and was the first reported instance of an *Enterococcus* organism containing a mutated CA gene [7]. This mutation effectively truncated the enzyme at position 209, eliminating twenty-three residues on the C-terminal strand that typically extends around the protein to form stabilizing interactions with other regions (Fig. 1). The loss of these stabilizing C-terminal residues likely resulted in a destabilized enzyme with diminished, if not completely destroyed, catalytic efficiency.

This mutant strain exhibited a gentamicin-hypersusceptible phenotype (16-fold lower MIC than in the isogenic wildtype strain), leading the researchers to investigate changes to the bacterial physiology that may have resulted from reduced α-CA activity and allowed for the observed gentamicin susceptibility. This susceptibility was linked to increased uptake of gentamicin driven by increased proton motive force (PMF) as confirmed by elevated intracellular levels of ATP detected in the mutant strain. Additionally, a propidium iodide assay found that the plasma membrane was more permeable, and the fatty acid profile of the membrane in the mutant strain revealed significantly lower levels of palmitic and cyclopropane fatty acids and higher levels of *cis*-vaccenic acid (an unsaturated fatty acid) which

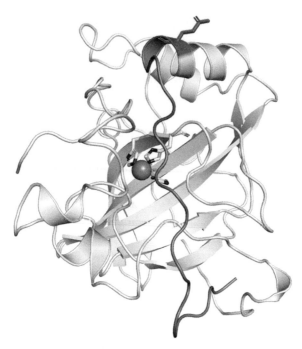

Fig. 1 Reported homology model (SWISS-model) of wildtype *E. faecalis* α-CA using *T. ammonificans* (PDB ID: 4COQ) as the template [7]. Three histidines (white sticks) coordinate the Zn²⁺ cation (gray sphere) in the catalytic active site. Glu209 (red sticks) and the remaining C-terminal residues (teal cartoon) indicate the truncated portion of the mutant *E. faecalis* α-CA reported by Chilambi et al. [7]. *Figure generated using PyMOL.*

further confirm the increased membrane permeability and fluidity phenotype. The authors proposed that the diminished α-CA activity results in a deficiency of protons within the cell, which leads to increased PMF and subsequent ATP production. However, the exact mechanistic link between *E. faecalis* α-CA activity and plasma membrane fatty acid composition remains unclear.

While the relationship between CA activity and membrane fatty acid composition is unexplored in *Enterococci* species, information may be gleaned from studies in other Gram-positive bacteria. The first step in fatty acid biosynthesis is the ATP–driven conversion of acetyl-CoA to malonyl-CoA as catalyzed by acetyl-CoA-carboxylase, in which a bicarbonate anion is utilized as the CO_2 source. Two studies have previously explored the relationship between CA activity and growth in Gram-positive *Streptococcus* species, finding direct relationships between CA-mediated bicarbonate levels and subsequent fatty acid biosynthesis and bacterial growth [8,9]. In 2010, Burghout et al. observed that when the putative β-class CA in *Streptococcus pneumoniae* was deleted (Δ*pca*), the cells were dependent on 5% atmospheric CO_2 for growth, whereas wildtype *S. pneumoniae* thrived in ambient air [8]. When the Δ*pca* strain was complemented with an inducible *pca* gene or an *E. coli* CA, the growth phenotype in ambient air was restored. Supplementing growth media with bicarbonate also allowed the Δ*pca* mutant to grow in CO_2-poor conditions, likely indicating that this CO_2-dependent growth in the mutant is related to the CA-dependent production of bicarbonate. Interestingly, supplementation of the growth media with the unsaturated fatty acid (UFA) oleic acid reversed the CO_2-dependent growth of the Δ*pca* strain, whereas supplementation with saturated fatty acids (SFA) did not. This observation suggests that the CO_2-dependent growth phenotype of the Δ*pca* strain is linked to a deficiency in UFAs rather than SFAs. Research conducted by Matsumoto et al. in 2019 further supported these CO_2-dependent, bicarbonate-dependent, and UFA-dependent growth observations in *Streptococcus anginosus,* a strain which lacks CA and requires CO_2 for growth [9]. When *S. agalactiae* CA was ectopically expressed in *S. anginosus*, the strain was no longer dependent on CO_2 for growth. Similarly, supplementation of growth media with bicarbonate or with the UFAs oleic acid, linoleic acid, and linolenic acid reversed the CO_2-dependent growth phenotype of *S. anginosus*, whereas SFAs actually had an inhibitory effect on growth [9]. Taking the study a step further, the group utilized RT-qPCR to investigate the effect of bicarbonate on transcription levels of UFA-synthetic genes *versus* SFA-synthetic genes [9]. They found that bicarbonate upregulated *fabM* expression and downregulated *fabK* expression, which are the enzymes that act on *trans*-2-decenoyl ACP at the branch point between the UFA and SFA biosynthesis pathways, respectively. This information suggests that in the absence of CA-mediated conversion of CO_2 to bicarbonate, UFA biosynthesis may be abrogated whereas SFA biosynthesis may be increased. Although both studies pertain to pH-dependent β-CA activity in *Streptococcus* organisms rather than to α-CA activity in *Enterococcus* organisms, both are Gram-positive and have similar biosynthetic gene clusters involved in FA biosynthesis [10]. Due to the difficulty of introducing foreign plasmids to *Enterococcus* organisms for genetic manipulation, parallel studies in *Enterococci* have not been pursued to study growth phenotypes in α-CA deletion strains [11]. However, other approaches utilizing enzyme inhibition rather than gene deletion mutants may be able to shed similar light on the role of α-CA on FA biosynthesis-related growth in *Enterococci*.

2.1.3 Essentiality

The essentiality of the α-CA gene to *Enterococci* organisms is largely unexplored. There have been no reports of genetic studies to knockout or knockdown the enzyme in *Enterococci*, likely due to the known difficulty in introducing foreign plasmids to *Enterococci* [11]. However, such studies have been successful in the Gram-negative organisms *Helicobacter pylori* and *Neisseria gonorrhoeae*. These studies have found that an α-CA is essential in both *H. pylori* and *N. gonorrhoeae* [2,12]. It should be noted that the essentiality of the enzyme in *H. pylori* was studied in acidic conditions, considering that *H. pylori* colonizes the gastric compartment. Due to the general role of CA in pH regulation of its microenvironment, it is possible that essentiality of CA in various organisms may be dependent on the pH of the media in which it is evaluated. It has been noted that many organisms that do not encode for a CA require elevated levels of CO_2 for growth [1,13]. This is believed to be due to the spontaneous conversion of CO_2 to bicarbonate under elevated CO_2 conditions, negating the need for CA enzymes to carry out the conversion [1,5,13]. Small molecules that are known to potently inhibit recombinant *Ef*α-CA *in vitro* have also been shown to have a bacteriostatic effect against *E. faecium* ATCC 700221, an effect which is abrogated in elevated CO_2 where the bicarbonate supply is produced by spontaneous equilibrium [5,14]. These observations suggest that α-CA activity may indeed be essential for *Enterococci* growth in environments with limited CO_2.

2.1.4 Structure

At present, there is no reported structural data for any *Enterococci* CAs. However, a homology model of the *E. faecium* α-CA (*Ef*α-CA, UniProt identifier Q3XYE8) has been published (Fig. 2) [5]. The *H. pylori* α-CA (*Hp*α-CA, UniProt identifier A0A0M3KL20; PDB ID: 4YGF) was utilized as a template for the homology model based on its 61.1% sequence identity + similarity to that determined from the UniProt sequence alignment with *Ef*α-CA [5,15]. The *Ef*α-CA homology model was generated using Prime (Schrodinger, LLC), obtaining a notable structural alignment score of 0.006 with a root mean square deviation (RMSD) of 0.397 Å. The active sites displayed excellent alignment in secondary and primary structure except for Ala192 and Asn108 in *Hp*α-CA which are replaced by Thr180 and Asp94, respectively, in the *Ef*α-CA model (Fig. 3) [5]. The active site is characterized by a Zn^{2+} cation coordinated by three histidine residues. The homology model of *Ef*α-CA also suggests that it lacks an alpha helix which is present in human CAs, including hCAII (Fig. 4) [16]. The absence of the alpha helix is a

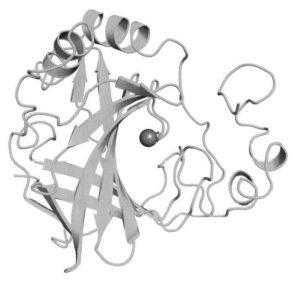

Fig. 2 Homology model of *Efa*-CA generated by Kaur et al., with the catalytic Zn^{2+} cation depicted as a gray sphere [5]. *Figure generated in PyMOL.*

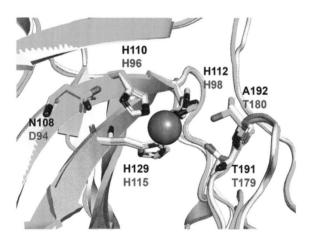

Fig. 3 Overlay of the active sites from the *Efa*-CA homology model (green) and the *Hpa*-CA crystal structure (yellow) (PDB ID: 4YGF) [5]. Key binding site residues are depicted as sticks. *Hpa*-CA residues are annotated in black and corresponding *Efa*-CA residues in red. The Zn^{2+} cation is depicted as a gray sphere. *Figure generated in PyMOL.*

commonly observed structural feature in bacterial α–CAs [5,15,17,18]. This common structural difference in the active sites between the human and bacterial α–CA isoforms may provide opportunity for design of small molecule agents that are selective for the bacterial isoforms over human counterparts.

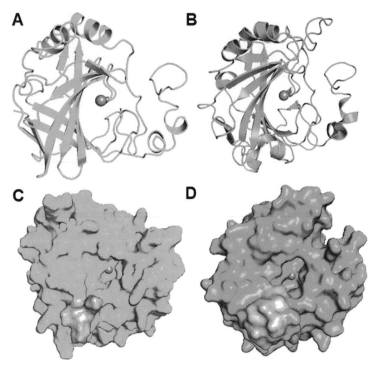

Fig. 4 Comparison of *Efa*-CA and hCAII structures. (A) *Efa*-CA homology model (green cartoon) with β-strand loop depicted in pink [5]. (B) The hCAII crystal structure (cyan cartoon) (PDB ID: 3HS4) with corresponding alpha helix in pink [16]. (C) Surface representation of the *Efa*-CA homology model (green) with β-strand loop in pink. (D) Surface representation of the hCAII crystal structure (cyan) with the alpha helix in pink. *Figures generated in PyMOL.*

2.1.5 Enzyme kinetics

Carbonic anhydrases are one of the most efficient classes of catalytic enzymes. The α–CA class is the most well-studied isoform as it is the only CA class present in vertebrates and is known to be widely distributed among both prokaryotes and eukaryotes, comprising all sixteen CA isoforms in humans [19–22]. The α–CA class typically exhibits the highest catalytic efficiencies among the families and includes the second-most catalytically efficient enzyme characterized thus far after superoxide dismutase [1,23]. Particularly, Gram-negative thermophiles living in extreme environments rely on their α-CAs to aid in rapid conversion of CO_2 to bicarbonate in the periplasmic and extra-cellular compartments [1]. The most efficient α-CA known is that of *Sulfur-ihydrogenibium azorense* (SazCA) with a K_{cat}/K_M of 3.5×10^8 M^{-1} s^{-1}, approaching the limit of diffusion [23]. While the kinetics of the α-CAs in *Enterococci* have not been well-studied to date, the Supuran laboratory recently determined the K_{cat}/K_M for recombinant *E. faecium* α-CA (*Ef*α-CA) to be $4.9{\times}10^7$ M^{-1} s^{-1} and the k_{cat} to be $8.1{\times}10^5$ s^{-1} (unpublished results) (Table 1).

The CA ping-pong mechanism for conversion of CO_2 to bicarbonate is well-characterized across CA isoforms and is a Zn^{2+}-hydroxide mediated process. Briefly, a Zn^{2+}-bound water lowers the pK_a of a water proton, allowing it to be extracted by a residue acting as a proton shuttle to transfer the proton to the bulk solvent, thereby generating the Zn^{2+}-bound hydroxide nucleophile. This hydroxide then attacks the electrophilic carbonyl center of CO_2, generating a Zn^{2+}-bound bicarbonate anion. A new water molecule then displaces the bicarbonate anion to repeat the cycle. In *Ef*α-CA, the residue participating as the proton shuttle has not been definitively determined. In the well-studied human CAII (hCAII), the proton shuttle is His64, which has shown the ability to flip back and forth between the binding pocket and the bulk solvent depending on pH which likely helps facilitate the proton transfer [26,27]. Human CAI (hCAI) also contains the proton shuttle at the same site according to sequence alignment and crystal structure (PDB ID: 3W6H) (Fig. 5). The corresponding His66 in *N. gonorrhoeae* α-CA (α-NgCA) was confirmed to serve as the proton shuttle, and it has been proposed that a similarly-positioned histidine in *Hp*α-CA also may serve as the proton shuttle for that enzyme [15,29]. However, *Ef*α-CA appears to lack this respective histidine and

Table 1 Kinetic parameters reported for bacterial α-CAs and hCAI and hCAII [23–25].

CA	k_{cat} (s^{-1})	k_{cat}/K_M $(M^{-1} s^{-1})$
E. faecium α-CA	$8.1{\times}10^5$	$4.9{\times}10^7$
H. pylori α-CA[a]	$2.5{\times}10^5$	$1.5{\times}10^7$
V. cholerae[b]	$8.23{\times}10^5$	$7.0{\times}10^7$
N. gonorrhoeae α-CA	$1.0{\times}10^6$	$7.4{\times}10^7$
S. azorense α-CA[c]	$4.4{\times}10^6$	$3.5{\times}10^8$
hCAI	$2.00{\times}10^5$	$5.0{\times}10^7$
hCAII	$1.4{\times}10^6$	$1.5{\times}10^8$

[a]Reported in Ref. [25].
[b]Reported in Ref. [24].
[c]Reported in Ref. [23].

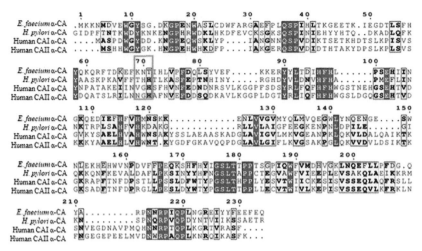

Fig. 5 Uniprot sequence alignments for *Efa*-CA, *Hpa*-CA, hCAI, and hCAII. The green box highlights the absence of histidine at the classic α-CA proton shuttle site. A red box with white characters indicates strict identity. Bold characters represent similarity within a group. A yellow box indicates similarity across the group. Similarities of *Hpa*-CA, hCAI, and hCAII to *Efa*-CA were 37.2%, 29.1%, and 26.1%, respectively. Uniprot accession codes: *E. faecium* α-CA – Q3XY38; *H. pylori* α-CA – A0A0M3KL20; hCAI – P00915; hCAII – P00918 [28]. *Image generated using ESPript3.0.*

according to sequence alignment and homology modeling instead contains Asn70 and Thr71, neither of which could facilitate the proton transfer (Figs. 5 and 6) [5]. Because the k_{cat} for *Efa*–CA is greater than that for water-mediated proton transfer (10^4 s^{-1}), an alternative proton shuttle must be at work [30]. Adjacent to Asn70 is Lys69, a flexible and basic residue also positioned at the interface of the binding site and bulk solvent, which may, in theory, also serve the role of proton shuttle (Fig. 6). However, the human CAIII contains a Lys64 in place of His64 in hCAII, and this enzyme is notably less efficient, with a k_{cat} below 10^4 s^{-1} [31]. Since no proton shuttle residues other than histidine have been characterized in the α–CA class, site-directed mutagenesis studies are needed to firmly identify which residue is serving as the proton shuttle in *Efa*–CA [5,15,16].

2.2 γ-Carbonic anhydrase

2.2.1 Subcellular location

No studies have specifically investigated the subcellular location of γ-CAs in *Enterococci*. The most well-studied γ-CAs are found in archaea, specifically methanogens, whose growth is dependent on energy generated from conversion of imported acetate to methane and CO_2. Ferry's group

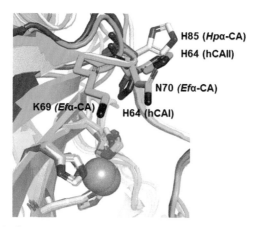

Fig. 6 Overlay of *Efα*-CA homology model (green), *Hpα*-CA (PDB ID: 4YGF) (yellow), hCAI (PDB ID: 3W6H) (dark gray), and hCAII (PDB ID: 3HS4) (cyan) active sites [5,15,16]. The Zn^{2+} cation (gray sphere) is coordinated by three conserved histidine residues in each enzyme (sticks). Histidine residues (sticks) serving as proton shuttles in *Hpα*-CA, hCAI, and hCAII are annotated, along with the corresponding *Efα*-CA residues at those sites. *Figure generated using PyMOL.*

proposed that a γ-CA in *Methanosarcina thermophila* is secreted to the extracellular side of the plasma membrane to regulate bicarbonate levels that affect acetate uptake [32,33]. However, the metabolic needs of *Enterococci* differ dramatically from those of methanogenic archaea. Additionally, sequence alignment of *E. faecium* γ-CA (*Efγ*-CA) (UniProt Accession Q3XYE8) with the full-length *M. thermophila* γ-CA (GenBank Accession AKB12346.1) reveals that *Efγ*-CA lacks an N-terminal signal peptide which is responsible for extracellular secretion of the enzyme in *M. thermophila*, likely precluding the *Efγ*-CA from being secreted [32–34]. Aside from archaea, the most well-studied γ-CAs in bacteria come from cyanobacteria which are photosynthetic organisms that concentrate CO_2 in their carboxysomes to overcome the low affinity of CO_2 for ribulose bisphosphate carboxylase/oxygenase in the CO_2 fixation step of photosynthesis [35]. The gene for γ-CA (CcmM) in *Synechocysis* organisms is located within a cluster of genes involved in its carbon concentration mechanism [36]. Western Blot analysis using antisera generated against recombinantly expressed CcmM identified that it is localized to the carboxysome [35,36]. However, *Enterococci* are facultative anaerobes and do not have CO_2-concentrating compartments such as carboxysomes. Thus, the subcellular locations of most bacterial γ-CAs, including that in *Enterococci*, are still under-explored.

2.2.2 Hypothesized role in bacterial physiology

Research regarding the physiological roles of γ-CA in various prokaryotes has been rather absent over the past two decades. Ferry's group, who has pioneered γ-CA research, has hypothesized that in methanogenic archaea, there may be both extracellular and cytosolic γ-CAs supporting an acetate-bicarbonate antiporter by regulating bicarbonate and CO_2 levels on either side of the plasma membrane [32–34]. While the enzyme may play a significant physiological role for acetotrophic methanogens, there is no known parallel role for facultative anaerobic bacteria such as *Enterococci*. Similarly, the known localization of γ-CA to the carboxysomes in cyanobacteria also has little physiological relevance for understanding the potential role of γ-CA in *Enterococci* [35]. Relative expression levels of α- and γ-CA in *Enterococci* in various growth conditions have not yet been probed; however, such a study could illuminate the importance of the enzymes in growth and metabolism of *Enterococci* and also distinguish between their respective physiological roles.

2.2.3 Essentiality

Similar to the aforementioned α-CA in *Enterococci*, the essentiality of the γ-CA enzyme to *Enterococci* is not known due to the weakness of the current tools to genetically manipulate *Enterococci* organisms [11]. The only indication thus far of any potential essentiality of the enzyme is a study from the Flaherty group showing that small molecule inhibitors with known *in vitro* inhibition against *Ef*γ-CA exhibit a bacteriostatic effect against *E. faecium* ATCC 700221, which can be abrogated when cultured in the presence of 5% CO_2 [5,14]. In cyanobacteria, the gene for γ-CA was shown to be essential for growth in low CO_2 and for appropriate formation of the carboxysome [35,36]. However, such physiological essentiality is irrelevant in the case of facultative anaerobic *Enterococci*.

2.2.4 Structure

No structures have been solved for a γ-CA in *Enterococci*. Generally, structurally characterized γ-CAs are homotrimers often with Zn^{2+} or Fe^{2+} (in some anaerobes) coordinated at a triad of histidine residues, with some activity studies suggesting they may functionally tolerate a variety of metal cations [32,37,38]. Kaur et al. have proposed a homology model for *E. faecium* γ-CA (*Ef*γ-CA) (Fig. 7, UniProt Accession Q3XX77) based on the 2.0 Å crystal structure of the *Clostridium difficile* γ-CA (*Cd*γ-CA) (PDB ID: 4MFG; Center for Structural Genomics of Infectious Disease) after running a UniProt sequence alignment that found it shared 77% sequence identity + similarity

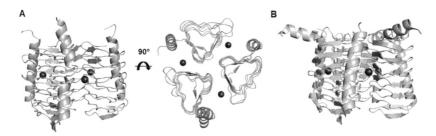

Fig. 7 (A) Homology model of *Efγ*-CA (gray cartoon) with Ni^{2+} cations (dark blue spheres), from both side and top perspectives [5]. (B) *C. difficile* γ CA (PDB ID. 4MFG) crystal structure (light blue cartoon) with Ni^{2+} cations (dark blue spheres). *Figure generated in PyMOL.*

with the *Efγ*-CA sequence (Fig. 8) [5]. The alignment of the *Efγ*-CA homology model with *Cdγ*-CA produced an alignment score of 0.015. Each catalytic active site is deeply embedded at the interface of one homotrimer with the next, containing a Ni^{2+} cation which would participate in catalysis. The electrostatic nature of the binding pocket is overall highly negatively charged with a neutral patch extending down the interior (Fig. 9).

3. Enzyme inhibition

3.1 FDA-approved carbonic anhydrase inhibitors

The concept of enterococcal CA inhibition remained overlooked until 2017 when Younis et al. performed a drug-repurposing screen against a panel of antibiotic-resistant pathogens and found that three FDA-approved human carbonic anhydrase inhibitors (CAIs) acetazolamide (AZM), methazolamide (MZM), and ethoxzolamide (EZM) had antimicrobial activity against a strain of vancomycin-resistant *Enterococci* (VRE), with AZM yielding the best minimum inhibitory concentration (MIC) among the three at 0.5 μg/mL (Table 2) [39]. It was only after this discovery that α- and γ-CA isoforms were noted in the *E. faecium* genome in 2020 and hypothesized to be novel antimicrobial drug targets worth exploring further for treatment of VRE [5]. Additionally, the observation that the antimicrobial potency of AZM was abrogated under an atmosphere of 5% CO_2 further supported the hypothesis of bacterial CA inhibition, considering that the excess CO_2 would spontaneously convert to bicarbonate and overcome the intracellular bicarbonate deficiency resulting from CA inhibition (Table 3) [5]. In 2021, another FDA-approved CAI dorzolamide (DZM) was also identified to have antimicrobial

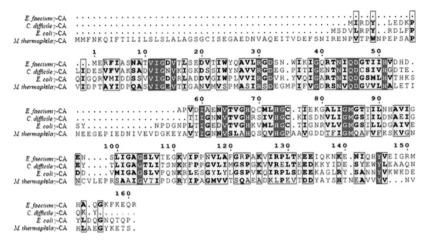

Fig. 8 Uniprot sequence alignments for *Efγ-CA, Cdγ-CA, Ecγ-CA,* and *Mtγ-CA.* A red box with white characters indicates strict identity. Bold characters represent similarity within a group. A yellow box indicates similarity across the group. Similarities of *Cdγ-CA, Ecγ-CA,* and *Mtγ-CA* to *Efα-CA* were 33.3%, 27.2%, and 25.9%, respectively. Uniprot accession codes: *E. faecium* γ-CA – Q3XX77; *C. difficile* γ-CA – Q18312; *E. coli* γ-CA – P0A9W9; *M. thermophila* γ-CA – P40881 [28]. *Image generated using ESPript3.0.*

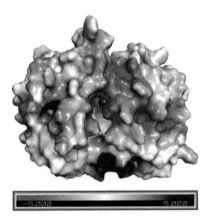

Fig. 9 Electrostatic surface of *Efγ-CA* homology model [5]. *Image generated in PyMOL.*

potency against several clinical isolates of VRE (Table 4) [40]. The *E. faecium* CAs were then recombinantly expressed for *in vitro* inhibition studies in which AZM was found to potently inhibit the CO$_2$ hydration activity of both *Efα*-CA and *Efγ*-CA (Table 5) [14]. The high activity of AZM against the *E. faecium* CAs, strengthens the hypothesis that these bacterial enzymes may be targets of FDA-approved CAIs [24].

Table 2 MICs of hit compounds from FDA-approved drug screen against VRE [39].

Compound	Structure	E. faecium ATCC 700221 MIC (µg/mL)
AZM		0.5
MZM		4
EZM		1

Table 3 MICs of AZM and linezolid against various E. faecium (E. fm) and E. faecalis (E. fs) VRE strains when incubated in ambient air and elevated atmospheric CO_2 [5].

Compound	E. fm ATCC 700221 MIC (µg/mL)		E. fs NR-31971 MIC (µg/mL)		E. fs NR-31972 MIC (µg/mL)		E. fm HM-334 MIC (µg/mL)	
	Ambient CO_2[a]		Ambient CO_2		Ambient CO_2		Ambient CO_2	
AZM	1	>64	2	>64	2	>64	2	>64
Linezolid[b]	0.5	0.5	1	1	0.5	0.5	1	1

[a]CO_2 concentration was 5%.
[b]Linezolid served as a negative control with no shift in MIC.

Homology modeling and molecular dynamics simulations using Desmond (Schrodinger, LLC, release 2020–1, [41]) were utilized to predict and rationalize the likely binding pose of AZM in Efα-CA based on the co-crystal structure of AZM bound to Hpα-CA (PDB ID:4YGF) [5,15]. The simulation quickly converged after 10 ns and generated an AZM binding pose very similar to that observed in the hCAII structure (PDB ID: 3HS4) (Fig. 10A and B) [16]. Most notably, the modeled pose in the Efα-CA active site maintained the ionic interaction of the deprotonated sulfonamide with the Zn^{2+} cation and indicated hydrogen bonding interactions with two threonine residues (Thr179 and Thr180) which are also present in the hCAII structure (Thr199 and Thr200) (Fig. 10A and B). These same interactions are also observed in the AZM-bound crystal structures of Hpα-CA

Table 4 MICs of DZM and linezolid against VRE strains reported by Abutaleb et al. [40].

VRE Strains	DZM MIC (µg/mL)	Linezolid MIC (µg/mL)
E. fm ATCC 700221	1	0.5
E. fs NR-31971	4	1
E. fs NR-31972	2	1
E. fm HM-334	2	1
MIC_{50}	4	1
MIC_{90}	4	1

and α-NgCA (Fig. 10C and D) [15,17]. As mentioned previously, the Efα-CA model seems to lack a particular alpha helix near the active site that is present in hCAII, which may provide an avenue for drug designers to gain selectivity for inhibition of the bacterial isoform. The parallels between the modeled binding pose of AZM in Efα-CA and the experimentally determined poses of AZM in other α-CAs further strengthen the hypothesis that α-CA is a likely enzyme target of AZM and other sulfonamide-based CAIs in *Enterococci*.

Similarly, a homology model of Efγ-CA based on the Cdγ-CA crystal structure (PDB ID: 4MFG; Center for Structural Genomics of Infectious Disease) was subject to Desmond molecular dynamics simulations to elucidate how AZM may bind to the Efγ-CA active site, considering that no ligand-bound structures exist for the γ- class of CAs [5]. The only well-defined interaction was that of the deprotonated sulfonamide moiety of the inhibitor coordinating with the Ni^{2+} cation (Fig. 11). All other interactions appeared to be due to somewhat transient water-mediated hydrogen bonding interactions. The simulation proposed that there may be some flexibility in the α-helix near the binding pocket to accommodate the inhibitor. However, these results are only highly suggestive without availability of an experimental ligand-bound structure for Efγ-CA. There also appear to be regions of hydrophobicity in proximity to the binding pocket toward which the acetamide moiety of AZM extends, suggesting room for inhibitor growth into that space (Fig. 12).

3.2 Published carbonic anhydrase inhibitors

While several classes of inhibitors have been developed and reported against other bacterial CAs such as for *Neisseria gonorrhoeae*, *Vibrio cholera*,

Table 5 Structure-activity-relationship studies with respect to antimicrobial activity and CA inhibition [5,14].

ID	R-group	E. faecium HM-965 MIC (µg/mL)[a]	K_i (nM)[b]			
			Efα-CA	Efγ-CA	hCAI	hCAII
AZM		2	56.7	322.8	250	12.5
1		1	37.5	250.1	235.4	37.2
2		1	23.7	218.4	179.7	30.9
3		0.25	29.8	440.8	167.3	9.5
4		0.5	50.1	325.3	212.5	22.3
5		0.125	62.7	279.0	215.1	55.6
6		0.015	69.6	389.5	230.9	7.3
7		2	36.2	192.7	156.9	24.6
8		0.5	27.2	285.9	125.2	41.5
9		0.25	9.8	194.9	76.5	26.2
10		0.25	49.3	131.1	63.3	20.2

(continued)

Table 5 Structure-activity-relationship studies with respect to antimicrobial activity and CA inhibition [5,14]. (*cont'd*)

ID	R-group	E. faecium HM-965 MIC (µg/mL)[a]	K_i (nM)[b]			
			Efα-CA	Efγ-CA	hCAI	hCAII
11		0.06	21.9	232.8	86.3	19.4
12		0.06	14.5	304.8	58.7	24.5
13		0.06	39.3	244.6	190.2	10.2
14		0.007	66.9	345.9	945.9	0.32
15		0.06	6.4	148.4	855.3	8.1
16		1	29.5	110.6	53.4	20.9
17		1	20.1	56.4	14	0.9
18	H₃C	>64	165.2	504.5	701	47.2

[a]Data reported by Kaur et al. [5].
[b]Data reported by An et al. [14].

Burkholderia organisms, *Helicobacter pylori*, and *Mycobacterium tuberculosis*, to name a few, only one class has been tested and reported against CAs in *Enterococci*, specifically [15,24,25,42–53]. The only reported series of CAIs evaluated against enterococcal CAs are based on derivatization of the AZM scaffold off the amide moiety that was initially shown to exhibit varying degrees of antimicrobial potency in VRE in 2020 [5,14]. The initial study reported robust structure–activity relationships (SAR) for the scaffold as it

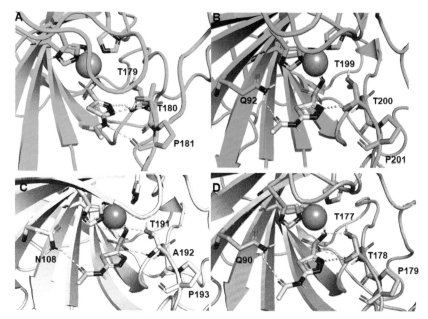

Fig. 10 Comparison of binding poses of AZM in different α-CAs. (A) Simulated binding pose of AZM in Efα-CA homology model (green) [5]. (B) Binding pose of AZM co-crystallized with hCAII (cyan) (PDB ID: 3HS4) [15]. (C) Binding pose of AZM co-crystallized with Hpα-CA (yellow) (PDB ID: 4YGF) [16]. (D) Binding pose of AZM co-crystallized with α-NgCA (orange) (PDB ID: 8DYQ) [17]. *Figures generated in PyMOL.*

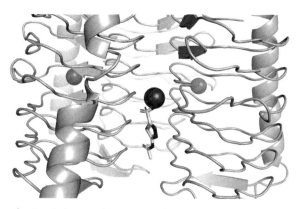

Fig. 11 Pose of AZM (white sticks) in the Efγ-CA homology model binding pocket after molecular dynamics simulation convergence (33 ns) [5]. *Image generated in PyMOL.*

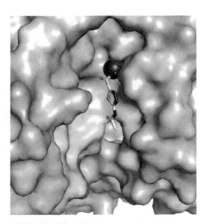

Fig. 12 Surface representation of AZM pose (white sticks) in the *Efγ*-CA binding pocket [5]. Hydrophobic patches are depicted in green. *Image generated in PyMOL.*

pertains to anti-*Enterococci* activity (Table 5). Among these were a trend of increased alkyl branching leading to increases in antibacterial activity from **1 − 3**. Additionally, extending the alkyl branch point away from the carbonyl of the amide by one methylene in **4 − 6** also improved anti-*Enterococci* activity for the scaffold, a trend that held true across multiple matched molecular pairs. Another notable trend was that increasing alkyl ring size improved antibacterial activity in **7 − 10**, with a parallel improvement again when the 5- and 6-membered rings were extended one and then two methylenes from the amide carbonyl in **11** and **12** and then **13** and **14**, respectively. In general, introducing heteroatoms into the rings in **16** and **17** reduced potencies against the strain of *Enterococci* with respect to the cyclohexyl **12**. Alternatively, removal of the carbonyl from the amide to yield an amine linker in **18** led to a loss of activity. Similarly, changes to the central thiadiazole heterocycle such as converting the 4–nitrogen to a carbon in **19** or swapping out the thiadiazole for a phenyl in **20** led to loss of activity (Table 6). The hypothesis from this study was that the molecules were acting on one or both *Ef*-CAs as the bacteria was less susceptible to the molecules in conditions of 5% CO_2 [5].

It was not until a follow-up study that the individual *E. faecium* CAs were cloned and molecules were subsequently assessed for activity [14]. In this study, the catalytic activity for both CAs was confirmed, and SAR was evaluated to determine the molecular features of the AZM scaffold that contribute to *Efα*-CA and *Efγ*-CA inhibition quantified as inhibition constant (K_i) (Tables 5 and 6) [14]. In parallel, the AZM derivatives were also

Table 6 Structure-activity-relationships between aromatic core and antimicrobial potency and CA inhibition [5,14].

ID	Structure	E. faecium HM-965 MIC (μg/mL)[a]	K_i (nM)[b]			
			Efα-CA	Efγ-CA	hCAI	hCAII
AZM		2	56.7	322.8	250	12.5
19		>64	347.7	1129	465.2	97.2
20		>64	179.7	528.6	1331	67.7

[a]Data reported by Kaur et al. [5].
[b]Data reported by An et al. [14].

evaluated for inhibition of both hCAI and hCAII to assess if the molecules were selective toward any of the CAs. Some features were found to be essential for potent *Efα*- and *Efγ*-CA inhibition, including the aromatic thiadiazole core and the amide carbonyl. When the thiadiazole in AZM was replaced with a simple phenyl ring in **20** or thiazole in **19** as the aromatic core, the inhibition of *Efα*-CA exhibited a 3-fold (56.7 nM to 179.7 nM) and 6-fold (56.7 nM to 347.7 nM) increase in K_i, respectively, with a similar but less extreme increase in K_i for *Efγ*-CA. These decreases in inhibition may have contributed to the reduced antibacterial activity for these analogs mentioned above. These changes also greatly reduced inhibition of hCAI and hCAII, suggesting that modifications to the aromatic core may disrupt key hydrogen bonding interactions, such as the known interactions between the threonine residues and the thiadiazole nitrogens described above in the previous section (Table 6). The absence of the carbonyl in **18** also displayed a greater than 2-fold reduction in inhibition of *Efα*-CA (165.2 nM) compared to AZM (56.7 nM) and displayed a similar reduction in inhibition of *Efγ*-CA (504.5 nM) and hCAI and II (701 nM and 47.2 nM, respectively), which also may have contributed to the loss in antibacterial potency noted above.

Most of the AZM analogs derivatized beyond the amide moiety were characterized by hydrophobic features. However, in a few analogs where heteroatoms were incorporated into the cycloalkyl rings, as in **16** and **17**, the inhibition of the *Ef*CA isoforms was reduced compared to that of their hydrophobic counterpart **12**, while interestingly improving inhibition of the hCAs (Table 6). When connected directly to the amide bond of the scaffold, increasing the ring size of the hydrophobic pendant from **7 – 10** seemed to improve the potency against *Ef*α-CA up through the cyclo-pentane ring in **9**. The cyclohexane moiety in **10** seemed to reverse the potency trend in this SAR series. No clear trend was noted for *Ef*γ-CA in this regard, although the K_is were generally less potent than for the α-isoform. When the cyclopentane and cyclohexane rings were extended one methylene from the amide carbonyl, only the cyclohexane in **12** increased in inhibition of *Ef*α-CA, whereas the cyclopentane ring in **11** produced a decrease in inhibition. Inhibition of the *Ef*γ-CA decreased for both **11** and **12**. Inhibition of hCAII increased for this series. Extending the cyclopentane and cyclohexane rings two methylenes from the amide car-bonyl in **13** and **14** resulted in a notable decrease in potency toward both *Ef*CA isoforms compared to the single methylene extensions in **11** and **12**, but inhibition of hCAII continued to improve to sub-nanomolar levels. Intriguingly, replacing the cyclohexane ring of **14** with a phenyl ring in **15** once again displayed increased activity against *Ef*α-CA by 10-fold from 66.9 nM to 6.4 nM and for *Ef*γ-CA by over 2-fold while decreasing activity against hCAII from 0.32 nM to 8.1 nM, respectively. Even though the phenyl pendant group was preferred for inhibition against the recombinant proteins, the opposite trend was preferred for antibacterial activity highlighting a disconnect between *in vitro* enzyme inhibition and cellular antibacterial activity [5,14].

Molecular dynamics simulations were once again utilized to gain an understanding of the differences in inhibitory activity observed between these cyclohexane and phenyl derivatives (**14** and **15**) [14]. For both compounds, the thiadiazole core was flipped relative to its position in the simulated AZM binding pose (Fig. 13) [5,14]. This may in fact be a realistic alternative thia-diazole orientation for AZM analogs with extended hydrophobic moieties, considering this kind of flip was experimentally observed in reported ligand-bound crystal structures of AZM derivatives (including **14** and **15**) bound to α-NgCA [17]. It is possible that the extended hydrophobicity of the molecules is drawn to a hydrophobic patch in the *Ef*α-CA which is partially obstructed by an alpha helix turn in the same location in hCAII structure (Fig. 4). However, no notable differences in binding site interactions could explain the differences in *in vitro* *Ef*α-CA inhibition between **14** and **15**.

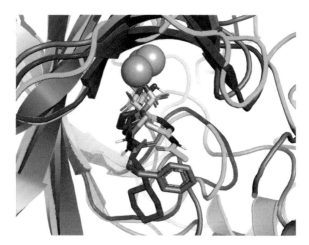

Fig. 13 Overlay of simulated binding poses of AZM (green sticks), **14** (purple sticks), and **15** (teal sticks) in the *Efα*-CA homology model active site [5,14]. *Figure generated using PyMOL.*

In 2023, 15 monothiocarbamates (MTCs) were reported to display inhibitory activity against both *Efα*-CA and *Efγ*-CA [54]. From this MTC cohort, three particular MTCs (**21**, **22**, and **23**) most strongly inhibited both *Efα*-CA and *Efγ*-CA (Table 7). MTC **22** interestingly demonstrated a greater than 2-fold improvement in inhibition of *Efγ*-CA compared to AZM (149 nM and 322 nM, respectively). Among these, **23** produced the greatest selectivity for both *Ef*CA isoforms over hCAII (Table 7). However, the promising initial *in vitro* inhibition of *Efα*-CA and *Efγ*-CA did not translate to anti-VRE efficacy for the MTCs tested against three *Enterococci* strains [54]. This lack of antimicrobial activity was hypothesized to be due to the hydrophilic nature of the MTCs which may prevent uptake by *Enterococci*, indicating an aspect for optimization in this scaffold.

3.3 Future of carbonic anhydrase inhibitors as anti-enterococcal agents

Targeting CAs as a novel therapeutic avenue for treating antibiotic-resistant *Enterococci* infections has produced promising work thus far. The CAIs presented herein have demonstrated *in vitro* inhibition to the target CAs, while also maintaining antimicrobial efficacy when dosed against various *Enterococci* strains [5,14,40]. This *in vitro* antimicrobial efficacy has also begun to translate to *in vivo* efficacy in which CAIs have demonstrated the ability to significantly reduce bioburden in both gastrointestinal and

Table 7 Inhibitory activity of select monothiocarbamates (MTCs) against *Efa*-CA and *Efy*-CA as well as against hCAI and hCAII [54].

ID	Structure	K_i (nM)			
		Efa-CA	*Efy*-CA	hCAI	hCAII
AZM		56	322	250	12
21		357	358	891	26
22		195	149	895	46
23		300	214	686	>2000

septicemic VRE infections in mice compared to the current standard of care, linezolid [40,55,56]. While the current progress suggests a bright future for CAIs in treatment of VRE, the mechanism of antimicrobial efficacy related to CA inhibition in *Enterococci* remains unelucidated due to a lack of genetic tools to confirm the role of CAs in *Enterococci* growth. The relative expression levels of the α- and γ-CAs in *Enterococci* during various stages of growth and in different physiological compartments is still uncharacterized but could shed light on the importance of each enzyme at different phases of growth and during different types of infection. While it seems likely that these CAIs are targeting CAs in *Enterococci*, studies confirming the on-target engagement of CAIs in the cellular context remain to be reported. As described above, the SAR trends for CA inhibition do not always correlate exactly with the SAR for antimicrobial potency [5,14]. Antimicrobial potency may be confounded by other factors outside of CA inhibition alone such as permeability or other unknown targets within *Enterococci*. Another aspect to consider when planning *in vivo* experiments with CAIs is that of off-target binding to hCAI and II isoforms which are highly concentrated in erythrocytes [57,58]. AZM itself is known to rapidly partition into erythrocytes, and it is likely this could be the case for other AZM-based CAIs with high affinity for hCAI or hCAII [59]. While this does not likely pose a toxicity concern, considering that the sulfonamide class of CAIs were originally designed to target hCAs, off-target binding to hCAI and II isoforms in red blood cells may reduce the amount of free CAI in circulation to reach the site of *Enterococci* infection. Thus, achieving CAI selectivity for *Enterococci* CA binding over hCA binding and sequestration into erythrocytes in the bloodstream may improve the pharmacokinetic profiles of CAIs, allowing them to distribute to satellite tissues more easily. With the reporting of new ligand-bound structures for CAIs to bacterial isoforms, there is a greater chance of structure-based design contributing to the achievement of bacterial selective inhibitors in the future. The field of research investigating CAs in *Enterococci* as therapeutic targets is still young, but the future holds great potential.

4. Conclusion

Much remains to be discovered regarding the biological mechanisms governing CA expression and activity in *Enterococci* and its resulting downstream effects. While it has proven difficult to study due to lack of genetic tools for manipulation of *Enterococci* organisms, powerful *in vitro* tools lie at the research community's disposal such as mutagenesis of recombinant proteins as well as structural biology for enabling drug discovery and elucidating key residues necessary for function and engagement by inhibitors. Likewise, utilizing common knowledge from CA activity and role in other organisms physiologically similar to *Enterococci* can direct further association of CA activity with other cellular processes which may then be investigated experimentally. The opportunities for advancing *Enterococci* CA research are many, and the antibiotic resistance crisis urges the research community to pursue alternative strategies for addressing drug-resistant infections. The *Enterococci* CA field is young, but with its possibilities for opening a new therapeutic avenue for treating drug-resistant *Enterococci* infections, it will surely continue to grow and develop into the future.

References

[1] C. Capasso, C.T. Supuran, An overview of the alpha-, beta- and gamma-carbonic anhydrases from Bacteria: can bacterial carbonic anhydrases shed new light on evolution of bacteria? J. Enzyme Inhib. Med. Chem. 30 (2015) 325–332, https://doi.org/10.3109/14756366.2014.910202

[2] E.A. Marcus, A.P. Moshfegh, G. Sachs, D.R. Scott, The periplasmic α-carbonic anhydrase activity of Helicobacter pylori is essential for acid acclimation, J. Bacteriol. 187 (2005) 729–738, https://doi.org/10.1128/JB.187.2.729-738.2005/ASSET/08B8E30E-7394-45F8-932F-AF4289987447/ASSETS/GRAPHIC/ZJB0020543740008.JPEG

[3] D.P. Flaherty, M.N. Seleem, C.T. Supuran, Bacterial carbonic anhydrases: underexploited antibacterial therapeutic targets, Future Med. Chem. 13 (2021) 1619–1622, https://doi.org/10.4155/FMC-2021-0207

[4] K.S. Smith, C. Jakubzick, T.S. Whittam, J.G. Ferry, Carbonic anhydrase is an ancient enzyme widespread in prokaryotes, Proc. Natl. Acad. Sci. U. S. A. 96 (1999) 15184–15189, https://doi.org/10.1073/PNAS.96.26.15184/ASSET/552AD14F-3D67-48F5-B473-5594D9A6FEBD/ASSETS/GRAPHIC/PQ2594492003.JPEG

[5] J. Kaur, X. Cao, N.S. Abutaleb, A. Elkashif, A.L. Graboski, A.D. Krabill, et al., Optimization of acetazolamide-based scaffold as potent inhibitors of vancomycin-resistant Enterococcus, J. Med. Chem. 63 (2020) 9540–9562, https://doi.org/10.1021/ACS.JMEDCHEM.0C00734/SUPPL_FILE/JM0C00734_SI_005.CSV

[6] Y. Pocker, J. Stone, The catalytic versatility of erythrocyte carbonic anhydrase: the enzyme-catalyzed hydrolysis of p-nitrophenyl acetate, J. Am. Chem. Soc. 87 (1965) 5497–5498.

[7] G.S. Chilambi, Y.-H. Wang, N.R. Wallace, C. Obiwuma, K.M. Evans, Y. Li, et al., Carbonic anhydrase inhibition as a target for antibiotic synergy in Enterococci, Microbiol. Spectr. (2023), https://doi.org/10.1128/SPECTRUM.03963-22

[8] Burghout, P. Cron, L.E. Gradstedt, H. Quintero, B. Simonetti, E. Bijlsma JJE, et al., Carbonic anhydrase is essential for Streptococcus pneumoniae growth in environmental ambient air, J. Bacteriol. 192 (2010) 4054–4062, https://doi.org/10.1128/JB.00151-10/SUPPL_FILE/TABLE_S1.ZIP

[9] Y. Matsumoto, K. Miyake, K. Ozawa, Y. Baba, T. Kusube, Bicarbonate and unsaturated fatty acids enhance capsular polysaccharide synthesis gene expression in oral Streptococci, Streptococcus anginosus, J. Biosci. Bioeng. 128 (2019) 511–517, https://doi.org/10.1016/J.JBIOSC.2019.04.010

[10] T.H. Eckhardt, D. Skotnicka, J. Kok, O.P. Kuipers, Transcriptional regulation of fatty acid biosynthesis in Lactococcus lactis, J. Bacteriol. 195 (2013) 1081–1089, https://doi.org/10.1128/JB.02043-12/SUPPL_FILE/ZJB999092460SO1.PDF

[11] A.L. Krause, T.P. Stinear, I.R. Monk, Barriers to genetic manipulation of Enterococci: Current approaches and future directions, FEMS Microbiol. Rev. 46 (2022) 1–14, https://doi.org/10.1093/FEMSRE/FUAC036

[12] C.W. Remmele, Y. Xian, M. Albrecht, M. Faulstich, M. Fraunholz, E. Heinrichs, et al., Transcriptional landscape and essential genes of Neisseria gonorrhoeae, Nucleic Acids Res. 42 (2014) 10579–10595, https://doi.org/10.1093/NAR/GKU762

[13] K. Ueda, H. Nishida, T. Beppu, Dispensabilities of carbonic anhydrase in proteobacteria, Int. J. Evol. Biol. 2012 (2012) 1–5, https://doi.org/10.1155/2012/324549

[14] W. An, K.J. Holly, A. Nocentini, R.D. Imhoff, C.S. Hewitt, N.S. Abutaleb, et al., Structure-activity relationship studies for inhibitors for vancomycin-resistant Enterococcus and human carbonic anhydrases, J. Enzyme Inhib. Med. Chem. 37 (2022) 1838–1844, https://doi.org/10.1080/14756366.2022.2092729

[15] J.K. Modakh, Y.C. Liu, M.A. Machuca, C.T. Supuran, A. Roujeinikova, Structural basis for the inhibition of Helicobacter pylori α-carbonic anhydrase by sulfonamides, PLoS One 10 (2015) e0127149, https://doi.org/10.1371/JOURNAL.PONE.0127149

[16] K.H. Sippel, A.H. Robbins, J. Domsic, C. Genis, M. Agbandje-Mckenna, R. McKenna, High-resolution structure of human carbonic anhydrase II complexed with acetazolamide reveals insights into inhibitor drug design, Acta Crystallogr. Sect. F. Struct. Biol. Cryst. Commun. 65 (2009) 992–995, https://doi.org/10.1107/S1744309109036665/HTTPS://JOURNALS.IUCR.ORG/SERVICES/TERMSOFUSE.HTML.

[17] A.K. Marapaka, A. Nocentini, M.S. Youse, W. An, K.J. Holly, C. Das, et al., Structural characterization of thiadiazolesulfonamide inhibitors bound to Neisseria gonorrhoeae α-carbonic anhydrase, ACS Med. Chem. Lett. 14 (2023) 103–109, https://doi.org/10.1021/ACSMEDCHEMLETT.2C00471/SUPPL_FILE/ML2C00471_SI_002.ZIP.

[18] G. De Simone, S.M. Monti, V. Alterio, M. Buonanno, V. De Luca, M. Rossi, et al., Crystal structure of the most catalytically effective carbonic anhydrase enzyme known, SazCA from the thermophilic bacterium Sulfurihydrogenibium azorense, Bioorg Med. Chem. Lett. 25 (2015) 2002–2006, https://doi.org/10.1016/J.BMCL.2015.02.068

[19] A. Aspatwar, Tolvanen MEE, S. Parkkila, Phylogeny and expression of carbonic anhydrase-related proteins, BMC Mol. Biol. 11 (2010) 1–19, https://doi.org/10.1186/1471-2199-11-25/FIGURES/11

[20] A. Nocentini, W.A. Donald, C.T. Supuran, Human carbonic anhydrases: tissue distribution, physiological role, and druggability, Carbonic Anhydrases: Biochem. Pharmacol. Evergr. Pharm. Target. (2019) 151–185, https://doi.org/10.1016/B978-0-12-816476-1.00008-3

[21] C.T. Supuran, Carbonic anhydrases: novel therapeutic applications for inhibitors and activators, Nat. Rev. Drug. Discov. 7 (2) (2008) 168–181, https://doi.org/10.1038/nrd2467

[22] K.M. Gilmour, Perspectives on carbonic anhydrase, Comp. Biochem. 157 (2010) 193–197, https://doi.org/10.1016/J.CBPA.2010.06.161

[23] V.De Luca, D. Vullo, A. Scozzafava, V. Carginale, M. Rossi, C.T. Supuran, et al., An α-carbonic anhydrase from the thermophilic bacterium Sulphurihydrogenibium azorense is the fastest enzyme known for the CO_2 hydration reaction, Bioorg Med. Chem. 21 (2013) 1465–1469, https://doi.org/10.1016/J.BMC.2012.09.047

[24] S. Del Prete, D. Vullo, V. De Luca, V. Carginale, P. di Fonzo, S.M. Osman, et al., Anion inhibition profiles of α-, β- and γ-carbonic anhydrases from the pathogenic bacterium Vibrio cholerae, Bioorg Med. Chem. 24 (2016) 3413–3417, https://doi.org/10.1016/J.BMC.2016.05.029

[25] I. Nishimori, T. Minakuchi, K. Morimoto, S. Sano, S. Onishi, H. Takeuchi, et al., Carbonic anhydrase inhibitors: DNA cloning and inhibition studies of the α carbonic anhydrase from Helicobacter pylori, a new target for developing sulfonamide and sulfamate gastric drugs, J. Med. Chem. 49 (2006) 2117–2126, https://doi.org/10.1021/JM0512600/SUPPL_FILE/JM0512600SI20060120_055150.PDF.

[26] I.D. Campbell, S. Lindskog, A.I. White, A study of the histidine residues of human carbonic anhydrase C using 270 MHz proton magnetic resonance, J. Mol. Biol. 98 (1975) 597–614, https://doi.org/10.1016/S0022-2836(75)80089-1

[27] S.K. Nair, D.W. Christianson, S. Eur, Unexpected pH-dependent conformation of His-64, the proton shuttle of carbonic anhydrase II. J. E Proc. Natl. Acad. Sci. U. S. A. 113 (1991) 4075–4081.

[28] X. Robert, P. Gouet, Deciphering key features in protein structures with the new ENDscript server, Nucleic Acids Res. 42 (2014) W320–W324, https://doi.org/10.1093/NAR/GKU316

[29] B. Elleby, L.C. Chirica, C. Tu, M. Zeppezauer, S. Lindskog, Characterization of carbonic anhydrase from Neisseria gonorrhoeae, Eur. J. Biochem. 268 (2001) 1613–1619, https://doi.org/10.1046/J.1432-1327.2001.02031.X

[30] F.F. Nord, F. Friedrich, Advances in Enzymology and Related Subjects of Biochemistry XXV Interscience/Wiley, 1963.

[31] P.V. LoGrasso, C. Tu, G.C. Wynns, D.N. Silverman, D.A. Jewell, P.J. Laipis, Catalytic enhancement of human carbonic anhydrase III by replacement of phenylalanine-198 with leucine, Biochemistry 30 (1991) 8463–8470, https://doi.org/10.1021/BI00098A025/ASSET/BI00098A025.FP.PNG_V03

[32] J.G. Ferry, The γ class of carbonic anhydrases, Biochimica et. Biophysica Acta (BBA)—Proteins Proteom. 1804 (2010) 374–381, https://doi.org/10.1016/J.BBAPAP.2009.08.026

[33] B.E. Alber, J.G. Ferry, Characterization of heterologously produced carbonic anhydrase from Methanosarcina thermophila, J. Bacteriol. 178 (1996) 3270–3274, https://doi.org/10.1128/JB.178.11.3270-3274.1996

[34] B.E. Alber, J.G. Ferry, A carbonic anhydrase from the archaeon Methanosarcina thermophila, Proc. Natl. Acad. Sci. U. S. A. 91 (1994) 6909–6913, https://doi.org/10.1073/PNAS.91.15.6909

[35] K.S. Smith, J.G. Ferry, Prokaryotic carbonic anhydrases, FEMS Microbiol. Rev. 24 (2000) 335–366, https://doi.org/10.1111/J.1574-6976.2000.TB00546.X

[36] G.D. Price, D. Sültemeyer, B. Klughammer, M. Ludwig, M.R. Badger, The functioning of the CO_2 concentrating mechanism in several cyanobacterial strains: a review of general physiological characteristics, genes, proteins, and recent advances, Canad. J. Bot. 76 (1998) 973–1002, https://doi.org/10.1139/B98-081

[37] C.T. Supuran, Structure and function of carbonic anhydrases, Biochem. J. 473 (2016) 2023–2032, https://doi.org/10.1042/BCJ20160115

[38] B.C. Tripp, C.B. Bell, F. Cruz, C. Krebs, J.G. Ferry, A role for iron in an ancient carbonic anhydrase, J. Biol. Chem. 279 (2004) 6683–6687, https://doi.org/10.1074/JBC.M311648200

[39] W. Younis, A. Abdelkhalek, A.S. Mayhoub, M.N. Seleem, In vitro screening of an FDA-approved library against ESKAPE pathogens HHS public access, Curr. Pharm. Des. 23 (2017) 2147–2157, https://doi.org/10.2174/1381612823666170209154745

[40] N.S. Abutaleb, A.E.M. Elhassanny, D.P. Flaherty, M.N. Seleem, In vitro and in vivo activities of the carbonic anhydrase inhibitor, dorzolamide, against vancomycin-resistant enterococci, PeerJ 9 (2021) e11059, https://doi.org/10.7717/PEERJ.11059/SUPP-5

[41] K.J. Bowers, E. Chow, H. Xu, R.O. Dror, M.P. Eastwood, B.A. Gregersen, et al., Scalable algorithms for molecular dynamics simulations on commodity clusters, Proc. 2006 ACM/IEEE Conf. Supercomputing SC'06 (2006), https://doi.org/10.1145/1188455.1188544

[42] A. Nocentini, C.S. Hewitt, M.D. Mastrolorenzo, D.P. Flaherty, C.T. Supuran, Anion inhibition Stud. α-carbonic anhydrases Neisseria gonorrhoeae, J. Enzyme Inhib. Med. Chem. 36 (2021) 1061–1066, https://doi.org/10.1080/14756366.2021.1929202

[43] Abutaleb, N.S. Elhassanny AEM, A. Nocentini, C.S. Hewitt, A. Elkashif, B.R. Cooper, et al., Repurposing FDA-approved sulphonamide carbonic anhydrase inhibitors for treatment of Neisseria gonorrhoeae, J. Enzyme Inhib. Med. Chem. 37 (2022) 51–61, https://doi.org/10.1080/14756366.2021.1991336/SUPPL_FILE/IENZ_A_1991336_SM8024.PDF.

[44] Hewitt, C.S. Abutaleb, N.S. Elhassanny AEM, A. Nocentini, X. Cao, D.P. Amos, et al., Structure-activity relationship studies of acetazolamide-based carbonic anhydrase inhibitors with activity against Neisseria gonorrhoeae, ACS Infect. Dis. 7 (2021) 1969–1984, https://doi.org/10.1021/ACSINFECDIS.1C00055/ASSET/IMAGES/LARGE/ID1C00055_0007.JPEG.

[45] S. Giovannuzzi, C.S. Hewitt, A. Nocentini, C. Capasso, D.P. Flaherty, C.T. Supuran, Coumarins effectively inhibit bacterial α-carbonic anhydrases, J. Enzyme Inhib. Med. Chem. 37 (2022) 333–338, https://doi.org/10.1080/14756366.2021.2012174

[46] S. Giovannuzzi, C.S. Hewitt, A. Nocentini, C. Capasso, G. Costantino, D.P. Flaherty, et al., Inhibition studies of bacterial α-carbonic anhydrases with phenols, J. Enzyme Inhib. Med. Chem. 37 (2022) 666–671, https://doi.org/10.1080/14756366.2022.2038592

[47] A. Angeli, G. Abbas, S. del Prete, C. Capasso, C.T. Supuran, Selenides bearing benzenesulfonamide show potent inhibition activity against carbonic anhydrases from pathogenic bacteria Vibrio cholerae and Burkholderia pseudomallei, Bioorg. Chem. 79 (2018) 319–322, https://doi.org/10.1016/J.BIOORG.2018.05.015

[48] V. De Luca, A. Petreni, A. Nocentini, A. Scaloni, C.T. Supuran, C. Capasso, Effect of sulfonamides and their structurally related derivatives on the activity of ι-carbonic anhydrase from Burkholderia territorii, Int. J. Mol. Sci. 22 (2021) 571, https://doi.org/10.3390/IJMS22020571

[49] R. Grande, S. Carradori, V. Puca, I. Vitale, A. Angeli, A. Nocentini, et al., Selective inhibition of helicobacter pylori carbonic anhydrases by carvacrol and thymol could impair biofilm production and the release of outer membrane vesicles, Int. J. Mol. Sci. 22 (2021) 11583, https://doi.org/10.3390/IJMS222111583/S1

[50] Y. Cau, M. Mori, C.T. Supuran, M. Botta, Mycobacterial carbonic anhydrase inhibition with phenolic acids and esters: kinetic and computational investigations, Org. Biomol. Chem. 14 (2016) 8322–8330, https://doi.org/10.1039/C6OB01477A

[51] A. Maresca, F. Carta, D. Vullo, C.T. Supuran, Dithiocarbamates strongly inhibit the β-class carbonic anhydrases from Mycobacterium tuberculosis, J. Enzyme Inhib. Med. Chem. 28 (2013) 407–411, https://doi.org/10.3109/14756366.2011.641015

[52] A. Maresca, D. Vullo, A. Scozzafava, G. Manole, C.T. Supuran, Inhibition of the β-class carbonic anhydrases from Mycobacterium tuberculosis with carboxylic acids, J. Enzyme Inhib. Med. Chem. 28 (2013) 392–396, https://doi.org/10.3109/14756366.2011.650168

[53] F. Carta, A. Maresca, A.S. Covarrubias, S.L. Mowbray, T.A. Jones, C.T. Supuran, Carbonic anhydrase inhibitors. Characterization and inhibition studies of the most active β-carbonic anhydrase from Mycobacterium tuberculosis, Rv3588c, Bioorg Med. Chem. Lett. 19 (2009) 6649–6654, https://doi.org/10.1016/J.BMCL.2009.10.009

[54] S. Giovannuzzi, A.K. Marapaka, N.S. Abutaleb, F. Carta, H.W. Liang, A. Nocentini, et al., Inhibition of pathogenic bacterial carbonic anhydrases by monothiocarbamates, J. Enzyme Inhib. Med. Chem. 38 (2023), https://doi.org/10.1080/14756366.2023.2284119

[55] N.S. Abutaleb, A. Elkashif, D.P. Flaherty, M.N. Seleem, In vivo antibacterial activity of acetazolamide, Antimicrob. Agents Chemother. 65 (2021), https://doi.org/10.1128/AAC.01715-20/FORMAT/EPUB

[56] N.S. Abutaleb, A. Shrinidhi, A.B. Bandara, M.N. Seleem, D.P. Flaherty, Evaluation of 1,3,4-thiadiazole carbonic anhydrase inhibitors for gut decolonization of vancomycin-resistant Enterococci, ACS Med. Chem. Lett. 14 (2023) 487–492, https://doi.org/10.1021/ACSMEDCHEMLETT.3C00032/ASSET/IMAGES/LARGE/ML3C00032_0004.JPEG

[57] C.T. Supuran, Emerging role of carbonic anhydrase inhibitors, Clin. Sci. 135 (2021) 1233–1249, https://doi.org/10.1042/CS20210040

[58] W.F. Boron, Evaluating the role of carbonic anhydrases in the transport of HCO_3^--related species, Biochimica et. Biophysica Acta (BBA)—Proteins Proteom. 1804 (2010) 410–421, https://doi.org/10.1016/J.BBAPAP.2009.10.021

[59] S.M. Wallace, S. Riegelman, Uptake of acetazolamide by human erythrocytes in vitro, J. Pharm. Sci. 66 (1977) 729–731, https://doi.org/10.1002/JPS.2600660532

CHAPTER ELEVEN

Carbonic anhydrases in bacterial pathogens

Reygan E. Braga[a], Fares Z. Najar[b], Chelsea L. Murphy[b], and Marianna A. Patrauchan[a,*]

[a]Department of Microbiology and Molecular Genetics, Oklahoma State University, Stillwater, OK, United States
[b]Bioinformatics Core, Oklahoma State University, Stillwater, OK, United States
*Corresponding author. e-mail address: m.patrauchan@okstate.edu

Contents

1. Introduction	313
2. Distribution and sequence conservation in carbonic anhydrases among bacterial pathogens	315
3. Importance of carbonic anhydrases in physiology of bacterial pathogens	328
4. Role in bacterial virulence and host-pathogen interactions	332
5. Potential in therapeutic implications	334
6. Remaining unknowns and challenges	335
References	336

Abstract

Carbonic anhydrases (CAs) catalyze the reversable hydration of carbon dioxide to bicarbonate placing them into the core of the biochemical carbon cycle. Due to the fundamental importance of their function, they evolved independently into eight classes, three of which have been recently discovered. Most research on CAs has focused on their representatives in eukaryotic organisms, while prokaryotic CAs received significantly less attention. Nevertheless, prokaryotic CAs play a key role in the fundamental ability of the biosphere to acquire CO_2 for photosynthesis and to decompose the organic matter back to CO_2. They also contribute to a broad spectrum of processes in pathogenic bacteria, enhancing their ability to survive in a host and, therefore, present a promising target for developing antimicrobials. This review focuses on the distribution of CAs among bacterial pathogens and their importance in bacterial virulence and host-pathogen interactions.

1. Introduction

Carbonic anhydrases (CAs) form a large family of metalloenzymes (EC 4.2.1.1). Despite their sequence and structural differences, they all catalyze the reversible hydration of carbon dioxide with the production of

The Enzymes, Volume 55
ISSN 1874-6047, https://doi.org/10.1016/bs.enz.2024.05.007

bicarbonate and proton [1]. Although defined as metalloenzymes, as most of them require metals for catalytic activity (zinc and less frequently, cadmium, iron or cobalt) [2,3], newly discovered CAs can be catalytically active without a metal cofactor [4–6]. In addition to conventional CO_2 hydratase activity, CAs can exhibit promiscuous esterase activity that probably stems from the mechanistic similarity between hydration of the carbonyl of CO_2 and that of an ester [7]. This feature has been discussed in detail in [8] and may have evolutionary implications [9]. Furthermore, a ζ-CA from marine diatom has been reported for its hydrolase activity towards CS_2, potentially extending sources of carbon acquisition [10] and further expanding the biological significance of this family of enzymes.

CAs are widespread in all three domains of life: Bacteria, Archaea and Eukarya [11]. According to their sequence and structural features, they are grouped into eight evolutionary distinct classes α, β, γ, δ, ζ, η, θ, and ι [12]. The last three classes have been discovered relatively recently [13–15], indicating that the potential of finding novel carbonic anhydrases is not yet exhausted. So far, bacteria have been shown to encode CA representatives of four classes α, β, γ, and ι [16], and most recently, ζ-class CAs have been bioinformatically projected in environmental microbiomes [17].

Due to their enzymatic activity balancing between CO_2 and HCO_3^-, CAs are involved in many important biochemical processes. Two of the most influential roles are the maintenance of cellular pH homeostasis and regulating metabolic flow of carbon [11,18]. Controlling the intracellular pH is essential for optimal activity of proteins, cellular transport and signaling, that collectively highlight the impact of CAs on cellular functions. The role of CAs in carbon assimilation is particularly profound in autotrophic bacteria, where they provide the necessary bicarbonate substrate for carboxylation reactions enabling carbon fixation and biosynthesis of organic matter [19,20]. Since CO_2 may also serve as an environmental signal, CAs can be involved in sensing the environment [11,21]. Additional roles include supplying HCO_3^- for cyanate decomposition [22], deposition of calcium [23], as well as de-novo biosynthesis of unsaturated fatty acids [24]. These roles define the importance of CAs in environmental adaptations and survival of bacteria in diverse niches.

One of the most challenging niches for bacteria to survive is a human host. There are two types of bacteria making a living in a human body: commensals that evolved to co-exist within human organs and tissues and pathogens that invade a host by using aggressive virulence factors that damage the host cells and cause infections. CAs of both bacterial and host origin have been implicated in various aspects of host-pathogen interactions, including bacterial adhesion to host tissues, invasion of host cells and modulation of the immune response [24–27]. By regulating pH and CO_2 levels, CAs enable the formation of micro-niches favorable for bacterial survival and enhance their adaptation to the host [28]. These and other aspects of CAs role in bacterial pathogens are discussed below.

Reflecting the importance of CAs in survival in a host and virulence of bacterial pathogens, they emerged as promising targets for developing antimicrobial therapies. Targeting these enzymes may impair the ability of pathogens to cause infections or/and sensitize bacteria increasing the efficacy of existing antibiotics and therefore, offer new avenues for combating bacterial infections. However, advancing in this direction requires new knowledge on functional and structural features of bacterial CAs, a better understanding of their role in bacterial physiology and interactions with the host, new insights about their regulation during infections, their apparent redundancy in bacterial genomes, and, finally, the distinct molecular characteristics of bacterial CAs that can guide the design of specific inhibitors that selectively target pathogens.

2. Distribution and sequence conservation in carbonic anhydrases among bacterial pathogens

The continuing growth of genomic information and computational analysis provides an excellent source and tools for discovering novel CAs and understanding their evolution. In 2014, a search within 656 completed prokaryotic genomes with the query "carbonic anhydrase" retrieved 1300 sequences [18]. Our search in 2024 detected about 229290 CA protein sequences within 23183 completed genomes, underscoring the widespread occurrence of these enzymes in diverse prokaryotes inhabiting the Earth. Such diversity is reflected in the variety of physiological roles that these enzymes play in different organisms, including respiration, pH homeostasis, CO_2/bicarbonate homeostasis and transport, as well as carbon fixation (reviewed in [11]). However, the occurrence and distribution of these enzymes in bacterial genomes vary widely. An overview of α-, β-, and γ-classes of CA enzymes present in bacteria [16] showed a broad spread of β- and γ-CA among diverse bacterial species and the predominant occurrence of α-CAs in Gram-negative bacteria. The authors related the distribution of the CAs, their subcellular localization and the geometry of their catalytic sites to the evolution and genetic variability of Gram-positive and Gram-negative bacteria. The distribution study of CAs among selected commensal bacteria concluded that the lack of α-CAs can be compensated by the presence of β- and γ-CAs [29].

Interestingly, some sequenced bacterial genomes appear to be missing CAs. For example, in 2012, it was shown that 39 strains of 20 genera of the genome-sequenced Proteobacteria do not retain any gene encoding CA [30]. An argument was made that the distribution of CAs among bacteria is guided by the adaptations to CO_2 levels in their immediate environment and that CAs are dispensable in high-CO_2. This was exemplified by intracellular bacterial genera *Buchnera* and *Rickettsia* containing CA-defective strains [30]. Therefore, understanding CAs' distribution in microbial genomes may provide an insight into the history of bacterial adaptation to their environments. Such understanding is also important, considering the role of CAs in bacterial pathogenicity and their potential as drug targets.

In addition to α-, β-, and γ-classes of CAs, recent research identified the representatives of two more CA classes in the prokaryotic world, ι and ζ. The ι-CAs have been discovered in the marine diatom *Thalassiosira pseudonana* [14] and then identified in *Burkholderia territorii* [13]. This group of CAs prefers Mn^{2+} to Zn^{2+} as a cofactor [14]. The ζ-class CA have been also discovered in marine diatoms, *T. weissflogii* [31] and, most recently, bioinformatically predicted in the environmental microbiomes [17]. The ζ-class CDCA1 is the first example of a Cd^{2+}-containing CA and can spontaneously exchange Zn^{2+} or Cd^{2+} at its active site, possibly as an adaptation to the metal-poor environments [32,33].

CAs represent an interesting group of enzymes, where on one hand, there is a certain degree of sequence conservation within each class that protects key catalytic residues and structural features involved in the enzymatic activity. This enables their mostly uniform catalytic activity and allows their prediction within diverse genomes. On the other hand, quite significant sequence variability exists among multiple CA homologs present either within the same or different bacterial genomes. This variability may reflect adaptation to specific environmental conditions or host niches and highlights the potential for functional diversity and specialization in CAs. In the case of bacterial pathogens, it is possible that CAs evolution reflects the co-evolution of pathogens with host responses, particularly nutritional immunity, when host limits the availability of nutrients, especially trace metals, required for enzymes activity [34]. This co-evolution can drive sequence variation in CAs or their transcriptional regulators.

Here we aimed to analyze the occurrence and distribution of α-, β-, γ-, ι- and ζ-CAs in 50 well-established bacterial pathogens, including 33 Gram-negative, 12 Gram-positive, and 5 Gram-neutral (termed in [35], included species from order Mycobacteriales containing a mycolic acid layer resembling an outer membrane [36]), species associated with three different lifestyles: opportunistic pathogens that do not require a host for survival, pathobionts that belong to commensal bacteria conditionally able to cause disease, and obligate human or animal pathogens that require a host for survival (Table 1). Retrieved CA sequences were searched against the InterPro Database [37] using default parameters to identify their characteristic families (Table 2). All reviewed protein sequences were downloaded and aligned with Clustal-Omega [38] and constructed into HMM profiles using hmmbuild from the HMMER suite [39]. Proteins from the selected bacterial genomes were scanned against this custom CA database using hmmscan [39], and results were filtered using a cutoff e-value of 1×10^{-5}. These proteins were considered homologous and their number in each genome is listed in Table 1 after the backslash. For a more stringent analysis, homologous CA sequences were then searched for motifs containing a catalytic triad using in-house scripts (Table 2). Sequences containing the motif were considered as probable CAs, their number in each genome is listed in Table 1 before the backslash. These sequences were carried into further analyses. For phylogenetic analysis, small subunit (SSU) rRNA sequences from each bacterial genome were obtained from the Genome Taxonomy Database (GTDB) [40] release 220. Sequences were aligned using Clustal-Omega [38] and a phylogenetic tree was constructed using MEGA11 [41]. The phylogenetic tree of pathogens was visualized and decorated with the information on Gram staining and lifestyle using iToL [42].

Out of the 50 selected genomes, 44 encoded at least one probable CA. Six genomes contained a single probable CA gene that encodes α-, β-, or γ-CA. The genomes of three Gram-positive and three Gram-negative species did not encode probable CAs. Overall, the largest number of CAs were present in the genomes of bacteria with opportunistic lifestyle. Since opportunistic bacterial pathogens transition between environment and host, this suggests that the enzymes evolved to support adaptation to a broader scope of conditions (Fig. 1; Table 1). On the other hand, on average, the smallest number of CAs were seen associated with obligate pathogens. Considering significantly higher levels of CO_2 within a human body than in the air, this may reflect the proposed role of low CO_2 as an evolutionary pressure supporting CAs [30]. We did not observe the

Table 1 The number of CA homologs present in diverse pathogenic bacteria.

Organism	CA Class					Lifestyle		
	α	β	γ	ι	ζ	Obl_H/A	Pathobi	Opport
Gram Negative								
Achromobacter xylosoxidans		2/2[a]	2/4					◀
Acinetobacter baumannii		2/1	4/7					◀
Aeromonas hydrophila	1	1/5	4/17	1				◀
Bordetella pertussis		1/3	1/5			◀		
Borrelia afzelii		0/1	0/1			◀		
Borrelia burgdorferi		0/1	0/1			◀		
Borrelia mayonii		0/1	0/1			◀		
Brucella suis		2/1	1/10	0/1				◀
Burkholderia cepacia		3/3	4/9	0/2				◀
Burkholderia pseudomallei		3/3	1/4	0/2	0/1			◀
Campylobacter jejuni		1	1/8					◀
Chlamydia pneumoniae		0/2	1/4			◀		

Species						
Chlamydia trachomatis	1	1[b]	0/3	▲		◁
Citrobacter koseri		1/3	4/13		◁	◁
Coxiella burnetii		1/1	1/8	▲		
Enterobacter cloacae	1	2/5	4/16	1	◁	◁
Escherichia coli K-12		2/1	4/8		◁	◁
Escherichia coli O157:H7		2/1	3/8			◁
Francisella tularensis		1/1	1/7	1	◁	
Haemophilus influenzae		1	1/4		◁	
Helicobacter pylori	1	1/1	1/3		◁	
Klebsiella aerogenes		4/5	4/12	0/1	◁	◁
Klebsiella pneumoniae		3/3	3/6		◁	◁
Legionella pneumophila		3[b]/3	5/3	0/1	▲	◁
Neisseria gonorrhoeae	1	1	1/4		◁	
Neisseria meningitidis		1	1/5	0/1	◁	
Porphyromonas gingivalis		1/1	1/3		◁	

(continued)

Table 1 The number of CA homologs present in diverse pathogenic bacteria. (*cont'd*)

Organism	CA Class					Lifestyle		
	α	β	γ	ι	ζ	Obl_H/A	Pathobi	Opport
Pseudomonas aeruginosa		3/4	4/5	0/1				◄
Rickettsia rickettsii		1	0/4			◄		
Salmonella enterica		2/1	3/5			◄		
Shigella flexneri		1/3	3/13			◄		
Vibrio cholerae	1	1/3	3/8	0/1				◄
Yersinia pestis		2/3	2/4			◄		
Gram Positive								
Bacillus anthracis		1/3	2/10					◄
Clostridioides difficile		1	1/9	0/1			◄	
Clostridium perfringens		1/2	1/7				◄	
Enterococcus faecalis	1	0/2	2/6				◄	◄
Enterococcus faecium	1	0/1	2/5				◄	◄
Listeria monocytogenes	1	0/7	0/8					◄

Paeniclostridium sordellii	1/1	2/8		◀
Staphylococcus aureus		0/6		◀ ◀
Staphylococcus epidermidis	0/1	0/3		◀
Streptococcus milleri		0/6		◀
Streptococcus mutans	1	0/3		◀
Streptococcus pneumoniae	1	0/5		◀
Gram Neutral[c]				
Corynebacterium diphtheriae	1	3/6		◀
Mycobacterium leprae	1/1	0/4		◀
Mycobacterium tuberculosis	3[b]/2	1/4	0/1	◀
Mycobacteroides abscessus	7[d]/1	1/4		◀
Nocardia asteroides	1	7[b]/8	2/9	0/2	◀

Accession numbers for genomes and genes are as listed: *A. xylosoxidans* (GCF_016728825): β-I WP_006386090.1, β-I WP_006390028.1, γ WP_006389349.1, γ WP_054480950.1; *A. baumannii* (GCF_008632635): β-I WP_011141695.1, β-I WP_004833634.1, γ WP_001107199.1, γ WP_024437954.1, γ WP_077166004.1, γ WP_104914775.1; *A. hydrophila* (GCF_020162255): α WP_073351461.1, β-II WP_005308055.1, γ WP_005030?4.1, γ WP_073348559.1, γ WP_073349014.1, γ WP_224481121.1, I WP_029301822.1; *B. pertussis* (GCF_004008975): β-I WP_003819076.1, γ WP_003813683.1; *B. suis* (GCF_000007505): β-I WP_006192575.1, β-II WP_004690295.1, γ WP_002964382.1; *B. cepacia* (GCF_009586235): β-I WP_021157399.1, β-I WP_027789886.1, β-I WP_060063574.1, γ WP_021157692.1, γ WP_119940000.1, γ WP_153488788.1, γ WP_153490518.1; *B. pseudomallei* (GCF_000756125): β-I WP_004186405.1, β-I WP_004189176.1, β-I WP_004522070.1, γ WP_004524734.1; *C. jejuni* (GCF_000009085): β-I YP_002343680.1, γ YP_002344006.1; *C. pneumoniae* (GCF_000007205):

(continued)

γ WP_010883386.1; *C. trachomatis* (GCF_000008725): β-TM NP_220378.1; *C. koseri* (GCF_000018045): α WP_012135396.1, β-II WP_012134030.1, γ WP_000944851.1, γ WP_012000826.1, γ WP_012134134.1, γ WP_012135410.1; *C. burnetii* (GCF_000007765): β-I NP_819189.1, γ NP_819848.1; *E. cloacae* (GCF_905331265): α WP_013094853.1, β-I WP_013094936.1, β-II WP_013095615.1, γ WP_013099110.1, γ WP_029883660.1, γ WP_230137142.1, ι WP_029883750.1; *E. coli K-12* (GCF_000005845): β-I NP_414668.1, β-I NP_414873.1, γ NP_414577.2, γ NP_415918.1, γ NP_417738.4, γ NP_418186.1; *E. coli O157:H7* (GCF_000008865): β-I NP_308157.1, β-I NP_308419.2, γ NP_308065.2, γ NP_312172.2, γ NP_312699.1; *F. tularensis* (GCF_000833355): β-I WP_003034182.1, γ WP_003034834.1, ι WP_003034056.1; *H. influenzae* (GCF_000931575): β-I WP_005631770.1, γ WP_044332992.1; *H. pylori* (GCF_017821535): α WP_209611489.1, β-I WP_209611813.1, γ WP_077645225.1; *K. aerogenes* (GCF_007632255): β-I WP_015366898.1, β-I WP_015704921.1, β-II WP_015368205.1, β-II WP_063445728.1, γ WP_047036359.1, γ WP_047075923.1, γ WP_047080914.1, γ WP_063446340.1; *K. pneumoniae* (GCF_000240185): β-I WP_005224839.1, β-I YP_005225161.1, β-I YP_005227295.1, γ WP_005226681.1, γ YP_005229129.1, γ YP_005229602.1; *L. pneumophila* (GCF_001753085): β-I WP_010947903.1, β-I WP_010948202.1, β-TM WP_010946434.1, γ WP_010946256.1, γ WP_010946314.1, γ WP_010948561.1, γ WP_011212755.1, γ WP_025862423.1; *N. gonorrhoeae* (GCF_013030075): α WP_003688976.1, β-I WP_003690382.1, γ WP_003687524.1; *N. meningitidis* (GCF_008330805): β-I WP_022191933.1, γ WP_022291164.1, ζ WP_002229295.1; *P. gingivalis* (GCF_000010505): β-I WP_012458351.1, γ WP_012457873.1; *P. aeruginosa* (GCF_000006765): β-I NP_248792.1, β-I NP_250743.1, β-I NP_253365.1, γ NP_248756.1, γ NP_252442.1, γ NP_254227.1, γ NP_254239.1; *R.rickettsii* (GCF_000018225): β-I WP_012150306.1; *S. enterica* (GCF_000006945): β-I NP_459176.1, β-I NP_490536.1, γ NP_459074.1, γ NP_462303.1, γ NP_462761.1; *S. flexneri* (GCF_000006925): β-II NP_706079.1, γ NP_705991.1, γ NP_709067.2, γ NP_709543.1; *V. cholerae* (GCF_008369605): α WP_114736123.1, β-I WP_001114117.1, γ WP_001154339.1, γ WP_001892013.1, γ WP_172626033.1; *Y. pestis* (GCF_000229975) β-I WP_002209342.1, β-I WP_002220090.1, γ WP_002215550.1, γ WP_002215709.1; *B. anthracis* (GCF_000008445): β-I WP_000838159.1, γ WP_000640195.1, γ WP_000679310.1; *C. difficile* (GCF_018885085): β-I WP_003423380.1, γ WP_004454132.1; *C. perfringens* (GCF_020138775): β-I WP_003461171.1, γ WP_003449781.1; *C. diphtheriae* (GCF_021560115): β-I WP_003852865.1, γ WP_003852749.1, γ WP_014318457.1, γ WP_196969453.1; *E. faecalis* (GCF_000393015): α WP_010819414.1, γ WP_002355129.1, γ WP_002356054.1; *E. faecium* (GCF_009734005): α WP_002321477.1, γ WP_002287446.1, γ WP_022289401.1; *L. monocytogenes* (GCF_000196035): α WP_464338.1; *P. sordelli* (GCF_002865995): β-I WP_057536998.1, γ WP_021122495.1, γ WP_101508621.1; *S. aureus* (GCF_000013425); *S. epidermidis* (GCF_006094875): β-I WP_001114117.1, γ WP_001154339.1, γ WP_009738105); *S. mutans* (GCF_009073883): β-I WP_002076863.1; *Borrelia afzelii* (GCF_026571835); *Borrelia burgdorferi* (GCF_000181575); *Borrelia mayonii* (GCF_001936295); *M. leprae* (GCF_003253775): β-I WP010908614.1; *M. tuberculosis* (GCF_000195955): β-I NP_215800.1, β-TM NP_217790.1, β-I NP_218105.1, γ NP_218042.1; *M. abscessus* (GCF_004028015): β-I WP_005057001.1, β-TM WP_005057131.1, β-I WP_005082077.1, β-I WP_005085383.1, β-I WP_005085913.1, β-I WP_005111190.1, β-I WP_021269287.1, γ WP_005085863.1; *N. asteroids* (GCF_900637185): α WP_019047934.1, β-I WP_019047047.1, β-I WP_019046717.1, β-I WP_019047241.1, β-I WP_019050575.1, β-TM WP_019047934.1, β-TM WP_022566411.1, β-TM WP_022567130.1, γ WP_019046020.1, γ WP_030201125 ∴

[a]The first number indicates the number of experimentally proven or probable CAs that contain a catalytic triad. The second number indicates the number of homologous CAs with no detected catalytic triad listed in Table 2.

[b]Includes 1 transmembrane CA.

[c]Included order *Mycobacteriales* containing a mycolic acid layer resembling an outer membrane [36].

[d]Includes 3 transmembrane CAs.

Table 2 The catalytic triad motifs used for predicting probable CAs.

CA class	Catalytic Triad Motif[a]	Selected families/proteins
Alpha	H[A-Z]H[A-Z]{5,20}H	IPR023561, IPR041891, IPR001148
Beta	C[A-Z]{5,60}H [A-Z]{2}C	IPR045066, IPR001765, cd00883, IPR001902
Gamma	H[A-Z]{5,20}H [A-Z]{4}H	IPR047324, cd04645, IPR011004
Iota	T[A-Z]{10,15}Y [A-Z]{30,60}HHSS	WP_063553346.1, 7C5V_A, WP_063553346.1, WP_089190700.1, WP_074176950.1, ECP0341226.1, WP_011036063.1, OYX05505.1
Zeta	C[A-Z]D[A-Z] R.ªH[A-Z]{6,14}C	AAX08632.1, A0A6S8C628, NBQ67882.1, A0A6U3G832, A0A6V2JW04, A0A1F5F3J8

[a]Sequence patterns used to extract the catalytic triad region from each CA predicted in the studied genomes. [A-Z] represents any amino acid, numbers in brackets represent the minimum and maximum number of amino acids.

reported earlier predominant occurrence of α-CAs in Gram-negative bacteria [16], and, instead, their distribution as well as the distribution of the predicted ι- CAs did not reflect phylogenetic relationship or lifestyle, at least in this set of bacteria. With the exception of *Nocardia asteroids* and *Mycobacteroides abscessus,* on average, fewer probable CAs were detected in the genomes of Gram-positive, Gram-neutral and the closely related Gram-negative species, indicating some phylogenetic bias. A larger data set would be necessary to identify its nature.

Both β- and γ-CAs showed the broadest distribution and were present in 40 and 38 genomes, respectively. Both types were absent in *Borellia, Staphylococcus,* and *Listeria* species that, except for *Listeria,* contained no other CAs. In addition, β-CAs were missing in both *Enterococcus* species and *Streptococcus milleri,* whereas γ-CAs were missing in all the included *Streptococcus* species and *Mycobacterium leprae.* So, overall, these enzymes were mostly missing from Gram-positive species and *Borellia,* which, although a Gram-negative, due to its unusual cell wall [43] is closer related to Gram-neutral species (Fig. 1). To investigate a possible relationship between taxonomy and evolution of β- and γ-CAs, we have performed phylogenetic analysis of all the

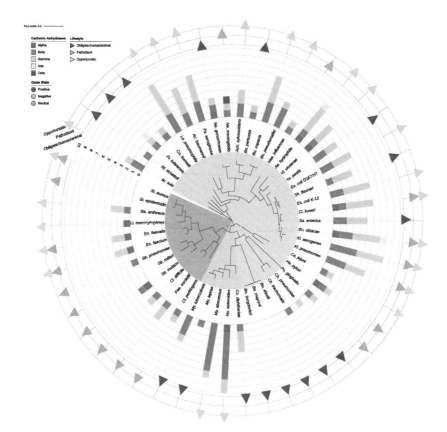

Fig. 1 A maximum-likelihood phylogenetic tree constructed with SSU rRNA sequences of selected bacterial pathogens. Branches are colored according to Gram Stain: Gram Positive shown in purple, Gram Negative—in pink, and Gram Neutral—in gray. As Gram-neutral, representatives of order Mycobacteriales containing a mycolic acid layer resembling an outer membrane [36] were included. The number and type of predicted probable CAs are displayed in bar graphs around the exterior of the tree (Red: Alpha, Blue: Beta, Green: Gamma, Yellow: Iota, Purple: Zeta). The colored triangles represent the lifestyle of each pathogen (Red: Obligate, Orange: Pathobiont, Green: Opportunistic).

predicted probable β- and γ-CAs (Fig. 2A and B). The results indicated that the β-CAs form fewer and more stable clades than the γ-CAs, suggesting that a closer phylogenetic relationship enhanced their distribution. The latter was likely mediated by the horizontal gene transfer, which was suggested in [44] based on the reported localization of β-CAs genes within bacterial genomic islands [45]. Further, β-CAs were shown to cross species boundaries [46] and even predicted to be horizontally transferred from a prokaryotic ancestor to

A

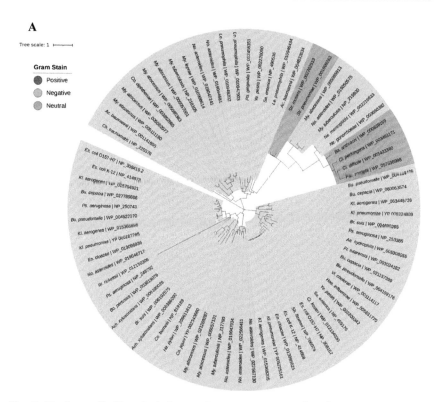

Fig. 2 Maximum-likelihood phylogenetic trees constructed with sequences of predicted probable CAs containing catalytic triads: β-CAs (A), γ-CAs (B). The accession numbers for each CA are provided next to the species names. The full names of the bacterial pathogens are listed in Table 1. Branches are colored according to Gram Stain: Gram Positive shown in purple, Gram Negative—in pink, and Gram Neutral—in gray. As Gram-neutral, representatives of order Mycobacteriales containing a mycolic acid layer resembling an outer membrane [36] were included.

several protists [45]. On the other hand, γ-CAs appeared to form smaller and less distinct clades (Fig. 2B), which may suggest that γ-CAs either cross the taxonomic boundaries easier or have longer evolutionary roots and evolved from a common ancestor. Another intriguing observation is that γ-CAs have the largest number of homologs with no catalytic triad (second number in Table 1), which may indicate that these homologous proteins are evolving either away or towards the CA activity. The second largest non–triad group of homologs consists of transmembrane β-CAs that mostly include sulfate transporters, suggesting their evolutionary relationship.

B

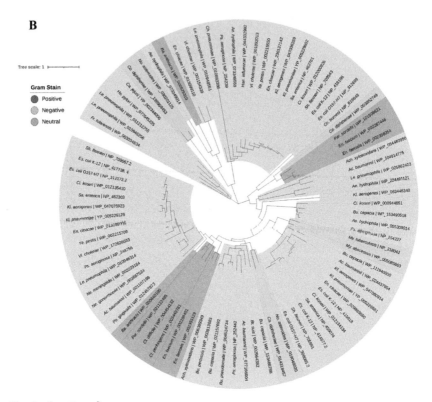

Fig. 2 (*continued*)

In addition to CAs, provision of the required CO_2/bicarbonate for carboxylation reactions in central metabolism and biosynthesis of small molecules can be achieved by the membrane potential-generating system (MpsAB). As observed in [47], CAs and Mps rarely coexist in a given species. It has been suggested that the bicarbonate transporter MpsAB may have an advantage in species where CO_2 diffusion is low, exemplified by mucus- and biofilm-forming bacteria, though the co-expression of MpsAB and CA did not confer any growth benefits, even under stress conditions [48]. Here, we have searched for homologs of MpsAB among the genomes of selected pathogens. In partial agreement with the earlier observation, two homologs of MpsAB were predicted in *Staphylococcus* species that do not contain CAs, however the other homologs were present in the genomes encoding at least one CA (Table 3; Fig. 1).

Overall, different structural and functional characteristics of CAs support their potentially different evolutionary origins and convergent evolution. A better understanding of the evolution and sequence conservation

Table 3 Predicted MpsAB homologs.

Organism	Genome Acc	Protein name	Protein Acc	e-value	%
MpsA					
Staphylococcus aureus	GCF_000013425	MpsA	YP_498998..	0	100
Staphylococcus epidermidis	GCF_006094375	NADH dehydrogenase	WP_001829413.1	0	61
Bacillus anthracis	GCF_000008445	NADH dehydrogenase	WP_000908909.1	1.00E-173	50
Legionella pneumophila	GCF_001753085	Proton-conducting transporter	WP_010945983.1	7.00E-49	39
Vibrio cholerae	GCF_008369605	NADH-quinone oxidoreductase	WP_149560317.1	2.00E-32	29
MpsB					
Staphylococcus aureus	GCF_000013425	MpsB	YP_498999.1	0	100
Staphylococcus epidermidis	GCF_006094375	MpsB	WP_002456832.1	0	53
Bacillus anthracis	GCF_000008445	Inorganic carbon transporter	WP_000026985.1	0	46
Legionella pneumophila	GCF_001753085	DUF2309 protein	WP_010945932.1	8.00E-124	41
Vibrio cholerae	GCF_008369605	YbcC family protein	WP_C00993432.1	2.00E-82	28
Rickettsia rickettsii	GCF_000018225	YdcF family protein	WP_012151464.1	2.00E-06	31

of bacterial CAs would provide an important insight into their functional roles and promote further studies of their potential as therapeutic targets. Elucidating specific functions of multiple CA homologs would guide our understanding of apparent functional redundancy. Furthermore, large-scale studies of the differences in the distribution of CAs among pathogenic and non-pathogenic microbes as well as eukaryotic hosts can be instrumental in targeting for therapeutics.

3. Importance of carbonic anhydrases in physiology of bacterial pathogens

Since CO_2 and HCO_3^- feed directly into the biogenesis of organic matter, the role of CAs in bacterial carbon metabolism cannot be over-estimated. This is particularly true for autotrophic bacteria. Cyanobacteria, the most abundant photosynthetic organism on the planet [49], rely on their ability to fix carbon, which requires raising their intracellular inorganic carbon concentration and demands the activity of β- and γ-class carbonic anhydrases residing within carboxysome [50]. In addition to their role in photosynthetic CO_2 fixation [50], bacterial CAs also contribute to CO_2 fixation performed by non-photosynthetic chemolithotrophs [51]. On the other hand, the decomposition of organic matter by prokaryotic hetero-trophs is just as essential for the global carbon cycle. The importance of several characterized CAs has been reported for aerobic and anaerobic het-erotrophs, exemplified by *Ralstonia eutropha* [52], *Geobacillus kaustophilus* [53], *Escherichia coli* [54], and *Clostridium perfringens* [55].

For human pathogens, the task of survival in a host is fundamentally similar to that in any other environment. It includes recognizing the environment they are in and adjusting their metabolic capacities to enable uptake of available nutrients and protection from threatening factors. Although the specifics of such conditions differ in a human body as a host, the basic requirements, such as carbon availability and pH remain essential. Due to their catalytic activity, CAs can enable both bicarbonate/CO_2 acquisition and removal. These molecules serve as metabolic substrates for central metabolism, as well as biosynthesis of various molecules, including purines, pyrimidines, fatty acids [56]. Although some small molecules could be acquired from the environment, the need for CO_2 for central metabolism and the function of several housekeeping enzymes such as phosphoe-nolpyruvate carboxylase, carbamoyl phosphate synthase, and acetyl-CoA carboxylase [57], cannot be replaced with supplements and the major known pathway for supplying bicarbonate is *via* hydration of CO_2 [56]. This also secures membrane-impermeable bicarbonate inside the cell. Clearly, the need is greater during growth at low CO_2 when spontaneous hydration is not sufficient. This agrees with the observed increased demand in CA activity during growth at ambient air exemplified by β-CAs Can in *E. coli* [56,58], psCA1 in *P. aeruginosa* [23], PCA in *S. pneumoniae* [59], Bca in *Corynebacterium glutamicum* [60], Can in *Ralstonia eutropha* [52], and supported by the differential regulation of CAs abundance at different levels of CO_2 [23,61]. Considering the requirement of CAs for survival at low CO_2, it was suggested that CAs can be dispensable in bacterial species residing at high CO_2, for example, soil, seawater, and intestine [30]. Nevertheless, the CAs' importance was illustrated for the physiology and survival of human pathogens residing at elevated CO_2 levels within a human body.

One example of the earlier established significance of CAs for bacterial survival in a human host relates to the role of membrane α-CA in *Helicobacter pylori* encoded by *hpαCA* in acid acclimation [28]. The expression of the gene is drastically increased during exposure to acidic pH and its deletion caused major loss of the pathogen's viability at pH 4 but not at neutral pH [62]. It was suggested that the CA converts CO_2 produced by urease to bicarbonate, buffering the periplasmic pH. Importantly, treatments of peptic ulcers with CA inhibitor acetazolamide reduced the reoccurrence of the infection to about 6%, indicating the necessity of CA for *H. pylori* colonization and infection [63,64].

Metabolic diversification is a potent strategy for bacterial survival in nutrient-challenging environments. The mechanisms enabling metabolism of a broader range of substrates significantly enhance fitness of bacteria. Several studies outlined the role of CAs in metabolic empowering of their hosts. For example, a β-CA CynT of *E. coli* enables the utilization of cyanate as a nitrogen source [22,65]. A γ-CA CaiE in *E. coli* contributes to carnitine metabolism by supporting activities of carnitine dehydratase and rasemase [66,67]. Another γ-CA, PaaY in *A. baumannii* is involved in the phenylacetic acid catabolic pathway [68]. If the metabolic needs are not met, bacteria begin to experience starvation. Considering nutrients withdrawal is a common host defense strategy, defined as nutritional immunity [69], invading pathogens frequently face starvation and therefore, evolved mechanisms to overcome it. There is evidence that β-CA Rv1284 in *M. tuberculosis* is significantly upregulated during starvation, suggesting its role in the pathogen's persistence within the host [70].

The increased abundance and activity of CA Can was shown to be required to provide bicarbonate for carboxylases in *de-novo* biosynthesis of fatty acids in *E. coli* [56]. In agreement, the observed essential role of β-CA PCA in *S. pneumoniae* in growth in ambient air was reversed when either bicarbonate or unsaturated fatty acids were supplemented into the media [24]. Similarly, the role of supplying bicarbonate to carboxylases involved in volatile fatty acids and amino acid production was attributed to β-CA Cpb *Clostridium perfringens* [55]. The deletion of *cpb* disabled growth on semi-defined medium in the absence of CO_2.

The observation that an increase in cell density, transition into stationary phase, and exposure to elevated temperature have a regulatory impact on the abundance of Can in *E. coli* [56] suggest that factors not directly reflecting the availability of bicarbonate/CO_2 also control the expression of the CA. The involvement of RpoS, a sigma factor regulating a general stress response [71], in this regulation indicates that the CA activity is important for *E. coli* survival during stresses commonly associated with stationary phase, including for example, nutrient deprivation, acid and osmotic shock. Since the physiological state of stationary phase cells is similar to that of cells within biofilms, surface-associated structured bacterial communities embedded into extracellular matrix [72], it is likely that CAs are in higher demand and therefore upregulated in biofilms. This can be explained by the reduced diffusion of CO_2 through biofilm matrix, exemplified in *S. aureus* biofilms involving polysaccharide intercellular adhesin (PIA) [48]. In agreement, according to the GEO profiles, the expression of *P. aeruginosa* psCA3 was significantly higher in biofilms (GDS3251), whereas psCA1 and psCA2 were significantly upregulated in alginate overproducing mucoid strain (GDS3174). Further supporting evidence showed an impairment of *H. pylori* biofilm growth in response to selective inhibition of CAs [73]. Furthermore, deletion of γ-CA PaaY impaired biofilm formation in *Acinetobacter baumannii* [68].

The role of CAs in enhancing formation of bacterial biofilms is important from several perspectives. Biofilm formation plays a particularly important role in host colonization and establishment of chronic infections. This impact can be mediated by CAs controlling the pH microenvironment and providing carbonate for calcium mineralization. The latter has been demonstrated for psCA1 in *P. aeruginosa* [23], where the formation of calcium deposits within biofilms likely supported the structural integrity and therefore enhanced biofilm development. In *Klebsiella pneumoniae*, CA activity was also recognized for its role in extracellular and intracellular depositing of calcium and magnesium [74]. Bacterial capsular poly-saccharides (CPs) are major components of biofilm matrix and play an important role in pathogenicity of oral streptococci [75]. CA activity was shown to enhance growth and CP production in *S. anginosus* likely due to forming bicarbonate [75]. It was also reported that bicarbonate increased pH-dependent export of eDNA, another major component of biofilm matrix, in *M. avium, M. abscessus,* and *M. chelonae* [76]. Besides, the β-CA MAVA5_02375 was among the most abundant proteins within surface-exposed proteome of *M. avium* biofilm. Significant reduction of eDNA release upon ethoxzolamide (EZA) treatment further supported the key role of CA activity in this process.

Finally, biofilm communities are recognized for their elevated resistance to different stresses and antibiotics [77], and CAs may serve as one of the contributing factors. In support, deletion of *paaY* decreased biofilm formation and impaired H_2O_2 resistance in *A. baumannii* during grwoth in the presence of phenylacetic acid [68]. Since increased antibiotic resistance can also be achieved by reducing a growth rate [78], the inverse relationship between β-CA Can expression and growth rate in *E. coli* [56] suggests at least an indirect link between CA activity and growth rate, and therefore, resistance. Another intriguing link between CA activity and antibiotic resistance was discovered in *A. baumannii,* where β-CA CanB plays an important role in survival of the strain overexpressing β-lactamase [79]. The authors also showed a decrease in imipenem resistance in the mutant lacking *canB* and related the impact to the role of CA in providing bicarbonate to synthesis of precursors required for peptidoglycan bio-synthesis, the target of β-lactams. Another link to antibiotic resistance was discovered for β-CA Mig5 in *Salmonella enterica* [80]. This protein enables a proper function of the proton motive force (PMF), which is important for a variety of cellular functions, including twin-arginine translocation (Tat) system secreting, among other molecules, enzymes remodeling pepti-doglycan during division. The deletion of *mig5* disrupted PMF and ren-dered cells sensitive to peptidoglycan-targeting antibiotics.

When studying the biology of newly identified CAs, understanding the genomic environment of the encoding genes may serve as potent predictor of the processes that CAs are functionally involved in. Several examples include the earlier discussed CAs: CynT of *E. coli* resides in the operon with cyanase and found to be important in utilizing cyanate as a nitrogen source [22,65]. CaiE, a γ-CA encoded by *E. coli* is situated in the operon encoding proteins involved in carnitine metabolism. It was shown to play a role in carnitine dehydratase and rasemase activities through either synthesis or activation of a required cofactor [66,67]. Another γ-CA, PaaY encoded within a *paa* gene cluster in *A. baumannii* was suggested to control, together with PaaX, the expression of all *paa* genes in a phenylacetic acid–CoA dependent manner [81]. PaaY has thioesterase activity that enables the breakdown of toxic metabolites through the phenylacetic acid pathway [68]. Finally, the encoding gene of β-CA Mig5 in *S. enterica* belongs to macrophage-induced genes and resides within a pathogenic island. Accordingly, its expression drastically increased during invasion into macrophages suggesting the role in pathogenicity [82].

Overall, the catalytic activity of CAs has broad implications in different physiological processes of bacterial pathogens, increasing their fitness and survival in the host. However, considering the abundance and diversity of CAs in bacterial pathogens, a much deeper understanding of their functional involvement is needed.

4. Role in bacterial virulence and host-pathogen interactions

Life-threatening pathogenic bacteria did not evolve to kill, but to survive, which may require damaging the host to silence its protection and provide nutrients. Therefore, the ability to survive in a host can be considered as one of the most potent virulence factors. Since CAs help bacterial pathogens to survive in specific host niches, they help establish and maintain infections and thus, play a key role in virulence.

Reflecting the importance of CAs for survival within a host, multiple evidence supports the increased expression of CAs during infection. For example, the expression of β-CA Mig5 was significantly increased in *S. typhimurium* internalized by macrophages as well as other cell lines [82]. It is regulated by PhoPQ, two component regulatory system controlling the expression of a large regulon in response to transition from an extracellular environment to a subcellular location [83]. Among major PhoPQ regulatory targets are the mechanisms of outer membrane permeability [84] and virulence [85]. The periplasmic α-CA in *H. pylori* is regulated by two component system ArsRS (HP0165/0166) in acidic environments [86]. This TCS was shown to be essential for competitive colonization of the pathogen in mice [87]. Further, according to the expression data available at the Gene Expression Omnibus (GEO) data repository, the transcription of two *P. aeruginosa* β-CAs psCA1 and psCA2 was increased in isolates from cystic fibrosis lung sputa grown *in vivo* (GDS2869). The two predicted γ-CAs encoded by PA3753 and PA5540 were upregulated during *P. aeruginosa* interactions with airway epithelia (GDS2502).

The activity of CAs may influence the expression of other potent virulence factors. One example is the case, where inhibition of CA activity in *Vibrio cholerae* by EZA inhibited bicarbonate-dependent expression of the genes encoding two key virulence factors, cholera toxin and toxin-regulated pilus [88]. Similarly, bicarbonate enhances the expression of the endocarditis- and biofilm-associated pilus locus *ebpR-ebpABC* in *Enterococcus faecalis,* which is required for a full scale of virulence in rat endocarditis and murine urinary tract infection models [89]. These findings suggest that conversion of CO_2 to bicarbonate by CAs enhances bacterial virulence. Another interesting example is the dual role of *E. coli* γ-CA, PaaY, with detected thioesterase activity involved in catabolism of phenylacetic acid [81]. This pathway plays an important role in virulence and antibiotic resistance [90]. At the metabolic level, PaaY is required for detoxification of some CoA derivatives potentially inhibiting the first steps of the *paa* pathway [91]. In addition, PaaY facilitates the induction of the *paa* genes whose expression is inhibited by the transcriptional repressor PaaX, possibly by reducing the amount of antagonistic CoA-derived intermediates produced during phenylacetic acid catabolism [81].

Several CAs have been reported to enhance bacterial survival during infection. The cytoplasmic β-CA and periplasmic α-CA of *H. pylori* are essential for acid acclimation that enables colonization and hence virulence of the pathogen [62,92,93]. In agreement, the deletion of one or both CAs showed decreased colonization as well as host inflammation [92]. The deletion of Mig5 in *S. typhimurium* also led to lower colonization in mice spleens [82]. The activity of PCA, the β-CA in *S. pneumoniae* was required for intracellular survival in host cell lines: J774 and human brain microvascular endothelial cells (HBMEC) [24].

CAs of bacterial pathogens have been implicated in various aspects of host-pathogen interactions, including modulation of the immune response that defines the ultimate outcome of the infection. One interesting case is the role of *H. pylori* CAs and urease in the detoxification by CO_2 of $ONOO^-$ produced by macrophages NO synthase [27]. However, as highlighted in [94], the formation of $ONOOCO_2^-$ may potentially be hazardous to *H. pylori,* and the benefits of CO_2-dependent survival of the pathogen [95] may be explained by CO_2 potentiating other mechanisms, for example, increasing the expression of α- and β-CAs as well as urease in *H. pylori* that act synergistically to adjust the periplasmic pH and enable colonization [96]. Another aspect of CA role in shaping the immune responses is their contributions to the physiology and survival of commensal host microbiota. Our understanding of the commensal microbiomes and their importance in the innate and adaptive immunity is rapidly growing [97–99]. However, no direct studies addressed the role of CAs in community homeostasis and relationship with the host of commensal bacteria. Interestingly, the analysis of CAs distribution among selected commensals suggested the lack of α-CAs [29], potentially indicating a particular evolutionary adaptation.

While a considerable body of research has been done to elucidate the role of bacterial CAs in host–pathogen interactions, there is still a significant lack of knowledge, and many CAs have not been studied at all. Among challenges, apparent redundancy with multiple homologs of the same or different CA classes encoded in bacterial genomes is a compounding factor. It is important for future research to focus on mechanistic and regulatory aspects of CAs contributing to pathogens' invasiveness and interactions with host and commensal communities. Additionally, the mechanistic interplay between CAs and other virulence factors in the context of both pathogen and host signaling and regulatory networks requires attention. A better understanding of the role of CAs in host–pathogen interactions will guide the development of therapeutic interventions.

5. Potential in therapeutic implications

It has been recognized that bacterial CAs emerged as promising targets for developing antimicrobial therapies [100,101]. This is primarily based on two key factors: evidence of the CAs importance for survival and virulence of several bacterial pathogens and failure of currently available antibiotics. Some distinct sequence, structural, and functional characteristics of bacterial CAs *vs* human homologs support the potential for selectivity of inhibition and minimizing the off-target effects [102].

It is understood that the reason for antibiotic failure is the strength of evolutionary adaptation leading to survival and causing a rapid spread of antibiotic-resistance [103], which shortens the lifespan of antibiotics and de-incentivizes the development of novel antibiotics [104]. An alternative strategy is to target, instead of essential mechanisms, virulence factors aiming to only "disarm" pathogens. This promises to not affect the overall fitness of bacterial pathogens and, therefore, to not trigger evolution of resistance [105,106]. However, as outlined in [107], the differential expression of virulence factors within a host can provide some fitness benefits, the reduction of which may trigger selection for resistance. This highlights the necessity of generating more extended knowledge on the regulation of virulence factors' expression at a site of infection and their role in *in vivo* fitness. This is particularly true for CAs of bacterial pathogens, where such knowledge is limited. Since we are currently in the era of RNA sequencing, a growing body of transcriptional data may serve well for this purpose.

As outlined above, CAs have been recognized for enhancing the ability of bacteria to survive host-related conditions and can be considered as virulence factors. Virulence factors are not expected to be highly expressed or essential in *in vitro* growth studies, especially those using regular lab media. Therefore, when choosing to target CAs as a virulence factor, a different screening procedure shall be in place. One step in the direction of clinical relevance is using the media formulated to mimic the nutritional composition of infection sites, for example, synthetic cystic fibrosis medium (SCFM) [108]. Considering the enzymatic specialization of CAs, the appropriate levels of CO_2 and pH shall reflect those in a human body, in general, or infection site, in particular. More experimentally challenging but insightful can be testing the role of CAs in host-pathogen interactions, using cell cultures or animal models.

A number of prokaryotic β-class CAs that share no sequence similarity with human CA isoenzymes have been recognized as promising drug targets for antimicrobial treatments [109,110]. Some of them have been tested in *in vitro* inhibition studies, including CAs from *E. coli* [111], *H. pylori* [112], *S.* Typhimurium [113], *S. pneumonia* [114], *H. influenzae* [115], *Brucella suis* [109,116–118], and *M. tuberculosis* [119]. However, as outlined in [100], the antibacterial effects of their inhibition were scarcely validated. Further research is needed to consider the expression pattern, optimize inhibitor design, improve selectivity, and assess the potential for resistance development. Since bacterial CAs are involved in multiple pathways, targeting them in combination with conventional antibiotics can offer a great potential to provide antibacterial efficacy and overcome the problem of resistance.

6. Remaining unknowns and challenges

Although many studies have been reported on bacterial CAs, several key aspects remain severely understudied, some of which are outlined below.

The functional redundancy of bacterial CAs is not fully understood and presents excellent examples of convergence and parallelism in protein evolution. It is unclear how different isoforms of these enzymes are regulated during infection and contribute to survival of their hosts at different stages and types of infections. The regulatory mechanisms controlling CA expression in bacterial pathogens are largely unknown. Elucidating the host cues that trigger their expression is an important prerequisite for understanding their functional roles during invasion and infection. Since the levels of CO_2 and pH play important roles in the physiology of a human body, their CA-dependent changes may trigger signaling cascades and drive host-pathogen interactions. The activity of bacterial CAs may also have a major impact on immune responses and pathogenic cellular processes in a human host defining the outcome of infection. Bearing in mind that commensal bacteria also encode CAs, their roles in keeping the host healthy shall not be overlooked, especially when designing inhibitors. The importance of the potential functional link between bacterial CAs and antibiotic resistance cannot be overestimated and requires future in-depth studies. Finally, detailed biochemical and structural studies of bacterial CAs are imperative for rationale design of highly selective and effective inhibitors.

References

[1] C.T. Supuran, Structure and function of carbonic anhydrases, Biochem. J. 473 (2016) 2023–2032.

[2] T.W. Lane, M.A. Saito, G.N. George, I.J. Pickering, R.C. Prince, F.M. Morel, Biochemistry: a cadmium enzyme from a marine diatom, Nature 435 (2005) 42.

[3] J.G. Ferry, The gamma class of carbonic anhydrases, Biochim. Biophys. Acta 1804 (2010) 374–381.

[4] G. Ren, Y. Wang, J. Qin, J. Tang, X. Zheng, Y. Li, Characterization of a novel carbonic anhydrase from freshwater pearl mussel Hyriopsis cumingii and the expression profile of its transcript in response to environmental conditions, Gene 546 (2014) 56–62.

[5] A. Nocentini, C.T. Supuran, C. Capasso, An overview on the recently discovered iota-carbonic anhydrases, J. Enzyme Inhib. Med. Chem. 36 (2021) 1988–1995.

[6] Y. Hirakawa, M. Senda, K. Fukuda, H.Y. Yu, M. Ishida, M. Taira, et al., Characterization of a novel type of carbonic anhydrase that acts without metal cofactors, BMC Biol. 19 (2021) 105.

[7] R.E. Tashian, D.P. Douglas, Y.S. Yu, Esterase and hydrase activity of carbonic anhydrase. I. From primate erythrocytes, Biochem. Biophys. Res. Commun. 14 (1964) 256–261.

[8] C.T. Supuran, Carbonic anhydrases, in: C.T. Supuran, W.A. Donald (Eds.), Metalloenzymes, Elsevier Inc, 2024, pp. 139–156.

[9] S.M. Gould, D.S. Tawfik, Directed evolution of the promiscuous esterase activity of carbonic anhydrase II, Biochemistry 44 (2005) 5444–5452.

[10] V. Alterio, E. Langella, M. Buonanno, D. Esposito, A. Nocentini, E. Berrino, et al., Zeta-carbonic anhydrases show CS(2) hydrolase activity: a new metabolic carbon acquisition pathway in diatoms? Comput. Struct. Biotechnol. J. 19 (2021) 3427–3436.

[11] K.S. Smith, J.G. Ferry, Prokaryotic carbonic anhydrases, FEMS Microbiol. Rev. 24 (2000) 335–366.

[12] A. Aspatwar, M.E.E. Tolvanen, H. Barker, L. Syrjanen, S. Valanne, S. Purmonen, et al., Carbonic anhydrases in metazoan model organisms: molecules, mechanisms, and physiology, Physiol. Rev. 102 (2022) 1327–1383.

[13] S. Del Prete, A. Nocentini, C.T. Supuran, C. Capasso, Bacterial iota-carbonic anhydrase: a new active class of carbonic anhydrase identified in the genome of the Gram-negative bacterium Burkholderia territorii, J. Enzyme Inhib. Med. Chem. 35 (2020) 1060–1068.

[14] E.L. Jensen, R. Clement, A. Kosta, S.C. Maberly, B. Gontero, A new widespread subclass of carbonic anhydrase in marine phytoplankton, ISME J. 13 (2019) 2094–2106.

[15] G. De Simone, A. Di Fiore, C. Capasso, C.T. Supuran, The zinc coordination pattern in the eta-carbonic anhydrase from Plasmodium falciparum is different from all other carbonic anhydrase genetic families, Bioorg Med. Chem. Lett. 25 (2015) 1385 1389.

[16] C. Capasso, C.T. Supuran, An overview of the alpha-, beta- and gamma-carbonic anhydrases from Bacteria: can bacterial carbonic anhydrases shed new light on evolution of bacteria? J. Enzyme Inhib. Med. Chem. 30 (2015) 325–332.

[17] F. Shalileh, M.S. Gheibzadeh, J.R. Lloyd, S. Fietz, H. Shahbani Zahiri, R. Zolfaghari Emameh, Evolutionary analysis and quality assessment of zeta-carbonic anhydrase sequences from environmental microbiome, J. Basic. Microbiol. 63 (2023) 1412–1425.

[18] R.S. Kumar, J.G. Ferry, Prokaryotic carbonic anhydrases of Earth's environment, Subcell. Biochem. 75 (2014) 77–87.

[19] G.D. Price, J.R. Coleman, M.R. Badger, Association of carbonic anhydrase activity with carboxysomes isolated from the Cyanobacterium Synechococcus PCC7942, Plant. Physiol. 100 (1992) 784–793.

[20] S. Bedu, F. Joset, Studies on the carbonic anhydrase activity in Synechocystis PCC6803 wild type and an acetazolamide-resistant mutant, Can. J. Bot. 69 (1991) 1103–1108.

[21] E. Soltes-Rak, M.E. Mulligan, J.R. Coleman, Identification and characterization of a gene encoding a vertebrate-type carbonic anhydrase in cyanobacteria, J. Bacteriol. 179 (1997) 769–774.

[22] E.I. Kozliak, J.A. Fuchs, M.B. Guilloton, P.M. Anderson, Role of bicarbonate/CO_2 in the inhibition of Escherichia coli growth by cyanate, J. Bacteriol. 177 (1995) 3213–3219.

[23] S.R. Lotlikar, B.B. Kayastha, D. Vullo, S.S. Khanam, R.E. Braga, A.B. Murray, et al., Pseudomonas aeruginosa beta-carbonic anhydrase, psCA1, is required for calcium deposition and contributes to virulence, Cell Calcium 84 (2019) 102080.

[24] P. Burghout, L.E. Cron, H. Gradstedt, B. Quintero, E. Simonetti, J.J. Bijlsma, et al., Carbonic anhydrase is essential for Streptococcus pneumoniae growth in environmental ambient air, J. Bacteriol. 192 (2010) 4054–4062.

[25] D.J. Culp, B. Robinson, S. Parkkila, P.W. Pan, M.N. Cash, H.N. Truong, et al., Oral colonization by Streptococcus mutans and caries development is reduced upon deletion of carbonic anhydrase VI expression in saliva, Biochim. Biophys. Acta 1812 (2011) 1567–1576.

[26] A. Walz, S. Odenbreit, K. Stuhler, A. Wattenberg, H.E. Meyer, J. Mahdavi, et al., Identification of glycoprotein receptors within the human salivary proteome for the lectin-like BabA and SabA adhesins of Helicobacter pylori by fluorescence-based 2-D bacterial overlay, Proteomics 9 (2009) 1582–1592.

[27] A.P. Gobert, K.T. Wilson, The immune battle against *Helicobacter pylori* infection: NO offense, Trends Microbiol. 24 (2016) 366–376.

[28] M.E. Compostella, P. Berto, F. Vallese, G. Zanotti, Structure of alpha-carbonic anhydrase from the human pathogen Helicobacter pylori, Acta Crystallogr. F. Struct. Biol. Commun. 71 (2015) 1005–1011.

[29] A. Amedei, C. Capasso, G. Nannini, C.T. Supuran, Microbiota, bacterial carbonic anhydrases, and modulators of their activity: Links to human diseases? Mediators Inflamm. 2021 (2021) 6926082.

[30] K. Ueda, H. Nishida, T. Beppu, Dispensabilities of carbonic anhydrase in proteo-bacteria, Int. J. Evol. Biol. 2012 (2012) 324549.

[31] T.W. Lane, F.M. Morel, Regulation of carbonic anhydrase expression by zinc, cobalt, and carbon dioxide in the marine diatom Thalassiosira weissflogii, Plant. Physiol. 123 (2000) 345–352.

[32] Y. Xu, L. Feng, P.D. Jeffrey, Y. Shi, F.M. Morel, Structure and metal exchange in the cadmium carbonic anhydrase of marine diatoms, Nature 452 (2008) 56–61.

[33] V. Alterio, E. Langella, F. Viparelli, D. Vullo, G. Ascione, N.A. Dathan, et al., Structural and inhibition insights into carbonic anhydrase CDCA1 from the marine diatom Thalassiosira weissflogii, Biochimie 94 (2012) 1232–1241.

[34] C.C. Murdoch, E.P. Skaar, Nutritional immunity: the battle for nutrient metals at the host-pathogen interface, Nat. Rev. Microbiol. 20 (2022) 657–670.

[35] M. Toyofuku, S. Schild, M. Kaparakis-Liaskos, L. Eberl, Composition and functions of bacterial membrane vesicles, Nat. Rev. Microbiol. 21 (2023) 415–430.

[36] P.J. Brennan, H. Nikaido, The envelope of mycobacteria, Annu. Rev. Biochem. 64 (1995) 29–63.

[37] T. Paysan-Lafosse, M. Blum, S. Chuguransky, T. Grego, B.L. Pinto, G.A. Salazar, et al., InterPro in 2022, Nucleic Acids Res. 51 (2023) D418–D427.

[38] F. Sievers, D.G. Higgins, Clustal Omega for making accurate alignments of many protein sequences, Protein Sci. 27 (2018) 135–145.

[39] S.C. Potter, A. Luciani, S.R. Eddy, Y. Park, R. Lopez, R.D. Finn, HMMER web server: 2018 update, Nucleic Acids Res. 46 (2018) W200–W204.

[40] D.H. Parks, M. Chuvochina, C. Rinke, A.J. Mussig, P.A. Chaumeil, P. Hugenholtz, GTDB: an ongoing census of bacterial and archaeal diversity through a phylogen-etically consistent, rank normalized and complete genome-based taxonomy, Nucleic Acids Res. 50 (2022) D785–D794.

[41] S. Kumar, G. Stecher, M. Li, C. Knyaz, K. Tamura, MEGA X: molecular evolu-tionary genetics analysis across computing platforms, Mol. Biol. Evol. 35 (2018) 1547–1549.

[42] I. Letunic, P. Bork, Interactive tree of life (iTOL) v6: recent updates to the phylo-genetic tree display and annotation tool, Nucleic Acids Res. (2024).

[43] T.G. DeHart, M.R. Kushelman, S.B. Hildreth, R.F. Helm, B.L. Jutras, The unusual cell wall of the Lyme disease spirochaete Borrelia burgdorferi is shaped by a tick sugar, Nat. Microbiol. 6 (2021) 1583–1592.

[44] C.T. Supuran, C. Capasso, New light on bacterial carbonic anhydrases phylogeny based on the analysis of signal peptide sequences, J. Enzyme Inhib. Med. Chem. 31 (2016) 1254–1260.

[45] R. Zolfaghari Emameh, H.R. Barker, V.P. Hytonen, S. Parkkila, Involvement of beta-carbonic anhydrase genes in bacterial genomic islands and their horizontal transfer to protists, Appl. Environ. Microbiol. 84 (2018).

[46] R. Zolfaghari Emameh, H.R. Barker, M.E. Tolvanen, S. Parkkila, V.P. Hytonen, Horizontal transfer of beta-carbonic anhydrase genes from prokaryotes to protozoans, insects, and nematodes, Parasit. Vectors 9 (2016) 152.

[47] S.H. Fan, P. Ebner, S. Reichert, T. Hertlein, S. Zabel, A.K. Lankapalli, et al., MpsAB is important for Staphylococcus aureus virulence and growth at atmospheric CO(2) levels, Nat. Commun. 10 (2019) 3627.

[48] S.H. Fan, M. Matsuo, L. Huang, P.M. Tribelli, F. Gotz, The MpsAB bicarbonate transporter is superior to carbonic anhydrase in biofilm-forming bacteria with limited CO(2) diffusion, Microbiol. Spectr. 9 (2021) e0030521.

[49] F. Partensky, W.R. Hess, D. Vaulot, Prochlorococcus, a marine photosynthetic prokaryote of global significance, Microbiol. Mol. Biol. Rev. 63 (1999) 106–127.

[50] M.S. Kimber, Carboxysomal carbonic anhydrases, Subcell. Biochem. 75 (2014) 89–103.

[51] S. Nakagawa, Z. Shtaih, A. Banta, T.J. Beveridge, Y. Sako, A.L. Reysenbach, Sulfurihydrogenibium yellowstonense sp. nov., an extremely thermophilic, facultatively heterotrophic, sulfur-oxidizing bacterium from Yellowstone National Park, and emended descriptions of the genus Sulfurihydrogenibium, Sulfurihydrogenibium subterraneum and Sulfurihydrogenibium azorense, Int. J. Syst. Evol. Microbiol. 55 (2005) 2263–2268.

[52] B. Kusian, D. Sültemeyer, B. Bowien, Carbonic anhydrase is essential for growth of Ralstonia eutropha at ambient CO(2) concentrations, J. Bacteriol. 184 (2002) 5018–5026.

[53] H. Takami, S. Nishi, J. Lu, S. Shimamura, Y. Takaki, Genomic characterization of thermophilic Geobacillus species isolated from the deepest sea mud of the Mariana Trench, Extremophiles 8 (2004) 351–356.

[54] H.M. Park, J.H. Park, J.W. Choi, J. Lee, B.Y. Kim, C.H. Jung, et al., Structures of the gamma-class carbonic anhydrase homologue YrdA suggest a possible allosteric switch, Acta Crystallogr. D. Biol. Crystallogr 68 (2012) 920–926.

[55] R.S. Kumar, W. Hendrick, J.B. Correll, A.D. Patterson, S.B. Melville, J.G. Ferry, Biochemistry and physiology of the beta class carbonic anhydrase (Cpb) from Clostridium perfringens strain 13, J. Bacteriol. 195 (2013) 2262–2269.

[56] C. Merlin, M. Masters, S. McAteer, A. Coulson, Why is carbonic anhydrase essential to Escherichia coli? J. Bacteriol. 185 (2003) 6415–6424.

[57] J. Aguilera, J.P. Van Dijken, J.H. De Winde, J.T. Pronk, Carbonic anhydrase (Nce103p): an essential biosynthetic enzyme for growth of Saccharomyces cerevisiae at atmospheric carbon dioxide pressure, Biochem. J. 391 (2005) 311–316.

[58] M.B. Guilloton, A.F. Lamblin, E.I. Kozliak, M. Gerami-Nejad, C. Tu, D. Silverman, et al., A physiological role for cyanate-induced carbonic anhydrase in Escherichia coli, J. Bacteriol. 175 (1993) 1443–1451.

[59] P. Burghout, L.E. Cron, H. Gradstedt, B. Quintero, E. Simonetti, J.J.E. Bijlsma, et al., Carbonic anhydrase is essential for Streptococcus pneumoniae growth in environmental ambient air, J. Bacteriol. 192 (2010) 4054–4062.

[60] S. Mitsuhashi, J. Ohnishi, M. Hayashi, M. Ikeda, A gene homologous to ß-type carbonic anhydrase is essential for the growth of Corynebacterium glutamicum under atmospheric conditions, Appl. Microbiol. Biot. 63 (2004) 592–601.

[61] S. Fujiwara, H. Fukuzawa, A. Tachiki, S. Miyachi, Structure and differential expression of two genes encoding carbonic anhydrase in Chlamydomonas reinhardtii, Proc. Natl. Acad. Sci. U. S. A. 87 (1990) 9779–9783.

[62] E.A. Marcus, A.P. Moshfegh, G. Sachs, D.R. Scott, The periplasmic alpha-carbonic anhydrase activity of Helicobacter pylori is essential for acid acclimation, J. Bacteriol. 187 (2005) 729–738.

[63] S. Valean, R. Vlaicu, I. Ionescu, Treatment of gastric ulcer with carbonic anhydrase inhibitors, Ann. N. Y. Acad. Sci. 429 (1984) 597–600.

[64] J.G. Penston, K.E. McColl, Eradication of Helicobacter pylori: an objective assessment of current therapies, Br. J. Clin. Pharmacol. 43 (1997) 223–243.

[65] M.B. Guilloton, J.J. Korte, A.F. Lamblin, J.A. Fuchs, P.M. Anderson, Carbonic anhydrase in Escherichia coli. A product of the cyn operon, J. Biol. Chem. 267 (1992) 3731–3734.

[66] H. Jung, K. Jung, H.P. Kleber, Purification and properties of carnitine dehydratase from Escherichia coli—a new enzyme of carnitine metabolization, Biochim. Biophys. Acta 1003 (1989) 270–276.

[67] K. Eichler, F. Bourgis, A. Buchet, H.P. Kleber, M.A. Mandrand-Berthelot, Molecular characterization of the cai operon necessary for carnitine metabolism in Escherichia coli, Mol. Microbiol. 13 (1994) 775–786.

[68] M. Jiao, W. He, Z. Ouyang, Q. Qin, Y. Guo, J. Zhang, et al., Mechanistic and structural insights into the bifunctional enzyme PaaY from Acinetobacter baumannii, Structure 31 (2023) 935–947 e934.

[69] S.R. Hennigar, J.P. McClung, Nutritional immunity: starving pathogens of trace minerals, Am. J. Lifestyle Med. 10 (2016) 170–173.

[70] J.C. Betts, P.T. Lukey, L.C. Robb, R.A. McAdam, K. Duncan, Evaluation of a nutrient starvation model of Mycobacterium tuberculosis persistence by gene and protein expression profiling, Mol. Microbiol. 43 (2002) 717–731.

[71] H.E. Schellhorn, Function, evolution, and composition of the RpoS regulon in Escherichia coli, Front. Microbiol. 11 (2020) 560099.

[72] K. Sauer, The genomics and proteomics of biofilm formation, Genome Biol. 4 (2003) 219.

[73] R. Grande, S. Carradori, V. Puca, I. Vitale, A. Angeli, A. Nocentini, et al., Selective inhibition of Helicobacter pylori carbonic anhydrases by carvacrol and thymol could impair biofilm production and the release of outer membrane vesicles, Int. J. Mol. Sci. 22 (2021).

[74] Y. Zhao, Z. Han, H. Yan, H. Zhao, M.E. Tucker, M. Han, et al., Intracellular and extracellular biomineralization induced by Klebsiella pneumoniae LH1 isolated from dolomites, Geomicrobiol. J. 37 (2020).

[75] Y. Matsumoto, K. Miyake, K. Ozawa, Y. Baba, T. Kusube, Bicarbonate and unsaturated fatty acids enhance capsular polysaccharide synthesis gene expression in oral streptococci, Streptococcus anginosus, J. Biosci. Bioeng. 128 (2019) 511–517.

[76] S.J. Rose, L.E. Bermudez, Identification of bicarbonate as a trigger and genes involved with extracellular DNA export in mycobacterial biofilms, mBio 7 (2016).

[77] C.W. Hall, T.F. Mah, Molecular mechanisms of biofilm-based antibiotic resistance and tolerance in pathogenic bacteria, FEMS Microbiol. Rev. 41 (2017) 276–301.

[78] M.H. Pontes, E.A. Groisman, Slow growth determines nonheritable antibiotic resistance in Salmonella enterica, Sci. Signal. 12 (2019).

[79] J.M. Colquhoun, M. Farokhyfar, A.C. Anderson, C.R. Bethel, R.A. Bonomo, A.J. Clarke, et al., Collateral changes in cell physiology associated with ADC-7 beta-lactamase expression in Acinetobacter baumannii, Microbiol. Spectr. 11 (2023) e0464622.

[80] A.M. Brauer, A.R. Rogers, J.R. Ellermeier, Twin-arginine translocation (Tat) mutants in Salmonella enterica serovar Typhimurium have increased susceptibility to cell wall targeting antibiotics, FEMS Microbes 2 (2021) xtab004.

[81] C. Fernandez, E. Diaz, J.L. Garcia, Insights on the regulation of the phenylacetate degradation pathway from Escherichia coli, Environ. Microbiol. Rep. 6 (2014) 239–250.

[82] R.H. Valdivia, S. Falkow, Fluorescence-based isolation of bacterial genes expressed within host cells, Science 277 (1997) 2007–2011.

[83] P. Monsieurs, S. De Keersmaecker, W.W. Navarre, M.W. Bader, F. De Smet, M. McClelland, et al., Comparison of the PhoPQ regulon in Escherichia coli and Salmonella typhimurium, J. Mol. Evol. 60 (2005) 462–474.

[84] T. Murata, W. Tseng, T. Guina, S.I. Miller, H. Nikaido, PhoPQ-mediated regulation produces a more robust permeability barrier in the outer membrane of Salmonella enterica serovar typhimurium, J. Bacteriol. 189 (2007) 7213–7222.

[85] E. Garcia Vescovi, F.C. Soncini, E.A. Groisman, The role of the PhoP/PhoQ regulon in Salmonella virulence, Res. Microbiol. 145 (1994) 473–480.

[86] Y. Wen, J. Feng, D.R. Scott, E.A. Marcus, G. Sachs, The HP0165-HP0166 two-component system (ArsRS) regulates acid-induced expression of HP1186 alpha-carbonic anhydrase in Helicobacter pylori by activating the pH-dependent promoter, J. Bacteriol. 189 (2007) 2426–2434.

[87] K. Panthel, P. Dietz, R. Haas, D. Beier, Two-component systems of Helicobacter pylori contribute to virulence in a mouse infection model, Infect. Immun. 71 (2003) 5381–5385.

[88] B.H. Abuaita, J.H. Withey, Bicarbonate induces Vibrio cholerae virulence gene expression by enhancing ToxT activity, Infect. Immun. 77 (2009) 4111–4120.

[89] A. Bourgogne, L.C. Thomson, B.E. Murray, Bicarbonate enhances expression of the endocarditis and biofilm associated pilus locus, ebpR-ebpABC, in Enterococcus faecalis, BMC Microbiol. 10 (2010) 17.

[90] M. Jiao, W. He, Z. Ouyang, Q. Shi, Y. Wen, Progress in structural and functional study of the bacterial phenylacetic acid catabolic pathway, its role in pathogenicity and antibiotic resistance, Front. Microbiol. 13 (2022) 964019.

[91] R. Teufel, T. Friedrich, G. Fuchs, An oxygenase that forms and deoxygenates toxic epoxide, Nature 483 (2012) 359–362.

[92] S. Bury-Mone, G.L. Mendz, G.E. Ball, M. Thibonnier, K. Stingl, C. Ecobichon, et al., Roles of alpha and beta carbonic anhydrases of Helicobacter pylori in the urease-dependent response to acidity and in colonization of the murine gastric mucosa, Infect. Immun. 76 (2008) 497–509.

[93] F.N. Stahler, L. Ganter, K. Lederer, M. Kist, S. Bereswill, Mutational analysis of the Helicobacter pylori carbonic anhydrases, FEMS Immunol. Med. Microbiol. 44 (2005) 183–189.

[94] D. Tsikas, E. Hanff, G. Brunner, Helicobacter pylori, its urease and carbonic anhydrases, and macrophage nitric oxide synthase, Trends Microbiol. 25 (2017) 601–602.

[95] H. Kuwahara, Y. Miyamoto, T. Akaike, T. Kubota, T. Sawa, S. Okamoto, et al., Helicobacter pylori urease suppresses bactericidal activity of peroxynitrite via carbon dioxide production, Infect. Immun. 68 (2000) 4378–4383.

[96] I. Nishimori, S. Onishi, H. Takeuchi, C.T. Supuran, The alpha and beta classes carbonic anhydrases from Helicobacter pylori as novel drug targets, Curr. Pharm. Des. 14 (2008) 622–630.

[97] Y. Gu, R. Bartolome-Casado, C. Xu, A. Bertocchi, A. Janney, C. Heuberger, et al., Immune microniches shape intestinal T(reg) function, Nature 628 (2024) 854–862.

[98] D.S. Spasova, C.D. Surh, Blowing on embers: commensal microbiota and our immune system, Front. Immunol. 5 (2014) 318.

[99] Y. Yao, W. Shang, L. Bao, Z. Peng, C. Wu, Epithelial-immune cell crosstalk for intestinal barrier homeostasis, Eur. J. Immunol. (2024) e2350631.

[100] C.T. Supuran, An overview of novel antimicrobial carbonic anhydrase inhibitors, Expert. Opin. Ther. Targets 27 (2023) 897–910.

[101] R. McKenna, C.T. Supuran, Carbonic anhydrase inhibitors drug design, Subcell. Biochem. 75 (2014) 291–323.

[102] C.T. Supuran, C. Capasso, An overview of the bacterial carbonic anhydrases, Metabolites 7 (2017).

[103] J. Davies, D. Davies, Origins and evolution of antibiotic resistance, Microbiol. Mol. Biol. Rev. 74 (2010) 417–433.

[104] A.R. Coates, G. Halls, Y. Hu, Novel classes of antibiotics or more of the same? Br. J. Pharmacol. 163 (2011) 184–194.

[105] A.E. Clatworthy, E. Pierson, D.T. Hung, Targeting virulence: a new paradigm for antimicrobial therapy, Nat. Chem. Biol. 3 (2007) 541–548.

[106] D.A. Rasko, V. Sperandio, Anti-virulence strategies to combat bacteria-mediated disease, Nat. Rev. Drug. Discov. 9 (2010) 117–128.

[107] R.C. Allen, R. Popat, S.P. Diggle, S.P. Brown, Targeting virulence: can we make evolution-proof drugs? Nat. Rev. Microbiol. 12 (2014) 300–308.

[108] K.L. Palmer, L.M. Aye, M. Whiteley, Nutritional cues control Pseudomonas aeruginosa multicellular behavior in cystic fibrosis sputum, J. Bacteriol. 189 (2007) 8079–8087.

[109] C.T. Supuran, Carbonic anhydrases as drug targets—an overview, Curr. Top. Medicinal Chem. 7 (2007) 825–833.

[110] C.T. Supuran, A. Scozzafava, Carbonic anhydrases as targets for medicinal chemistry, Bioorganic & Medicinal Chem. 15 (2007) 4336–4350.

[111] J. Cronk, R. Rowlett, K. Zhang, C. Tu, J. Endrizzi, J. Lee, et al., Identification of a novel noncatalytic bicarbonate binding site in eubacterial beta-carbonic anhydrase, Biochemistry 45 (2006) 4351–4361.

[112] I. Nishimori, T. Minakuchi, T. Kohsaki, S. Onishi, H. Takeuchi, D. Vullo, et al., Carbonic anhydrase inhibitors: the [beta]-carbonic anhydrase from Helicobacter pylori is a new target for sulfonamide and sulfamate inhibitors, Bioorg. Med. Chem. Lett. 17 (2007) 3585–3594.

[113] I. Nishimori, T. Minakuchi, D. Vullo, A. Scozzafava, C.T. Supuran, Inhibition studies of the [beta]-carbonic anhydrases from the bacterial pathogen Salmonella enterica serovar Typhimurium with sulfonamides and sulfamates, Bioorg. Med. Chem. 19 (2011) 5023–5030.

[114] P. Burghout, D. Vullo, A. Scozzafava, P.W.M. Hermans, C.T. Supuran, Inhibition of the β-carbonic anhydrase from Streptococcus pneumoniae by inorganic anions and small molecules: Toward innovative drug design of antiinfectives? Bioorg. Med. Chem. 19 (2011) 243–248.

[115] K.M. Hoffmann, D. Samardzic, K. v.d. Heever, R.S. Rowlett, Co(II)-substituted Haemophilus influenzae β-carbonic anhydrase: spectral evidence for allosteric regulation by pH and bicarbonate ion, Arch. Biochem. Biophys. 511 (2011) 80–87.

[116] P. Joseph, F. Turtaut, S. Ouahrani-Bettache, J. Montero, I. Nishimori, T. Minakuchi, et al., Cloning, characterization, and inhibition studies of a beta-carbonic anhydrase from Brucella suis, J. Med. Chem. 53 (2010) 2277–2285.

[117] J.Y. Winum, P. Joseph, S. Ouahrani-Bettache, J.L. Montero, I. Nishimori, T. Minakuchi, et al., A new beta-carbonic anhydrase from Brucella suis, its cloning, characterization, and inhibition with sulfonamides and sulfamates, leading to impaired pathogen growth, Bioorg. Med. Chem. 19 (2011) 1172–1178.

[118] J.Y. Winum, S. Kohler, C.T. Supuran, Brucella carbonic anhydrases: new targets for designing anti-infective agents, Curr. Pharm. Des. 16 (2010) 3310–3316.

[119] C.T. Supuran, I. Nishimori, T. Minakuchi, A. Maresca, F. Carta, A. Scozzafava, The beta-carbonic anhydrases from Mycobacterium tuberculosis as drug targets, Curr. Pharm. Des. 16 (2010) 3300–3309.

CHAPTER TWELVE

Mycobacterial β-carbonic anhydrases: Molecular biology, role in the pathogenesis of tuberculosis and inhibition studies

Jenny Parkkinen[a], Ratul Bhowmik[a], Martti Tolvanen[b], Fabrizio Carta[c], Claudiu T. Supuran[c], Seppo Parkkila[a,d], and Ashok Aspatwar[a,*]

[a]Faculty of Medicine and Health Technology, Tampere University, Tampere, Finland
[b]Department of Computing, University of Turku, Finland
[c]Neurofarba Department, Sezione di Chimica Farmaceutica e Nutraceutica, Università degli Studi di Firenze, Firenze, Italy
[d]Fimlab Ltd. and Tampere University Hospital, Tampere, Finland
*Corresponding author. e-mail address: ashok.aspatwar@tuni.fi

Contents

1. Introduction   344
2. Computational tools for studying *M. tuberculosis* β-carbonic anhydrases   350
3. Molecular biology   355
4. β-CA1 is induced during environmental stress   357
5. β-CA2 is essential for mycobacterial virulence   358
6. β-CA3 serves as a dual-function transmembrane protein   360
7. Inhibition of *M. tuberculosis* β-carbonic anhydrases   361
8. Inhibition with sulfonamides   362
9. Inhibition with dithiocarbamates   368
10. Inhibition with natural products and carboxylic acids   368
11. Inhibition of *M. tuberculosis* in culture   369
    11.1 *In vitro* inhibition   369
12. Summary and future directions   374
Acknowledgments   375
References   375

Abstract

Mycobacterium tuberculosis (Mtb), which causes tuberculosis (TB), is still a major global health problem. According to the World Health Organization (WHO), TB still causes more deaths worldwide than any other infectious agent. Drug-sensitive TB is treatable using first-line drugs; treatment of multidrug-resistant (MDR) and extensively drug-resistant

The Enzymes, Volume 55
ISSN 1874-6047, https://doi.org/10.1016/bs.enz.2024.05.012

(XDR) TB requires second- and third-line drugs. However, due to the long duration of treatment, the noncompliance of patients with different levels of resistance of Mtb to these drugs has worsened the situation. Previously developed anti-TB drugs targeted the replication machinery, protein synthesis, and cell wall biosynthesis pathways of Mtb. Therefore, novel drugs targeting alternate pathways crucial for the survival and pathogenesis of Mtb in the human host are needed. The genome of Mtb encodes three β-carbonic anhydrases (CAs) that are fundamental for pH homeostasis, hypoxia, survival, and pathogenesis. Recently, several studies have shown that the β-CAs of Mtb could be inhibited both *in vitro* and *in vivo* using small chemical molecules, suggesting that these enzymes could be novel targets for developing anti-TB compounds that are devoid of resistance by Mtb. In addition, homologs of β-CAs are absent in humans; therefore, drugs developed to target these enzymes might have minimal off-target effects. In this work, we describe the roles of β-CAs in Mtb and discuss bioinformatics and cheminformatics tools used in development and discovery of novel inhibitors of these enzymes. In addition, we summarize the *in vitro* and *in vivo* studies demonstrating that the β-CAs of Mtb are indeed druggable targets.

1. Introduction

Mycobacterium tuberculosis (Mtb) is a rod-shaped, nonmotile, acid–fast bacterium that contains a high amount of $G + C$ in its genome [1,2]. Mtb is normally transmitted through the air by droplet contact. In humans, tuberculosis (TB) caused by Mtb affects many organs, but the main target organ is the lung. Consequently, pulmonary TB is observed in 80% of patients. In the host, Mtb resides in granulomas, which are lesions involving differentiated macrophages and other immune cells. Within granulomas Mtb bacteria exist in a state of nonreplicating persistence [3–5]. These properties make humans a natural host and reservoir of Mtb. Only a small percentage of people have active TB, and the remaining Mtb carriers have latent TB infection (LTBI) [6].

TB caused by Mtb is a leading cause of morbidity and mortality after SARS-CoV-2 infection in 2022 [7,8]. According to the latest WHO report, 10.6 million people develop TB worldwide, causing 1.3 million deaths [7]. In addition, the global burden of LTBI was estimated to be 23.0% or approximately 1.7 billion people in 2014 [9]. This creates an important public health problem because people with LTBI act as reservoirs for TB disease. TB is a treatable disease in which patients with drug-susceptible TB can be treated with the first line anti–TB compounds for six months. The treatment of TB has side effects, and there is noncompliance by many TB patients, reducing the efficacy of the treatment (Table 1) [39].

Table 1 Details of the drugs used for the treatment of drug-sensitive and drug-resistant TB.

Level of treatment	TB-drugs (Year of development)	Mechanism of action	Resistance mechanism	Side effects	References
First line of treatment	Isoniazid (1951)	Block mycolic acid synthesis by inhibiting NADH–dependent enoyl acyl carrier protein reductase	KatG suppression causing decreased prodrug activation, and a mutation in the promoter region of InhA causing an overexpression of InhA	Hepatotoxicity, peripheral neuritis dermatological, gastrointestinal, hypersensitivity, hematological and renal reaction	[10,11]
	Rifampicin (1957)	Inhibit β subunit of DNA–dependent RNA polymerase	Mutation of rpoB induces a change at β–subunit of RNA polymerase, causing decrease in binding affinity	Epigastric distress, thrombocytopenia, leukopenia, hemolytic anemia, menstrual disturbances	[10,12]
	Ethambutol (1962)	Inhibits arabinogalactan biosynthesis in cell wall	Mutations in embB and ubiA	Optic neuritis nausea, rashes, gastrointestinal disturbance	[13,14]
	Pyrazinamide (1952)	Inhibits coenzyme A and pantothenate synthesis	Mutations in pncA reducing conversion to active acid	Hepatotoxicity, hyperuricemia arthralgia, GI disturbances, thrombocytopenia, sideroblastic anemia	[15,16]
Second line of treatment	Kanamycin[a]	Protein synthesis inhibition	16S rRNA target site modulation (1400 and 1401 rrs gene), increased drug inactivation via over expression of aminoglycoside acetyltransferase	Pain or irritation diarrhea, hearing loss, nephrotoxicity	[17,18]

(continued)

Table 1 Details of the drugs used for the treatment of drug-sensitive and drug-resistant TB. (*cont'd*)

Level of treatment	TB-drugs (Year of development)	Mechanism of action	Resistance mechanism	Side effects	References
	Streptomycin	Inhibits protein synthesis	Mutations in rpsl and rrs confer binding site modulation	Itching, numbness, ototoxicity, nephrotoxicity	[19]
	Capreomycin	Inhibits protein synthesis	Cross-resistance with aminoglycoside plus mutation of tlyA which decreases rRNA methyltransferase activity	Ototoxicity nephrotoxicity eosinophilia, rashes, fever, and injection site pain	[18]
	Amikacin	Protein synthesis inhibition	16S rRNA target site modulation (1400 and 1401 rrs gene), increased drug inactivation *via* over expression of aminoglycoside acetyltransferase	Pain or irritation, diarrhea, hearing loss, nephrotoxicity	[17,18]
	Fluroquinol drugs Levofloxacin Ofloxacin (1982) Moxifloxacin Gatifloxacin	DNA gyrase and topoisomerase IV inhibitor	Mutations in gyrA and gyrB causing alteration to DNA Gyrase A/B binding site (later generations not always cross-resistant with first generation) and increased ABC-type efflux pump expression	Tendonitis and tendon rupture, photosensitivity, seizure, QT prolongation, nausea, diarrhea, constipation, gas, vomiting, skin disturbances	[20–22]

Drug	Mechanism of action	Resistance mechanism	Adverse effects	References
Cycloserine (1955)	Peptidoglycan synthesis inhibition	Overexpression of alrA decreasing drug efficiency	Headache, drowsiness, depression	[23]
Thiacetazone	Inhibits methyltransferases in mycolic acid biosynthesis	ethA mutation decrease prodrug activation and mutations to had ABC operon affecting dehydratase activity	GI intolerance, rash, ototoxicity, hypersensitivity reactions, such as hepatitis, bore marrow aplastic syndromes	[24]
β-Lactam/β-lactamase inhibitor: Meropenem Imipenem	Cell wall disruption via PG modulation	Overexpression of β-lactamases, (BlaC), point mutations at target site altering deacylation rate and binding affinity, cell permeability, and increased efflux		[25,26]
Para-aminosalicylic acid	Inhibits thymine nucleotide, iron and folic acid metabolism	Mutations in the thyA causing a decrease in activated drug concentrations and folC mutations which cause binding site mutations	GI intolerance lupuslike reactions	[27,28]
Bedaquiline	Inhibition of mitochondrial ATP synthase	atpE mutations induces binding site modulation. Noted efflux via mmpl5 (cross-resistance with Clofazimine)	QT-prolongation, unexplained excess mortality	[29]
MDR–TB treatment (Less effective and more toxic than first-line drugs)				

(continued)

Table 1 Details of the drugs used for the treatment of drug-sensitive and drug-resistant TB. (*cont'd*.)

Level of treatment	TB-drugs (Year of development)	Mechanism of action	Resistance mechanism	Side effects	References
	Linezolid	Protein synthesis inhibitor (50S subunit)	T460C mutation in rplC, encoding the 50S ribosomal 13 protein and possible efflux mechanisms	Myelosuppression, neuropathies, lactic acidosis, rhabdomyolysis, diarrhea	[30]
	Delamanid	Inhibits mycolic acid synthesis	Mutation of reductive activating Rv3547 gene fgd1, fbiA	QT-prolongation	[31,32]
	Pretomanid	Inhibits synthesis of mycolic acid	Mutations in ddn, fgd, and fbiA–C genes	Nerve damage, vomiting, headache, low blood sugar, diarrhea, liver inflammation	[29,33]
	Ethionamide (1956)	Inhibits synthesis of mycolic acids in cell wall	Mutations in ethA and inhA cause decreased prodrug activation and InhA overexpression (cross-resistance with Isoniazid)	Nausea, vomiting	[34–36]
	Clofazimine	Release of Reactive Oxygen Species (ROS) and cell membrane disruption	Mutation to Rv0678 causes upregulation of Mmpl5, a multisubstrate efflux pump (cross-resistance with Bedaquiline)	Red–brown skin discoloration, ichthyosis, acute abdominal pain	[37,38]

aDiscontinued use in the USA.

This situation is exacerbated by the emergence of multidrug-resistant (MDR) and extensively drug-resistant (XDR) strains of Mtb and totally drug-resistant (TDR) strains, which require longer durations of treatment with second-line and third-line anti-TB drugs and are becoming more expensive (Table 1) [40,41]. The currently used treatment methods are effective only against Mtb strains that are actively growing [42]. LTBI, in which Mtb bacilli are metabolically inactive, does not respond to treatment methods, complicating treatment efforts.

Therefore, novel drugs that can target new cellular pathways of Mtb and that are effective against both LTBI and drug-resistant Mtb strains are needed (Fig. 1). In addition, these drugs can reduce the duration of treatment, have minimal off-target effects, and can be combined with currently available anti-TB compounds. In the last 25 years, research on Mtb has made substantial progress with the sequencing of the Mtb genome and the discovery of many new drug targets, which may lead to the development of innovative drugs possessing a novel mechanism of action, thus helping to solve the problem of drug resistance [44,45]. Recent progress in the structural and functional analyses of genomes and proteomes has opened new possibilities for the design of mechanism-based drugs targeting proteins crucial for the pathogenesis of myco-bacteria [46]. Among many such proteins, β-carbonic anhydrases (β-CAs) of mycobacteria could be possible targets for developing novel anti-mycobacterial agents with the potential to treat persistent and drug-resistant TB (Fig. 1).

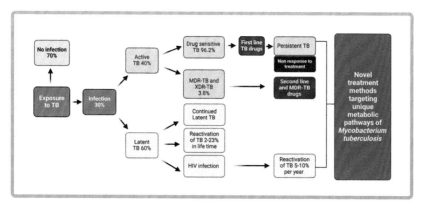

Fig. 1 Tuberculosis infection by *M. tuberculosis* and disease outcomes worldwide and treatment options for drug-sensitive and drug-resistant TB [7,43].

Carbonic anhydrases (EC 4.2.1.1) (CAs) are a group of metalloenzymes containing mainly zinc as a metal ion that is involved in the reversible hydration of carbon dioxide to bicarbonate and protons in the reaction $CO_2 + H_2O \leftrightharpoons HCO_3^- + H^+$ [47–50]. There are eight CA families, namely, α, β, γ, δ, ζ, η, θ, and ι CAs [47]. The α, β and γ families are widespread in all kingdoms of life, but human and other vertebrates only have α-CAs [51]. The other families are more restricted in their distribution. α-CAs have been thoroughly investigated from the perspective of drug development [52]. However, the β-CAs that are found in pathogenic bacteria, including Mtb, and in parasites, have recently been considered targets for the development of drugs [52–54]. In Mtb, there are three β-CA genes in the genome, namely, Rv1284, Rv3588c, and Rv3273, encoding CanA, CanB, and bifunctional SulP family inorganic anion transporter/carbonic anhydrase, also known as β-CA1, β-CA2, and β-CA3, respectively, and henceforth referred to as such in this chapter [55]. The β-CAs of Mtb have been cloned and characterized, and the crystal structures of two β-CAs (β-CA1 and β-CA2) have been resolved [55–59]. Inhibition studies using different classes of inhibitors have demonstrated that the activity of Mtb β-CAs can be efficiently inhibited in vitro [55,60]. In addition, in vivo studies using *Mycobacterium marinum*, a model bacterium, and zebrafish, its natural host, have shown that these inhibitors impair the growth of the bacterium, demonstrating proof-of-concept that β-CAs of TB are druggable targets [61].

In this chapter, we present comprehensive information on the three β-CAs that play crucial physiological functions in *M. tuberculosis*, a medically important pathogen that causes difficult-to-treat TB disease in humans. This chapter mainly focuses on cloning, characterization, and inhibition studies using different classes of chemical inhibitors of β-CAs both *in vitro* and *in vivo*. We also present the latest information on the metabolic roles of Mtb β-CAs that are crucial for the survival and pathogenesis of TB in humans, including future directions for developing anti-TB drugs targeting β-CAs that have the potential to treat not only drug-sensitive TB but also persistent and drug-resistant TB.

2. Computational tools for studying *M. tuberculosis* β-carbonic anhydrases

Computational methods concerning chemoinformatics, including quantitative structure–activity relationship (QSAR), pharmacophore modeling, molecular docking, and molecular dynamics (MD) simulations, are crucial for the development and identification of novel drugs against tuberculosis (TB) that have improved therapeutic effects compared with traditional FDA-approved medications. Cheminformatics and molecular modeling methods have been used for many years to discover and enhance new drugs with improved therapeutic properties in many areas. At present, *in silico* modeling is an integral component of the standard drug discovery process [62,63]. These methods are typically utilized to identify new medications or enhance the therapeutic efficacy of a chemical series during the early stages of drug development Fig. 2.

The quantitative structure–activity relationship (QSAR), an integral part of the *in silico* modeling approach, is a robust computational methodology employed to establish quantitative associations between the chemical structures of molecules and their corresponding biological activities. Traditional QSAR models have been constructed for anti-TB drug development by utilizing the structures of established anti-TB drugs and their efficacy against Mtb Mathematical techniques leveraging several machine learning and deep learning algorithms have been employed by these QSAR models to establish a correlation between the physicochemical attributes and structural characteristics of small molecules and their corresponding biological effects. The activity of novel compounds may be predicted using these QSAR models through the process of extrapolation,

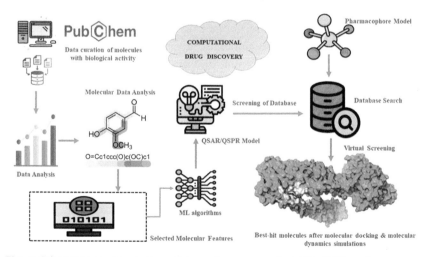

Fig. 2 Schematic representation of drug discovery against Mtb β-CAs using computational tools.

making use of the established structure–activity connections of preexisting molecules. QSAR models have been traditionally utilized to prioritize compounds for synthesis and optimize their structures to maximize potency, selectivity, and other pharmacological features. This has been achieved by the analysis of molecular descriptors and attributes, including physicochemical features, atom indices, and functional groups, along with quantum chemical features of molecules [64]. In addition, QSAR strategies have also significantly assisted in the virtual screening of extensive chemical databases to find prospective lead compounds that possess the requisite anti-TB properties. This process effectively decreases the time and expenses often involved in experimental screening.

Despite the success of traditional approaches to the QSAR model, recent chemogenomic strategies have significantly revolutionized the drug discovery process concerning predictive modeling using machine learning and artificial intelligence strategies. The concept of chemogenomics encompasses methodologies that use extensive bioactivity data, including drug and target information, to identify unidentified interactions between novel compounds and their respective targets. Proteochemometrics (PCM) or hybridization of QSAR modeling refers to a set of techniques that integrate a computational representation of the ligand (small molecules/drugs) side of a system with a description of the biological side under investigation. These two aspects are interconnected and pertain to a specific readout of interest. On the other hand, the biological parameters inside the model may include protein binding locations, as well as data on gene expression levels specific to particular cell lines. The provided information outlines the biological impact of a specific ligand on the protein or cell line under investigation, including the IC50 value associated with the chemical molecule and its interaction with the biological system. Furthermore, PCM is relevant to precision medicine treatments because it can predict the impact of a ligand on an intricate biological system, such as a cell line, based on genotypic data. The unique approach of PCM, which takes into account the integration of chemical and target space, enables a more detailed comprehension and prediction of the impact of target variability on compound activity [65,66]. For example, by predicting the activity of compounds on a panel of Mtb cell lines, it is possible to find compounds that specifically target a particular cell line. Likewise, the impact of changes in different mycobacterial proteins on the mechanism of action of compounds may be measured. Hence, PCM provides novel opportunities to extract drug affinity databases to develop models that encompass multiple targets and species, along with the incorporation of toxicogenomic and phenotypic data into predictive models. PCM also helps in the systematic prediction of synthetic or naturally occurring molecules for orphan receptors, a process known as receptor deorphanization. Additionally, genotypic information on PCM can be used to develop individualized therapies for multi-drug-resistant tuberculosis. In future prospective research, PCM models can be employed to unravel the mechanism of action of beta-carbonic anhydrase inhibitors (CAIs) by integrating the molecular features of the inhibitors with fold changes against a cell line for each Mtb beta-carbonic anhydrase (MtbCAs: β-CA1, β-CA2, β-CA3) to predict biological/toxicological activity, which can help us elucidate the drug susceptibility and toxicogenomic mechanism involved in the biological process. The mechanistic interpretation of the PCM model for different gene expression profiles of a drug concerning TB cell lines can further provide us with a proper link with different metabolic pathways, along with associations with key biological processes, cellular components, and molecular functions involved in the physiological mechanism of MtbCAs.

In addition to QSAR techniques, pharmacophore models have also served as invaluable tools in drug discovery, aiding in the identification and optimization of potential drug candidates for treating tuberculosis. Pharmacophore models have been mostly employed to ascertain the essential structural and chemical characteristics required for a molecule to effectively bind to its designated receptor and manifest its biological pharmacological properties [66,67]. Several researchers have previously constructed pharmacophore hypotheses that depict the critical connections necessary for biological activity by studying the pharmacophore properties present in known active anti-TB drugs. The optimization of lead molecules has also been guided by pharmacophore models, which employ structure–activity relationship (SAR) investigations and rational drug design methodologies. Researchers can increase the potency, selectivity, and pharmacokinetic aspects of hit drugs by studying the spatial arrangement of pharmacophore features and their interactions with the target receptor [67]. This modification allows for the preservation of critical pharmacophore components. The utilization of a pharmacophore-based hit optimization strategy can be further implemented for the creation of MtbCA inhibitor analogs and derivatives that possess enhanced drug-like characteristics and therapeutic capabilities. In prospective studies, pharmacophore models can also be used to predict the biological activity of novel MtbCA inhibitors by showing their correspondence with essential pharmacophore characteristics. Through the utilization of quantitative analysis, researchers can easily assess the conformity of novel molecules to the pharmacophore model and estimate their resemblance to established active anti-TB MtbCA inhibitors. This will further enable them to effectively prioritize compounds for experimental evaluation and make predictions about their probable effectiveness against MtbCAs. Pharmacophore-based activity prediction can further add to the evaluation of significant insights into the connections between structure and activity that regulate novel MtbCA inhibitor binding mechanisms, hence aiding in the prioritization of drugs for further optimization and development.

In addition to pharmacophore modeling, the field of molecular docking is of great significance in the realm of anti-TB drug design. It involves computer prediction and analysis of the interactions that occur between small-molecule ligands and target proteins. During this procedure, the three-dimensional configurations of both the ligands and the target proteins are generated and subsequently subjected to docking simulations, whereby the ligands are freely positioned within the binding site of the protein [66]. Docking algorithms are utilized to explore different ligand conformations and orientations to identify the binding posture that is most energetically advantageous. This pose is then assessed using scoring functions that measure complementarity, hydrogen bonding, hydrophobic interactions, and steric conflicts. The obtained binding information offers significant insights into the mechanisms of binding and the preferences of ligands, hence informing lead optimization, investigations on the link between structure and activity, and virtual screening of chemical libraries for the identification of prospective therapeutic candidates [68]. In future studies, molecular docking strategies can play a key role in gaining invaluable insights into the binding mechanism of novel inhibitors of MtbCAs through visualization of the amino acid residues in the active site, analysis of different hydrophilic and hydrophobic chemical interactions formed during the binding process, and determination of their associations with different structural attributes of the inhibitors. Preestablished knowledge concerning the binding mechanism of inhibitors will further accelerate the drug development process by utilizing computational predictions and analysis to assist in the logical design and optimization of new drugs that can exhibit improved effectiveness and selectivity against MtbCAs.

Molecular dynamics (MD) simulations in conjunction with molecular docking also act as a vital tool in anti-TB drug design, offering insights into the dynamic behavior of biomolecular systems at the atomic level [69]. With respect to MtbCA inhibitor design, MD simulations can be employed to investigate the interactions between drug candidates and their target proteins or nucleic acids over time, providing information on binding kinetics, conformational changes, and stability [70]. By integrating MD simulations with experimental data, proper refinement of the understanding of ligand—receptor interactions, optimization of lead compounds, and prediction of the effects of structural modifications on drug binding and efficacy of MtbCAs can be achieved. MD simulations can also be employed to explore drug-protein dynamics under different physiological conditions and aid in the identification of allosteric binding sites and off-target effects of MtbCAs. Through their ability to capture the dynamic nature of biomolecular interactions, MD simulations can enhance the rational design of novel anti-TB therapeutics with improved potency, selectivity, and pharmacokinetic properties, thereby accelerating the drug discovery process for multidrug-resistant tuberculosis.

3. Molecular biology

In recent decades, advances in molecular biology and genetics have enabled the study of full genomes. The genomes of various mycobacterial species, including Mtb, *M. bovis* (the causative agent of bovine TB and the progenitor strain used in the BCG vaccine), *M. marinum* (Mmr, the causative agent of TB-like disease in fish) and several other nontuberculous mycobacteria (NTMs), such as *M. avium*, have been sequenced [44,71]. All these species have three β-class carbonic anhydrases encoded in their genomes. In the NCBI sequence databases there are more than 100 reference genomes for named species of genus *Mycobacterium* (https://www.ncbi.nlm.nih.gov/datasets/genome/?taxon=1763&reference_only=true) in the controversial amended genus [72–74] and more than 200 reference genomes for the pre-2018 *Mycobacterium*, now seen as family *Mycobacteriaceae* at NCBI (https://www.ncbi.nlm.nih.gov/datasets/genome/?taxon=1762&reference_only=true). A full analysis of the BCAs in all of *Mycobacteriaceae* has been completed in our research group and will be published separately.

Mycobacterial β-CAs are zinc-containing metalloenzymes that catalyze the well-known reaction involving the hydration of membrane-permeable carbon dioxide to a proton and membrane-impermeable bicarbonate [75]. The genome of Mtb encodes β-CA1 (*Rv1284*), β-CA2 (*Rv3588c*) and β-CA3 (*Rv3273*). Both β-CA1 and β-CA2 are cytosolic enzymes, whereas β-CA3 is predicted to be a membrane-bound protein [56–58].

CAs are commonly composed of structural oligomers, and the homodimerization process is essential for creating a specific microenvironment for the active site [59]. All β-CAs share a conserved α/β-fold structure, with the core being a five-stranded β-sheet. Zinc in the active site is bound to either three or four amino acid side chains [59]. One histidine and two cysteines are constant, and the fourth coordination position can be occupied by either a water (accessible conformation) or the aspartate side chain of the CxDxR motif (blocked conformation). The X-ray crystallographic structures of β-CA1 and β-CA2 have been previously resolved, and interestingly both of them can exist in either the accessible or blocked conformation [59,76,77].

All three Mtb β-CAs have previously quantified activities that can be inhibited by classical CA inhibitors (Table 2). Of these, β-CA2 has the highest activity, which is quantified to fall between the activity levels of human CA I and II (hCAI and hCA II) [58].

In contrast to β-CAs, α- and γ-CA active site zinc is bound by three histidine residues, and the three families display very different tertiary structures. Accordingly, the most common CA classes (α, β and γ) are believed to have evolved separately [78]. Hence, β-CA-selective inhibitors could be an interesting avenue for drug development. Linking existing TB treatments and mycobacterial β-CAs, Srinivas et al. studied the changes in Mtb gene expression in a large superset of transcriptional analyses when treated with TB drugs. In this study, *in silico* drug response predictions were made based on transcriptomic data [79]. Pyrazinoic acid, a metabolite of pyrazinamide, was the only drug that upregulated β-CA gene expression, specifically the expression of *Rv1284* [79]. Other treatments either had no transcriptional effect or downregulated the genes. Interestingly, moxifloxacin downregulated the expression of all three CAs. Bedaquiline, linezolid and

Table 2 Activity and inhibition properties of Mtb β-CAs compared to those of human CA I and CA II.

CA	Gene ID	Kcat (s^{-1})	Kcat/km (M^{-1} s^{-1})	Activity	KI (nM)[a]	References
β-CA1	Rv1284	3.9×10^5	3.7×10^7	Moderate	480	[57]
β-CA2	Rv3588c	9.8×10^5	9.3×10^7	High	9.8	[58]
β-CA3	Rv3273	4.3×10^5	4.0×10^7	Moderate	104	[56]
hCA I	CA1	2.0×10^5	5.0×10^7	Moderate	250	[56]
hCAII	CA2	1.4×10^6	1.5×10^8	Very high	12	[56]

[a]Inhibition activity using acetazolamide.

rifampicin downregulated both *Rv3588c* and *Rv3273* [79]. Reduced transcription of *Rv3273* was also achieved with isoniazid and pretomanid [79].

The biological characteristics of mycobacterial β-CAs remain largely unknown, but predictions about their functions exist. According to the literature, Mtb β-CAs are involved in various processes, such as virulence, eDNA transport, biofilm formation and pH regulation [80,81]. Both β-CA1 and β-CA2 have been reported to be essential for both mycobacterial growth and *in vivo* virulence [82,83]. However, a more recent transposon mutagenesis screening suggested that all Mtb β CAs may be nonessential for bacterial viability [84], but this is the expected result for proteins with a shared catalytic activity.

4. β-CA1 is induced during environmental stress

The protein β-CA1, encoded by the *Rv1284* gene, is 163 amino acid residues in length. Among other mycobacteria, an identical protein (UniProt ID A0A679LF87) is encoded by the *M. bovis* genome and compared to the protein encoded by MMAR_4135 in Mmr, the sequence identity is 84.66%. The mycobacterial β-CA1 proteins match the domain signature of beta_CA_cladeD (CDD:239474) in the NCBI Conserved Domain Database (CDD) [85]. This is the clade of prokaryotic BCA proteins recognized to be more distant from the other three clades in the seminal article on BCA phylogenetics [86]. In practice this means that the orthologs of β-CA1 will be missed in sequence similarity based database searches (*e.g.* BLAST, https://blast.ncbi.nlm.nih.gov/Blast.cgi) if the query sequence is β-CA2 or β-CA3.

The Mtb protein has entries in the Protein Data Bank (PDB) described in two publications: the original X-ray crystallography structures PDB ID 1YLK [76] and 4YF4, 4YF5 and 4YF6 [77]. The biological assembly of the β-CA1 protein structure is a dimer in all of these structures, even if the crystal unit cells contain two dimers. In 1YLK the active site is in the blocked conformation, with Asp 37 binding to the fourth coordination position of zinc, whereas in 4YF4, 4YF5 and 4YF6 the conformation is open.

The subcellular localization of β-CA1 has been shown to be plasma membrane in a large-scale membrane protein proteomics screen [87]. As there are no hydrophobic parts or known lipid modifications in β-CA1, it is to be assumed that the association is through protein-protein binding to other membrane proteins.

Using a transposon site hybridization (TraSH) assay, Rv1284 was identified as an essential gene for both mycobacterial *in vitro* growth and *in vivo* virulence in mice [82,83]. Another recent study revealed that β-CA1 is dispensable for Mtb virulence in macrophages [80]. β-CA1 is less stable than β-CA2, and it is subject to oxidization; thus, environmental redox conditions are hypothesized to regulate the activity of this enzyme [77]. Oxidation can release zinc ions through the reversible formation of an alternative disulfide Cys35–Cys61, which removes Cys35 from zinc coordination. Tyr120 is suggested to mediate the electron transfer reactions from protein surface to the active site which cause Cys35 to change position depending on the redox potential of the protein surroundings [77]. Furthermore, shifts in the microenvironment, such as nutritional starvation or hypoxia, can trigger the expression of *Rv1284,* as shown in Table 3 [83]. With respect to Mtb transcriptional profiling and whether there are insights into the function and regulation of β-CA1, several studies have reported that *Rv1284* is upregulated upon environmental stress (Table 3). Avican et al. (2021) characterized bacterial stress responses mimicking human host conditions. The transcriptomes of several pathogens, including Mtb, were analyzed, and *Rv1284* was shown to be upregulated under several stress conditions, such as acidic pH and hypoxia [88]. Similarly, Vilchèze et al. (2022) studied stress responses in human macrophages. The Mtb transcriptome yields, once again, *Rv1284* upregulation and *Rv3588c* and *Rv3273* downregulation upon stress [89]. Similarly, interfering with the expression of other genes, such as *mtrA* and *esx-1,* can upregulate *Rv1284* expression. Downregulation of *Rv1284* is rare; however, DosRST–regulon inhibitors and mutants seem to cause this phenomenon (Table 3). The DosRST regulon includes more than 50 genes that are regulated upon dormancy after encountering environmental stresses such as hypoxia, acidic pH or nutrient starvation [93].

5. β-CA2 is essential for mycobacterial virulence

The β-CA2 protein encoded by the Mtb *Rv3588c* gene is 207 amino acid residues in length. Again, the ortholog (UniProt ID A0A0H3MII2) in *M. bovis* is identical to that in B-CA2. The orthologs of Mmr (MMAR_5088) and *Mycobacterium leprae* (UniProt ID Q9CBJ1) are 86.34% and 84.54% identical to those of Mtb. β-CA2 belongs to clade C of β-CAs, matching the CDD domain signature beta_CA_cladeC (CDD:239473) [86].

Table 3 Environmental and genetic factors that induce differential expression of the three mycobacterial carbonic anhydrases.

Gene	Regulation	Environmental factor	Genetic factor	References
β-CA1	Upregulation	Acidic pH, high chloride levels, high temperature, hypoxia	mtrA KD, (rv0950c(GSE134574), rv1096 (GSE134574), rv1432 (GSE134574), rv1565c (GSE134574), esx-1) KO	[88–92]
	Downregulation	DosR-inhibitors	dosRS KO	[93]
β-CA2	Upregulation	Biofilm (GSE70718), phosphate starvation (GSE183469) increased CO_2 (5% *vs.* 0.5%)		[80]
	Downregulation	Hypoxia	sigE KO	[89,94]
β-CA3	Downregulation	Acidic pH, biofilm (GSE70718), hypoxia, nitrosative stress, nutrient starvation	whiB3 KO	[88,89,95]

In the inhibition studies, KO=knockout, KD=knockdown, and OE=overexpression.

The protein has been crystallized in different conditions to yield structures PDB ID 1YM3 [76] and 2A5V [59], which are a dimer and tetramer, respectively. β-CA2 was the first β-CA to exhibit a two-state conformation. In the 1YM3 crystal dimer the protein is in a so-called 'blocked' state (zinc coordinated by four amino acid residues), whereas in the 2AV5 tetramer, crystallized in the presence of thiocyanate, the fourth coordination position of zinc is occupied by thiocyanate instead of the asparate side chain, giving the accessible conformation of the active site. β-CA2 is inactive at neutral and acidic pH levels [59], but enzymatic activity is observed at pH 8.4, and above this pH, the maximum activity was reached at pH 9.0. In solution, the upward pH shift was also seen to lead to a dimer-to-tetramer conversion, based on dynamic light scattering measurements [59]. As the blocked state leaves no room for the water molecule or substrate in the active site, it inactivates the enzyme [59]. So, in summary, the accessible state and β-CA2 activity require the pH-dependent dimer-to-tetramer conversion, creating more room in the active site core [59]. Additionally, the net charge of the tetramer is positive, possibly explaining why activity is preferred at higher pH values [59]. In the active site core, the aspartate residue forms a salt bridge with the arginine residue and is replaced with a water molecule that can act as a nucleophile that enables enzymatic activity [58,76]. It is to be noted that in β-CA1 structures both the blocked and accessible conformation are present as dimers, so there is no evidence for a similar activation mechanism through tetramerization for β-CA1.

β-CA2 exhibits the highest level of CO_2 hydration activity of the three enzymes [58]. The active site is more adaptable and open in β-CA2 and is therefore also more accessible to inhibitors; thus, β-CA2 is an interesting target for drug development [96]. Additionally, Rv3588c has been identified as an essential gene for both mycobacterial *in vitro* growth and *in vivo* virulence in mice *via* transposon site hybridization (TraSH) assays and macrophage infections [80,82,83,97]. β-CA2 seems to be enriched under biofilm conditions (GSE70718) and upregulated during phosphate starvation (GSE183469). Most of the studied environmental stresses had no effect on Rv3588c expression, although the expression of this gene decreased during hypoxia (Table 3).

6. β-CA3 serves as a dual-function transmembrane protein

The β-CA3 protein encoded by the *Rv3273* gene is 764 amino acid residues in length [56]. Orthologs in *M. bovis* (UniProt ID A0A0H3MED5) and Mmr (MMAR_1266) have 100% and 73.09% identical protein sequences, respectively. In contrast to the two cytosolic mycobacterial β-CAs, β-CA3 is directly attached to the plasma membrane. This transmembrane protein has a dual function as both a carbonic anhydrase and an inorganic ion sulfate/bicarbonate transporter [56]. The hydrophobic N-terminus contains the SLC26A/SulP transporter domain (Pfam: PF00916) between amino acid residues 31–398. The ten transmembrane helices are annotated between residues 35–424 The β-CA domain (Pfam: SM00947) is predicted to be located between amino acid residues 572–743 at the C terminus of the protein (domain and region data from UniProt: P96878, https://www.uniprot.org/uniprotkb/P96878/entry). In CDD, the CA domain does not match any of the specific beta CA clades, just the more generic CynT signature (CDD:440057). Similar CA-sulfate transporter combination proteins have also been found in other bacterial species, and the authors suggest that this arrangement provides a built-in transport metabolon in one polypeptide [98]. The structure of β-CA3 has not been resolved experimentally, but AlphaFold predictions of the structure can be found under UniProt ID P96878. β-CA3 also possesses all the conserved residues of the active site motifs CxDxR and HxxC, but as ligands are missing from AlphaFold models, we cannot infer the precise metal coordination.

In contrast to the two other mycobacterial β-CAs, this enzyme is not necessary for Mtb growth or virulence. Moreover, *Rv3273* was not included in the TraSH assays, which indicated that the two other enzymes are more essential [82,83]. Other studies also indicate that *Rv3273* is not essential [80,84]. β-CA3 seems to have some type of role in mycobacterial virulence, but the specific mechanisms involved remain unknown [56]. The enzyme also seems to be more vulnerable to environmental stress, and its expression is strongly downregulated upon, *e.g.*, acidic pH, hypoxia, and nutrient starvation (Table 3). Furthermore, *Rv3273* expression is downregulated when the cytoplasmic redox sensor *whiB3* is inactivated, which interferes with the transcriptional response to acidic pH [95].

7. Inhibition of *M. tuberculosis* β-carbonic anhydrases

Mycobacterial β-CAs are interesting targets for drug development because they are involved in essential mechanisms for both viability and virulence. CA inhibitors (CAIs) can be used to treat tuberculosis alone or in combination with preexisting treatments as well as in the treatment of other mycobacterial diseases [99]. There are already several α–CAIs in clinical practice that are commonly used for the treatment of glaucoma, epilepsy and diuretics [100]. The most common inhibition mechanism utilized by CAIs is to anchor the enzyme active site zinc, and CAIs can, for instance, interfere with dimer-to-tetramer changes, keeping specific CAs inactive [100]. As mycobacterial β–CAs have rather small active cavities and CAIs need to pass through the hydrophobic mycobacterial cell wall, it is challenging to discover β–CA-selective inhibitors that are functional *in vivo* [59,101].

When examining classical CAIs in clinical practice, highly active hCA II can be inhibited at low nanomolar concentrations (Table 4). Some classical CAIs have good inhibitory effects on mycobacterial enzymes, and acetazolamide can inhibit β–CA2 even more efficiently than hCA II. However, these classical CAIs have mainly moderate-to-poor inhibitory effects on mycobacterial CAs. Similarly, some of the other known CAIs, such as sulpiride and sulthiame, have good inhibitory effects on human isozymes but lower effects on mycobacterial enzymes. The most common inhibitor classes and their results are reviewed below.

8. Inhibition with sulfonamides

Sulfonamides, RSO_2NH_2, and their derivatives are the most classical and abundant group of CAIs. They were discovered in 1940, and ever since, there have been more than 20 sulfonamides inhibiting α–CAs in clinical practice and several in clinical trials [100]. Minakuchi et al. were the first to evaluate β–CA1 as a druggable target and to determine whether sulfonamides could inhibit the activity of this enzyme [57]. From the panel of sulfonamides, 3-bromosulfanilamide (β–CA1 K_i = 186 nM) and indisulam (β–CA1 K_i = 97 nM) had nanomolar inhibitory effects [57]. However, specific β–CA1 inhibition is difficult to achieve (Table 5). Maresca et al. reported a series of halogenated sulfanilamides with the greatest activity toward β–CA1, the two top inhibitors being bromo–chloro–derivative 8 (β–CA1 K_i = 580 nM) and dibromosulfanilamide 10 (β–CA1 K_i = 410 nM) [107]. In the same study, halogenated benzolamides (monochloro derivative 1, bromo–fluoro-substituted benzolamide 2 and iodo–fluoro-substituted compound 3) showed nanomolar inhibitory effects on all three β–CAs [107]. Another CAI inhibiting β–CA1 at nanomolar concentrations is halogen-substituted sulfanilamide 9 (β–CA1 K_i = 186 nM) [56].

Table 4 Inhibition constants ($_{KIs}$) (nM) of classical clinically relevant CAIs.

CAI	hCA I	hCA II	β-CA1	β-CA2	β-CA3	References
Acetazolamide	250	12.5	480	9.8	104	[75]
Benzolamide	15	9	810	460	340	[101]
Brinzolamide	45,000	3	839	201	127	[56]
Celecoxib	50,000	21	10,350	7760	713	[56]
Dichlorophenamide	1200	38	870	2010	611	[101]
Dorzolamide	50,000	9	744	137	99	[56,58]
Ethoxzolamide	25	8	1030	594	27	[56,58]
Methazolamide	50	14	781	562	660	[56,58]
Saccharin	18,540	5950	7960	7150	792	[56,58]
Sulpiride	12,000	40	2300	7920	266	[56,58]
Sulthiame	374	9	5160	6720	664	[56,58]
Topiramate	250	10	610	na	3020	[102]
Zonisamide	56	35	286,800	na	210	[102]

na = data not available.

In search of novel CAIs, *para-* and *meta*-substituted diazenylbenzene-sulfonamides were found to inhibit mycobacterial enzymes very poorly in contrast to hCA II [110]. However, prontosil, the first clinically used sulfonamide, was found to inhibit both β-CA1 (K_i = 126 nM) and β-CA3 (K_i = 148 nM) at far lower concentrations than hCA II [110]. Additionally, Pacchiano et al. (2011) prepared ureido-sulfonamides by combining sulfanilamides with aryl/alkyl isocyanates [113]. Derivates 7–10, all of which included 4-halogenosubstituted phenyl or pentafluorophenyl R moieties, inhibited β-CA1 (K_i range 4.8–7.5 nM), whereas derivatives 17–21 with alkyl/alkoxyphenyl or cyanophenyl moieties inhibited β-CA3 (K_i range 57.5–81.2 nM) [113].

Similarly, β-CA2 was proven to be a target for inhibition [58]. According to the data from Nishimori et al. (2009) and Carta et al. (2009), a series of diazenylbenzenesulfonamides have inhibitory effects on all Mtb β-CAs [56,58]. The inhibition ranges mostly at micromolar concentrations,

Table 5 A list of CAIs with inhibition constants (K_{Is}) (nM) in the nanomolar range and selectivity toward mycobacterial CAs.

CAI	hCA I	hCA II	β-CA1	β-CA2	β-CA3	References
2-Amino-pyrimidin-4-yl-sulfanilamide (18)	109	33	750	2630	91	[56,58]
Aromatic/heterocyclic carboxylates	na	na	2250–7130	590–8100	110–970	[103]
Benzenesulphonamide derivatives (5f,h,i,k,o,p and 6i-j)	276–898	240–3461	na	na	127–275	[104]
3-Bromosulfanilamide	6500	40	186	na	na	[57]
BTD-CAIs	547–ns	140–2940	556–ns	38.1–4480	15.1–2250	[105]
C-Cinnamoyl glycosides	na	3100–8800	140–4500	130–1150	3250–19,000	[106]
Diazenylbenzenesulfonamides 23e-f	na	104–105	6780–8710	45–59	8900–9030	[58]
Dithiocarbamates (8–10, 12)	0.88–33.5	0.92–33.0	na	na	2.4–43.0	[81]
Halogenated benzolamides	na	na	120–290	410–450	170–300	[107]
Halogenated sulfanilamides	na	na	410–1270	4140–4740	2030–3190	[107]
Halogeno-substituted sulfanilamide (9)	6500	40	186	31,600	7320	[56,58]
2-(Hydrazinocarbonyl)-3-aryl-1H-indole-5-Sulfonamides	3.2–113	0.93–3380	0.92–48	na	0.88–31	[108]

Indisulam	31	15	97	7340	717	[56,58]
Monothiocarbamates (4, 5, 7)	891–ns	26.7–43.7	na	na	83.3–97.9	[81]
N-Mono- and N,N-disubstituted dithiocarbamates (1–12)	3.5–33.5	0.7–325	4.2–89.4	na	2.4–87.3	[109]
N-Mono- and N,N-disubstituted dithiocarbamates (13–27)	0.88–1838	0.92–6910	0.94–893	na	0.91–659	[109]
Prontosil	na	13,000	126	na	148	[110]
Pyrazole-tethered sulfamoylphenyl acetamides (meta-aminobenzene)	545–1327	270–845	404–710	61.7–97.8	197–399	[75]
Pyrazole-tethered sulfamoylphenyl acetamides (para-aminobenzene)	36.6–170	31.0–114	924–1241	232–513	488–652	[75]
Pyrazole-tethered sulfamoylphenyl benzamides (9)	701–918	383–764	477–780	69.5–93.6	169–366	[75]
Pyrazole-tethered sulfamoylphenyl benzamides (10)	6617–9857	4919–8409	6683–8968	115–165	3270–5510	[75]
Seleno/sulfur-acetamides (3e, 7a–b)	na	na	na	na	202.5–282.8	[111]
Sulfanilyl-sulfonamide (17)	185	50	853	2290	170	[56,58]

(continued)

Table 5 A list of CAIs with inhibition constants (K_{Is}) (nM) in the nanomolar range and selectivity toward mycobacterial CAs. (*cont'd*)

CAI	hCA I	hCA II	β-CA1	β-CA2	β-CA3	References
Triazinyl-sulfonamides (2, 4, 6)	90–250	5.0–7.1	42–610	8.1–9.6	2.1–1000	[101]
1,2,3-triazole-bearing benzenesulfonamides (13–17)	92.8–ns	1.0–15.7	na	na	28.2–192.5	[112]
Ureido-sulfonamides 7–10	na	96–2634	4.8–7.5	na	53.0–87.1	[113]
Ureido-sulfonamides 17–21	na	2.1–9600	35.2–6500	na	57.5–81.2	[113]

na=data not available, ns=nonsignificant.

which is still insufficient compared to the low nanomolar range of hCA II inhibition [58]. Interestingly, two diazenylbenzenesulfonamides, amino-methylene sodium sulfonate derivative 23e (β-CA2 K_i = 59 nM) and N-methylated analog 23 f (β-CA2 K_i = 45 nM), inhibited β-CA2 at nanomolar concentrations lower than those of hCA II [58]. Meta-aminobenzene pyrazole-tethered sulfamoylphenyl acetamides were found to inhibit β-CA2 (K_i = 61.7–97.8 nM) at much lower concentrations than either hCA I or hCA II [75]. Similar results were obtained with pyrazole-tethered sulfamoylphenyl benzamides (β-CA2 K_i = 69.6–93.6 nM), which also inhibited β-CA3 (K_i = 169–366 nM) [75]. A panel of N-((4-sulfamoylphenyl)carbamothioyl) amides showed low nanomolar inhibition of β-CA2, with most compounds being more selective toward β CA2 than hCA I and II [114]. Moreover, Ceruso et al. (2013) screened triazinyl sulfonamides and identified three interesting CAIs that inhibit all Mtb β-CAs [101]. The difluorotriazine derivative of 4–aminoethyl benzenesulfonamide (2) had the greatest activity (β-CA1 K_i = 4.2 nM, β-CA2 K_i = 8.6 nM, β-CA3 K_i = 2.1 nM), and the difluoro derivative of sulfanilamide with a methyl amine (4) and glycine derivative of 4–aminoethylbenzenesulfonamide (6) inhibited the activity, especially β-CA2 [101]. These compounds could be interesting for future bacterial experiments to determine whether enzyme kinetics are correlated with in vivo efficacy.

As already stated above, β-CA3 inhibition is also achievable with sulfonamides. CAIs that inhibit β-CA3 to a lesser extent than hCA I and II include sulfanilyl-sulfonamide 17 (β-CA3 K_i = 170 nM) and 2-amino-pyrimidin-4-yl-sulfanilamide 18 (β-CA3 K_i = 91 nM) [56]. Bua et al. tested two sulfonamide subsets against β-CA3 and found that 1,2,3-triazole-bearing benzenesulfonamides 13–17 inhibited β-CA3 better than did acetazolamide 30716618 [112]. Wani et al. [104] screened novel sulfonamide derivatives against β-CA3, with most compounds having medium-to-poor potency. However, some CAIs had better efficacy against β-CA3 than hCA I and II, which are a series of benzenesulfonamide derivatives [104]. 2-thio- and 2-seleno-acetamides were tested against several pathogen (Vibrio cholerae, Burkholderia pseudomallei, and Salmonella enterica serovar Typhimurium) CAs, including β-CA3 [111]. A few selenium- or sulfur-containing acetamides bearing a benzenesulfonamide moiety inhibited β-CA3 in the K_i-range of 202.5–282.8 nM [111]. Interestingly, many compounds showed low nanomolar inhibition of Salmonella enterica serovar Typhimurium β-CA2, and hence, it would be fascinating to determine whether these sulfonamides inhibit the two other Mtb β-CAs. Güzel et al. identified a series of 2-(hydrazino-carbonyl)-3-aryl-1H-indole-5-sulfonamides in which the inhibitory activities against β-CA1 and β-CA3 were phenomenal, reaching subnanomolar K_i-ranges [108]. In particular, β-CA1 reached K_i < 1.0 nM with compounds 13–17, and for β-CA3, the same was achieved with compounds 9–12 and 17 [108]. Finally, Bua et al. (2024) screened a panel of twenty 2H-benzo[e] [1,2,4]thiadiazin-3(4H)-one 1,1-dioxide (BTD)-based CAIs (BTD-CAIs), which proved to have good inhibition and selectivity toward β-CA2 and β-CA3, in addition to compounds containing diCH3 moieties. These BTD-CAIs were designed to target the narrow active site of β-CAs, a strategy for seeking selective inhibitors [105].

9. Inhibition with dithiocarbamates

Dithiocarbamate (R_2N-CS_2^-) derivatives can coordinate the transition of metal ions and inhibit CAs by replacing the fourth zinc ligand, which is not an amino acid residue [109]. These compounds have pre-existing applications in agriculture [109]. Maresca et al. screened a series of N-mono- and N,N-disubstituted dithiocarbamates [109]. The best overall inhibition was achieved with compounds obtained from primary amines and carbon disulfide (1−12), whereas compounds obtained from secondary amines and carbon disulfide (13−27) occasionally had higher K_is but were also the best hit (β-CA1 K_i = 0.94 nM, β-CA3 K_i = 0.91 nM) with morpholine-dithiocarbamate (24) [109]. Similarly, Aspatwar et al. studied the inhibitory effects of a series of mono- and dithiocarbamates [81]. Several compounds from both groups had β-CA3 inhibition constants within the low-nanomolar range, and N-methyl piperazine (compound 12) showed β-CA3 selectivity over human enzymes [81].

10. Inhibition with natural products and carboxylic acids

To increase the repertoire of nonsulfonamide CAIs, C-cinnamoyl glycosides containing a phenol moiety were synthesized [106]. Seven out of the eight compounds inhibited β-CA2 at nanomolar concentrations, two compounds inhibited β-CA1 at nanomolar concentrations, and none of the compounds inhibited β-CA3 at the same level [106]. Notably, all the compounds inhibited hCA II in the micromolar range, indicating selectivity toward mycobacterial enzymes [106]. It was also shown that phenol alone is not capable of inhibiting Mtb enzymes [106]. Additionally, Dallaston et al. screened a library of phenol-based natural products. Of the 15 compounds tested, 2,3-dichloronaphthoquinone had the best efficacy, inhibiting approximately 80% of β-CA2 activity [115]. Interestingly, these compounds seem to target sites other than the active cavity [115]. Another class of nonsulfonamides with potential CAIs is carboxylic acids. Maresca et al. screened carboxylic acids incorporating various aromatic/heterocyclic scaffolds against Mtb β-CAs and reported nanomolar K_i values against B-CA3 [103].

In addition to the compounds reviewed above, several studies have reported inhibition studies in which the K_i values were greater for Mtb CAs than for human enzymes. Such CAIs are likely to be ineffective because they are more selective toward human isozymes. These compounds include para- and meta-substituted diazenylbenzenesulfonamides [110], sulfanilamide [102], natural product-based phenols [102], phenolic acids [96] and dopamine-related sulfonamides [116].

11. Inhibition of *M. tuberculosis* in culture
11.1 *In vitro* inhibition

The numerous studies described above show CA inhibition from the perspective of enzyme kinetics; however, *in vitro* and *in vivo* proof of CAI efficacy is still lacking [107]. For instance, several β-CA2 inhibitors have great potential for use in CO_2 hydration assays but fail to reduce mycobacterial growth or survival, emphasizing the need for more efficient inhibitors to reduce this bottleneck effect [97]. Some success stories clearly exist, as described below (Table 6). The series of pyrazole-tethered sulfamoyl phenyl acetamides and pyrazole-tethered sulfamoyl phenyl benzamides described in the previous section were further tested against the Mtb H37Rv (MC^2 6230) strain. Minimum inhibitory concentrations (MICs) were measured from liquid cultures after ten days of growth, and the greatest inhibition was observed with *para*-aminobenzene sulfonamide compounds, followed by compounds containing a phenyl reverse carboxamide linker and *meta*-aminobenzene sulfonamides [75]. In total, six of the tested CAIs reached an MIC of 4 µg/mL [75]. Scozzafava et al. [120] tested the inhibitory activities of a panel of eleven 9-sulfenylated/

Table 6 CAIs that inhibit mycobacteria *in vitro*.

CAI	Mycobacteria	Result	References
BTD-CAIs (2a, 2c, 2d, 2h, 2t)	Mtb	MIC < 2 µg/mL	[106]
C-glycosides (1, 2, 6, 7, 10)	Mtb H37Rv	MIC 50 µg/mL	[117]
β-CA2-knockdown	Mtb CDC1551	Macrophage survival assay, Mtb growth reduced by 1-log	[80]
β-CA2-knockdown	Mtb H37Rv	Reduced growth and survival	[97]
Dithiocarbamates (Fc14-584b and Fc14-594a)	Mmr ATCC 927	MIC 17–18 µg/mL, bactericidal	[61]
(E)-1-(2′,3′,4′,6′-tetra-O-acetyl-β-d-glucopyranosyl)-4-(3-hydroxyphenyl)but-3-en-2-one	Mtb H37Rv	MIC 3.125–6.25 µg/mL.	[106]
Ethoxzolamide	Mtb CDC1551	Macrophage survival assay, Mtb growth reduced by 0.25–0.5-log	[80]
Ethoxzolamide	Mtb CDC1551	Macrophage survival assay, Mtb growth reduced by 1-log	[118]
Ethoxzolamide	*M. avium* subsp. hominis. suis A5	Reduced eDNA transport and biofilm formation	[119]
N,N′-Diphenylterephthalamide derivatives (7 derivates)	Mtb H37Rv	MIC < 1 µg/mL	[97]
pyrazole-tethered sulfamoylphenyl acetamides (meta-aminobenzene)	Mtb H37Rv (MC2 6230)	MIC 8–128 µg/mL	[75]

Pyrazole-tethered sulfamoylphenyl acetamides (para-aminobenzene)	Mtb H37Rv (MC2 6230)	MIC 4 μg/mL	[75]
Pyrazole-tethered sulfamoylphenyl benzamides	Mtb H37Rv (MC2 6230)	MIC 4–64 μg/mL	[75]
9-sulfenylated/ sulfonylated-6-mercaptopurines	Mtb H37Rv	MIC 0.39–3.39 μg/mL	[120]
9-sulfenylated/ sulfonylated-6-mercaptopurines	M. avium	40–81% inhibition at 12.5 μM	[120]
9-sulfenylated/ sulfonylated-6-mercaptopurine (with 4-O2NC6H4 moiety)	Mtb H37Rv (drug-resistant strains)	MIC 0.78 μg/mL (RMP, EMB) MIC 1.56 μg/mL (KM) MIC 6.25 μg/mL (INH)	[120]

In vivo inhibition.

sulfonylated-6-mercaptopurines against Mtb H37Rv. Mercaptopurines inhibited growth at low concentrations, and similar inhibition was achieved with *M. avium* [120]. Notably, a compound from this series inhibited the growth of several drug-resistant Mtb strains [120]. Compound 8b contained a $4\text{-}O_2NC_6H_4$ moiety and inhibited mycobacterial growth at 0.78–6.25 μg/mL in rifampicin (RMP)-, ethambutol (EMB)-, kanamycin (KM)- and isoniazid (INH)-resistant strains as well as in a wild-type, drug-susceptible strain [120]. Similarly, five BTD-CAIs selective for β-CAs (Table 4) were screened against drug-susceptible and drug-resistant strains of Mtb [105]. Interestingly, all the compounds were extremely effective, halting bacterial growth at concentrations less than 2 μg/mL in both drug-susceptible and rifampicin- and isoniazid-resistant strains [105].

Among the *C*-cinnamoyl glycosides discussed in the previous chapter, four compounds (1, 2, 6, 7) inhibited Mtb H37Rv growth completely at 100 μg/mL and partially at 50 μg/mL [106]. Additionally, compound 1, ((*E*)-1-(2′,3′,4′,6′-tetra-O-acetyl-β-d-glucopyranosyl)-4-(3-hydroxyphenyl) but-3-en-2-one), inhibited over 99.99% of the growth on solid media during a 30-day observation period, with MIC values ranging from 3.125 to 6.25 μg/mL. Similar MIC tests were performed on C-glycosides previously known to inhibit α-CAs [117,121]. Peracetylated C-glycosides (1−2) and fully deprotected derivatives (6−7) containing a phenol moiety, as well as per-O-acetylated C-glycoside (10), inhibited the growth of Mtb H37Rv at 50 and 100 μg/mL in liquid culture after seven days of growth. Recently, 25 N,N′-diphenylterephthalamide derivative compounds were proven to be ligands of β-CA2 by molecular docking studies, and the affinity of the hit compounds was verified with a thermal shift assay [97]. Forty-eight percent of the tested derivates inhibited Mtb growth at 10 μg/mL; furthermore, seven compounds with MICs ≤ 1 μg/mL were identified [97]. The same study also confirmed the essentiality of β-CA2 by studying the growth and survival of the conditionally $Rv3588c$-knockdown (TetR-Pip-OFF repressible system) Mtb H37Rv strain [97].

In a recent study, Dechow et al. disrupted the function of all three Mtb β-CAs. Both $Rv1284$ and $Rv3588c$ were knocked down via CRISPRi, whereas $Rv3273$ was knocked out via ORBIT [80]. CA-modified Mtb CDC1551 strains were used to infect bone marrow-derived macrophages, and after a 9-day assay, the survival of these strains was evaluated [80]. Neither $Rv1284$ knockdown nor $Rv3273$ knockout reduced growth, indicating that these enzymes are not essential for Mtb virulence [80]. However, $Rv3588c$ knockdown results in a 1-log decrease in growth, strengthening the hypothesis that β-CA2 plays an essential role in mycobacterial virulence [80]. Additionally, ethoxzolamide treatment led to a 0.25–0.5-log decrease in bacterial growth, highlighting the fact that CAIs could be used to treat mycobacterial infections [80]. Moreover, in a different study, ethoxzolamide (100 nM) was found to reduce eDNA transport [119]. The same study screened $M.$ $avium$ transposon mutants and found a CA among the most eDNA-deficient mutants, linking CA activity to biofilm formation processes [119]. CAs were also found to be enriched in wild-type biofilms [119].

In addition to $M.$ $avium$, Mmr is used as a model organism to study tuberculosis. Mmr is a natural pathogen of zebrafish ($Danio$ $rerio$) and causes TB-like disease in fish, in addition to being safer and faster to grow in laboratories [61]. Accordingly, the presence of the three β-CAs has been verified and quantified in several Mmr laboratory strains (ATCC 927, E11 and M) [61]. Dithiocarbamates screened by Maresca et al. were tested against Mmr [61,109]. Fc14-584b and Fc14-594a were identified as potential hits because they inhibited bacterial growth at 17–18 µg/mL, and growth could be eliminated at high concentrations [61]. Mmr did not recover after CAI treatment, indicating that these compounds have bactericidal effects [61].

Only a few publications have described the *in vivo* antimycobacterial applications of CAIs (Table 7). Johnson et al. were the first to describe the effects of a CAI, ethoxzolamide, against Mtb [118]. The treatment significantly reduced reporter fluorescence, indicating successful enzyme inhibition, but had no effect on *in vitro* growth [118]. However, ethoxzolamide has several phenotypic effects, including the suppression of the PhoPR regulon, virulence-associated lipid production, and Esx-1 secretion [118]. Despite not affecting *in vitro* bacterial growth, Mtb CDC1551-infected macrophages were assayed for nine days and showed an approximately 1-log decrease in growth when treated with ethoxzolamide (80 µM) [118]. Similarly, ethoxzolamide-treated Erdman-Mtb-infected mice showed reduced bacterial loads, indicating that CAI treatment attenuates Mtb virulence [118]. Another study utilized a zebrafish larval model as an infection host for Mmr [61]. Two dithiocarbamates previously shown to inhibit β-CAs and be bactericidal *in vitro* were first evaluated for safe administration [61,109]. Following the protocol for *in vivo* safety analysis described by Aspatwar et al., Fc14-594A was shown to be toxic, and the safer dithiocarbamate Fc14-584B (LC_{50} = 498.1 µM) was chosen for zebrafish infections [61]. Zebrafish embryos (1 day postfertilization) were infected with reporter-fluorescent Mmr, and after five days, the bacterial load between treatment groups was quantified, revealing a strikingly reduced (2.9–8.9-fold) bacterial load with Fc14-584B-treated (300 µM) larvae [61].

Table 7 CAIs that inhibit mycobacteria *in vivo*.

CAI	Mycobacteria	Host	Result	References
Ethoxzolamide	Mtb Erdman	C57BL/6 mice	0.72-log reduction of bacteria	[118]
Dithiocarbamate	Mmr ATCC 927	Zebrafish larvae	2.9–8.9-fold reduction of bacteria	[61]

12. Summary and future directions

This chapter provides the currently available information on the three β-carbonic anhydrases of *Mycobacterium tuberculosis,* a pathogen responsible for more than a billion deaths to date. We presented data on the molecular biology and kinetics of the three β-carbonic anhydrase enzymes and described the possible physiological roles of each enzyme reported thus far. In addition, we described the use of computational tools to discover compounds targeting these enzymes that accelerated the process of drug development against the TB pathogen.

Studies using different classes of carbonic anhydrase inhibitors have shown that the enzymatic activity of Mtb β-CAs can be efficiently inhibited. These inhibitions impaired the growth of both laboratory and clinical strains of Mtb. These compounds were effective in reducing the growth of biofilms in *M. marinum,* a close relative of Mtb. In addition, *in vivo* studies using zebrafish and *M. marinum,* a natural TB pathogen of zebrafish, demonstrated that β-CA inhibitors inhibit the growth of the bacterium. The available studies on the roles of Mtb β-CAs demonstrated that these enzymes play crucial metabolic roles unique to Mtb, helping them survive and establish infection in the human host. Similarly, *in vitro* and *in vivo* studies are proof-of-concept that the β-CAs of Mtb are novel targets for developing anti–TB drugs that are potentially safe, as humans lack β-CAs, are cost-effective, and are devoid of resistance by the tuberculin bacillus. Few studies have also shown that when these drugs are combined with rifampicin, a first-line TB drug, the growth of the bacterium can be reduced more efficiently than with either β-CAI or rifampicin alone. These findings suggest that some compounds could be combined with currently used TB drugs to reduce the duration of TB treatment.

Some Mtb bacilli that become refractory to treatment with antibiotics transiently and are otherwise sensitive to first-line drugs are known as persister cells that are difficult to kill. We need to study the specific molecular changes that occur during antibiotic treatment. To obtain better insights into the active site cavity and structural basis of the role of β-CA3, the crystal structure of this protein in complex with CA inhibitors is needed to help us design and synthesize specific inhibitors against each CA and understand their modes of action. It is well known that the active site cavities of Mtb β-CAs are narrower than those of α-CAs. In the future, it is important to consider this when designing inhibitors so that these compounds specifically inhibit the activity of Mtb β-CAs. In addition, the newly discovered compounds should be able to pass through the bacterial cell walls on their own or use innovative methods such as nanoemulsification of the inhibitor to deliver the inhibitors through the cell wall to inhibit the activity of intracellular β-CAs.

Acknowledgments

The authors thank Research Council of Finland, Jane & Aatos Erkko Foundation, Tampere Tuberculosis Foundation, Finnish Cultural Foundation, and Finnish Tuberculosis Resistance Association Foundation for granting research funding to the members of our research group.

References

[1] D.A. Stahl, J.W. Urbance, The division between fast- and slow-growing species corresponds to natural relationships among the mycobacteria, J. Bacteriol. 172 (1) (1990) 116–124.

[2] C.J. Kim, N.H. Kim, K.H. Song, P.G. Choe, E.S. Kim, S.W. Park, et al., Differentiating rapid- and slow-growing mycobacteria by difference in time to growth detection in liquid media, Diagn. Microbiol. Infect. Dis. 75 (1) (2013) 73–76.

[3] I. Comas, M. Coscolla, T. Luo, S. Borrell, K.E. Holt, M. Kato-Maeda, et al., Out-of-Africa migration and Neolithic coexpansion of Mycobacterium tuberculosis with modern humans, Nat. Genet. 45 (10) (2013) 1176–1182.

[4] M.C. Gutierrez, S. Brisse, R. Brosch, M. Fabre, B. Omaïs, M. Marmiesse, et al., Ancient origin and gene mosaicism of the progenitor of Mycobacterium tuberculosis, PLoS Pathog. 1 (1) (2005) e5.

[5] M.J. Fenton, M.W. Vermeulen, Immunopathology of tuberculosis: roles of macrophages and monocytes, Infect. Immun. 64 (3) (1996) 683–690.

[6] C.E. Barry 3rd, H.I. Boshoff, V. Dartois, T. Dick, S. Ehrt, J. Flynn, et al., The spectrum of latent tuberculosis: rethinking the biology and intervention strategies, Nat. Rev. Microbiol. 7 (12) (2009) 845–855.

[7] World Health Organization and (WHO), Global Tuberculosis Report, 2023, pp. 1–232. ISBN 978-92-4-008385-1.

[8] W. Gong, J. Xie, H. Li, A. Aspatwar, Editorial: Research advances of tuberculosis vaccine and its implication on COVID-19, Front. Immunol. 14 (2023) 1147704.

[9] R.M. Houben, P.J. Dodd, The global burden of latent tuberculosis infection: a re-estimation using mathematical modelling, PLoS Med. 13 (10) (2016) e1002152.

[10] M. Combrink, D.T. Loots, I. du Preez, Metabolomics describes previously unknown toxicity mechanisms of isoniazid and rifampicin, Toxicol. Lett. 322 (2020) 104–110.

[11] L.Y. Hsu, L.Y. Lai, P.F. Hsieh, T.L. Lin, W.H. Lin, H.Y. Tasi, et al., Two novel katG mutations conferring isoniazid resistance in Mycobacterium tuberculosis, Front. Microbiol. 11 (2020) 1644.

[12] B.P. Goldstein, Resistance to rifampicin: a review, J. Antibiot. (Tokyo) 67 (9) (2014) 625–630.

[13] D. Saroha, D. Garg, A.K. Singh, R.K. Dhamija, Irreversible neuropathy in extremely-drug resistant tuberculosis: an unfortunate clinical conundrum, Indian J. Tuberc. 67 (3) (2020) 389–392.

[14] M.C. Li, R. Chen, S.Q. Lin, Y. Lu, H.C. Liu, G.L. Li, et al., Detecting ethambutol resistance in Mycobacterium tuberculosis isolates in China: a comparison between phenotypic drug susceptibility testing methods and DNA sequencing of embAB, Front. Microbiol. 11 (2020) 781.

[15] Q. Sun, X. Li, L.M. Perez, W. Shi, Y. Zhang, J.C. Sacchettini, The molecular basis of pyrazinamide activity on Mycobacterium tuberculosis PanD, Nat. Commun. 11 (1) (2020) 339.

[16] B.S. Kwon, Y. Kim, S.H. Lee, S.Y. Lim, Y.J. Lee, J.S. Park, et al., The high incidence of severe adverse events due to pyrazinamide in elderly patients with tuberculosis, PLoS One 15 (7) (2020) e0236109.

[17] M.M. Islam, Y. Tan, H.M.A. Hameed, Y. Liu, C. Chhotaray, X. Cai, et al., Prevalence and molecular characterization of amikacin resistance among Mycobacterium tuberculosis clinical isolates from southern China, J. Glob. Antimicrob. Resist. 22 (2020) 290–295.

[18] A. Sowajassatakul, T. Prammananan, A. Chaiprasert, S. Phunpruch, Molecular characterization of amikacin, kanamycin and capreomycin resistance in M/XDR-TB strains isolated in Thailand, BMC Microbiol. 14 (2014) 165.

[19] D. Shrestha, B. Maharjan, N.A. Thida Oo, N. Isoda, C. Nakajima, Y. Suzuki, Molecular analysis of streptomycin-resistance associating genes in Mycobacterium tuberculosis isolates from Nepal, Tuberculosis (Edinb.) 125 (2020) 101985.

[20] R. Shi, N. Itagaki, I. Sugawara, Overview of anti-tuberculosis (TB) drugs and their resistance mechanisms, Mini Rev. Med. Chem. 7 (11) (2007) 1177–1185.

[21] A.S. Ginsburg, J.H. Grosset, W.R. Bishai, Fluoroquinolones, tuberculosis, and resistance, Lancet Infect. Dis. 3 (7) (2003) 432–442.

[22] A.D. Pranger, T.S. van der Werf, J.G.W. Kosterink, J.W.C. Alffenaar, The role of fluoroquinolones in the treatment of tuberculosis in 2019, Drugs 79 (2) (2019) 161–171.

[23] J. Chen, S. Zhang, P. Cui, W. Shi, W. Zhang, Y. Zhang, Identification of novel mutations associated with cycloserine resistance in Mycobacterium tuberculosis, J. Antimicrob. Chemother. 72 (12) (2017) 3272–3276.

[24] G.D. Coxon, D. Craig, R.M. Corrales, E. Vialla, L. Gannoun-Zaki, L. Kremer, Synthesis, antitubercular activity and mechanism of resistance of highly effective thiacetazone analogues, PLoS One 8 (1) (2013) e53162.

[25] D. Jaganath, G. Lamichhane, M. Shah, Carbapenems against Mycobacterium tuberculosis: a review of the evidence, Int. J. Tuberc. Lung Dis. 20 (11) (2016) 1436–1447.

[26] P. Kumar, A. Kaushik, D.T. Bell, V. Chauhan, F. Xia, R.L. Stevens, et al., Mutation in an unannotated protein confers carbapenem resistance in Mycobacterium tuberculosis, Antimicrob. Agents Chemother. 61 (3) (2017).

[27] T. Zhang, G. Jiang, S. Wen, F. Huo, F. Wang, H. Huang, et al., Para-aminosalicylic acid increases the susceptibility to isoniazid in clinical isolates of Mycobacterium tuberculosis, Infect. Drug. Resist. 12 (2019) 825–829.

[28] Y. Minato, J.M. Thiede, S.L. Kordus, E.J. McKlveen, B.J. Turman, A.D. Baughn, Mycobacterium tuberculosis folate metabolism and the mechanistic basis for para-aminosalicylic acid susceptibility and resistance, Antimicrob. Agents Chemother. 59 (9) (2015) 5097–5106.

[29] A.D. Bendre, P.J. Peters, J. Kumar, Tuberculosis: past, present and future of the treatment and drug discovery research, Curr. Res. Pharmacol. Drug. Discov. 2 (2021) 100037.

[30] S.M.R. Hashemian, T. Farhadi, M. Ganjparvar, Linezolid: a review of its properties, function, and use in critical care, Drug. Des. Devel. Ther. 12 (2018) 1759–1767.

[31] T.V.A. Nguyen, R.M. Anthony, T.T.H. Cao, A.L. Bañuls, V.A.T. Nguyen, D.H. Vu, et al., Delamanid resistance: update and clinical management, Clin. Infect. Dis. 71 (12) (2020) 3252–3259.

[32] J.M. Lewis, D.J. Sloan, The role of delamanid in the treatment of drug-resistant tuberculosis, Ther. Clin. Risk Manag. 11 (2015) 779–791.

[33] D. Rifat, S.Y. Li, T. Ioerger, K. Shah, J.P. Lanoix, J. Lee, et al., Mutations in fbiD (Rv2983) as a novel determinant of resistance to pretomanid and delamanid in Mycobacterium tuberculosis, Antimicrob. Agents Chemother. 65 (1) (2020).

[34] G. Rajalakshmi, M.S. Pavan, P. Kumaradhas, Charge density distribution and electrostatic interactions of ethionamide: an inhibitor of the enoyl acyl carrier protein reductase (inhA) enzyme of Mycobacterium tuberculosis, RSC Adv. 4 (101) (2014) 57823–57833.

[35] C. Vilchèze, W.R. Jacobs Jr., Resistance to isoniazid and ethionamide in Mycobacterium tuberculosis: genes, mutations, and causalities, Microbiol. Spectr. 2 (4) (2014) Mgm2-0014-2013.

[36] E. Narmandakh, O. Tumenbayar, T. Borolzoi, B. Erkhembayar, T. Boldoo, N. Dambaa, et al., Genetic mutations associated with isoniazid resistance in Mycobacterium tuberculosis in Mongolia, Antimicrob. Agents Chemother. 64 (7) (2020).

[37] J.A.M. Stadler, G. Maartens, G. Meintjes, S. Wasserman, Clofazimine for the treatment of tuberculosis, Front. Pharmacol. 14 (2023) 1100488.

[38] S. Kadura, N. King, M. Nakhoul, H. Zhu, G. Theron, C.U. Köser, et al., Systematic review of mutations associated with resistance to the new and repurposed Mycobacterium tuberculosis drugs bedaquiline, clofazimine, linezolid, delamanid and pretomanid, J. Antimicrob. Chemother. 75 (8) (2020) 2031–2043.

[39] E. Sumartojo, When tuberculosis treatment fails. A social behavioral account of patient adherence, Am. Rev. Respir. Dis. 147 (5) (1993) 1311–1320.

[40] N.R. Gandhi, P. Nunn, K. Dheda, H.S. Schaaf, M. Zignol, D. van Soolingen, et al., Multidrug-resistant and extensively drug-resistant tuberculosis: a threat to global control of tuberculosis, Lancet 375 (9728) (2010) 1830–1843.

[41] Z.F. Udwadia, R.A. Amale, K.K. Ajbani, C. Rodrigues, Totally drug-resistant tuberculosis in India, Clin. Infect. Dis. 54 (4) (2012) 579–581.

[42] L.G. Wayne, H.A. Sramek, Metronidazole is bactericidal to dormant cells of Mycobacterium tuberculosis, Antimicrob. Agents Chemother. 38 (9) (1994) 2054–2058.

[43] N.M. Parrish, J.D. Dick, W.R. Bishai, Mechanisms of latency in Mycobacterium tuberculosis, Trends Microbiol. 6 (3) (1998) 107–112.

[44] S.T. Cole, R. Brosch, J. Parkhill, T. Garnier, C. Churcher, D. Harris, et al., Deciphering the biology of Mycobacterium tuberculosis from the complete genome sequence, Nature 393 (6685) (1998) 537–544.

[45] E.C. Rivers, R.L. Mancera, New anti-tuberculosis drugs with novel mechanisms of action, Curr. Med. Chem. 15 (19) (2008) 1956–1967.

[46] J. Yang, L. Zhang, W. Qiao, Y. Luo, Mycobacterium tuberculosis: pathogenesis and therapeutic targets, MedComm 4 (5) (2020) e3532023.

[47] A. Aspatwar, M.E.E. Tolvanen, H. Barker, L. Syrjänen, S. Valanne, S. Purmonen, et al., Carbonic anhydrases in metazoan model organisms: molecules, mechanisms, and physiology, Physiol. Rev. (2022).

[48] A. Aspatwar, M.E. Tolvanen, S. Parkkila, Phylogeny and expression of carbonic anhydrase-related proteins, BMC Mol. Biol. 11 (2010) 25.

[49] A. Aspatwar, K.A. Qureshi, A. Parvez, J. Parkkinen, S. Parkkila, Bicarbonate and bacteria: friends or foes? PLoS Pathog. (2024), https://doi.org/10.1371/journal.ppat.1012029

[50] A. Aspatwar, C.T. Supuran, A. Waheed, W.S. Sly, S. Parkkila, Mitochondrial carbonic anhydrase VA and VB: properties and roles in health and disease, J. Physiol. 601 (2) (2023) 257–274.

[51] A. Aspatwar, L. Syrjänen, S. Parkkila, Roles of carbonic anhydrases and carbonic anhydrase related proteins in zebrafish, Int. J. Mol. Sci. 23 (8) (2022).

[52] C.T. Supuran, Carbonic anhydrases: novel therapeutic applications for inhibitors and activators, Nat. Rev. Drug. Discov. 7 (2) (2008) 168–181.

[53] A. Aspatwar, H. Barker, M. Tolvanen, R.Z. Emameh, S. Parkkila, Chapter 20—Carbonic anhydrases from pathogens: protozoan CAs and related inhibitors as potential antiprotozoal agents, in: C.T. Supuran, A. Nocentini (Eds.), Carbonic Anhydrases, Academic Press, 2019, pp. 449–475.

[54] K.A. Qureshi, A. Parvez, N.A. Fahmy, B.H. Abdel Hady, S. Kumar, A. Ganguly, et al., Brucellosis: epidemiology, pathogenesis, diagnosis and treatment-a comprehensive review, Ann. Med. 55 (2) (2023) 2295398.

[55] A. Aspatwar, V. Kairys, S. Rala, M. Parikka, M. Bozdag, F. Carta, et al., Mycobacterium tuberculosis beta-carbonic anhydrases: novel targets for developing antituberculosis drugs, Int. J. Mol. Sci. 20 (20) (2019).

[56] I. Nishimori, T. Minakuchi, D. Vullo, A. Scozzafava, A. Innocenti, C.T. Supuran, Carbonic anhydrase inhibitors. Cloning, characterization, and inhibition studies of a new beta-carbonic anhydrase from Mycobacterium tuberculosis, J. Med. Chem. 52 (9) (2009) 3116–3120.

[57] T. Minakuchi, I. Nishimori, D. Vullo, A. Scozzafava, C.T. Supuran, Molecular cloning, characterization, and inhibition studies of the Rv1284 beta-carbonic anhydrase from Mycobacterium tuberculosis with sulfonamides and a sulfamate, J. Med. Chem. 52 (8) (2009) 2226–2232.

[58] F. Carta, A. Maresca, A.S. Covarrubias, S.L. Mowbray, T.A. Jones, C.T. Supuran, Carbonic anhydrase inhibitors. Characterization and inhibition studies of the most active beta-carbonic anhydrase from Mycobacterium tuberculosis, Rv3588c, Bioorg Med. Chem. Lett. 19 (23) (2009) 6649–6654.

[59] A.S. Covarrubias, T. Bergfors, T.A. Jones, M. Hogbom, Structural mechanics of the pH-dependent activity of beta-carbonic anhydrase from Mycobacterium tuberculosis, J. Biol. Chem. 281 (8) (2006) 4993–4999.

[60] A. Aspatwar, S. Haapanen, S. Parkkila, An update on the metabolic roles of carbonic anhydrases in the model alga Chlamydomonas reinhardtii, Metabolites 8 (1) (2018).

[61] A. Aspatwar, M. Hammaren, S. Koskinen, B. Luukinen, H. Barker, F. Carta, et al., beta-CA-specific inhibitor dithiocarbamate Fc14-584B: a novel antimycobacterial agent with potential to treat drug-resistant tuberculosis, J. Enzyme Inhib. Med. Chem. 32 (1) (2017) 832–840.

[62] B.K. Chung, T. Dick, D.Y. Lee, In silico analyses for the discovery of tuberculosis drug targets, J. Antimicrob. Chemother. 68 (12) (2013) 2701–2709.

[63] R. Mehra, I.A. Khan, A. Nargotra, Anti-tubercular drug discovery: in silico implications and challenges, Eur. J. Pharm. Sci. 104 (2017) 1–15.

[64] S. Ahamad, S. Rahman, F.I. Khan, N. Dwivedi, S. Ali, J. Kim, et al., QSAR based therapeutic management of M. tuberculosis, Arch. Pharm. Res. 40 (6) (2017) 676–694.

[65] I. Cortés-Ciriano, Q.U. Ain, V. Subramanian, E.B. Lenselink, O. Méndez-Lucio, A.P. Ijzerman, et al., Polypharmacology modelling using proteochemometrics (PCM): recent methodological developments, applications to target families, and future prospects, MedChemComm 6 (1) (2015) 24–50.

[66] R. Bhowmik, R. Kant, A. Manaithiya, D. Saluja, B. Vyas, R. Nath, et al., Navigating bioactivity space in anti-tubercular drug discovery through the deployment of advanced machine learning models and cheminformatics tools: a molecular modeling based retrospective study, Front. Pharmacol. 14 (2023) 1265573.

[67] S. Ekins, P.B. Madrid, M. Sarker, S.G. Li, N. Mittal, P. Kumar, et al., Combining metabolite-based pharmacophores with bayesian machine learning models for Mycobacterium tuberculosis drug discovery, PLoS One 10 (10) (2015) e0141076.

[68] S. Rajasekhar, R. Karuppasamy, K. Chanda, Exploration of potential inhibitors for tuberculosis via structure-based drug design, molecular docking, and molecular dynamics simulation studies, J. Comput. Chem. 42 (24) (2021) 1736–1749.

[69] W.F. de Azevedo Jr., Molecular dynamics simulations of protein targets identified in Mycobacterium tuberculosis, Curr. Med. Chem. 18 (9) (2011) 1353–1366.

[70] M. Karplus, G.A. Petsko, Molecular dynamics simulations in biology, Nature 347 (6294) (1990) 631–639.

[71] T. Garnier, K. Eiglmeier, J.C. Camus, N. Medina, H. Mansoor, M. Pryor, et al., The complete genome sequence of Mycobacterium bovis, Proc. Natl. Acad. Sci. U. S. A. 100 (13) (2003) 7877–7882.

[72] R.S. Gupta, B. Lo, J. Son, Phylogenomics and comparative genomic studies robustly support division of the genus Mycobacterium into an emended genus Mycobacterium and four novel genera, Front. Microbiol. 9 (2018) 67.

[73] C.J. Meehan, R.A. Barco, Y.E. Loh, S. Cogneau, L. Rigouts, Reconstituting the genus Mycobacterium, Int. J. Syst. Evol. Microbiol. 71 (9) (2021).

[74] E.W. Sayers, J. Beck, E.E. Bolton, J.R. Brister, J. Chan, D.C. Comeau, et al., Database resources of the National Center for Biotechnology Information, Nucleic Acids Res. 52 (D1) (2024) D33–D43.

[75] O. Ommi, N. Paoletti, A. Bonardi, P. Gratteri, H.A. Bhalerao, S. Sau, et al., Exploration of 3-aryl pyrazole-tethered sulfamoyl carboxamides as carbonic anhydrase inhibitors, Arch. Pharm. (Weinh.) 356 (11) (2023) e2300309.

[76] A. Suarez Covarrubias, A.M. Larsson, M. Hogbom, J. Lindberg, T. Bergfors, C. Bjorkelid, et al., Structure and function of carbonic anhydrases from Mycobacterium tuberculosis, J. Biol. Chem. 280 (19) (2005) 18782–18789.

[77] L. Nienaber, E. Cave-Freeman, M. Cross, L. Mason, U.M. Bailey, P. Amani, et al., Chemical probing suggests redox-regulation of the carbonic anhydrase activity of mycobacterial Rv1284, FEBS J. 282 (14) (2015) 2708–2721.

[78] A. Liljas, M. Laurberg, A wheel invented three times. The molecular structures of the three carbonic anhydrases, EMBO Rep. 1 (1) (2000) 16–17.

[79] V. Srinivas, R.A. Ruiz, M. Pan, S.R.C. Immanuel, E.J.R. Peterson, N.S. Baliga, Transcriptome signature of cell viability predicts drug response and drug interaction in Mycobacterium tuberculosis, Cell Rep. Methods 1 (8) (2021) (None).

[80] S. Dechow, R. Goyal, B.J. Johnson, R.B. Abramovitch, Carbon dioxide regulates Mycobacterium tuberculosis PhoPR signaling and virulence, bioRxiv (2022).

[81] A. Aspatwar, M. Hammaren, M. Parikka, S. Parkkila, F. Carta, M. Bozdag, et al., In vitro inhibition of Mycobacterium tuberculosis beta-carbonic anhydrase 3 with Mono- and dithiocarbamates and evaluation of their toxicity using zebrafish developing embryos, J. Enzyme Inhib. Med. Chem. 35 (1) (2020) 65–71.

[82] C.M. Sassetti, D.H. Boyd, E.J. Rubin, Genes required for mycobacterial growth defined by high density mutagenesis, Mol. Microbiol. 48 (1) (2003) 77–84.

[83] C.M. Sassetti, E.J. Rubin, Genetic requirements for mycobacterial survival during infection, Proc. Natl. Acad. Sci. U. S. A. 100 (22) (2003) 12989–12994.

[84] M.A. DeJesus, E.R. Gerrick, W. Xu, S.W. Park, J.E. Long, C.C. Boutte, et al., Comprehensive essentiality analysis of the Mycobacterium tuberculosis genome via saturating transposon mutagenesis, mBio 8 (1) (2017).

[85] J. Wang, F. Chitsaz, M.K. Derbyshire, N.R. Gonzales, M. Gwadz, S. Lu, et al., The conserved domain database in 2023, Nucleic Acids Res. 51 (D1) (2023) D384–D388.

[86] K.S. Smith, C. Jakubzick, T.S. Whittam, J.G. Ferry, Carbonic anhydrase is an ancient enzyme widespread in prokaryotes, Proc. Natl. Acad. Sci. U. S. A. 96 (26) (1999) 15184–15189.

[87] S. Gu, J. Chen, K.M. Dobos, E.M. Bradbury, J.T. Belisle, X. Chen, Comprehensive proteomic profiling of the membrane constituents of a Mycobacterium tuberculosis strain, Mol. Cell Proteom. 2 (12) (2003) 1284–1296.

[88] K. Avican, J. Aldahdooh, M. Togninalli, A. Mahmud, J. Tang, K.M. Borgwardt, et al., RNA atlas of human bacterial pathogens uncovers stress dynamics linked to infection, Nat. Commun. 12 (1) (2021) 3282.

[89] C. Vilchèze, B. Yan, R. Casey, S. Hingley-Wilson, L. Ettwiller, W.R. Jacobs Jr., Commonalities of Mycobacterium tuberculosis transcriptomes in response to defined persisting macrophage stresses, Front. Immunol. 13 (2022) 909904.

[90] A.M. Abdallah, E.M. Weerdenburg, Q. Guan, R. Ummels, S. Borggreve, S.A. Adroub, et al., Integrated transcriptomic and proteomic analysis of pathogenic mycobacteria and their esx-1 mutants reveal secretion-dependent regulation of ESX-1 substrates and WhiB6 as a transcriptional regulator, PLoS One 14 (1) (2019) e0211003.

[91] D. Giacalone, R.E. Yap, A.M.V. Ecker, S. Tan, PrrA modulates Mycobacterium tuberculosis response to multiple environmental cues and is critically regulated by serine/threonine protein kinases, PLoS Genet. 18 (8) (2022) e1010331.

[92] E.J.R. Peterson, A.N. Brooks, D.J. Reiss, A. Kaur, J. Do, M. Pan, et al., MtrA modulates Mycobacterium tuberculosis cell division in host microenvironments to mediate intrinsic resistance and drug tolerance, Cell Rep. 42 (8) (2023) 112875.

[93] H. Zheng, C.J. Colvin, B.K. Johnson, P.D. Kirchhoff, M. Wilson, K. Jorgensen-Muga, et al., Inhibitors of Mycobacterium tuberculosis DosRST signaling and persistence, Nat. Chem. Biol. 13 (2) (2017) 218–225.

[94] G. Baruzzo, A. Serafini, F. Finotello, T. Sanavia, L. Cioetto-Mazzabò, F. Boldrin, et al., Role of the extracytoplasmic function sigma factor SigE in the stringent response of Mycobacterium tuberculosis, Microbiol. Spectr. 11 (2) (2023) e0294422.

[95] R. Mishra, S. Kohli, N. Malhotra, P. Bandyopadhyay, M. Mehta, M. Munshi, et al., Targeting redox heterogeneity to counteract drug tolerance in replicating Mycobacterium tuberculosis, Sci. Transl. Med. 11 (518) (2019).

96] Y. Cau, M. Mori, C.T. Supuran, M. Botta, Mycobacterial carbonic anhydrase inhibition with phenolic acids and esters: kinetic and computational investigations, Org. Biomol. Chem. 14 (35) (2016) 8322–8330.

[97] G. Degiacomi, B. Gianibbi, D. Recchia, G. Stelitano, G.I. Truglio, P. Marra, et al., CanB, a druggable cellular target in Mycobacterium tuberculosis, ACS Omega 8 (28) (2023) 25209–25220.

[98] J. Felce, M.H. Saier Jr., Carbonic anhydrases fused to anion transporters of the SulP family: evidence for a novel type of bicarbonate transporter, J. Mol. Microbiol. Biotechnol. 8 (3) (2004) 169–176.

[99] A. Aspatwar, J.Y. Winum, F. Carta, C.T. Supuran, M. Hammaren, M. Parikka, S. Parkkila, Carbonic anhydrase inhibitors as novel drugs against mycobacterial beta-carbonic anhydrases: an update on in vitro and in vivo studies, Molecules 23 (11) (2018).

[100] C.T. Supuran, How many carbonic anhydrase inhibition mechanisms exist? J. Enzyme Inhib. Med. Chem. 31 (3) (2016) 345–360.

[101] M. Ceruso, D. Vullo, A. Scozzafava, C.T. Supuran, Sulfonamides incorporating fluorine and 1,3,5-triazine moieties are effective inhibitors of three beta-class carbonic anhydrases from Mycobacterium tuberculosis, J. Enzyme Inhib. Med. Chem. 29 (5) (2014) 686–689.

[102] R.A. Davis, A. Hofmann, A. Osman, R.A. Hall, F.A. Muhlschlegel, D. Vullo, et al., Natural product-based phenols as novel probes for mycobacterial and fungal carbonic anhydrases, J. Med. Chem. 54 (6) (2011) 1682–1692.

[103] A. Maresca, D. Vullo, A. Scozzafava, G. Manole, C.T. Supuran, Inhibition of the beta-class carbonic anhydrases from Mycobacterium tuberculosis with carboxylic acids, J. Enzyme Inhib. Med. Chem. 28 (2) (2013) 392–396.

[104] T.V. Wani, S. Bua, P.S. Khude, A.H. Chowdhary, C.T. Supuran, M.P. Toraskar, Evaluation of sulphonamide derivatives acting as inhibitors of human carbonic anhydrase isoforms I, II and Mycobacterium tuberculosis β-class enzyme Rv3273, J. Enzyme Inhib. Med. Chem. 33 (1) (2018) 962–971.

[105] S. Bua, A. Bonardi, G.R. Mük, A. Nocentini, P. Gratteri, C.T. Supuran, Benzothiadiazinone-1,1-dioxide carbonic anhydrase inhibitors suppress the growth of drug-resistant Mycobacterium tuberculosis strains, Int. J. Mol. Sci. 25 (2024), https://doi.org/10.3390/ijms25052584

[106] M.V. Buchieri, L.E. Riafrecha, O.M. Rodriguez, D. Vullo, H.R. Morbidoni, C.T. Supuran, et al., Inhibition of the beta-carbonic anhydrases from Mycobacterium tuberculosis with C-cinnamoyl glycosides: identification of the first inhibitor with anti-mycobacterial activity, Bioorg. Med. Chem. Lett. 23 (3) (2013) 740–743.

[107] A. Maresca, A. Scozzafava, D. Vullo, C.T. Supuran, Dihalogenated sulfanilamides and benzolamides are effective inhibitors of the three beta-class carbonic anhydrases from Mycobacterium tuberculosis, J. Enzyme Inhib. Med. Chem. 28 (2) (2013) 384–387.

[108] O. Guzel, A. Maresca, A. Scozzafava, A. Salman, A.T. Balaban, C.T. Supuran, Discovery of low nanomolar and subnanomolar inhibitors of the mycobacterial beta-carbonic anhydrases Rv1284 and Rv3273, J. Med. Chem. 52 (13) (2009) 4063–4067.

[109] A. Maresca, F. Carta, D. Vullo, C.T. Supuran, Dithiocarbamates strongly inhibit the beta-class carbonic anhydrases from Mycobacterium tuberculosis, J. Enzyme Inhib. Med. Chem. 28 (2) (2013) 407–411.

[110] A. Maresca, F. Carta, D. Vullo, A. Scozzafava, C.T. Supuran, Carbonic anhydrase inhibitors. Inhibition of the Rv1284 and Rv3273 beta-carbonic anhydrases from Mycobacterium tuberculosis with diazenylbenzenesulfonamides, Bioorg. Med. Chem. Lett. 19 (17) (2009) 4929–4932.

[111] A. Angeli, M. Pinteala, S.S. Maier, B.C. Simionescu, A. Milaneschi, G. Abbas, et al., Evaluation of thio- and seleno-acetamides bearing benzenesulfonamide as inhibitor of carbonic anhydrases from different pathogenic bacteria, Int. J. Mol. Sci. 21 (2) (2020).

[112] S. Bua, S.M. Osman, S. Del Prete, C. Capasso, Z. AlOthman, A. Nocentini, et al., Click-tailed benzenesulfonamides as potent bacterial carbonic anhydrase inhibitors for targeting Mycobacterium tuberculosis and Vibrio cholerae, Bioorg. Chem. 86 (2019) 183–186.

[113] F. Pacchiano, F. Carta, D. Vullo, A. Scozzafava, C.T. Supuran, Inhibition of beta-carbonic anhydrases with ureido-substituted benzenesulfonamides, Bioorg. Med. Chem. Lett. 21 (1) (2011) 102–105.

[114] M. Abdoli, A. Bonardi, N. Paoletti, A. Aspatwar, S. Parkkila, P. Gratteri, et al., Inhibition studies on human and mycobacterial carbonic anhydrases with N-((4-sulfamoylphenyl)carbamothioyl) amides, Molecules 28 (10) (2023).

[115] M.A. Dallaston, S. Rajan, J. Chekaiban, M. Wibowo, M. Cross, M.J. Coster, et al., Dichloro-naphthoquinone as a non-classical inhibitor of the mycobacterial carbonic anhydrase Rv3588c, MedChemComm 8 (6) (2017) 1318–1321.

[116] K. Aksu, M. Nar, M. Tanc, D. Vullo, I. Gülçin, S. Göksu, et al., Synthesis and carbonic anhydrase inhibitory properties of sulfamides structurally related to dopamine, Bioorg. Med. Chem. 21 (11) (2013) 2925–2931.

[117] M.J. Zaro, A. Bortolotti, L.E. Riafrecha, A. Concellón, H.R. Morbidoni, P.A. Colinas, Anti-tubercular and antioxidant activities of C-glycosyl carbonic anhydrase inhibitors: towards the development of novel chemotherapeutic agents against Mycobacterium tuberculosis, J. Enzyme Inhib. Med. Chem. 31 (6) (2016) 1726–1730.

[118] B.K. Johnson, C.J. Colvin, D.B. Needle, F. Mba Medie, P.A. Champion, R.B. Abramovitch, The carbonic anhydrase inhibitor ethoxzolamide inhibits the Mycobacterium tuberculosis PhoPR regulon and Esx-1 secretion and attenuates virulence, Antimicrob. Agents Chemother. 59 (8) (2015) 4436–4445.

[119] S.J. Rose, L.E. Bermudez, Identification of bicarbonate as a trigger and genes involved with extracellular DNA export in mycobacterial biofilms, MBio 7 (6) (2016).

[120] A. Scozzafava, A. Mastrolorenzo, C.T. Supuran, Antimycobacterial activity of 9-sulfonylated/sulfenylated-6-mercaptopurine derivatives, Bioorg. Med. Chem. Lett. 11 (13) (2001) 1675–1678.

[121] L.E. Riafrecha, O.M. Rodríguez, D. Vullo, C.T. Supuran, P.A. Colinas, Synthesis of C-cinnamoyl glycosides and their inhibitory activity against mammalian carbonic anhydrases, Bioorg. Med. Chem. 21 (6) (2013) 1489–1494.

Challenges for developing bacterial CA inhibitors as novel antibiotics

Claudiu T. Supuran*

Neurofarba Department, Pharmaceutical and Nutraceutical Section, University of Florence, Sesto Fiorentino, Florence, Italy
*Corresponding author. e-mail address: claudiu.supuran@unifi.it

Contents

1. Introduction   384
2. CA inhibitors (CAIs) classes   385
3. Drug repurposing of human CAIs as antiinfectives   386
4. Drug design studies of bacterial CAIs   389
   4.1 *H. pylori* CA   389
   4.2 *N. gonorrhoeae* CAs   392
5. Vacomycin resistant *Enterococci* CAs   395
   5.1 *Vibrio cholerae* CAs   397
   5.2 *Mycobacterium tuberculosis* CAs   399
   5.3 *Pseudomonas aeruginosa* CAs   400
6. Conclusions   400
Funding   402
Declaration of interest   402
References   402

Abstract

Acetazolamide, methazolamide, ethoxzolamide and dorzolamide, classical sulfonamide carbonic anhydrase (CA) inhibitors (CAIs) designed for targeting human enzymes, were also shown to effectively inhibit bacterial CAs and were proposed for repurposing as antibacterial agents against several infective agents. CAs belonging to the α-, β- and/or γ-classes from pathogens such as *Helicobacter pylori*, *Neisseria gonorrhoeae*, vacomycin resistant enterococci (VRE), *Vibrio cholerae*, *Mycobacterium tuberculosis*, *Pseudomonas aeruginosa* and other bacteria were considered as drug targets for which several classes of potent inhibitors have been developed. Treatment of some of these pathogens with various classes of such CAIs led to an impairment of the bacterial growth, reduced virulence and for drug resistant bacteria, a resensitization to clinically used antibiotics. Here I will discuss the strategies and challenges for obtaining CAIs with enhanced selectivity for inhibiting bacterial versus human enzymes, which may constitute an important weapon for addressing the drug resistance to β-lactams and other clinically used antibiotics.

The Enzymes, Volume 55
ISSN 1874-6047, https://doi.org/10.1016/bs.enz.2024.05.006

1. Introduction

The enzymes denominated carbonic anhydrases (CAs, EC 4.2.1.1) are widely spread in almost all investigated organisms (prokaryotes and eukaryotes), and are encoded by at least eight genetically unrelated families, the α-, β-, γ-, δ-, ζ-, η-, θ- and ι-CAs [1–6]. They catalyze a simple, yet fundamental chemical reaction: the interconversion between CO_2 and bicarbonate, through the nucleophilic attack of an activated water molecule to the weak electrophile CO_2 [1,2,7,8]. Among the 8 CA classes reported to date [6–10], bacteria encode for at least 4 of them, the α-, β-, γ- and ι-CAs, of which the first three are metalloenzymes [1,2,4,5], whereas the ι CAs do not have metal ions within their active sites [6,8]. In all metallo-CAs, the metal ion has an essential role in catalysis, being prevalently Zn (II), but enzymes with other metal ions such as Fe(II), Cd(II) or Co(II) are also known [2,4,6,7,9]. Indeed, the metal ion from the CA active site coordinates a water molecule (apart 3 amino acid residues), which is involved in the catalytic process: enhancement of its nucleophilicity due to the positive charge of the cation and the hydrophobic environment of the active site thereafter promotes catalysis [9–11]. Some members of the CAs are among the catalytically most effective enzymes known in nature, being second only to superoxide dismutase as turnover numbers, which however varies between the various CA classes, but for the effective ones reaches k_{cat}/K_M values in the range of 1.5×10^8–3.5×10^8 $M^{-1} \times s^{-1}$ [12]. This extreme catalytic power is probably connected to the fact that CO_2 is generated in high amounts in most vital processes and being a gas, must be transformed efficiently into soluble products [1,3,4]. The hydration of CO_2 leads to the formation of bicarbonate and protons and occurs also non-catalytically, but it is too slow a process at physiological pH values (i.e., pH in the range of 6.0–7.5, typical of most tissues/organisms) [2–4,10–12]. CO_2 is generated in relevant amounts in many physiological processes, as mentioned above, and it is crucial for the growth of many organisms, including bacteria, plants, algae, etc. [1,2,10,11]. However, this reaction also induces a pH imbalance, since H^+ is a strong acid whereas bicarbonate is a weak base (both of them generated from two neutral molecules, carbon dioxide and water), and thus, the system CO_2/bicarbonate/CA is one of the most common biological buffers, which probably explains the wide occurrence of these enzymes in organisms all over the phylogenetic tree and the many genetic families discovered so far [1–4,7,9–12]. However, CO_2/bicarbonate may also be a source of one carbon atom for biosynthetic processes involving carboxylating reactions [11,13], and for rapidly growing organisms such as bacteria, this may be an essential opportunity for promoting the growth, colonization and virulence of pathogenic species, but also for their acclimation in particular niches in which they live, for example the highly acidic environment of the stomach in the case of *Helicobacter pylori* [14]. Altered levels of either CO_2 or bicarbonate, elicit the activation of multiple adaptive pathways both in prokaryotes and eukaryotes [3]. For example, in bacterial and fungal pathogens, CO_2/bicarbonate sensing is correlated with an increased virulence and/or pathogenicity of several relevant human pathogens [15–17]. Thus, interfering with prokaryotic CAs has been proposed already a decade ago [11] to be a potential strategy for obtaining antiinfectives with a different mechanism of action compared to that of clinically used such drugs [9–11,18]. Indeed, the increasing drug resistance phenomenon to clinically used antibiotics/antifungals [19–22] and the scarcity of new classes of antibiotics approved in the last decades [23–25] led to a diffuse lack of clinical efficacy of the presently available agents, with serious problems faced by clinicians and governments all over the world. Although there was some initial reluctance from the scientific community to accept the idea that a pH regulating enzyme may be an antibiotic drug target (although as mentioned here and in the entire volume, CAs have many other functions not only the one related with pH balance) data accumulated over the last decade demonstrated that bacterial CA inhibition may be an additional and valuable strategy to design antibacterials. In this chapter I will discuss the main strategies, challenges and results obtained in the field.

2. CA inhibitors (CAIs) classes

The inhibition mechanism of CAs is well understood and there are several distinct inhibition mechanisms known to date, as determined by X-ray crystallography and other biophysical/biochemical techniques, as well as a rather ample variety of CAI classes belonging to many different chemotypes [4,9,10,26,27]. CAIs known to date possess four main inhibitory mechanisms: (i) the classical inhibitors coordinate to the metal ion, which as mentioned above, except for ι-CAs [9,26,27], is crucial for the catalytic activity. Many classes of CAIs were investigated to date which possess this inhibitory mechanism, among which (in)organic anions, sulfonamides and their isosteres (sulfamates, sulfamides, *N*-substituted sulfonamides, etc.), carboxylates, hydroxamates, thiols, selenols, phosphonamidates, benzoxaboroles, ninhydrins, oxazolidinediones, etc. [9,26,27]; (ii) inhibitors that anchor to the metal-ion coordinated nucleophile (water or hydroxide ion), among which phenols, polyamines, some carboxylates, sulfocoumarins, 2-thioxo-coumarins, etc. [26–30]; (iii) compounds which occlude the active site entrance, e.g., coumarins/isocoumarins and many of their derivatives, which act as prodrug inhibitors, undergoing a CA-catalyzed hydrolysis of the lactone ring which leads to the formation of the real inhibitor [31,32]; (iv) inhibitors that bind out of the active site, in a hydrophobic pocket adjacent to the entrance to the cleft, and inhibit the enzyme by blocking a His residue involved in the catalytic cycle in α-CAs [33]. Up until now, inhibition of bacterial CAs was investigated with a large number of derivatives acting as inhibitors via mechanisms (i)–(iii) mentioned above, which are valid not only for α-CAs but also for at least the β- and γ-CAs (among the enzymes present in bacteria) [1,9]. A special note on the ι-CAs, which although do not possess metal ions, were shown by X-ray crystallography [8] and kinetic measurements [34] to bind CAIs belonging to the anions and sulfonamide classes, which inhibit their activity in a non-metal ion dependent manner [6,8].

As mentioned above, bacterial CAs furnish CO_2 and HCO_3^-/H^+ ions which are indispensable for several biosynthetic pathways, in pH regulation, represent virulence and survival factors for bacteria in various niches, and their inhibition with modulators of activity as those mentioned above impair survival, growth and virulence of the pathogens. As thus, bacterial CAIs show multiple antibacterial effects and might constitute a new class of antibacterials in the presently available armamentarium of such drugs, strongly compromised by the emergence of drug resistance phenomena [1,7,9–11].

3. Drug repurposing of human CAIs as antiinfectives

There are at least six sulfonamides in clinical use as CAIs, which have been designed for targeting various human enzymes over the last several decades [9]: acetazolamide **1**, methazolamide **2**, ethoxzolamide **3**, dichlorophenamide **4**, dorzolamide **5** and brinzolamide **6** (Fig. 1).

These FDA/EMEA-approved drugs are used nowadays as diuretics, antiglaucoma, or antiepileptic drugs [9,18], but also as tools for understanding a range of processes in which these enzymes from different organisms not only mammals, are involved [35–37]. They were also assayed as inhibitors of bacterial CAs from many pathogenic (as well as non-pathogenic) species, and most of the times showed significant inhibitory

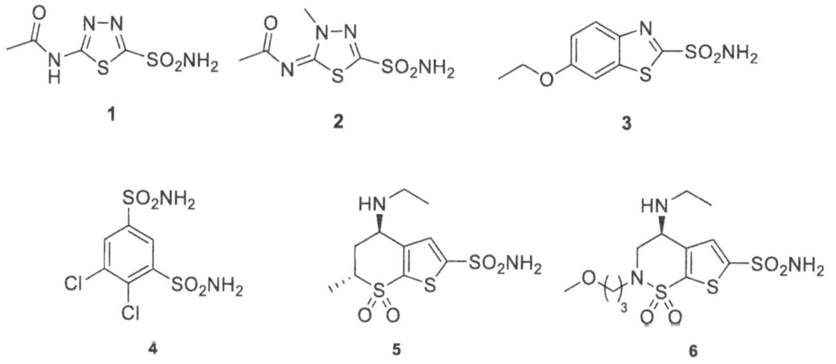

Fig. 1 Sulfonamide CAIs of type **1–6** originally designed for targeting human enzymes.

effects [1,7]. Their well-known pharmacology, toxicology, safety and lack of relevant toxicity [38,39] led to investigations on their possible repurposing as antibacterial agents, alone or in combination with other drugs [40]. Most such studies have been performed with acetazolamide **1**, ethoxzolamide **3** and dorzolamide **5**, but in principle all other sulfonamides of Fig. 1 might be useful for such an approach. The most investigated pathogens were on the other hand *Helicobacter pylori* and *Neisseria gonorrhoeae*. The available data for these two pathogens will be discussed here.

Helicobacter pylori is a Gram-negative neutralophilic bacterium which is considered to be the principal cause of gastric ulcer, gastritis and gastric cancer [41,42]. It is adapted to live in the highly acidic environment of the stomach, and it was observed already more than two decades ago to encode for two CAs, an α- and a β-class enzyme, known as HpCAα and HpCAβ, respectively [43–46]. Both enzymes have been characterized in detail, both biochemically and for their inhibition profiles with a large number of inhibitors [44–46]. Furthermore, acetazolamide **1** has been used for at least two decades as an antiulcer drug in Romania by Puscas and his group, with the rationale that by inhibiting gastric mucosa CAs a reduction of gastric acid secretion is achieved [47,48]. Indeed, in the period before the discovery of the histamine H_2 antagonists or proton pump inhibitors, this therapy based on acetazolamide or ethoxzolamide as CAIs was quite effective in curing ulcers, although associated with a range of side effects due to inhibition of CAs in various other tissues and organs, mainly in the kidneys and red blood cells [49]. However, decades later, in 2005, it has been demonstrated by Sachs' group [14,50] that the antiulcer effects of these CAIs is due to the inhibition of the bacterial enzymes HpCAα (and presumably also HpCAβ), not of the human ones, which leads to the inability of the bacterium to survive in the acidic environment within the stomach. Indeed, these CAs are involved together with the urease of *H. pylori* to the acclimation of the bacterium in the acidic environment in which it lives, whereas the inhibition of one or more such enzymes, interferes with the process, leading to the death and decolonization of the stomach from this pathogen [49,50]. Several other more recent studies [51–53] confirmed the fact that most CAIs of the sulfonamide type, especially **1–3** (Fig. 1), might be repurposed as anti-ulcer agents, also considering the fact that there is significant resistance of the clinically used antibiotics for many *H. pylori* clinical isolates worldwide [54]. Furthermore, Roujeinikova's group evaluated ethoxzolamide on various *H. pylori* strains *ex vivo*, in cell cultures, in order to understand whether drug resistance to this sulfonamide may emerge after repeated passages [52,53]. For strains SS1 and 26695, it has been observed that resistance did not develop easily, and a quite low rate of spontaneous resistance acquisition of $< 10^{-8}$ has been reported [77]. Acquisition of resistance was associated with mutations in 3 genes in strain SS1, and 6 different genes in strain 26695, but the bacterial CA genes were not among them. In fact, the resistant isolates had mutations in genes involved in bacterial cell wall synthesis, gene expression control but remained susceptible to inhibition by ethoxzolamide [52,53].

The second pathogen for which interesting data are available regarding a repurposing of sulfonamide CAIs used clinically (and originally designed as mammalian enzymes inhibitors), is *Neisseria gonorrhoeae* [55]. *N. gonorrhoeae* is also a Gram-negative, coffee-bean-shaped, facultative intracellular diplococcus bacterium responsible for the third-most frequently reported sexually-transmitted disease in the world, gonorrhoea [56]. The pathogen has developed resistance to all antimicrobials previously used as first-line therapeutic agents, and to many second line antibiotics too, creating wide concern all over the world [57,58]. As the previously discussed pathogen, also *N. gonorrhoeae* encodes for two CAs, an α- and a β-class enzyme [59,60], of which the first one was the most investigated one, being nominated NgCAα [61]. Acetazolamide **1** was shown by Lindskog's group to act as an in vitro inhibitor of NgCAα with a K_I of 74 nM [59] whereas Flaherty's group later demonstrated *ex vivo* inhibition of the bacterial growth, with MIC values of around 2 μg/mL for different wild type and drug resistant strains [62]. A recent study thereafter showed that other FDA-approved sulfonamides, apart **1**, such as compounds **3–6** (Fig. 1) showed effective in vitro inhibition of NgCAα and MIC values in the range of 0.015–4 μg/mL against a range of various drug-resistant strains of the bacterium [62], leading to the proposal of repurposing these sulfonamides as antibacterials for the management of *N. gonorrhoea* infection. Seleem 's group [63] thereafter demonstrated the in vivo efficacy of **1**, using a mouse model infection of *N. gonorrhoeae*. Administration of acetazolamide **1** for three days reduced by 90% the gonococcal burden in vagina of infected mice when a dose of 50 mg/kg drug was used [63]. As this drug is a safe, inexpensive and well-known sulfonamide CAI, these data represent an excellent basis for using such agents in the management of this disease, although the compound also inhibits effectively human CAs and as thus, might lead to some side effects [9].

Although repurposing already approved drugs belonging to the CAI class as antibacterials may be an inexpensive and rather safe approach, there are also inconveniences due to the possible off-target effects, due to inhibition of human enzymes, which as mentioned above, are widespread in many tissues and organs in humans [3,9,10]. The two examples considered above, for the management of *H. pylori* and *N. gonorrhoeae* infections with FDA/EMEA-approved drugs are indeed of great interest as they provide the proof-of-concept that bacterial CAs may indeed be considered as antibiotic drug targets, and they encountered a good level of success, as presented above. However, *de novo* drug design studies of bacterial CAIs might be an even better such approach, and the successful (and less successful) stories in this field will be discussed in the next paragraphs.

4. Drug design studies of bacterial CAIs

4.1 *H. pylori* CA

A large number of sulfonamides, sulfamates and sulfamides have been assayed in vitro for their inhibitory effects against recombinant HpCAα and HpCAβ [44,45,64], confirming that many such derivatives are effective inhibitors of both these enzymes. Inhibition data for sulfonamides **1–3** (Fig. 1), which are clinically used human CAIs, and derivatives **7–9**, some of which were designed as bacterial CAIs [44,66], are presented in Table 1, where the inhibition of the offtarget human isoforms hCA I and II is also given.

Acetazolamide **1** and compound **9** are the most effective inhiibtors of both bacterial enzymes, (K_Is < 50 nM) but it may be observed that only compound **9** is a more effective HpCAα than hCA II inhibitor (hCA II is the dominant human isoform, widespread in many tissues and presumably

Table 1 HpCAα, HpCAβ, hCA I and hCA II inhibition data for sulfonamides **1–3** and **7–9**.

Compound	K_I (nM)			
	HpCAα	**HpCAβ**	**hCA I**	**hCA II**
1	21	40	250	12
2	225	176	50	14
3	193	33	25	8
7	323	2590	8600	60
8	315	54	15	9
9	8	40	338	24

responsible for the side effects of sulfonamide pan-CAIs [9]). The compounds were high nanomolar bacterial CA inhibitors, apart deacetylated acetazolamide (compound **7**) which was the most ineffective inhibitor (high nM for the α-class enzyme and low micromolar for the β-CA) [44]. These differences in inhibition between these sulfonamides were in fact rationalized by the report of their X-ray crystal structures in adduct with the bacterial (and human) enzymes (Fig. 2) [65,66].

It may be observed that only the sulfonamide moiety and the 1,3,4-thiadiazole/thiadiazoline rings are rather superimposable in the adducts of HpCAα with sulfonamides **1–3** and **7–9** (Fig. 2). The tails present in these derivatives (except ethoxzolamide), especially in benzolamide **8** and the structutally similar compound **9**, extend towards the nydrophilic half of the active site, towards its exit, and make contacts/interactions with residues Val141, Lys133 and Leu139 (Fig. 2). This stabilizes especially the interactions of the *tert*-butyl-phenyl tail of **9**, making it the best HpCAα inhibitor detected so far. Furthermore, in the X-ray crystal structiure of **9** bound to hCA II (data not shown), the *tert*-butyl-phenyl tail of **9** makes a clash with a short α-helix of hCA II (whoch is not present in the bacterial enzyme), denominated helix 130–136, which is involved in hydrophobic stabilizing interactions with the scaffold of bulkier

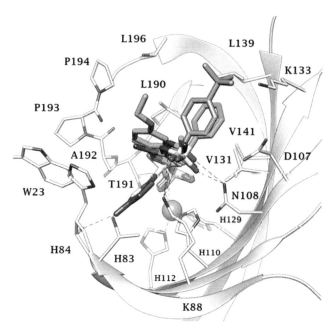

Fig. 2 X-ray crystal structures of adducts of HpCAα with sulfonamides **1–3** and **7–9**. Green, acetazolamide **1** adduct (PDB code 4YGF); dark blue, methazolamide **2** adduct (PDB code 4YHA); orange, ethoxzolamide **3** adduct (PDB code 5TT3); violet, 5-amino-1,3,4-thiadiazole-2-sulfonamide **7** adduct (PDB code 5TUO); sky blue, benzolamide **8** adduct (PDB code 5TT8); metal blue, (E)-5-(((4-(*tert*-butyl)phenyl)sulfonyl)imino)-4-methyl-4,5-dihydro-1,3,4-thiadiazole-2-sulfonamide **9** adduct (PDB code 5TV3). The zinc ion from the enzyme active site is shown as a gray sphere, whereas its three protein ligands (His83, His110 and His112) as sticks. Amino acid residues involved in the binding of inhibitors are also shown as stick representations. Dotted lines are H bonds.

inhibitors, such as ethoxzolamide **3** or benzolamide **8** (in their hCA II not bacterial enzyme complexes) [66]. Indeed, as seen from Fog. 2, the scaffold of ethoxzolamide **3** is olrientated toeards the hydrophobic half of the enzyme active site, in an orientation totally dofferent from the other inhibitors bound to HpCAα. This constitutes thus a very interesting and rare example of drug design which afforded bacterial CA—selective inhibitors (over hCA II, the main offtarget human isoform) with also a very good efficacy in inhibiting HpCAα over the main human isoforms.

Few phenols were investigated as HpCAα and HpCAβ inhibitors, but carvacrol **10** and thymol **11** were investigated in some detail recently (Table 2) [67–69].

Carvacrol **10** acted as a medium potency, micromolar inhibitor of both bacterial enzymes, and was ineffective as hCA I and II inhibitor

Table 2 HpCAα and HpCAβ inhibition data for phenols **10** and **11**.

Structure		K_I (μM)			
		hCA I	hCA II	HpCAα	HpCAβ
Carvacrol **10**		>100	>100	8.4	13.3
Thymol **11**		>100	>100	>100	3.4
1		0.25	0.012	0.021	0.040

hCA I, hCA II inhibitory effects and acetazolamide (1) inhibition data are also shown for comparison.

(K_Is > 100 μM), whereas thymol **11** inhibited in the micromolar range only HpCAβ [67–69]. For both phenols **7** and **8**, the *ex vivo* anti-*H. pylori* activity in terms of growth impairment, biofilm production, and release of outer membrane vesicles were also assessed [67–69]. The outer membrane vesicles were shown earlier to be rich in HpCAα, with a higher amount of the enzyme in the planktonic than biofilm cultures of the bacterium [70]. It has been thus hypothesized that outer membrane vesicles enriched in this enzyme play a role in colonization, survival, persistence, and pathogenesis of *H. pylori* [70]. Although phenols **10** and **11** are relatively ineffective HpCAα and HpCAβ inhibitors, in *ex vivo* experiments it has been observed that they inhibit the growth of planktonic bacteria and the biofilm formation, leading also to a decreased release of outer membrane vesicles, effects which are synergistic with sub-MIC (minimal inhibitory concentration) of amoxicillin, a widely used beta-lactam antibiotic [80], or salicylhydroxamic acid, an urease inhibitor [69]. Phenols remain however an under-investigated class of bacterial CAIs, and the data presented above show that their potential for developing alternative classes of antibacterials is quite high.

4.2 *N. gonorrhoeae* CAs

As mentioned above, acetazolamide **1** acts as an efficient in vitro NgCAα inhibitor, has an MIC of around 4 μg/mL for inhibiting various strains of the bacterium, some of which resistant to other antibiotics [40,55,61,62], and as thus has been used as lead compound for drug design studies of even more effective bacterial CAIs.

Flaherty's group designed analogs of **1** incorporating diverse amides instead of the acetamido group of the lead, some of which are shown in Table 3 [61]. They contain aliphatic, cycloalkyl, aromatic and heterocyclic moieties, and showed significant inhibitory effects against the bacterial enzyme, with in vitro inhibition constants ranging between 0.7 and 79 nM (Table 3), also inhibiting the off-target isoform hCA II. However, they showed highly effective MICs against wild-type and drug-resistant *N. gonorrhoea* strains, between 0.25 and 4 µg/mL (some of the acetazolamide analogs are more effective ex vivo than the lead compound **1**, e.g., **13**, **16** and **18**). Not all the in vitro active

Table 3 Inhibitory effects against the bacterial enzyme NgCAα and offtarget enzyme hCA II with sulfonamides structurally related to acetazolamide **1** (**AAZ**), of types **7**, **12–19**, and their minimum inhibitory concentration (MIC) against strain CDC 181 of *N. gonorrhoeae*.

1 (AAZ)   **12-19**

Compound	R	K_I (nM)		MIC
		hCA II	NgCAα	(µg/mL)
1 (AAZ)	MeCO	12.5	74	4
7	H	60	73	4
12	Et	47	74	>64
13	Cyclohexyl-CO	20	9.8	2
14	Cyclohexyl-CH_2	78	42	>64
15	PhCO	29	79	4
16	Cyclohexyl-CH_2CO	24.5	37	0.5
17	Cyclohexyl-CH_2CH_2	38	66	>64
18	$PhCH_2CH_2CO$	8.1	8.3	0.25
19	Cyclohexyl-CH_2CH_2CO	0.32	0.7	1

compounds also showed in vivo activity and this could be not explained in this interesting study. However, in another study from the same group [71], some of the compounds of Table 3, more exactly **1**, **13**, **16** and **18** were co–crystallized with the bacterial enzyme, showing interesting interactions with the zinc ion from the active site (to which they are coordinated as deprotonated sulfonamidate anions, as for all other similar adducts bound to other CAs [9]) and with different active site amino acid residues (evidenced in Fig. 3), which explain their good inhibitory effects and also the structure–activity relationship (SAR) differences which emerged in the previous study (Fig. 3) [61,71]. It may be observed that as for the *Helicobacter* enzyme discussed above, the tail of the inhibitor makes various interactions with different amino acid residues, which may explain the differences in the inhibition pattern of these derivatives, but unlike the compounds shown in Fig. 2, those of Fig. 3 when bound to the

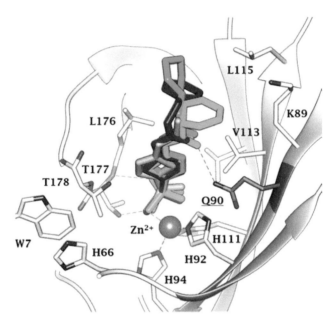

Fig. 3 X-ray crystal structure showing the NgCAα complexed with different sulfonamides, superimposed in the active site: **1 (AAZ)** (tan, PDB 8DYQ), **13** (green, PDB 8DQF), **16** (cyan, PDB 8DR2), **18** (magenta, PDB 8DRB). The active site details are as follows: zinc(II) ion shown as a gray sphere; zinc protein ligands (His92, His94, His111) blue and white stick model. Residue Gln90 is shown in green, being involved in H-bonds with the amide group of several inhibitors. Residues Thr177, Thr178, Val113, Leu176, Lys 89 and Leu115 (in blue and white) are also shown, as they are in contact with some of the bound inhibitors. The H-bonds are shown as dashed lines.

Neisseria enzyme have the 1,3,4-thiadiazole ring flipped by 180°, which constitutes a relevant and diverse orientation compared to their binding to hCA II or HpCAα [71].

Subsequent studies showed efficient in vitro inhibition of NgCAα and sometimes also in vivo antibacterial activity with dithiocarbamates [72], monothiocarbamates [73], anions [74], phenols [75] and coumarins [76], but no crystallographic or drug design studies were thereafter performed with these classes of CAIs.

5. Vacomycin resistant *Enterococci* CAs

Various species belonging to the genus *Enterococcus*, among which *E. faecium* and *E. faecalis*, gram-positive bacteria commonly found in the gastrointestinal tract (GIT) microbioma, may cause disease in hospitalized/ immunosupressed patients, ranging from mild to serious ones [77], with few therapeutic options for their treatment, as the pathogens became highly multi-drug resistant to most antibiotics used clinically, including vanco-mycin [78–80]. Thus, these bacteria are commonly reffered to as vanco-mycin-resistant enterococci (VRE) [77,81]. FDA approved only two therapies for the treatment of VRE, i.e., linezolid, which is only marginally effective, and a combination of quinupristin and dalfopristin, which leads to serious toxicity problems in the treated patients [77–81].

Flaherty's group proposed the CAs present in these bacteria as alternative antimicrobial drug targets, obtaining remarkable results by inhibiting them [81,82]. *E. faecium* CAs (an α- and a γ-class enzyme are present in its genome, denominated EfCAα and EfCAγ, respectively) have been used as model enzymes, being cloned, purified and characterized biochemically and for their inhibition profiles [83]. Initially, as for the validation of NgCAα as an antibacterial target, acetazolamide **1** has been used as investigational inhibitor, being observed that it possesses a rather effective MIC against *E. faecium* strain HM-965 (MIC of 2 μg/mL) as well as against several multidrug-resistant strains of the bacterium [81,82]. Analogs of acetazolamide in which the acetamide moiety of the lead molecule has been modified to a variety of other amides (compounds **13–22**), led to CAIs which showed both effective MIC values for the inhibition of *E. faecium* HM-965 strain, but also inhibited in vitro the two CAs from the pathogen (Table 4).

Among the many acetazolamide **1**derivatives investigated, which incor-porated aliphatic, cycloaliphatic, aromatic and heterocyclic tails, some of the

Table 4 Inhibitory effects of hCA II, EfCAα and EfCAγ with sulfonamides **1**, **13–22** incorporating the 1,3,4-thiadiazole-2-sulfonamide scaffold.

1, 13-22

Compound	R	K_I (nM)		MIC	
	hCA II	EfCAα	EfCAγ	(μg/mL)	
1 (AAZ)	MeCO	12.5	56.7	323	2
13	Cyclohexyl-CO	20.2	49.3	131	0.25
15	PhCO	29.0	11.7	310	2
16	Cyclohexyl-CH₂CO	24.5	14.5	305	0.06
18	PhCH₂CH₂CO	8.1	6.4	148	0.06
19	Cyclohexyl-CH₂CH₂CO	0.32	66.9	346	0.007
20	EtCO	37.2	37.5	250	1
21	t-Bu-CH₂CO	7.3	69.6	390	0.015
22	O[CH₂CH₂]₂NCH₂CO	0.9	20.1	56.4	1

The minimum inhibitory concentration (MIC) against E. faecium strain HM-965 is also shown.

most effective anti-VRE derivatives (considering their MIC and K_Is against the two CAs form the pathogen) are shown in Table 4. The lowest MIC values against E. faecium strain HM-965 were observed for the tert-butylacetamido (**21**), cyclohexylcarboxamido (**13**), cyclohexylmethylcarboxamido (**16**), cyclohexylethylcarboxamido (**19**), and phenethylcarboxamido (**18**) derivatives, for which MIC-s in the range of 0.007–0.25 μg/mL were measured. Acetazolamide **1**, its ethyl congener (**20**), the benzoyl derivative **15** or the morpholinyl-methyl- derivative **22** were also active, with MIC-s in the range of 1–2 μg/mL [82,83]. Many other investigated derivatives (with R groups such as ethylamino, benzyl, etc.) structurally related to **13–22**, did not show any anti-VRE effects, with MIC-s > 64 μg/mL—data not shown [82]. All amides shown in Table 4 were also effective EfCAα inhibitors (K_Is in the range of 6.4–69.6 nM) and rather effective EfCAγ inhibitors

(K_Is in the range of 56.4–390 nM), but there was not a straightforward correlation between EfCAα/EfCAγ inhibition and the measured MIC, although generally the most effective in vivo anti-VRE agents also showed effective inhibition of the two enzymes. It should also be mentioned that an offtarget effect of these sulfonamides on other enzymes than the CAs was also excluded, since the derivatives of acetazolamide incorporating a SO_2Me instead of the SO_2NH_2 moiety (which is not at all a CAI) [9], as well as the corresponding thiazole-sulfonamide (a much weaker CAI compared to **1**) did not show any anti-VRE effects in vivo [40,82,83]. The two CAs from this pathogen EfCAα and EfCAγ were not yet crystallized, but by using computational techniques, some of the effective inhibitors were docked in the active sites of the two enzymes, showing favorable interactions which can rationalize the observed inhibitory data and anti-VRE effects [82,83]. Flaherty's group also demonstrated the possibility of using such bacterial CAIs for the gut decolonization of VRE in animal model of the disease [84]. Acetazolamide and three of its derivatives (**19**, **21** and **22**) were used in such experiments, all of which had MIC-s in the range of 0.007–0.060 μg/mL and were also effective inhibitors of the two CAs present in *E. faecium*. Mice infected with *E. faecium* strain HM-952 have been used in the experiments and their GIT bioburden was quantified in the fecal pellets after treatment for 8 days with 20 mg/kg CAIs (**1**, **19**, **21** and **22**). The treated animals with CAIs showed significantly reduced *E. faecium* fecal bioburden as compared to those treated with the vehicle (dimethylacetamide, propylene glycol, polyethylene glycol 400, in phosphate buffer) and linezolid as control drug [84]. After three days of treatment with CAIs, the *E. faecium* fecal bioburden was already reduced by 90%, whereas the highest effect was observed after 5 days of treatment, with the bioburden being reduced by 92.6–99.3%, with **1** being the most effective inhibitor among the four investigated compounds [84]. Of note, the effect of linezolid was of only 86.5% reduction of the fecal bioburden (after 8 days of treatment), demonstrating thus that sulfonamide CAIs were by far more effective as decolonization agents against VRE as compared to the standard, clinically used drug. The improved VRE decolonization of **1** has been attributed to its reduced GIT absorption compared to the other more lipophilic CAIs used in the study [84].

5.1 *Vibrio cholerae* CAs

Vibrio cholerae is a Gram-negative bacillus possessing several serogroups, of which two, the O1 and O139, cause cholera. The disease is spread in poor countries, being characterized by a massive loss of water and electrolytes, and leads to death if untreated [85]. *V. cholerae* encodes for three CAs, belonging to the α-, β- and γ-classes, denominated VchCAα, VchCAβ and VchCAγ, which were cloned and characterized biochemically over the last decade [86–93]. Cholera treatment is nowaday based on intravenous rehydration and administration of antibiotics (erythromycin, doxycycline, norfloxacin, ciprofloxacin, azithromycin, and trimethoprim-sulfamethoxazole combinations), but an increasing drug resistance was reported, making the search for new anti-cholera drugs a medical emergency [94–96].

There are three CAs present in this pathogen, which were proposed as alternative antibacterial drug targets [85] and many sulfonamides [86,88,90,93], anionic [87], phenolic [91], or benzoxaborole [92] effective in vitro inhibitors have been detected, although only the X-ray crystal structure of the tetrameric VchCAβ is known at the moment, without inhibitors bound to it [89]. Among the effective in vitro VchCAα inhibitors reported so far are clinically used sulfonamides such as **1** (K_I of 6.8 nM), **2** (K_I of 0.69 nM), **8** (K_I of 4.2 nM), and many others [86]. VchCAβ and VchCAγ have also been investigated for their inhibition profile with sulfonamides and anions [88,90], and they generally showed one–two orders of magnitude lower affinity for these classes of inhibitors. It is presently not known whether a pharmacologic, antibacterial effect may be observed by inhibiting only VchCAα (for which highly effective, low nanomolar inhibitors are available), or whether the other two enzymes should be also targeted. Bicarbonate, which is generated by the action of the three CAs on CO_2, has been demonstrated earlier to be a virulence factor which enhances the expression of the cholera toxin ToxT, whereas **2**, a potent VchCAα inhibitor inhibited this bicarbonate-mediated virulence induction [97]. Only recently, some sulfonamide CAIs of type **23** and **24**, which act efficiently in inhibiting the three *V. cholerae* CAs, were investigated in vivo for their antibacterial effects on different clinical isolates of the bacterium, i.e., SI-Vc22, SI-Vc71, and SI-Vc912 [93] (Fig. 4).

Fig. 4 Structure of SLC-0111 **23**, used as lead molecule for obtaining anti-*Vibrio cholerae* derivatives **24**.

These sulfonamides have been designed considering derivative **23**, also known as SLC-0111, an antitumor/antimetastatic sulfonamide CAI in Phase Ib/II clinical trials [98] as lead molecule. Compounds **24** possess the sulfamoyl moiety in *para* or *meta* with respect to the cyclic urea functionality, which induces rigidity to the scaffold of **24**, compared to the flexible ureido-derivative **23** [93]. These sulfonamides indeed showed effective in vitro VchCAα, VchCAβ and VchCAγ inhibitory activity, in some cases in the low nanomolar range, but their MICs were $> 64 \, \mu g/mL$ (although inhibition of growth of the bacterium has been observed, but without the possibility to determine the MIC value) [93]. Thus, more detailed studies of the three CAs present in this pathogen and better inhibitors are needed in order to validate/invalidate these enzymes as antibacterial drug targets.

5.2 *Mycobacterium tuberculosis* CAs

M. tuberculosis, the pathogen provoking tuberculosis, encodes for three β-CAs which have been rather well characterized [99–103], and presumably also a γ-class enzyme, which has not yet been cloned and characterized [104]. The three β-CAs, sometimes called with names derived from the genes encoding them as Rv1284, Rv3588c and Rv3273, or also as mtCA1, mtCA2 and mtCA3, have been investigated in detail for their catalytic activity and inhibition with sulfonamides [99–103] and other classes of compounds, such as phenol and carboxylic acid natural products [105], dithiocarbamates [106], aromatic and heterocyclic carboxylic acids [107], C-cinnamoyl glycosides [108], boronic acids [109], nitroimidazole based sulfonamides, sulfamides and sulfamates [110], with many effective, in some cases low nanomolar in vitro inhibitors detected. However, such compounds were ineffective in vivo/ex vivo, especially for the sulfonamides, presumably due to their low penetrability through the complex membranes of the bacterium [99–104]. Only the dithiocarbamates, especially the derivatives Fc14–594A Bis and Fc14–584B Bis (Fig. 5) showed significant anti-mycobacterial activity in vivo, in zebra fish larvaeinfected with *M. marinum* [111,112].

Both these compounds acted as potent, nanomolar inhibitors in vitro against all three β-CAs from the pathogen, and inhibited the growth of *M. marinum* at a concentration of 75 μM. In vivo inhibition studies using 300 μM Fc14–584B showed significant impairment of bacterial growth in zebrafish larvae at 6 days post infection, allowing for a certain degree of optimism regarding the potential development of even more effective anti-mycobacterial CAIs [104].

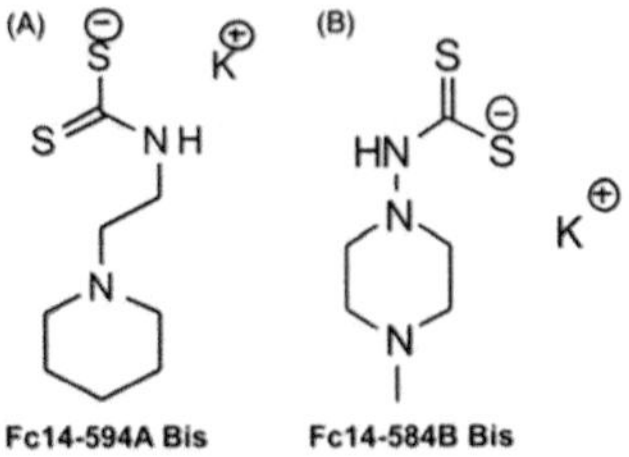

Fig. 5 Dithiocarbamates with *M. tuberculosis* CAs inhibitory action in vitro and in vivo.

5.3 *Pseudomonas aeruginosa* CAs

P. aeruginosa also encodes for at least three β-CAs and several γ-class enzymes, but at the moment only the first ones have been investigated to some detail, especially the isoforms psCA3 and psCA1 for which sulfonamide and anion inhibition data were generated [113–115]. The X-ray structure of psCA3 was also reported, alone [113] or in adduct with sulfamide, imidazole, 4-methylimidazole, and thiocyanate [114], which inhibit the enzyme with efficiencies in the micro- to millimolar range and bind to the metal ion from the enzyme cavity as for most anion/sulfonamide adducts with CAs studied to date [114,116]. The organic sulfonamides act as much more effective inhibitors and as some of these enzymes were shown to be required for calcium carbonate deposition in various tissues infected with the pathogen and to contribute to its virulence [115], it is probable that the development of more selective/potent psCA inhibitors may have relevance in clinical settings. However, no drug design studies of such a type have been reported up until now.

6. Conclusions

Apart the enzymes discussed above, many other CAs have been described in other pathogenic bacterial species, such as *Brucella suis, Salmonella enterica* serovar *Typhimurium, Legionella pneumophila, Porphyromonas gingivalis, Clostridium perfringens, Streptococcus mutans, Burkholderia pseudomallei, Francisella tularensis, Escherichia coli, Mammaliicoccus (Staphylococcus) sciuri*, etc. [117–142]. For many of them the detailed catalytic activity and inhibition profiles with anions and sulfonamides were also published, allowing the discovery of effective in vitro inhibitors [117–142]. However, no in vivo data were obtained and although the need of new antibacterials is stringent, no appropriate models were developed to investigate in more detail such compounds, except for *Escherichia coli*, for which an elegant and simple model system has recently been proposed by Capasso's group [143]. It has been proposed to investigate *E. coli* growth in the presence/absence of CAIs of sulfonamide type, by monitoring the glucose consumption spectro-photometrically during the growth of the bacteria in standard conditions [143]. This method is simple, can be run in a normal laboratory without the need of extensive safety biohazard measures, and allowed to determine an MIC for acetazolamide of 31.2 μg/mL (other compounds and standard antibiotics were also used as controls) [143]. Extending this method to other classes of inhibitors may lead to a facile and inexpensive preliminary screening method of antibacterials based on CAIs, before starting experiments with pathogenic species in high safety laboratories.

Thus, what are the main challenges for obtaining antibacterials based one CAIs in this almost "post-antibiotic" era ? I will try to propose some answers here:

i. although there is extensive knowledge regarding the molecular biology, cloning and purification of bacterial CAs, only in the last two decades saw important contributions in the field, and only a rather limited number of around 15 species have been analyzed in details and their enzymes characterized both for catalytic activity and inhibition with the main classes of inhiibtors, anions and sulfonamides;

ii. few X-ray crystal structures of bacterial CAs and especially of enzymes complexed with inhibitors were so far reported, and most such examples have been presented in the chapter;

iii. few other classes of inhibitors apart the anions and the sulfonamides have been investigated in detail for the inhibition of bacterial enzymes, of the rather important armamentarium of such compounds available;

iv. *ex vivo* experiments for measuring the MICs of bacterial CAIs have emerged only in the last 5 years and they led to significant success, as exemplified here for *H. pylori*, *N. gonorrhoeae* and VRE among others.

v. in vivo data in animal models are so far available only for *N. gonorrhoeae* and VRE, and probably also *H. pylori*, if one considers the clinical data from the 80s and 90s in Romania, which have been mentioned in the chapter.

Probably by intensifying all steps (i)–(v) mentioned above one can obtain the validation of many other bacterial CAs as antiinfective drug targets, and allow the discovery of novel antibiotics devoid of the drug resistance problems affecting all known clinically used classes. Another approach may be that of hybridizing CAIs with various classes of antibiotic chemotypes leading to hybrid compounds which may be more effectiuve than single agents due to their dual nature. All these approaches will probably greatly be expanded in the future.

Funding

Work from the author's laboratory was financed by the Italian Ministry for University and Research (MIUR), project FISR2019_04819 BacCAD and EU–Horizon2020, for the project Springboard (grant agreement No. 951883).

Declaration of interest

The author declares no conflict of interest.

References

[1] C. Capasso, C.T. Supuran, An overview of the alpha-, beta- and gamma-carbonic anhydrases from Bacteria: can bacterial carbonic anhydrases shed new light on evolution of bacteria? J. Enzyme Inhib. Med. Chem. 30 (2015) 325–332.

[2] K.S. Smith, C. Jakubzick, T.S. Whittam, J.G. Ferry, Carbonic anhydrase is an ancient enzyme widespread in prokaryotes, Proc. Natl. Acad. Sci. U. S. A. 96 (26) (1999) 15184–15189.

[3] C.T. Supuran, Carbonic anhydrase versatility: from pH regulation to CO_2 sensing and metabolism, Front. Mol. Biosci. 10 (2023) 1326633.

[4] C.T. Supuran, Structure and function of carbonic anhydrases, Biochem. J. 473 (14) (2016) 2023–2032.

[5] J.G. Ferry, The gamma class of carbonic anhydrases, Biochim. Biophys. Acta 1804 (2) (2010) 374–381.

[6] A. Nocentini, C.T. Supuran, C. Capasso, An overview on the recently discovered iota-carbonic anhydrases, J. Enzyme Inhib. Med. Chem. 36 (1) (2021) 1988–1995.

[7] C.T. Supuran, An overview of novel antimicrobial carbonic anhydrase inhibitors, Expert. Opin. Ther. Targets 27 (10) (2023) 897–910.

[8] Y. Hirakawa, M. Senda, K. Fukuda, H.Y. Yu, M. Ishida, M. Taira, K. Kinbara, T. Senda, Characterization of a novel type of carbonic anhydrase that acts without metal cofactors, BMC Biol. 19 (1) (2021) 105.

[9] C.T. Supuran, Carbonic anhydrases: novel therapeutic applications for inhibitors and activators, Nat. Rev. Drug. Discov. 7 (2) (2008) 168–181.

[10] C.T. Supuran, A simple yet multifaceted 90 years old, evergreen enzyme: carbonic anhydrase, its inhibition and activation, Bioorg. Med. Chem. Lett. 93 (2023) 129411.

[11] C.T. Supuran, Bacterial carbonic anhydrases as drug targets: toward novel antibiotics? Front. Pharmacol. 2 (2011) 34.

[12] V.D. Luca, D. Vullo, A. Scozzafava, V. Carginale, M. Rossi, C.T. Supuran, C. Capasso, An α-carbonic anhydrase from the thermophilic bacterium *Sulphurihydrogenibium azorense* is the fastest enzyme known for the CO_2 hydration reaction, Bioorg. Med. Chem. 21 (6) (2013) 1465–1469.

[13] T. Matsumoto, M. Hashimoto, W.C. Huang, C.H. Teng, T. Niwa, M. Yamada, T. Negishi, Molecular characterization of a carbon dioxide-dependent Proteus mirabilis small-colony variant isolated from a clinical specimen (in press): S1341-321X(24)00072-2, J. Infect. Chemother. (2024), https://doi.org/10.1016/j.jiac.2024.02.031

[14] E.A. Marcus, A.P. Moshfegh, G. Sachs, D.R. Scott, The periplasmic alpha-carbonic anhydrase activity of *Helicobacter pylori* is essential for acid acclimation, J. Bacteriol. 187 (2) (2005) 729–738.

[15] B.H. Abuaita, J.H. Withey, Bicarbonate induces *Vibrio cholerae* virulence gene expression by enhancing ToxT activity, Infect. Immun. 77 (9) (2009) 4111–4120.

[16] F. Cottier, W. Leewattanapasuk, L.R. Kemp, M. Murphy, C.T. Supuran, O. Kurzai, F.A. Mühlschlegel, Carbonic anhydrase regulation and CO(2) sensing in the fungal pathogen Candida glabrata involves a novel Rca1p ortholog, Bioorg. Med. Chem. 21 (6) (2013) 1549–1554.

[17] Y.S. Bahn, G.M. Cox, J.R. Perfect, J. Heitman, Carbonic anhydrase and CO_2 sensing during *Cryptococcus neoformans* growth, differentiation, and virulence, Curr. Biol. 15 (2005) 2013–2020.

[18] C.T. Supuran, Emerging role of carbonic anhydrase inhibitors, Clin. Sci. (Lond.) 135 (10) (2021) 1233–1249.

[19] M. Castanheira, R.E. Mendes, A.C. Gales, Global epidemiology and mechanisms of resistance of *Acinetobacter baumannii*-calcoaceticus complex, Clin. Infect. Dis. 76 (Suppl. 2) (2023) S166–S178.

[20] K.R. Hummels, S.P. Berry, Z. Li, A. Taguchi, J.K. Min, S. Walker, D.S. Marks, T.G. Bernhardt, Coordination of bacterial cell wall and outer membrane biosynthesis, Nature 615 (7951) (2023) 300–304.

[21] F. Benz, A.R. Hall, Host-specific plasmid evolution explains the variable spread of clinical antibiotic-resistance plasmids, Proc. Natl. Acad. Sci. U. S. A. 120 (15) (2023) e2212147120.

[22] R. Magnano San Lio, G. Favara, A. Maugeri, M. Barchitta, A. Agodi, How anti-microbial resistance is linked to climate change: an overview of two intertwined global challenges, Int. J. Environ. Res. Public Health 20 (3) (2023) 1681.

[23] A. Plotniece, A. Sobolev, C.T. Supuran, F. Carta, F. Björkling, H. Franzyk, J. Yli-Kauhaluoma, K. Augustyns, P. Cos, L. De Vooght, M. Govaerts, J. Aizawa, P. Tammela, R. Žalubovskis, Selected strategies to fight pathogenic bacteria, J. Enzyme Inhib. Med. Chem. 38 (1) (2023) 2155816.

[24] A.G. Atanasov, S.B. Zotchev, V.M. Dirsch, C.T. Supuran, Natural products in drug discovery: advances and opportunities, Nat. Rev. Drug. Discov. 20 (3) (2021) 200–216.

[25] C. Lepore, L. Silver, U. Theuretzbacher, J. Thomas, D. Visi, The small-molecule antibiotics pipeline: 2014–2018, Nat. Rev. Drug. Discov. 18 (10) (2019) 739.

[26] C.T. Supuran, How many carbonic anhydrase inhibition mechanisms exist? J. Enzyme Inhib. Med. Chem. 31 (3) (2016) 345–360.

[27] A. Nocentini, A. Angeli, F. Carta, J.Y. Winum, R. Zalubovskis, S. Carradori, C. Capasso, W.A. Donald, C.T. Supuran, Reconsidering anion inhibitors in the general context of drug design studies of modulators of activity of the classical enzyme carbonic anhydrase, J. Enzyme Inhib. Med. Chem. 36 (1) (2021) 561–580.

[28] S.K. Nair, P.A. Ludwig, D.W. Christianson, Two-site binding of phenol in the active site of human carbonic anhydrase II: structural implications for substrate association, J. Am. Chem. Soc. 116 (1994) 3659–3660.

[29] F. Carta, C. Temperini, A. Innocenti, A. Scozzafava, K. Kaila, C.T. Supuran, Polyamines inhibit carbonic anhydrases by anchoring to the zinc-coordinated water molecule, J. Med. Chem. 53 (15) (2010) 5511–5522.

[30] K. Tars, D. Vullo, A. Kazaks, J. Leitans, A. Lends, A. Grandane, R. Zalubovskis, A. Scozzafava, C.T. Supuran, Sulfocoumarins (1,2-benzoxathiine-2,2-dioxides): a class of potent and isoform-selective inhibitors of tumor-associated carbonic anhydrases, J. Med. Chem. 56 (2013) 293–300.

[31] A. Maresca, C. Temperini, H. Vu, N.B. Pham, S.A. Poulsen, A. Scozzafava, R.J. Quinn, C.T. Supuran, Non-zinc mediated inhibition of carbonic anhydrases: coumarins are a new class of suicide inhibitors, J. Am. Chem. Soc. 131 (8) (2009) 3057–3062.

[32] M. Onyılmaz, M. Koca, A. Bonardi, M. Degirmenci, C.T. Supuran, Isocoumarins: a new class of selective carbonic anhydrase IX and XII inhibitors, J. Enzyme Inhib. Med. Chem. 37 (1) (2022) 743–748.

[33] K. D'Ambrosio, S. Carradori, S.M. Monti, M. Buonanno, D. Secci, D. Vullo, C.T. Supuran, G. De Simone, Out of the active site binding pocket for carbonic anhydrase inhibitors, Chem. Commun. (Camb.) 51 (2) (2015) 302–305.

[34] V. De Luca, A. Petreni, A. Nocentini, A. Scaloni, C.T. Supuran, C. Capasso, Effect of sulfonamides and their structurally related derivatives on the activity of ι-carbonic anhydrase from Burkholderia territorii, Int. J. Mol. Sci. 22 (2) (2021) 571.

[35] A. García-Llorca, F. Carta, C.T. Supuran, T. Eysteinsson, Carbonic anhydrase, its inhibitors and vascular function, Front. Mol. Biosci. 11 (2024) 1338528.

[36] D. Tsikas, Acetazolamide and human carbonic anhydrases: retrospect, review and discussion of an intimate relationship, J. Enzyme Inhib. Med. Chem. 39 (1) (2024) 2291336.

[37] C.T. Supuran, Carbonic anhydrase inhibitors and their potential in a range of therapeutic areas, Expert. Opin. Ther. Pat. 28 (10) (2018) 709–712.

[38] C.T. Supuran, Drug interaction considerations in the therapeutic use of carbonic anhydrase inhibitors, Expert. Opin. Drug. Metab. Toxicol. 12 (4) (2016) 423–431.

[39] C.T. Supuran, An update on drug interaction considerations in the therapeutic use of carbonic anhydrase inhibitors, Expert. Opin. Drug. Metab. Toxicol. 16 (4) (2020) 297–307.

[40] C.T. Supuran, C. Capasso, Antibacterial carbonic anhydrase inhibitors: an update on the recent literature, Expert. Opin. Ther. Pat. 30 (12) (2020) 963–982.

[41] J.R. Warren, Helicobacter: the ease and difficulty of a new discovery (Nobel lecture), ChemMedChem 1 (2006) 672–685.

[42] G.M. Buzas, P. Birinyi, Newer, older, and alternative agents for the eradication of Helicobacter pylori infection: a narrative review, Antibiotics 12 (2023) 946.

[43] C. Campestre, V. De Luca, S. Carradori, R. Grande, V. Carginale, A. Scaloni, C.T. Supuran, C. Capasso, Carbonic anhydrases: new perspectives on protein functional role and inhibition in Helicobacter pylori, Front. Microbiol. 12 (2021) 629163.

[44] I. Nishimori, T. Minakuchi, K. Morimoto, S. Sano, S. Onishi, H. Takeuchi, D. Vullo, A. Scozzafava, C.T. Supuran, Carbonic anhydrase inhibitors: DNA cloning and inhibition studies of the alpha-carbonic anhydrase from Helicobacter pylori, a new target for developing sulfonamide and sulfamate gastric drugs, J. Med. Chem. 49 (6) (2006) 2117–2126.

[45] I. Nishimori, T. Minakuchi, T. Kohsaki, S. Onishi, H. Takeuchi, D. Vullo, A. Scozzafava, C.T. Supuran, Carbonic anhydrase inhibitors: the beta-carbonic anhydrase from Helicobacter pylori is a new target for sulfonamide and sulfamate inhibitors, Bioorg. Med. Chem. Lett. 17 (13) (2007) 3585–3594.

[46] I. Nishimori, S. Onishi, H. Takeuchi, C.T. Supuran, The alpha and beta classes carbonic anhydrases from Helicobacter pylori as novel drug targets, Curr. Pharm. Des. 14 (7) (2008) 622–630.

[47] I. Puşcaş, Treatment of gastroduodenal ulcers with carbonic anhydrase inhibitors, Ann. N. Y. Acad. Sci. 429 (1984) 587–591.

[48] I. Puscas, G. Buzas, Treatment of duodenal ulcers with ethoxzolamide, an inhibitor of gastric mucosa carbonic anhydrase, Int. J. Clin. Pharmacol. Ther. Toxicol. 24 (2) (1986) 97–99.

[49] G.M. Buzás, C.T. Supuran, The history and rationale of using carbonic anhydrase inhibitors in the treatment of peptic ulcers. In memoriam Ioan Puşcaş (1932–2015), J. Enzyme Inhib. Med. Chem. 31 (4) (2016) 527–533.

[50] G. Sachs, Y. Wen, D.R. Scott, Gastric infection by *Helicobacter pylori*, Curr. Gastroenterol. Rep. 11 (6) (2009) 455–461.

[51] S. Morishita, I. Nishimori, T. Minakuchi, S. Onishi, H. Takeuchi, T. Sugiura, D. Vullo, A. Scozzafava, C.T. Supuran, Cloning, polymorphism, and inhibition of beta-carbonic anhydrase of *Helicobacter pylori*, J. Gastroenterol. 43 (11) (2008) 849–857.

[52] M.M. Rahman, A. Tikhomirova, J.K. Modak, M.L. Hutton, C.T. Supuran, A. Roujeinikova, Antibacterial activity of ethoxzolamide against *Helicobacter pylori* strains SS1 and 26695, Gut Pathog. 12 (2020) 20.

[53] J.K. Modak, A. Tikhomirova, R.J. Gorrell, M.M. Rahman, D. Kotsanas, T.M. Korman, J. Garcia-Bustos, T. Kwok, R.L. Ferrero, C.T. Supuran, A. Roujeinikova, Anti-*Helicobacter pylori* activity of ethoxzolamide, J. Enzyme Inhib. Med. Chem. 34 (2019) 1660–1667.

[54] T. Rokkas, K. Ekmektzoglou, Advances in the pharmacological and regulatory management of multidrug resistant *Helicobacter pylori*, Expert. Rev. Clin. Pharmacol. 16 (12) (2023) 1229–1237.

[55] D.P. Flaherty, M.N. Seleem, C.T. Supuran, Bacterial carbonic anhydrases: under-exploited antibacterial therapeutic targets, Future Med. Chem. 13 (19) (2021) 1619–1622.

[56] S.J. Quillin, H.S. Seifert, Neisseria gonorrhoeae host adaptation and pathogenesis, Nat. Rev. Microbiol. 16 (4) (2018) 226–240.

[57] G.L. Aitolo, O.S. Adeyemi, B.L. Afolabi, A.O. Owolabi, *Neisseria gonorrhoeae* anti-microbial resistance: past to present to future, Curr. Microbiol. 78 (3) (2021) 867–878.

[58] M. Unemo, R.A. Nicholas, Emergence of multidrug-resistant, extensively drug resistant and untreatable gonorrhoea, Future Microbiol. 7 (2012) 1401–1422.

[59] S. Huang, Y. Xue, E. Sauer-Eriksson, L. Chirica, S. Lindskog, B.H. Jonsson, Crystal structure of carbonic anhydrase from *Neisseria gonorrhoeae* and its complex with the inhibitor acetazolamide, J. Mol. Biol. 283 (1) (1998) 301–310.

[60] B. Elleby, L.C. Chirica, C. Tu, M. Zeppezauer, S. Lindskog, Characterization of carbonic anhydrase from *Neisseria gonorrhoeae*, Eur. J. Biochem. 268 (6) (2001) 1613–1619.

[61] C.S. Hewitt, N.S. Abutaleb, A.E.M. Elhassanny, A. Nocentini, X. Cao, D.P. Amos, M.S. Youse, K.J. Holly, A.K. Marapaka, W. An, J. Kaur, A.D. Krabill, A. Elkashif, Y. Elgammal, A.L. Graboski, C.T. Supuran, M.N. Seleem, D.P. Flaherty, Structure-activity relationship studies of acetazolamide-based carbonic anhydrase inhibitors with activity against *Neisseria gonorrhoeae*, ACS Infect. Dis. 7 (7) (2021) 1969–1984.

[62] N.S. Abutaleb, A.E.M. Elhassanny, A. Nocentini, C.S. Hewitt, A. Elkashif, B.R. Cooper, C.T. Supuran, M.N. Seleem, D.P. Flaherty, Repurposing FDA-approved sulphonamide carbonic anhydrase inhibitors for treatment of *Neisseria gonorrhoeae*, J. Enzyme Inhib. Med. Chem. 37 (1) (2022) 51–61.

[63] N.S. Abutaleb, A.E.M. Elhassanny, M.N. Seleem, In vivo efficacy of acetazolamide in a mouse model of *Neisseria gonorrhoeae* infection, Microb. Pathog. 164 (2022) 105454.

[64] I. Nishimori, T. Minakuchi, T. Kohsaki, S. Onishi, H. Takeuchi, D. Vullo, A. Scozzafava, C.T. Supuran, Carbonic anhydrase inhibitors: the beta-carbonic anhydrase from Helicobacter pylori is a new target for sulfonamide and sulfamate inhibitors, Bioorg. Med. Chem. Lett. 17 (13) (2007) 3585–3594.

[65] J.K. Modak, Y.C. Liu, M.A. Machuca, C.T. Supuran, A. Roujeinikova, Structural basis for the inhibition of *Helicobacter pylori* α-carbonic anhydrase by sulfonamides, PLoS One 10 (5) (2015) e0127149.

[66] J.K. Modak, Y.C. Liu, C.T. Supuran, A. Roujeinikova, Structure-activity relationship for sulfonamide inhibition of *Helicobacter pylori* α-carbonic anhydrase, J. Med. Chem. 59 (24) (2016) 11098–11109.

[67] C. Campestre, V. De Luca, S. Carradori, R. Grande, V. Carginale, A. Scaloni, C.T. Supuran, C. Capasso, Carbonic anhydrases: new perspectives on protein functional role and inhibition in *Helicobacter pylori*, Front. Microbiol. 12 (2021) 629163.

[68] R. Grande, S. Carradori, V. Puca, I. Vitale, A. Angeli, A. Nocentini, A. Bonardi, P. Gratteri, P. Lanuti, G. Bologna, P. Simeone, C. Capasso, V. De Luca, C.T. Supuran, Selective inhibition of *Helicobacter pylori* carbonic anhydrases by carvacrol and thymol could impair biofilm production and the release of outer membrane vesicles, Int. J. Mol. Sci. 22 (21) (2021) 11583.

[69] V. Puca, G. Turacchio, B. Marinacci, C.T. Supuran, C. Capasso, P. Di Giovanni, I. D'Agostino, S. Carradori, R. Grande, Antimicrobial and antibiofilm activities of carvacrol, amoxicillin and salicylhydroxamic acid alone and in combination vs. *Helicobacter pylori*: towards a new multi-targeted therapy, Int. J. Mol. Sci. 24 (5) (2023) 4455.

[70] M. Ronci, S. Del Prete, V. Puca, S. Carradori, V. Carginale, R. Muraro, G. Mincione, A. Aceto, F. Sisto, C.T. Supuran, R. Grande, C. Capasso, Identification and characterization of the α-CA in the outer membrane vesicles produced by Helicobacter pylori, J. Enzyme Inhib. Med. Chem. 34 (1) (2019) 189–195.

[71] A.K. Marapaka, A. Nocentini, M.S. Youse, W. An, K.J. Holly, C. Das, R. Yadav, M.N. Seleem, C.T. Supuran, D.P. Flaherty, Structural characterization of thiadiazolesulfonamide inhibitors bound to *Neisseria gonorrhoeae* α-carbonic anhydrase, ACS Med. Chem. Lett. 14 (1) (2022) 103–110.

[72] S. Giovannuzzi, N.S. Abutaleb, C.S. Hewitt, F. Carta, A. Nocentini, M.N. Seleem, D.P. Flaherty, C.T. Supuran, Dithiocarbamates effectively inhibit the α-carbonic anhydrase from *Neisseria gonorrhoeae*, J. Enzyme Inhib. Med. Chem. 37 (1) (2022) 1–8.

[73] S. Giovannuzzi, A.K. Marapaka, N.S. Abutaleb, F. Carta, H.W. Liang, A. Nocentini, L. Pisano, M.N. Seleem, D.P. Flaherty, C.T. Supuran, Inhibition of pathogenic bacterial carbonic anhydrases by monothiocarbamates, J. Enzyme Inhib. Med. Chem. 38 (1) (2023) 2284119.

[74] A. Nocentini, C.S. Hewitt, M.D. Mastrolorenzo, D.P. Flaherty, C.T. Supuran, Anion inhibition studies of the α-carbonic anhydrases from *Neisseria gonorrhoeae*, J. Enzyme Inhib. Med. Chem. 36 (1) (2021) 1061–1066.

[75] S. Giovannuzzi, C.S. Hewitt, A. Nocentini, C. Capasso, D.P. Flaherty, C.T. Supuran, Coumarins effectively inhibit bacterial α-carbonic anhydrases, J. Enzyme Inhib. Med. Chem. 37 (1) (2022) 333–338.

[76] S. Giovannuzzi, C.S. Hewitt, A. Nocentini, C. Capasso, G. Costantino, D.P. Flaherty, C.T. Supuran, Inhibition studies of bacterial α-carbonic anhydrases with phenols, J. Enzyme Inhib. Med. Chem. 37 (1) (2022) 666–671.

[77] R.F. O'Toole, K.W.C. Leong, V. Cumming, S.J. Van Hal, Vancomycin-resistant *Enterococcus faecium* and the emergence of new sequence types associated with hospital infection, Res. Microbiol. 174 (4) (2023) 104046.

[78] R.P.J. Willems, K. van Dijk, M.J.G.T. Vehreschild, L.M. Biehl, J.C.F. Ket, S. Remmelzwaal, C.M.J.E. Vandenbroucke-Grauls, Incidence of infection with multidrug-resistant Gram-negative bacteria and vancomycin-resistant enterococci in carriers: a systematic review and meta-regression analysis, Lancet Infect. Dis. 23 (6) (2023) 719–731.

[79] C. Geraldes, L. Tavares, S. Gil, M. Oliveira, *Enterococcus virulence* and resistant traits associated with its permanence in the hospital environment, Antibiotics (Basel) 11 (7) (2022) 857.

[80] C. Horner, S. Mushtaq, M. Allen, R. Hope, S. Gerver, C. Longshaw, R. Reynolds, N. Woodford, D.M. Livermore, Replacement of *Enterococcus faecalis* by *Enterococcus faecium* as the predominant enterococcus in UK bacteraemias, JAC Antimicrob. Resist. 3 (4) (2021) dlab185.

[81] J. Kaur, X. Cao, N.S. Abutaleb, A. Elkashif, A.L. Graboski, A.D. Krabill, A.H. AbdelKhalek, W. An, A. Bhardwaj, M.N. Seleem, D.P. Flaherty, Optimization of acetazolamide-based scaffold as potent inhibitors of vancomycin-resistant *Enterococcus*, J. Med. Chem. 63 (17) (2020) 9540–9562.

[82] N.S. Abutaleb, A. Elkashif, D.P. Flaherty, M.N. Seleem, In vivo antibacterial activity of acetazolamide, Antimicrob. Agents Chemother. 65 (4) (2021) e01715-20.

[83] W. An, K.J. Holly, A. Nocentini, R.D. Imhoff, C.S. Hewitt, N.S. Abutaleb, X. Cao, M.N. Seleem, C.T. Supuran, D.P. Flaherty, Structure-activity relationship studies for inhibitors for vancomycin-resistant *Enterococcus* and human carbonic anhydrases, J. Enzyme Inhib. Med. Chem. 37 (1) (2022) 1838–1844.

[84] N.S. Abutaleb, A. Shrinidhi, A.B. Bandara, M.N. Seleem, D.P. Flaherty, Evaluation of 1,3,4-thiadiazole carbonic anhydrase inhibitors for gut decolonization of vancomycin resistant *Enterococci*, ACS Med. Chem. Lett. 14 (4) (2023) 487–492.

[85] C.T. Supuran, Inhibition of bacterial carbonic anhydrases and zinc proteases: from orphan targets to innovative new antibiotic drugs, Curr. Med. Chem. 19 (6) (2012) 831–844.

[86] S. Del Prete, S. Isik, D. Vullo, V. De Luca, V. Carginale, A. Scozzafava, C.T. Supuran, C. Capasso, DNA cloning, characterization, and inhibition studies of an α-carbonic anhydrase from the pathogenic bacterium *Vibrio cholerae*, J. Med. Chem. 55 (23) (2012) 10742–10748.

[87] D. Vullo, S. Isik, S. Del Prete, V. De Luca, V. Carginale, A. Scozzafava, C.T. Supuran, C. Capasso, Anion inhibition studies of the α-carbonic anhydrase from the pathogenic bacterium *Vibrio cholerae*, Bioorg. Med. Chem. Lett. 23 (6) (2013) 1636–1638.

[88] S. Del Prete, D. Vullo, V. De Luca, V. Carginale, M. Ferraroni, S.M. Osman, Z. AlOthman, C.T. Supuran, C. Capasso, Sulfonamide inhibition studies of the β-carbonic anhydrase from the pathogenic bacterium *Vibrio cholerae*, Bioorg. Med. Chem. 24 (5) (2016) 1115–1120.

[89] M. Ferraroni, S. Del Prete, D. Vullo, C. Capasso, C.T. Supuran, Crystal structure and kinetic studies of a tetrameric type II β-carbonic anhydrase from the pathogenic bacterium *Vibrio cholerae*, Acta Crystallogr. D. Biol. Crystallogr. 71 (Pt 12) (2015) 2449–2456.

[90] S. Del Prete, D. Vullo, V. De Luca, V. Carginale, S.M. Osman, Z. AlOthman, C.T. Supuran, C. Capasso, Comparison of the sulfonamide inhibition profiles of the α-, β- and γ-carbonic anhydrases from the pathogenic bacterium *Vibrio cholerae*, Bioorg. Med. Chem. Lett. 26 (8) (2016) 1941–1946.

[91] A. Nocentini, S.M. Osman, S. Del Prete, C. Capasso, Z.A. AlOthman, C.T. Supuran, Extending the γ-class carbonic anhydrases inhibition profiles with phenolic compounds, Bioorg. Chem. 93 (2019) 103336.

[92] A. Bonardi, A. Nocentini, R. Cadoni, S. Del Prete, P. Dumy, C. Capasso, P. Gratteri, C.T. Supuran, J.Y. Winum, Benzoxaboroles: new potent inhibitors of the carbonic anhydrases of the pathogenic bacterium *Vibrio cholerae*, ACS Med. Chem. Lett. 11 (11) (2020) 2277–2284.

[93] M. Fantacuzzi, I. D'Agostino, S. Carradori, F. Liguori, F. Carta, M. Agamennone, A. Angeli, F. Sannio, J.D. Docquier, C. Capasso, C.T. Supuran, Benzenesulfonamide derivatives as *Vibrio cholerae* carbonic anhydrases inhibitors: a computational-aided insight in the structural rigidity-activity relationships, J. Enzyme Inhib. Med. Chem. 38 (1) (2023) 2201402.

[94] T. Mashe, D. Domman, A. Tarupiwa, P. Manangazira, I. Phiri, K. Masunda, P. Chonzi, E. Njamkepo, M. Ramudzulu, S. Mtapuri-Zinyowera, A.M. Smith, F.X. Weill, Highly resistant cholera outbreak strain in Zimbabwe, N. Engl. J. Med. 383 (7) (2020) 687–689.

[95] M.H. Ahmadi, Global status of tetracycline resistance among clinical isolates of *Vibrio cholerae*: a systematic review and meta-analysis, Antimicrob. Resist. Infect. Control. 10 (1) (2021) 115.

[96] B. Das, J. Verma, P. Kumar, A. Ghosh, T. Ramamurthy, Antibiotic resistance in *Vibrio cholerae*: understanding the ecology of resistance genes and mechanisms, Vaccine 38 (Suppl 1) (2020) A83–A92.

[97] B.H. Abuaita, J.H. Withey, Bicarbonate induces *Vibrio cholerae* virulence gene expression by enhancing ToxT activity, Infect. Immun. 77 (2009) 4111–4120.

[98] P.C. McDonald, S. Chia, P.L. Bedard, Q. Chu, M. Lyle, L. Tang, M. Singh, Z. Zhang, C.T. Supuran, D.J. Renouf, S. Dedhar, A phase 1 study of SLC-0111, a novel inhibitor of carbonic anhydrase IX, in patients with advanced solid tumors, Am. J. Clin. Oncol. 43 (7) (2020) 484–490.

[99] T. Minakuchi, I. Nishimori, D. Vullo, A. Scozzafava, C.T. Supuran, Molecular cloning, characterization, and inhibition studies of the Rv1284 beta-carbonic anhydrase from *Mycobacterium tuberculosis* with sulfonamides and a sulfamate, J. Med. Chem. 52 (8) (2009) 2226–2232.

[100] I. Nishimori, T. Minakuchi, D. Vullo, A. Scozzafava, A. Innocenti, C.T. Supuran, Carbonic anhydrase inhibitors. Cloning, characterization, and inhibition studies of a new beta-carbonic anhydrase from *Mycobacterium tuberculosis*, J. Med. Chem. 52 (9) (2009) 3116–3120.

[101] F. Carta, A. Maresca, A.S. Covarrubias, S.L. Mowbray, T.A. Jones, C.T. Supuran, Carbonic anhydrase inhibitors. Characterization and inhibition studies of the most active beta-carbonic anhydrase from *Mycobacterium tuberculosis*, Rv3588c, Bioorg. Med. Chem. Lett. 19 (23) (2009) 6649–6654.

[102] I. Nishimori, T. Minakuchi, A. Maresca, F. Carta, A. Scozzafava, C.T. Supuran, The β-carbonic anhydrases from *Mycobacterium tuberculosis* as drug targets, Curr. Pharm. Des. 16 (29) (2010) 3300–3309.

[103] A. Maresca, F. Carta, D. Vullo, A. Scozzafava, C.T. Supuran, Carbonic anhydrase inhibitors. Inhibition of the Rv1284 and Rv3273 beta-carbonic anhydrases from *Mycobacterium tuberculosis* with diazenylbenzenesulfonamides, Bioorg. Med. Chem. Lett. 19 (17) (2009) 4929–4932.

[104] A. Aspatwar, V. Kairys, S. Rala, M. Parikka, M. Bozdag, F. Carta, C.T. Supuran, S. Parkkila, *Mycobacterium tuberculosis* β-carbonic anhydrases: novel targets for developing antituberculosis drugs, Int. J. Mol. Sci. 20 (20) (2019) 5153.

[105] R.A. Davis, A. Hofmann, A. Osman, R.A. Hall, F.A. Mühlschlegel, D. Vullo, A. Innocenti, C.T. Supuran, S.A. Poulsen, Natural product-based phenols as novel probes for mycobacterial and fungal carbonic anhydrases, J. Med. Chem. 54 (6) (2011) 1682–1692.

[106] A. Maresca, F. Carta, D. Vullo, C.T. Supuran, Dithiocarbamates strongly inhibit the β-class carbonic anhydrases from *Mycobacterium tuberculosis*, J. Enzyme Inhib. Med. Chem. 28 (2) (2013) 407–411.

[107] A. Maresca, D. Vullo, A. Scozzafava, G. Manole, C.T. Supuran, Inhibition of the β-class carbonic anhydrases from *Mycobacterium tuberculosis* with carboxylic acids, J. Enzyme Inhib. Med. Chem. 28 (2) (2013) 392–396.

[108] M.V. Buchieri, L.E. Riafrecha, O.M. Rodríguez, D. Vullo, H.R. Morbidoni, C.T. Supuran, P.A. Colinas, Inhibition of the β-carbonic anhydrases from *Mycobacterium tuberculosis* with C-cinnamoyl glycosides: identification of the first inhibitor with antimycobacterial activity, Bioorg. Med. Chem. Lett. 23 (3) (2013) 740–743.

[109] C.T. Supuran, Bortezomib inhibits bacterial and fungal β-carbonic anhydrases, Bioorg. Med. Chem. 24 (18) (2016) 4406–4409.

[110] A. Aspatwar, H.M. Becker, N.K. Parvathaneni, M. Hammaren, A. Svorjova, H. Barker, C.T. Supuran, L. Dubois, P. Lambin, M. Parikka, S. Parkkila, J.Y. Winum, Nitroimidazole-based inhibitors DTP338 and DTP348 are safe for zebrafish embryos and efficiently inhibit the activity of human CA IX in Xenopus oocytes, J. Enzyme Inhib. Med. Chem. 33 (1) (2018) 1064–1073.

[111] A. Aspatwar, M. Hammarén, S. Koskinen, B. Luukinen, H. Barker, F. Carta, C.T. Supuran, M. Parikka, S. Parkkila, β-CA-specific inhibitor dithiocarbamate Fc14-584B: a novel antimycobacterial agent with potential to treat drug-resistant tuberculosis, J. Enzyme Inhib. Med. Chem. 32 (1) (2017 Dec) 832–840.

[112] A. Aspatwar, M. Hammaren, M. Parikka, S. Parkkila, F. Carta, M. Bozdag, D. Vullo, C.T. Supuran, In vitro inhibition of *Mycobacterium tuberculosis* β-carbonic anhydrase 3 with Mono- and dithiocarbamates and evaluation of their toxicity using zebrafish developing embryos, J. Enzyme Inhib. Med. Chem. 35 (1) (2020) 65–71.

[113] M.A. Pinard, S.R. Lotlikar, C.D. Boone, D. Vullo, C.T. Supuran, M.A. Patrauchan, R. McKenna, Structure and inhibition studies of a type II beta-carbonic anhydrase psCA3 from *Pseudomonas aeruginosa*, Bioorg. Med. Chem. 23 (15) (2015) 4831–4838.

[114] A.B. Murray, M. Aggarwal, M. Pinard, D. Vullo, M. Patrauchan, C.T. Supuran, R. McKenna, Structural mapping of anion inhibitors to β-carbonic anhydrase psCA3 from *Pseudomonas aeruginosa*, ChemMedChem 13 (19) (2018) 2024–2029.

[115] S.R. Lotlikar, B.B. Kayastha, D. Vullo, S.S. Khanam, R.E. Braga, A.B. Murray, R. McKenna, C.T. Supuran, M.A. Patrauchan, *Pseudomonas aeruginosa* β-carbonic anhydrase, psCA1, is required for calcium deposition and contributes to virulence, Cell Calcium 84 (2019) 102080.

[116] G. De Simone, C.T. Supuran, (In)Organic anions as carbonic anhydrase inhibitors, J. Inorg. Biochem. 111 (2012) 117–129.

[117] P. Joseph, F. Turtaut, S. Ouahrani-Bettache, J.L. Montero, I. Nishimori, T. Minakuchi, D. Vullo, A. Scozzafava, S. Köhler, J.Y. Winum, C.T. Supuran, Cloning, characterization, and inhibition studies of a beta-carbonic anhydrase from *Brucella suis*, J. Med. Chem. 53 (5) (2010) 2277–2285.

[118] D. Vullo, I. Nishimori, T. Minakuchi, A. Scozzafava, C.T. Supuran, Inhibition studies with anions and small molecules of two novel β-carbonic anhydrases from the bacterial pathogen *Salmonella enterica* serovar *Typhimurium*, Bioorg. Med. Chem. Lett. 21 (12) (2011) 3591–3595.

[119] I. Nishimori, T. Minakuchi, D. Vullo, A. Scozzafava, C.T. Supuran, Inhibition studies of the β-carbonic anhydrases from the bacterial pathogen *Salmonella enterica* serovar *Typhimurium* with sulfonamides and sulfamates, Bioorg. Med. Chem. 19 (16) (2011) 5023–5030.

[120] I. Nishimori, D. Vullo, T. Minakuchi, A. Scozzafava, S.M. Osman, Z. AlOthman, C. Capasso, C.T. Supuran, Anion inhibition studies of two new β-carbonic anhydrases from the bacterial pathogen *Legionella pneumophila*, Bioorg. Med. Chem. Lett. 24 (4) (2014) 1127–1132.

[121] I. Nishimori, D. Vullo, T. Minakuchi, A. Scozzafava, C. Capasso, C.T. Supuran, Sulfonamide inhibition studies of two β-carbonic anhydrases from the bacterial pathogen *Legionella pneumophila*, Bioorg. Med. Chem. 22 (11) (2014) 2939–2946.

[122] S. Del Prete, D. Vullo, V. De Luca, V. Carginale, A. Scozzafava, C.T. Supuran, C. Capasso, A highly catalytically active γ-carbonic anhydrase from the pathogenic anaerobe *Porphyromonas gingivalis* and its inhibition profile with anions and small molecules, Bioorg. Med. Chem. Lett. 23 (14) (2013) 4067–4071.

[123] D. Vullo, S. Del Prete, S.M. Osman, V. De Luca, A. Scozzafava, Z. Alothman, C.T. Supuran, C. Capasso, Sulfonamide inhibition studies of the γ-carbonic anhydrase from the oral pathogen *Porphyromonas gingivalis*, Bioorg. Med. Chem. Lett. 24 (1) (2014) 240–244.

[124] S. Del Prete, D. Vullo, V. De Luca, Z. AlOthman, S.M. Osman, C.T. Supuran, C. Capasso, Biochemical characterization of recombinant β-carbonic anhydrase (PgiCAb) identified in the genome of the oral pathogenic bacterium *Porphyromonas gingivalis*, J. Enzyme Inhib. Med. Chem. 30 (3) (2015) 366–370.

[125] S.D. Prete, D. Vullo, S.M. Osman, A. Scozzafava, Z. AlOthman, C. Capasso, C.T. Supuran, Sulfonamide inhibition study of the carbonic anhydrases from the bacterial pathogen *Porphyromonas gingivalis*: the β-class (PgiCAb) versus the γ-class (PgiCA) enzymes, Bioorg. Med. Chem. 22 (17) (2014) 4537–4543.

[126] D. Vullo, R.S. Sai Kumar, A. Scozzafava, C. Capasso, J.G. Ferry, C.T. Supuran, Anion inhibition studies of a β-carbonic anhydrase from *Clostridium perfringens*, Bioorg. Med. Chem. Lett. 23 (24) (2013) 6706–6710.

[127] D. Vullo, R.S.S. Kumar, A. Scozzafava, J.G. Ferry, C.T. Supuran, Sulphonamide inhibition studies of the β-carbonic anhydrase from the bacterial pathogen *Clostridium perfringens*, J. Enzyme Inhib. Med. Chem. 33 (1) (2018) 31–36.

[128] N. Dedeoglu, V. DeLuca, S. Isik, H. Yildirim, F. Kockar, C. Capasso, C.T. Supuran, Sulfonamide inhibition study of the β-class carbonic anhydrase from the caries producing pathogen *Streptococcus mutans*, Bioorg. Med. Chem. Lett. 25 (11) (2015) 2291–2297.

[129] N. Dedeoglu, V. De Luca, S. Isik, H. Yildirim, F. Kockar, C. Capasso, C.T. Supuran, Cloning, characterization and anion inhibition study of a β-class carbonic anhydrase from the caries producing pathogen *Streptococcus mutans*, Bioorg. Med. Chem. 23 (13) (2015) 2995–3001.

[130] D. Vullo, S. Del Prete, S.M. Osman, Z. AlOthman, C. Capasso, W.A. Donald, C.T. Supuran, *Burkholderia pseudomallei* γ-carbonic anhydrase is strongly activated by amino acids and amines, Bioorg. Med. Chem. Lett. 27 (1) (2017) 77–80.

[131] D. Vullo, S. Del Prete, P. Di Fonzo, V. Carginale, W.A. Donald, C.T. Supuran, C. Capasso, Comparison of the sulfonamide inhibition profiles of the β- and γ-carbonic anhydrases from the pathogenic bacterium *Burkholderia pseudomallei*, Molecules 22 (3) (2017 Mar 7) 421.

[132] A. Angeli, M. Ferraroni, M. Pinteala, S.S. Maier, B.C. Simionescu, F. Carta, S. Del Prete, C. Capasso, C.T. Supuran, Crystal structure of a tetrameric type II β-carbonic anhydrase from the pathogenic bacterium *Burkholderia pseudomallei*, Molecules 25 (10) (2020) 2269.

[133] A. Di Fiore, V. De Luca, E. Langella, A. Nocentini, M. Buonanno, S.M. Monti, C.T. Supuran, C. Capasso, G. De Simone, Biochemical, structural, and computational studies of a γ-carbonic anhydrase from the pathogenic bacterium *Burkholderia pseudomallei*, Comput. Struct. Biotechnol. J. 20 (2022) 4185–4194.

[134] S. Del Prete, D. Vullo, S.M. Osman, Z. AlOthman, C.T. Supuran, C. Capasso, Sulfonamide inhibition profiles of the β-carbonic anhydrase from the pathogenic bacterium *Francisella tularensis* responsible of the febrile illness tularemia, Bioorg. Med. Chem. 25 (13) (2017) 3555–3561.

[135] S. Del Prete, D. Vullo, S.M. Osman, Z. AlOthman, W.A. Donald, J.Y. Winum, C.T. Supuran, C. Capasso, Anion inhibitors of the β-carbonic anhydrase from the pathogenic bacterium responsible of tularemia, *Francisella tularensis*, Bioorg. Med. Chem. 25 (17) (2017) 4800–4804.

[136] S. Del Prete, V. De Luca, A. Nocentini, A. Scaloni, M.D. Mastrolorenzo, C.T. Supuran, C. Capasso, Anion inhibition studies of the beta-carbonic anhydrase from *Escherichia coli*, Molecules 25 (11) (2020) 2564.

[137] S. Del Prete, V. De Luca, S. Bua, A. Nocentini, V. Carginale, C.T. Supuran, C. Capasso, The effect of substituted benzene-sulfonamides and clinically licensed drugs on the catalytic activity of CynT2, a carbonic anhydrase crucial for *Escherichia coli* life cycle, Int. J. Mol. Sci. 21 (11) (2020) 4175.

[138] S. Del Prete, S. Bua, C.T. Supuran, C. Capasso, *Escherichia coli* γ-carbonic anhydrase: characterisation and effects of simple aromatic/heterocyclic sulphonamide inhibitors, J. Enzyme Inhib. Med. Chem. 35 (1) (2020) 1545–1554.

[139] L.J. Urbanski, S. Bua, A. Angeli, M. Kuuslahti, V.P. Hytönen, C.T. Supuran, S. Parkkila, Sulphonamide inhibition profile of *Staphylococcus aureus* β-carbonic anhydrase, J. Enzyme Inhib. Med. Chem. 35 (1) (2020) 1834–1839.

[140] A. Angeli, L.J. Urbański, C. Capasso, S. Parkkila, C.T. Supuran, Activation studies with amino acids and amines of a β-carbonic anhydrase from *Mammaliicoccus (Staphylococcus) sciuri* previously annotated as *Staphylococcus aureus* (SauBCA) carbonic anhydrase, J. Enzyme Inhib. Med. Chem. 37 (1) (2022) 2786–2792.

[141] S. Giovannuzzi, V. De Luca, C. Capasso, C.T. Supuran, Inhibition studies with simple and complex (in)organic anions of the γ-carbonic anhydrase from *Mammaliicoccus (Staphylococcus) sciuri*, MscCAγ, J. Enzyme Inhib. Med. Chem. 38 (1) (2023) 2173748.

[142] V. De Luca, S. Giovannuzzi, C.T. Supuran, C. Capasso, May sulfonamide inhibitors of carbonic anhydrases from *Mammaliicoccus sciuri* prevent antimicrobial resistance due to gene transfer to other harmful *Staphylococci*? Int. J. Mol. Sci. 23 (22) (2022) 13827.

[143] V. De Luca, V. Carginale, C.T. Supuran, C. Capasso, The gram-negative bacterium *Escherichia coli* as a model for testing the effect of carbonic anhydrase inhibition on bacterial growth, J. Enzyme Inhib. Med. Chem. 37 (1) (2022) 2092–2098.

Printed and bound by CPI Group (UK) Ltd, Croydon, CR0 4YY

08/05/2025

01864966-0003